This is an introduction to Lie Algebras and their applications in physics.

The first three chapters show how Lie algebras arise naturally from symmetries of physical systems and illustrate through examples much of their general structure. Chapters 4 to 13 give a detailed introduction to Lie algebras and their representations, covering the Cartan–Weyl basis, simple and affine Lie algebras, real forms and Lie groups, the Weyl group, automorphisms, loop algebras and highest weight representations. Chapters 14 to 22 cover specific further topics, such as Verma modules, Casimirs, tensor products and Clebsch–Gordan coefficients, invariant tensors, subalgebras and branching rules, Young tableaux, spinors, Clifford algebras and supersymmetry, representations on function spaces, and Hopf algebras and representation rings. A detailed reference list is provided, and many exercises and examples throughout the book illustrate the use of Lie algebras in real physical problems.

The text is written at a level accessible to graduate students, but will also provide a comprehensive reference for researchers.

CAMBRIDGE MONOGRAPHS ON MATHEMATICAL PHYSICS

General editors: P. V. Landshoff, D. R. Nelson, D. W. Sciama, S. Weinberg

SYMMETRIES, LIE ALGEBRAS AND REPRESENTATIONS

CAMBRIDGE MONOGRAPHS ON MATHEMATICAL PHYSICS

†Issued as a paperback

Symmetries, Lie Algebras and Representations

A graduate course for physicists

JÜRGEN FUCHS

DESY

CHRISTOPH SCHWEIGERT

Institut des Hautes Études Scientifiques

CAMBRIDGE
UNIVERSITY PRESS

PUBLISHED BY THE PRESS SYNDICATE OF THE UNIVERSITY OF CAMBRIDGE
The Pitt Building, Trumpington Street, Cambridge, United Kingdom

CAMBRIDGE UNIVERSITY PRESS
The Edinburgh Building, Cambridge CB2 2RU, UK
40 West 20th Street, New York NY 10011-4211, USA
477 Williamstown Road, Port Melbourne, VIC 3207, Australia
Ruiz de Alarcón 13, 28014 Madrid, Spain
Dock House, The Waterfront, Cape Town 8001, South Africa

http://www.cambridge.org

First published 1997
First paperback edition 2003

Set in Computer Modern

A catalogue record for this book is available from the British Library

Library of Congress Cataloguing in Publication data

Fuchs, Jürgen, 1957–
Symmetries, Lie algebras and representations: a graduate course for
physicists / Jürgen Fuchs, Christoph Schweigert.
p.   cm. (Cambridge monographs on mathematical physics)
Includes bibliographical references and index.
ISBN 0 521 56001 2 hardback
1. Lie algebras.   2. Symmetry (Physics)   3. Representations of groups.
4. Mathematical physics.
I. Schweigert, Christoph.   II. Title.   III. Series.
QC20.7.L54F83 1997
512'.55-dc21 96-38842 CIP

ISBN 0 521 56001 2 hardback
ISBN 0 521 54119 0 paperback

Contents

Table I. *Table of tables*

Preface

There is hardly any student of physics or mathematics who will never come across symbols like $\mathfrak{su}(2)$ and $\mathfrak{su}(3)$. Indeed, many of them even need quite a good working knowledge about the objects which are denoted by these symbols and about many related structures. These structures, which are known as *Lie algebras*, are so ubiquitous because they are closely connected to one of the central themes of physics: to *symmetries*.

Symmetries are well-known in science in general and in physics in particular both for their power and for their beauty. Here, we would like to emphasize the following aspect. The most convenient description of continuous symmetries of physical systems is in terms of Lie algebras (and, when they exist, the corresponding Lie groups); e.g. rotations are described in terms of the Lie algebra of angular momentum. Lie algebras provide a handle to apply *algebraic* tools to analyze the properties of physical systems. Applying these tools, one can make quantitative theoretical predictions which can be confronted with nature.

Lie algebras have found extensive use in concrete applications, in such diverse areas as the formulation of symmetries of Hamiltonian systems in terms of moment maps, the description of atomic, molecular and nuclear spectra, aberration phenomena in optics, gauge theories, theories of gravity, and superstring theory. Another aspect of Lie algebras is that they are closely connected to many other algebraic structures that appear in mathematics and mathematical physics, like group algebras, Hopf algebras, quantum groups, vertex operator algebras, or fusion rings. Many of these structures have already been used to gain new insight in physical theories, and it can be expected that their impact on physics will grow in the future. Lie algebras provide an optimal initiation to this rich reservoir of mathematical concepts.

There are of course also purely mathematical reasons to investigate Lie algebras, and indeed the study of Lie algebras and related structures is an active field of contemporary mathematics. As examples let us mention the relation between differential Lie algebras and vector fields on a manifold, the applications of Lie algebras and gauge theories in differential geometry, e.g. in the description of Donaldson and Seiberg–Witten invariants, and other links to low-dimensional topology, e.g. via the knot invariants which are obtained from quantum groups. The solutions to various differential equations can also be understood in terms of Lie algebras. Moreover, the theory of Lie algebras provides crucial tools for the study of topological groups, including also infinite-dimensional ones, and enters in the classification of finite groups. It also plays a rôle in the *A-D-E* classification of simple singularities. Finally, deep connections have been unravelled between affine Lie algebras and number theory, e.g. the theory of lattices and of automorphic and modular forms.

In accordance with the considerations above, we have two main objectives in our presentation. One goal is to provide a thorough, self-contained introduction to the theory of Lie algebras and their representations which can serve as a textbook for a basic graduate course as well as for independent study. On the other hand, we also present more advanced concepts like Verma modules, null vectors, Hopf algebras and quantum groups. Using these tools we can explain rather deep results like the e.g. the Poincaré–Birkhoff–Witt theorem, the Weyl–Kac character formula, the Haag–Lopuszanski–Sohnius theorem, or the Kac–Walton formula for fusion rule multiplicities. Yet another example is the notion of an intertwiner; it allows us e.g. to present a very simple and conceptual proof of Schur's lemma.

At this point it is appropriate to address the question why one should publish yet another book about Lie algebras. Let us describe our main concern in writing this book which, we think, distinguishes it from the many other books on Lie algebras that have been written for physicists or mathematicians. Namely, our treatment incorporates modern developments both at a conceptual and at a technical level.

Conceptually, we make consequent use of the triangular decomposition of Lie algebras, the existence of a basis of simple roots, and the Dynkin basis of the weight space. This is not only the most transparent way of analyzing finite-dimensional semisimple Lie algebras, but also has numerous computational advantages in applications, in particular when combined with the use of computers. Moreover, it also has the virtue that we can treat a large class of infinite-dimensional Lie algebras on the same footing as finite-dimensional Lie algebras. The triangular decomposition also gives rise to powerful methods for analyzing representation spaces, e.g. via the notion of highest weight vectors, and it allows us to visualize various data of these Lie algebras by means of Dynkin diagrams. To have such a triangular decomposition at our disposal, we start with Lie algebras over the complex numbers, rather than over the real numbers. Accordingly the reader will e.g. often encounter the notation $\mathfrak{sl}(2)$ or $\mathfrak{sl}(2,\mathbb{C})$ in place of the possibly more familiar symbol $\mathfrak{su}(2)$.

Besides these more conceptual aspects, another concern in writing this book was to present the tools and algorithms one actually uses in practice. We have also included tables and formulæ that usually cannot be found in textbooks, but that we found useful in our own work (see e.g. the list of tables on page xiii). It is therefore our hope that this book will not only serve as a textbook for a graduate course, but that many readers will even use it as reference tool for their whole professional life. One particular aspect of this concern is that nowadays one should be aware of the existence of computers and of the computational power of modern algebraic manipulation programs. These computer programs typically allow to perform tasks like the calculation of weight multiplicities or the decomposition of tensor products by employing transparent and powerful general algorithms, rather than case by case studies with a limited range of applicability which were often advocated previously.

After having referred to the disciplines mathematics and physics several times, it is appropriate to point out that there seems to be no general agreement on what is to be called, respectively, 'physics' and 'mathematics'. We think that this is due to the quite simple, and at the same time fundamental, reason that human knowledge does not naturally fall into different isolated compartments. In particular, the fact that individuals have to delimit borders for their areas of interest must not be mixed up with the rather sterile debate on what should be called 'physics' and 'mathematics'. One should resist the temptation to impose one's own choices on other people and possibly even declare them to 'real physics' or 'good mathematics'. Like in any other human community, tolerance and open-mindedness should be considered of high value in science, too. It is in this spirit that we also try to rephrase – where corresponding notions exist – mathematical statements in physicist's terms. We hope that this helps to resolve language problems between mathematics and mathematical physics on one side and physics in a broader sense on the other.

However, it is not only for this reason that we strongly advocate open-mindedness. Rather, we are confident that new structures and notions which emerge in a natural manner in mathematics or mathematical physics have good chances to make their way also into other areas of physics which are more directly accessible to experiments. ('*There seems to be no part of (so-called pure) mathematics that is not in immediate danger of being applied.*' [M. Hazewinkel]) In particular this will happen for many of the new concepts associated with symmetry structures alluded to above, and they may even become standard tools in much the same way as the notion of a Lie algebra has already become a familiar concept for anyone dealing with continuous symmetries. We also believe that many of these new concepts are accessible to the working physicist and to students who wish to specialize in the more theoretical aspects of physics. Unfortunately there often exist prejudices in the physics community against introducing unfamiliar terminology. One aim of this book is therefore to help people realize that such concepts are helpful and that in fact they often already use them implicitly. As an illustration let us mention that until recently the notion of a coproduct was not familiar in theoretical physics, although in fact it is implicitly used whenever one is considering tensor products of representations of Lie algebras, and thereby it is also intimately linked to the physical concept of additivity of a quantum number. It is also in this spirit that we freely use more mathematical terminology, e.g. the term 'module', which is a synonym for the notion 'representation space' that is more familiar to physicists.

To overcome the prejudices against the unfamiliar terminology we mentioned, we think that students should better learn these concepts at an early stage of their education, where they are less handicapped by the rising degree of infallibility that scientists typically develop in the course of their career. We are, however, also confident that any other physicist will benefit in getting familiar with these more recent developments; on the other hand, applications in physics are often a particularly valuable source of intuition for mathematicians.

We would now like to make a few comments concerning the style in which this book is written. To enhance readability, we do not adhere to a lemma – proposition – theorem – proof type presentation. Nonetheless we think that it is mandatory as well as convenient to adhere to a rather high level of mathematical rigor. Here we understand rigor not as a formalistic attitude towards mathematics, but rather as the struggle for *conceptual* clarity. Our general experience is that mathematical precision in the definition of notions and accuracy in the description of tools tends to help people rather than confuse or discourage them. This striving for precision does not, in our opinion, require an unpleasantly formal presentation. Correspondingly, various statements are presented without any proof, but rather we often resort to phrases like 'one can show that' or 'it turns out that'. We do, however, outline the main ideas of proofs whenever we think this provides additional insight in the structures and concepts or uses techniques that have turned out to be useful in practice. Those parts of the proofs which we do present are mathematically rigorous and do not involve any hand waving.

The following comments are intended to help the readers to work with this book according to their own needs.

Structure of the book.

• We start in chapter 1 by explaining how Lie algebras arise naturally in the study of symmetries of physical systems. This is followed by two chapters which examine simple examples of Lie algebras which, however, already illustrate much of the general structure.

• The general description of Lie algebras and their representation theory starts in chapter 4, with the main parts of the structure theory being developed in chapters 6 – 13. More mathematically minded readers may prefer to start a first reading in chapter 4.

• The remaining chapters 14 – 22 are devoted to more specialized issues. Having mastered the chapters 4 – 13, they can be read more or less independently.

• We focus on Lie *algebras* rather than groups to describe symmetries, because they are conceptually and technically far more accessible than Lie groups. They are linear spaces, so that one can apply tools from linear algebra. This becomes particularly important as soon as one is also interested in infinite-dimensional symmetries. But we provide some information on Lie groups, too, in chapter 9 and also in chapter 21.

• We conclude the book with a short Epilogue in which we try to provide a survey of the available literature and a guide to topics which require more advanced knowledge.

Presentation of the material.

• Within each chapter, part of the material is written in small print. The contents of those parts, including e.g. the details of some proofs or possible generalizations, is not, or at least not heavily, used in the rest of the book. When a whole section appears in small print, it is in addition marked by an asterisk '*' (also in the table of contents).

• Examples and applications are usually included in the main text. When an application requires considerably more background than we are able to provide, it is included in a special environment which we call '*Information*'. In this case also references are given; we hope that this encourages the reader to independent study of more advanced topics.

• At the end of the text of each chapter we present a list of keywords and a short summary. Most of the keywords are the new notions explained in the chapter. Any item in this list should ring a bell in the reader's mind, so that it can be used for a simple and efficient check on whether one has absorbed the main points of the chapter.

• Each chapter is supplemented by number of exercises.

The basic information.

• For a 'crash course' on (complex) Lie algebras, one should start with chapter 6, and use chapter 4 as a dictionary when notions occur with which one is not familiar. Chapters 5, 6 and 7 should be read in any case (those who are not interested in infinite-dimensional Lie algebras at all may skip sections 8 – 10 of chapter 7). One should proceed with sections 1 – 5 of chapter 10 and sections 1 – 6 of chapter 13. After this quick reading one should be able to work through any other part without major problems.

Citations.

• This book does not claim to present any novel results. Indeed most of the results are known in the literature and even covered in monographs and reviews. Let us explain our attitude on how to refer to the existing literature. In the body of the book we usually do not cite any sources. Exceptions are made when we think that some statements require additional information for a profound understanding, as is e.g. often the case in the '*Information*' parts mentioned above.

• When citations do appear, we do not make any attempt to indicate the historical development or to give credit to the original work. We even do not necessarily quote the most widely used textbooks. Rather, we prefer to mention those sources which, to our personal taste and judgement, are most easily accessible, either bibliographically or because they try to be particularly pedagogical, and which can serve as a guide to further literature.

In writing this book we have benefited from the help and criticism of K. Blaubär, R. Borcherds, P. Cartier, W. de Graaf, N. Dragon, O. Gabber, T. Gannon, R. Gebert, M. Geck, K. Jünemann, W. Kalau, I. Kausz, M.L. Kontsevich, M. Kreuzer, R. Kühn, M. van Leeuwen, M. Lüscher, J. Neubüser, K. Peeters, H.J. Pirner, U. Ray, H. Samtleben, F. Scheck, A.N. Schellekens, M.G. Schmidt, V. Schomerus, J. Stembridge, P. Stevens, S. Theisen, C. Unkmeir, H. Wagner and B. Wybourne.

We are particularly grateful to M.G. Schmidt who, apart from encouraging us to write a book of this kind at all, made numerous suggestions on the conception, the contents and the presentation. We also thank T. Gannon, K. Jünemann, M.G. Schmidt and C. Unkmeir for a careful reading of a preliminary version of the book, M.-C. Vergne for drawing several PostScript figures, and R. Neal for the kind assistance concerning the technical details of the production.

Despite our efforts, it is unavoidable that this book will contain errors and misprints. However, we take advantage of the possibilities the World Wide Web offers nowadays: Updated information on all errors and misprints can be obtained from the home page of our book on the World

Wide Web, which has the address

$$\text{http://norma.nikhef.nl/lie}$$

(this information is also available from the Cambridge World Wide Web server, see

$$\text{http://www.cup.cam.ac.uk}$$

for further details). Any reader who detects an error is urged to contact the authors via this address. On the same WWW page we also provide information about computer programs and software packages which allow to perform symbolic and numerical calculations with Lie algebras and related structures.

We believe that this book can convey to its readers an impression of the elegance of the algebraic structures which underly symmetries and give them some flavor of their applications. It is our hope that the book may even contribute to the goal [Dyson 1972] that less opportunities will be missed for a fruitful interaction between mathematicians and physicists.

1
Symmetries and conservation laws

1.1 Symmetries

Symmetry is one of the most fruitful principles in science. In fact, nowadays acquaintance with the structures that describe symmetries is an indispensable prerequisite for anybody who wants to become familiar with the deep and beautiful concepts of modern physics. Besides their aesthetic appeal, the virtue of symmetries is that they enable us to make precise quantitative predictions about various physical systems, which can be compared with experiment.

The concept of symmetry is extremely general, and correspondingly the precise meaning of the term depends to a large extent on the setting. For the purposes of the present book, we reserve the word *symmetry* for mappings of the physical states of a system which leave the dynamics invariant. These operations must be either invertible or infinitesimal analogues of invertible mappings. What is meant by the terms 'invariant' and 'dynamics' still depends on the framework which is used to study the system. In the rest of this chapter several such frameworks will be presented, and each time the specific meaning of 'symmetry' will be explained. In all these examples the structure of Lie algebras emerges in the description of the symmetries. In chapter 2 and chapter 3 we will analyze specific examples of such Lie algebra symmetries which are still quite simple but nevertheless already display much of the general structure. Only afterwards, from chapter 4 on, we investigate the general theory of Lie algebras and related structures. Those readers who already have enough background or who are exclusively interested in the mathematical aspects of the description of symmetries may proceed directly to chapter 4.

Already before the restriction to a particular framework some fundamental observations can be made. Namely, first, the composition of mappings is always associative; second, there is the identity map which triv-

1

ially respects the dynamics; and third, the mappings are invertible by assumption. This way the mathematical structure of a *group* of symmetries arises in a natural manner.

1.2 Continuous parameters and local one-parameter groups

In many examples, such as translations or rotations, the group elements that describe symmetries of physical systems can be labelled by a continuous parameter, like the angle of a rotation or the translation vector, on which the group elements depend in a differentiable manner. In this situation, one can also study 'infinitesimal' symmetries. Such an infinitesimal symmetry is also called a *generator of a symmetry*, because often there are methods to recover group elements from these infinitesimal symmetries. A basic ingredient of a group is the operation of multiplication, which associates to any two group elements a third one. This operation has an infinitesimal counterpart as well, which associates to two infinitesimal symmetries a third infinitesimal symmetry. However, this operation has no longer the properties of a group multiplication; rather, it will lead us to introduce the mathematical structure of an *algebra*.

To characterize an infinitesimal symmetry, it is actually sufficient to know how the group elements look in some small neighborhood of the identity element of the group. Suppose we are given a continuous parametrization $t \mapsto \gamma(t)$ of group elements, where the parameter t is a real number, and assume for simplicity that $\gamma(0)$ is the unit element of the group. This parametrization should be such that the product of two group elements $\gamma(t_1)$ and $\gamma(t_2)$ satisfies $\gamma(t_1)\gamma(t_2) = \gamma(t_1 + t_2)$. An infinitesimal symmetry will be described by the derivative of the function $\gamma(t)$ at $t = 0$. To construct such a generator, it is not necessary to know the function $\gamma(t)$ for all values of t, and it is not even necessary that such a function is defined for all values of t. Rather, to obtain an infinitesimal symmetry it is sufficient to have a function γ obeying $\gamma(t_1)\gamma(t_2) = \gamma(t_1 + t_2)$ that is only defined on some small interval around 0. Such a function is sometimes called a *local one-parameter group*; the qualification 'local' refers to the fact that the function is only defined near zero.

Already this simple consideration indicates that there may exist symmetries of physical systems which cannot be described by groups. We will see shortly that, just like symmetry groups, such symmetries can give rise to conserved quantities and therefore contain valuable information.

1.3 Classical mechanics: Lagrangian description

In the Lagrangian description of classical mechanics, a system is characterized by the Lagrangian L, which is a function of positions q_i and of the tangent vectors to position space, i.e. the velocities $\dot{q}_i$. To describe the characteristic features of this framework, it suffices to restrict attention to the case of a single position q and velocity $\dot{q}$. The solutions to the equations of motion are precisely the extrema of the action, i.e. of the time integral $S = \int L(t)\, dt$ of the Lagrangian (subject to the requirement that the boundary conditions are kept fixed). Hence the equations of motion are given by the *Euler–Lagrange equations*

$$\frac{\mathrm{d}}{\mathrm{d}t}\frac{\partial L}{\partial \dot{q}} = \frac{\partial L}{\partial q}. \tag{1.1}$$

Now suppose that for any real number s in some small interval around 0, $s \in (-\varepsilon, \varepsilon)$, there is a mapping h_s of the position variable according to

$$h_s: \quad q \mapsto h_s(q), \tag{1.2}$$

which is accompanied by an induced linear mapping $\hat{h}_s$ of the velocity $\dot{q}$, i.e.

$$\hat{h}_s: \quad \dot{q} \mapsto \hat{h}_s(\dot{q}) = \frac{\partial h_s(q)}{\partial q}\,\dot{q}. \tag{1.3}$$

The Lagrangian L is said to be *invariant* under the transformation (1.2) and (1.3) if for any fixed s there exists a function $F_s(q, \dot{q}, t)$ that can also depend on time, such that

$$L(h_s(q), \hat{h}_s(\dot{q})) = L(q, \dot{q}) + \frac{\mathrm{d}}{\mathrm{d}t}F_s. \tag{1.4}$$

Note that we allowed for a total time derivative of an arbitrary function F_s; such an additional term is allowed because it does not change the equations of motion. Upon integration over some time interval $[t_1, t_2]$, the total derivative gives rise to a 'surface term' $F_s(q(t_2), \dot{q}(t_2), t_2) - F_s(q(t_1), \dot{q}(t_1), t_1)$ which only receives contributions from the boundary of the time interval. Surface terms play an important rôle in several applications. For instance, Lagrangians can be invariant under a *supersymmetry* (see section 20.9) only up to such terms.

1.4 Conservation laws

Any invariance of the type (1.4) gives rise to a conservation law. This statement, known as *Noether's theorem*, can be derived as follows. A mapping

$$\phi: \quad t \mapsto q = \phi(t) \tag{1.5}$$

from a time interval to the position space is called a *path*. We can now use any symmetry h_s to map such a path ϕ to a new path ϕ_s:

$$\phi_s = h_s \circ \phi: \quad t \mapsto q = h_s(\phi(t)). \tag{1.6}$$

This way we obtain a whole family of paths, one path for every value of s. The invariance (1.4) tells us that the value of the action of the path ϕ_s is independent of s. In particular, if the path (1.5) is a solution of the equations of motion (1.1), the action is minimal on the path ϕ and, by the invariance, it is minimal on any other path ϕ_s as well. In other words, any path ϕ_s is a solution of the equations of motion.

To investigate this whole family of paths, it is convenient to treat the variable t parametrizing the individual path and the variable s that labels the different paths on an equal footing. Thus we introduce a function Φ of two variables,

$$\Phi(s,t) := h_s \circ \phi(t), \tag{1.7}$$

i.e. for fixed value of s one just obtains the path ϕ_s. With the help of the invariance (1.4) it follows that Φ satisfies

$$0 = \frac{\partial}{\partial s} \left(L(\Phi, \frac{\partial}{\partial t}\Phi) - \frac{\mathrm{d}}{\mathrm{d}t}F_s \right) = \frac{\partial L}{\partial q}\frac{\partial \Phi}{\partial s} + \frac{\partial L}{\partial \dot{q}}\frac{\partial^2 \Phi}{\partial t\,\partial s} - \frac{\mathrm{d}}{\mathrm{d}t}\frac{\partial F_s}{\partial s}. \tag{1.8}$$

Using the equations of motion (1.1), this can be rewritten as

$$0 = \left(\frac{\mathrm{d}}{\mathrm{d}t}\frac{\partial L}{\partial \dot{q}} \right)\frac{\partial \Phi}{\partial s} + \frac{\partial L}{\partial \dot{q}}\frac{\partial^2 \Phi}{\partial t\partial s} - \frac{\mathrm{d}}{\mathrm{d}t}\frac{\partial F_s}{\partial s} = \frac{\mathrm{d}}{\mathrm{d}t}\left(\frac{\partial L}{\partial \dot{q}}\frac{\partial \Phi}{\partial s} - \frac{\partial F_s}{\partial s} \right). \tag{1.9}$$

This shows that the quantity

$$Q := \frac{\partial L}{\partial \dot{q}}\frac{\partial \Phi}{\partial s} - \frac{\partial F_s}{\partial s} \tag{1.10}$$

does not depend on time, i.e. is a *conserved quantity* (also called a conserved charge or Noether charge). Note that in the derivation of this result we used the equations of motion; correspondingly the statement that the quantity Q does not depend on time means that it is constant in time for any solution to the equations of motion.

As an illustration, let us employ Noether's theorem to derive the conservation of momentum for the motion of a point particle in $\mathbb{R}^3$. Suppose that the Lagrangian L is of the form $L = \frac{1}{2}m\,\dot{\vec{x}}^2 - V(\vec{x})$, with $\vec{x} \in \mathbb{R}^3$, and that the potential $V(\vec{x})$ is invariant under a shift $\vec{x} \mapsto \vec{x} + s\vec{a}$, i.e. $V(\vec{x} + s\vec{a}) = V(\vec{x})$ for all s. Then also L is invariant under the transformation

$$h_s: \quad \vec{x} \mapsto \vec{x} + s\vec{a}, \tag{1.11}$$

while the mapping $\hat{h}_s$ is the identity map and $F_s \equiv 0$ for all s. By computing

$$\frac{\partial \Phi}{\partial s} = \frac{\mathrm{d}h_s(\vec{x})}{\mathrm{d}s} = \vec{a} \tag{1.12}$$

we find that the conserved quantity (1.10) is

$$Q_{\vec{a}} = m\,\dot{\vec{x}} \cdot \vec{a}\,. \tag{1.13}$$

That is, the conserved charge associated to the invariance under the translation $\vec{x} \mapsto \vec{x} + s\vec{a}$ is the momentum in $\vec{a}$-direction.

Noether's theorem shows that any differentiable family of symmetry transformations yields a conservation law and a conserved charge Q. We have just seen that, if the system is invariant under translations, one obtains conservation of momentum; similarly (see exercise 1.1), invariance under rotations leads to conservation of angular momentum. The theorem can also be easily generalized from mechanics to classical field theories. However, we warn the reader that the converse of Noether's theorem does *not* hold: There can be conserved quantities, such as the so-called *topological charges*, which are not Noether charges. For example, it can happen that the *configuration space*, i.e. the space spanned by the (generalized) coordinates of the system, is not connected. As the motion is continuous and hence can never connect configurations in different components, it takes place in a single connected component. Thus when we attach a label, say an integer, to each component, we can interpret this label as the value of a conserved quantity.

To choose a sensible labelling corresponding to a topological charge can often be a complicated task. Typically the counting of connected components of the configuration space is most conveniently described by relating it to topological properties of other objects. An example for this is provided by four-dimensional Yang–Mills theories, where a suitable labelling of the connected components is given by the instanton number or Pontryagin index of the gauge field configuration (see e.g. chapter 9.4.1 of [Nakahara 1990]).

Information

Noether's theorem can in particular be applied to any continuous, differentiable *group* of symmetries. However, in the derivation of the theorem only a *local* one-parameter family of symmetries was used; thus the conserved quantities are actually associated not so much to the group, but rather to the infinitesimal version of the symmetry transformations; these form an algebra rather than a group. Algebras and groups are related by a certain exponential mapping (which will be studied in chapter 9). However, not any local family of symmetries can be integrated up to a group by exponentiation. In other words, there can be symmetries which can only be described by algebras, and these symmetries imply additional conservation laws. In fact, this situation is encountered rather frequently in modern physics; examples are provided by 'supersymmetry' (see section 20.9) and by 'quantum groups' (see chapter 22). Another important situation where this phenomenon occurs is conformal symmetry in a two-dimensional space: While the dimension of the conformal group (over $\mathbb{R}$) is six, i.e. in particular finite, the conformal *algebra* in two dimensions is the so-called Virasoro algebra (see section 12.12) which is infinite-dimensional; thus only a tiny part of the symmetries can be regarded as the infinitesimal version of a group transformation. An additional virtue of infinitesimal symmetries is that they are described by a *linear* structure, whereas the structure of groups is in general inherently non-linear.

1.5 Classical mechanics: Hamiltonian description

In the Hamiltonian framework of classical mechanics, all information is encoded in the Hamiltonian, a function of the positions q_i and momenta p_i (and possibly of the time t). The space spanned by the positions and momenta is called the *phase space*, to be distinguished from the configuration space, i.e. the space spanned only by the positions, which is the space in which Lagrangian dynamics takes place. Considering again the case of a single position variable q, the equations of motion read

$$\frac{\mathrm{d}p}{\mathrm{d}t} = -\frac{\partial H}{\partial q}, \qquad \frac{\mathrm{d}q}{\mathrm{d}t} = \frac{\partial H}{\partial p}. \tag{1.14}$$

For dynamical systems which can be treated both in the Hamiltonian and in the Lagrangian description, the relation between the two is given by a Legendre transformation, $H(q,p) = p\dot{q} - L$, where $\dot{q}$ is implicitly defined by $p = \partial L/\partial \dot{q}$.

However, the relation between Hamiltonian and Lagrangian classical dynamical systems is not one-to-one. On one hand, not any phase space of a Hamiltonian system (i.e. in mathematical terms, a symplectic manifold with a Hamiltonian vector field) can be described as the cotangent bundle of some Lagrangian submanifold; this is e.g. definitely not possible if the symplectic manifold is compact. On the other hand, in many interesting examples of Lagrangian dynamics some canonical momenta vanish, so that it is not possible to express $\dot{q}$ in terms of p and q as needed in the Legendre transform. A handle on this situation is provided by the theory of constrained systems; for details about constrained dynamics see e.g. [Henneaux and Teitelboim 1992] and [Sundermeyer 1982].

One of the fundamental tools employed in the Hamiltonian formulation of mechanics is the *Poisson bracket*. This maps any pair f, g of differentiable functions of the dynamical variables p and q to a single function of p and q which is denoted by $\{f,g\}$ and is obtained as

$$\{f,g\} := \frac{\partial f}{\partial p}\frac{\partial g}{\partial q} - \frac{\partial f}{\partial q}\frac{\partial g}{\partial p}. \tag{1.15}$$

It is easy to check (see exercise 1.2) that the Poisson bracket possesses the following three characteristic properties:

First, it is bilinear, i.e. satisfies

$$\text{a)} \quad \{\lambda f + \mu g, h\} = \lambda \{f, h\} + \mu \{g, h\} \tag{1.16}$$

for all real numbers λ, μ and an analogous formula for $\{f, \lambda g + \mu h\}$. Second, it is antisymmetric:

$$\text{b)} \quad \{f, g\} = -\{g, f\}. \tag{1.17}$$

And third, the so-called *Jacobi identity*

$$\text{c)} \quad \{f, \{g, h\}\} + \{g, \{h, f\}\} + \{h, \{f, g\}\} = 0 \tag{1.18}$$

holds.

These three properties can be summarized by the statement that the Poisson bracket $\{\cdot, \cdot\}$ turns the space of functions of p and q into what is called a 'Lie algebra'.

Let us now assume that the Hamiltonian H does not depend on time, $\partial H/\partial t = 0$. (Sometimes this is described by saying that H does not depend *explicitly* on time, since after evaluating H on some specific trajectory, say a solution to the equations of motion, H would depend implicitly on t through $p(t)$ and $q(t)$. However, the correct point of view is to regard H just as a function of the two *indeterminates* p and q, so that this qualification is unnecessary.) The time derivative of any function $f(p(t), q(t))$ on a trajectory that is a solution to the equations of motions can then be written as the Poisson bracket with the Hamiltonian (see exercise 1.2):

$$\frac{d}{dt} f(p(t), q(t)) = \{H, f\}. \tag{1.19}$$

A function f describes a conserved quantity if for any solution to the equations of motion the time derivative on the left hand side vanishes, and hence if it 'Poisson-commutes' with the Hamiltonian,

$$\{H, f\} = 0. \tag{1.20}$$

Such functions are said to be *in involution* with the Hamiltonian. Using the Jacobi identity (1.18), it follows (see exercise 1.2) that, if f and g are conserved, then $\{f, g\}$ is a conserved quantity, too. In other words, the space of functions that describe conserved quantities closes under the Poisson bracket. In fact, it constitutes in itself a Lie algebra, a subalgebra of the space of all functions. Hence, to learn more on the conserved charges of a system described by classical Hamiltonian mechanics, it is necessary to investigate the structure of Lie algebras. The Jacobi identity which ensures that together with f and g also $\{f, g\}$ is conserved, is a crucial and non-trivial input required for having the structure of a Lie algebra.

The presence of conserved quantities puts severe constraints on the dynamics of a system. Any trajectory that fulfills the equations of motion remains inside a subspace of the phase space in which each conserved quantity takes a fixed value. If there are, for a phase space of dimension $2n$, n conserved quantities which all Poisson commute, this subspace is typically an n-dimensional torus, if it is compact.

In the Hamiltonian framework the relation between conserved quantities and local one-parameter groups can be described as follows. The gradient $\mathrm{d}f$ of a function f on phase space that Poisson-commutes with the Hamilton function H is a one-form on phase space. The symplectic form Ω on the phase space is non-degenerate and allows us to identify vector fields and one-forms. In particular, we can associate to $\mathrm{d}f$ the unique vector field $I(\mathrm{d}f)$ for which $\Omega(I(\mathrm{d}f), X) = \mathrm{d}f(X)$ for any vector field X on phase space.

Given the vector field $I(\mathrm{d}f)$ one can obtain a *Hamiltonian phase flow*, i.e. a local one-parameter group of diffeomorphisms φ_t of the phase space which obey $\frac{\mathrm{d}}{\mathrm{d}t}\big|_{t=0}\varphi_t = I(\mathrm{d}f)$. (A special case of this construction is the time evolution of the system, which is obtained by taking f equal to the Hamilton function H.) The maps φ_t are indeed symmetries of the system: φ_t and the time evolution commute precisely if f and H Poisson-commute.

A more detailed investigation of symmetries in classical Hamiltonian systems or, more generally, group actions on symplectic manifolds, can be performed with the help of so-called moment maps. For a description of this concept we refer the reader to appendix 5 of [Arnold 1978] and to chapter 11 of [Marsden and Ratiu 1994].

1.6 Quantum mechanics

To analyze symmetries in quantum mechanics, we will work with canonical quantization in the Heisenberg picture. The basic idea of canonical quantization is to associate to any classical dynamical system in the Hamiltonian formulation a quantum dynamical system, which is accomplished by a certain quantization prescription. In short, this prescription amounts to declaring, first, that self-adjoint linear operators on some Hilbert space $\mathcal{H}$ over the complex numbers take over the rôle of the observables of the system from the functions on phase space; and second, to replacing the Poisson bracket of functions by the commutator of the associated operators on $\mathcal{H}$, according to

$$\{\cdot,\cdot\} \mapsto \tfrac{\mathrm{i}}{\hbar}[\cdot,\cdot]\,. \tag{1.21}$$

This leads in particular to the Heisenberg commutation relation between position and momentum operators:

$$\{p,q\} = 1 \mapsto \tfrac{\mathrm{i}}{\hbar}[p,q] = 1\,. \tag{1.22}$$

The number $\hbar$ appearing in the prescription (1.21) is Planck's constant, which has the dimension of an action, and is present for dimensional reasons. The factor of $\mathrm{i} \equiv \sqrt{-1}$ is needed in (1.21) because the Poisson bracket of two real functions is real again, while the commutator of two self-adjoint operators is anti-self-adjoint.

It should be noted that the formula

$$[A, B] := A \circ B - B \circ A \tag{1.23}$$

defining the *commutator* of two operators A and B on $\mathcal{H}$ is not as innocent as it may seem. The reason is that in quantum mechanics, observables are typically described by operators which are not defined on the whole Hilbert space $\mathcal{H}$, but only on a dense subspace. As a consequence, the composition of operators, and hence also their commutator, is not necessarily defined. However, assuming that the commutator of the operators in question does exist, it is easy to check that it fulfills the same three properties that characterize the Poisson bracket of classical mechanics: it is bilinear,

$$\text{a)} \quad [\lambda A + \mu B, C] = \lambda [A, C] + \mu [B, C] \,; \tag{1.24}$$

it is antisymmetric,

$$\text{b)} \quad [A, B] = -[B, A] \,; \tag{1.25}$$

and the Jacobi identity

$$\text{c)} \quad [A, [B, C]] + [B, [C, A]] + [C, [A, B]] = 0 \tag{1.26}$$

holds. Again we can summarize these properties by the statement that the commutator $[\cdot, \cdot]$ endows the space of linear operators on $\mathcal{H}$ with the structure of a Lie algebra.

In the Heisenberg picture of quantum mechanics the physical states, i.e. the elements of the Hilbert space $\mathcal{H}$, do not depend on time, but the observables do. Their time dependence is described by the Heisenberg equation

$$\tfrac{\mathrm{d}}{\mathrm{d}t} A(t) = \tfrac{i}{\hbar} [H, A] \tag{1.27}$$

which is obtained by applying (1.22) to the classical formula (1.19). As a consequence, in quantum mechanics observables A that describe conservation laws are characterized by the fact that they commute with the Hamiltonian H,

$$[H, A] = 0 \,. \tag{1.28}$$

Again it follows from the Jacobi identity that the operators which satisfy (1.28) form a subalgebra. These statements are valid in the Schrödinger picture as well, as can be seen by employing the relation between operators $A \equiv A_{(H)}$ in the Heisenberg picture and $A_{(S)}$ in the Schrödinger picture, which reads

$$A_{(H)}(t) = e^{iHt/\hbar} \, A_{(S)} \, e^{-iHt/\hbar} \,. \tag{1.29}$$

In quantum mechanics, too, the existence of conserved quantities constrains the dynamics; it implies selection rules for the matrix elements of the Hamiltonian. As an example, consider a system in which orbital angular momentum is conserved, as happens e.g. for the electron in the hydrogen atom when the spin degree of freedom is neglected. The eigenstates of H can then be labelled by their angular momentum l and the projection m of angular momentum on some coordinate axis, say the z-axis, together with some other quantum numbers, in the case of the hydrogen atom the principal quantum number n. The fact that H and L_z commute then implies that

$$
\begin{aligned}
0 &= \langle n', l', m' | \, [L_z, H] \, | n, l, m \rangle \\
&= (m' - m) \, \langle n', l', m' | \, H \, | n, l, m \rangle
\end{aligned}
\tag{1.30}
$$

for all eigenstates $|n, l, m\rangle$, $|n', l', m'\rangle$. As a consequence, the matrix element $\langle n', l', m' | \, H \, | n, l, m \rangle$ can be non-zero only if $m' = m$. This simple observation proves to be very powerful, and is the basis for many selection rules in physics (for some details, see chapter 16).

Suppose now that we are given a self-adjoint operator A which commutes with the Hamiltonian H. By exponentiation, one obtains operators

$$
U_A(t) := \mathrm{e}^{\mathrm{i}tA} \equiv 1 + \mathrm{i}tA + \tfrac{1}{2!}\left(\mathrm{i}tA\right)^2 + \dots
\tag{1.31}
$$

for any $t \in \mathbb{R}$. These operators are unitary and form a group (see exercise 1.5), with multiplication law

$$
U_A(t_1) \circ U_A(t_2) = U_A(t_1 + t_2) \qquad \text{for all } t_1, t_2 \in \mathbb{R}.
\tag{1.32}
$$

Together all these unitary one-parameter groups generate a group of symmetries which is unitarily represented on the Hilbert space. The construction of this group is commonly referred to as 'integrating up' the 'infinitesimal symmetries' described by the operators A. As an example consider the momentum operator p which generates infinitesimal translations; in a position space description it is described by the differential operator $p := \frac{\hbar}{\mathrm{i}} \frac{\mathrm{d}}{\mathrm{d}x}$. One then finds (see exercise 1.4) that finite translations by a shift a are described by the unitary operator

$$
\mathrm{e}^{\mathrm{i}pa/\hbar} = \exp\!\left(a \tfrac{\mathrm{d}}{\mathrm{d}x}\right).
\tag{1.33}
$$

It must be noted that an exponentiation of infinitesimal symmetries to unitary operators as in (1.31) is not always possible, and that in the case of unbounded self-adjoint operators the exponential in (1.31) has to be defined using functional calculus (e.g. if the power series does not naively make sense). Particular care is required if there are infinitely many independent infinitesimal symmetries. The translations (1.33) are in this respect rather special even among finite-dimensional symmetries, because partial derivatives $\frac{\partial}{\partial x_j}$ commute. (Thus all commutators $[p_i, p_j]$

vanish, so that the axioms of a Lie algebra are trivially satisfied; such Lie algebras are called abelian.) In this case the exponentiation can be immediately generalized to an arbitrary number of dimensions of position space.

But even those symmetry transformations U of a quantum mechanical system which are not infinitesimal are not necessarily implemented by *unitary* operators. Rather, it follows from the conservation of probability under the symmetry U, i.e. from $|\langle Ux\,|\,Uy\rangle| = |\langle x\,|\,y\rangle|$, that, choosing a suitable convention for the phases, any non-infinitesimal symmetry of a quantum mechanical system can be implemented either by unitary or by anti-unitary operators (see e.g. chapter XV, §2 of [Messiah 1986]). An operator Θ on a complex vector space is called anti-linear if it obeys $\Theta(\xi_1 v_1 + \xi_2 v_2) = \bar{\xi}_1 \Theta v_1 + \bar{\xi}_2 \Theta v_2$ for all complex numbers ξ_1, ξ_2. (Throughout this book, we denote complex conjugation by a bar.) An anti-unitary operator is an anti-linear operator that obeys in addition $(\Theta v_1, \Theta v_2) = \overline{(v_1, v_2)} = (v_2, v_1)$.

Thus in physicists' language, an anti-unitary operator interchanges the rôle of 'in-states' and 'out-states'. Accordingly, it should not come as a surprise that anti-unitary symmetries are intimately linked with time reversal (for details see e.g. chapter 12 of [Tung 1985]). In fact, the Schrödinger equation is a first order differential equation in time (whereas the equations of motion in classical mechanics are typically second order in time). This implies that for a time independent Hamiltonian along with $\psi(x,t)$ also $\overline{\psi(x,-t)}$ (rather than $\psi(x,-t)$) is a solution. On the other hand, other discrete symmetries like parity or charge conjugation are implemented on the Hilbert space of physical states by unitary operators, both in non-relativistic quantum mechanics and in the relativistic description of fermions via the Dirac equation.

Let us also note that in quantum physics there can be more general symmetries which are not described by groups anymore. The presence of such symmetries can heuristically be understood as follows. In quantum physics, the commutative algebra of functions on the configuration space is traded for a non-commutative algebra of operators. Now for a classical dynamical system the symmetries can be viewed as acting on the points of the configuration space; hence they form a group, with the group multiplication provided by the composition of maps. In contrast, in quantum physics the configuration space is no longer present, so that this argument does not apply any more. Actually, the question on how to implement the most general quantum symmetry has not been fully answered yet; for a few further remarks and a guide to the literature, see the Epilogue of the book.

1.7 Observation of symmetries

Given a physical system, a natural question to ask is how one can *measure* what its symmetries are. Besides the observation of selection rules there are two other possibilities. First, one may detect the multiplets of the symmetric theory by perturbing the symmetric system in a controlled manner so as to arrive at a system for which the symmetry is valid only approximately. In other words, one constructs a family of systems that generically have approximate symmetry which, however, is enhanced to an exact symmetry for one member of the family. A typical example for this procedure is the measurement of the energy levels of the hydrogen atom in a homogeneous magnetic field. While at zero field strength the system is symmetric with respect to rotations, this symmetry is broken by the presence of the magnetic field. As a consequence the energy levels which are degenerate at zero field split off (the Zeeman effect) so that one can count the dimension of the multiplets.

The second possibility for detecting the presence of a symmetry is that the degeneracy implied by the symmetry has some influence on the statistics of the excitations. For instance, in quantum mechanics the total wave function of multi-particle states of fermions must be antisymmetric. For example, in the quark model of hadrons where the baryons are interpreted as bound states of three quarks (which are fermions), this requirement can only be fulfilled by including in the total wave function also a factor corresponding to the color degree of freedom of the quarks. This observation was in fact one of the sources of the introduction of the color $\mathfrak{su}(3)$ symmetry (compare the end of section 3.7) in the theory of strong interactions.

Typically the last mentioned method provides only limited information, while the former requires that the symmetry breaking perturbations can be realized in an experimental setup, which generically need not be possible. Therefore selection rules provide the only generic handle to observe symmetries. A refinement of the result about selection rules that was obtained above, the so-called Wigner–Eckart theorem, will be derived in section 16.9.

1.8 Gauge 'symmetries'

While symmetries play an important rôle for many aspects of physics, in different contexts the term 'symmetry' can have rather different meanings. By a *symmetry* in the strict sense, we understand a mapping of the space of physical states of a system which respects the dynamics of the system, but maps states to physically distinct states. The symmetries analyzed in the previous sections are all of this type.

Also, we have always described the mappings of physically different states from the so-called *active* point of view. This is to be distinguished from the *passive* point of view which is conceptually rather different, although superficially many manipulations look almost the same. In the passive picture one analyzes the effects of coordinate changes on the description of a system. The difference between the two pictures is illustrated conveniently by the example of general relativity. When taking the active point of view, one considers diffeomorphisms of the space-time manifold (which are defined everywhere), while when using the passive point of view one deals with changes of (local) coordinates. The two descriptions are linked by the fundamental postulate of general relativity, the equivalence principle, which states that the *form* of the laws of physics is invariant under local changes of the coordinates. (It is worth noting that this principle is sometimes misstated by saying that 'the physics' or 'the theory' is invariant under coordinate changes. But that physical phenomena do not depend on the particular coordinate system that is adopted to describe them is a tautological statement. What *is* non-trivial is the assertion that the equations which govern the theory look the same in all coordinate systems.)

Of a quite different nature than the symmetries in the strict sense, i.e. mappings between distinct physical states, are the so-called *gauge symmetries*. For a gauge symmetry, all configurations which are related by symmetry transformations are to be regarded as corresponding to a *single* physical state. In other words, this kind of symmetry is part of the definition of what one really understands by physical states. A symmetry transformation of this type is thus *not* a mapping from a physical state to another physical state; rather, it realizes the redundancy in the description of the configurations of the system. (This has nothing to do with the passive point of view described above. The basic feature of gauge theories is that in any convenient description of the system each physical state is represented by several different configurations.)

The term gauge symmetry is also often used in a more specific context. Namely, in field theory it refers to transformations which correspond to automorphisms of a so-called principal fiber bundle over space-time and hence can be presented locally (though in general not globally) by maps from space-time to a group of internal transformations, the *structure group* of the theory. For example, in electrodynamics the structure group is the group U(1) of complex numbers of absolute value 1; gauge transformations act on the vector potential $A_\mu(x)$ as

$$A_\mu \;\mapsto\; A_\mu + \partial_\mu \Lambda \tag{1.34}$$

with some function $\Lambda(x)$, so that the physical states are not labelled by the vector potential, but rather by the vector potential modulo arbitrary gradients.

While in this specific context gauge group elements are described as 'local' transformations which vary smoothly in space-time, this is *not* the characteristic feature of a gauge symmetry. (For instance, when a gauge theory is formulated on a space-time lattice rather than in the continuum, discrete gauge symmetries can play the same rôle as continuous gauge symmetries; see e.g. chapter 16 of [Creutz 1983].) Rather, the distinctive property of gauge symmetries is that they represent redundancies. Also, the present use of the qualification 'local' (which refers to space-time) must not be confused with the locality concept that appears in the notion of local one-parameter groups of symmetries, where 'local' refers to the space of parameters of the symmetries.

1.9 Duality symmetries

Sometimes the term symmetry is also used in yet another context. One is given a collection of distinct theories, and a priori different theories in this set are related by suitable 'symmetry' relations. An example for this kind of symmetries is the so-called Kramers–Wannier duality of the Ising model, which relates quantities in the Ising model at high temperature to other quantities at low temperature (see e.g. chapter 5.4 of [Feynman 1972]). Another class of examples is provided by the so-called space-time dualities in string theory and 'electric-magnetic' dualities in supersymmetric Yang–Mills theories (for a review, see [Intriligator and Seiberg 1995]). Expressed in terms of the fundamental fields of the theory, duality symmetries in quantum field theory are realized as highly non-local transformations; in some cases (such as the electric-magnetic dualities just mentioned) this has the puzzling consequence that even the interpretation of a conserved charge as being a Noether charge or a topological charge is changed by the transformation.

Admittedly, there is no clear distinction between symmetries which relate the states of one and the same theory and those which relate different theories, since in order to distinguish between the two possibilities one first has to specify what one understands by 'a theory'. In fact, the attitude as to whether two situations correspond to distinct theories or are merely two realizations of one and the same theory has in many cases changed with time. In the example of the Ising model, one might wish to consider not the Ising model at a fixed temperature as one 'theory', but rather the family of Ising models at all different temperatures, thereby promoting the Kramers–Wannier duality to a symmetry of the first type.

In practice, there is frequently a natural notion of what one should consider as a theory, so that any further analysis of the difference between these types of symmetries would be a rather academic exercise. On the other hand, it can be a major breakthrough in science to discover a symmetry between a priori different theories and then promote it to a symmetry of the first type. Such a paradigm change is currently advocated in supersymmetric gauge theories and superstring theory with extended supersymmetry.

Summary:

Symmetries in physical systems lead to the structure of groups and Lie algebras. The Noether theorem describes how in a Lagrangian setting symmetries yield conservation laws. To apply the Noether theorem only a local one-parameter group of symmetries is necessary; accordingly algebras can be more comprehensive than groups of symmetries.

In the Hamiltonian description of classical mechanics, conserved quantities are described by functions on phase space that Poisson-commute with the Hamiltonian. Similarly, in quantum mechanics, any observable commuting with the Hamilton operator is conserved in time; conserved quantities imply selection rules on the matrix elements of the Hamiltonian.

Keywords:

Symmetry, (local) one-parameter family of symmetries, symmetry group, infinitesimal symmetry;
Noether's theorem, conserved charge, topological charge, Poisson bracket, Jacobi identity;
selection rule, active and passive point of view, gauge symmetry.

Exercises:

Use Noether's theorem to calculate the conserved quantity associated to a rotation in a three-dimensional space $\mathbb{R}^3$ around the axis given by the vector $\vec{a}$.

Exercise 1.1

Hint: In this case we have

$$\frac{\mathrm{d}h_s}{\mathrm{d}s}(\vec{x}) = \vec{a} \times \vec{x}, \qquad (1.35)$$

where '$\times$' denotes the cross product.

Check the three characteristic properties (1.16) – (1.18) of the Poisson bracket

Exercise 1.2

$$\{f, g\} = \sum_{i=1}^{n} \left(\frac{\partial f}{\partial q_i} \frac{\partial g}{\partial p_i} - \frac{\partial g}{\partial q_i} \frac{\partial f}{\partial p_i} \right) \qquad (1.36)$$

of smooth functions f, g on the phase space (with canonical coordinates q_i and conjugate momenta p_i ($i = 1, \ldots, n$)).
Deduce the relation (1.19) between time derivatives and Poisson brackets with H.

Show that with f and g also $\{f, g\}$ Poisson-commutes with the Hamilton function H.

Consider the bracket

Exercise 1.3

$$\{\{f, g\}\} := \sum_{i,j=1}^{n} \Omega^{ij} \left(\frac{\partial f}{\partial q_i} \frac{\partial g}{\partial p_j} - \frac{\partial g}{\partial q_i} \frac{\partial f}{\partial p_j} \right) \qquad (1.37)$$

in place of (1.36).

Which property must the matrix Ω possess in order that this bracket satisfies the Jacobi identity? Can it depend on p and q?

(A manifold like (p, q)-space, for which any tangent space is endowed with such a structure Ω which depends differentiably on p and q, is called a symplectic manifold. Such manifolds play an important rôle in the description of Hamiltonian systems and in differential geometry. For an introduction, we refer the reader to chapter 8 of [Arnold 1978], to chapter 5 of [Marsden and Ratiu 1994], and to chapter 5 of [Scheck 1994].)

Check that a finite translation can be represented by an operator of the form (1.33). Hint: Expand in a series.

Exercise 1.4

Verify that the operators (1.31) are unitary, i.e. obey $(U_A)^{\dagger} = (U_A)^{-1}$, and that by the multiplication law (1.32) they form a unitary one-parameter group.

Exercise 1.5

Show that unitary operators preserve the inner product of Hilbert space vectors,

$$\langle U_A \psi \mid U_A \psi' \rangle = \langle \psi \mid \psi' \rangle. \qquad (1.38)$$

2

Basic examples

2.1 Angular momentum

Before we start to develop the theory of Lie algebras systematically, let us have a look at two examples with which many of the readers will have some familiarity: First, the algebra of angular momentum in quantum mechanics – this is a finite-dimensional algebra; and second, the algebra that arises when one describes a free scalar field on a one-dimensional space in the formalism of 'second quantization' – this algebra is infinite-dimensional. We will see that the structure of both algebras exhibits a rather similar pattern. The goal of this chapter is not to give a thorough treatment of these matters, but rather to describe the pattern in an informal way. In the subsequent chapters we will develop the theory of these structures in a much more complete manner and in particular carefully state the relevant definitions. So if in the present description the reader encounters terminology with which he or she is not yet familiar, it is still a good idea to continue nonetheless, and possibly to reconsider the problematic point after the relevant notion has been properly introduced in one of the chapters that follow. Conversely, after having learned more about the general structure of Lie algebras, the reader might wish to return to this chapter to see how things work in the special cases treated here; this might be a particularly good idea for the more mathematically minded reader.

Similarly as translations are generated by momentum, rotations in three-dimensional space are generated by the components of angular momentum (compare exercise 1.1). To describe angular momentum in non-relativistic quantum mechanics, one introduces self-adjoint operators L_i on the Hilbert space, one for each of the three components of angular momentum. They can be expressed through the position operator $\vec{q}$ and the momentum operator $\vec{p}$ (which in the position space picture acts as

$p_j = -\mathrm{i}\,\frac{\partial}{\partial q_j}$) by the formula $\vec{L} = \vec{q} \times \vec{p}$, or in components,

$$L_i = \sum_{j,k=1}^{3} \epsilon_{ijk}\, q_j\, p_k\,, \tag{2.1}$$

where ϵ_{ijk} is the totally antisymmetric tensor in three dimensions. (Here and below we follow – unlike in chapter 1, compare e.g. equation (1.22) – the convention to work with units such that Planck's constant $\hbar$ is unity, $\hbar = 1$, i.e. in particular dimensionless.) It is then easy to see (see exercise 2.2) that the canonical commutation relations

$$[q_i, p_j] \equiv q_i p_j - p_j q_i = \mathrm{i}\,\delta_{ij} \tag{2.2}$$

between position and momentum operators imply the following commutation relations for the components of angular momentum:

$$[L_i, L_j] = \mathrm{i}\sum_{k=1}^{3} \epsilon_{ijk} L_k\,. \tag{2.3}$$

More generally, in quantum mechanics the commutator of two operators leads again to operators in the considered set (provided that the commutator can be properly defined at all, compare the remarks after equation (1.23)). In mathematical terms, this means that the commutator defines a bilinear and antisymmetric 'product' on the space of operators; in addition, this operation satisfies the relation

$$[A, [B, C]] + [B, [C, A]] + [C, [A, B]] = 0 \tag{2.4}$$

between double 'products', which is the *Jacobi identity* which we already encountered in formula (1.26).

For commutators, the identity (2.4) is immediate – just write out each of the three double commutators in terms of ordinary products ABC etc. The reason why we nevertheless state it explicitly here is that later on, in the more abstract description, the Jacobi identity will no longer be automatic and correspondingly play a fundamental rôle. The product provided by the commutator turns the vector space that is spanned by the operators of interest, say by the angular momentum operators L_i, into what is called a Lie algebra.

2.2 Step operators

In the standard treatment of angular momentum the next step is to consider instead of the operators L_i the linear combinations

$$L_\pm := L_1 \pm \mathrm{i}L_2\,, \quad \text{and} \quad L_0 := 2L_3\,. \tag{2.5}$$

Note that here we admit *complex* linear combinations of the original generators L_i. As a consequence, the operators $L_\pm$ are not hermitian any more; rather, we have

$$(L_+)^\dagger = L_-. \tag{2.6}$$

Also, while the vector space that the operators (2.5) span over the complex numbers $\mathbb{C}$ is the same as the one spanned by the operators L_i, the vector spaces spanned by the respective sets of operators over the real numbers $\mathbb{R}$ are different. In more mathematical terms, in the first case we are dealing with one and the same complex Lie algebra for which the relation between the L_i and $L_\pm$, L_0 is just a basis transformation, while in the second case the two sets $\{iL_j \mid j = 1, 2, 3\}$ and $\{L_\pm, L_0\}$ of operators generate two distinct real Lie algebras. The relation between complex and real Lie algebras involves quite a few subtleties, which we will have to explain later (see chapter 8).

For the moment, the only feature the reader must be aware of is that, since the relation (2.5) involves complex numbers, the respective Lie algebras are equivalent as complex Lie algebras, but not as real Lie algebras. When one considers instead of the L_j the multiples iL_j ($j = 1, 2, 3$) as generators, then according to (2.3) all structure constants are real, and the vector space over the real numbers spanned by these generators is a real Lie algebra. This real Lie algebra spanned by the iL_j is referred to as $\mathfrak{su}(2)$. (In physics it is common to consider instead of the iL_j directly the generators L_j. When doing so, the 'structure constants' $i\epsilon_{ijk}$ are purely imaginary rather than real.) The real Lie algebra spanned by the generators (2.5) is not the same as $\mathfrak{su}(2)$; it is known as $\mathfrak{sl}(2, \mathbb{R})$. For the algebra over the complex numbers one writes $\mathfrak{sl}(2, \mathbb{C})$, or briefly, $\mathfrak{sl}(2)$.

We also note that a common realization of Lie algebras is through matrices (for an example, see e.g. equation (3.19); more general Lie algebras will be dealt with in section 5.6). It proves to be convenient to write the generators of certain *real* Lie algebras in terms of complex matrices, because using the multiplication rules for complex numbers then allows us to simplify expressions considerably. Hence it is in particular *not* true that the entries of the elements of a matrix Lie algebra over the real numbers are all real.

The elements (2.5) of the new basis obey the commutation relations

$$[L_0, L_\pm] = \pm 2 \, L_\pm, \qquad [L_+, L_-] = L_0. \tag{2.7}$$

The first of these two relations follows from the formulæ (2.3) as

$$\begin{aligned}
[L_0, L_\pm] &\equiv [2L_3, L_1 \pm iL_2] = 2i \cdot L_2 \pm 2i \cdot (-iL_1) \\
&= \pm 2 \, (L_1 \pm iL_2) \equiv \pm 2 \, L_\pm,
\end{aligned} \tag{2.8}$$

and the second relation is obtained in a similar manner (see exercise 2.3).

To make use of the symmetry algebra in the concrete quantum mechanical problem, one would like to obtain more information on how these operators act on the Hilbert space of physical states. To this end, the operators $L_\pm$ and L_0 are used quite differently. According to the relations (2.3), the generators L_k do not commute for different values of k. However, each L_k commutes with the operator

$$\vec{L}^2 \equiv L_1^2 + L_2^2 + L_3^2 = \tfrac{1}{4}L_0^2 + \tfrac{1}{2}\left(L_+L_- + L_-L_+\right) \qquad (2.9)$$

for total angular momentum (compare exercise 2.2). Therefore one can diagonalize simultaneously the operators L_0 and $\vec{L}^2$. Accordingly, one uses for the Hilbert space a basis of eigenvectors of L_0 and $\vec{L}^2$. Note that the operator $\vec{L}^2$ is quadratic rather than linear in the generators L_i (or $L_\pm$, L_0); as it turns out, it is *not* an element of the Lie algebra. Nevertheless it is a well-defined operator; for more complicated Lie algebras, it is often helpful to consider arbitrary powers of the generators, too, but it is important to keep in mind that such objects do not belong to the Lie algebra itself.

Let us denote by v_λ an eigenvector of L_0 to the eigenvalue, or *weight*, λ:

$$L_0\, v_\lambda = \lambda \cdot v_\lambda\,. \qquad (2.10)$$

Next we consider the Hilbert space vectors $L_\pm v_\lambda$ that are obtained from v_λ by applying to it the two other basis elements $L_\pm$; one finds that acting on $L_\pm v_\lambda$ with L_0 yields a scalar multiple of $L_\pm v_\lambda$:

$$L_0(L_\pm v_\lambda) = L_\pm L_0 v_\lambda \pm 2L_\pm v_\lambda = (\lambda \pm 2) \cdot L_\pm v_\lambda\,. \qquad (2.11)$$

This is again an eigenvalue equation, i.e. the vectors $L_\pm v_\lambda$ are again eigenvectors of L_0, with eigenvalues $\lambda \pm 2$, respectively. Thus the non-hermitian operators $L_\pm$ raise and lower, respectively, the L_0-eigenvalue; they are therefore called *step operators*, or more specifically, *raising* and *lowering* operators, respectively. Symbolically, we can depict this situation as follows:

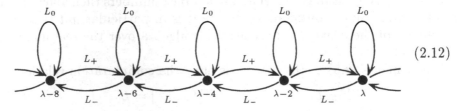

$$(2.12)$$

2.3 Irreducible representations

The set of all vectors in the Hilbert space which are connected to some given vector v_λ by the action of the algebra forms a subspace of the Hilbert space. As far as the action of the algebra is concerned, we can deal with each such subspace separately. A more careful analysis shows that in the situation of our present interest we can decompose the total Hilbert space into finite-dimensional subspaces, the so-called irreducible representations. (Later on, in chapter 13, we will see that at this point it is crucial that we deal with Hilbert spaces, i.e. that a positive definite scalar product can be defined on the relevant representation spaces, and that there exists a basis (namely the L_i) which consists of *hermitian* generators of the symmetry algebra.)

Since the irreducible representations are finite-dimensional, there must exist an element v_Λ in the basis of eigenvectors of L_0 which has the highest eigenvalue within the representation. This special eigenvalue is denoted by Λ and is called the *highest weight* of the representation. When acting with L_- several times on v_Λ, we obtain more vectors in the same irreducible representation,

$$v_{\Lambda-2n} := (L_-)^n v_\Lambda \qquad (2.13)$$

for any positive integer n. Using again the fact that irreducible representations are finite-dimensional, it follows that all these vectors except for finitely many actually have to vanish. So let $v_{\Lambda-2N}$ be the last non-vanishing element in this string of Hilbert space vectors. This vector is itself non-zero, but it is annihilated by L_-:

$$L_- v_{\Lambda-2N} = 0. \qquad (2.14)$$

The string of vectors appearing in (2.13) is constructed in such a way that the action of L_0 and L_- closes on it; to get more information, we also have to analyze the action of L_+. Using the fact that $L_+ v_\Lambda = 0$, it follows that

$$L_+ v_{\Lambda-2} = L_+ L_- v_\Lambda = [L_+, L_-] v_\Lambda = L_0 v_\Lambda = \Lambda \cdot v_\Lambda, \qquad (2.15)$$

i.e. that $L_+ v_{\Lambda-2}$ is a scalar multiple of v_Λ. By induction one can show that, more generally, $L_+ v_\lambda$ is proportional to $v_{\lambda+2}$. Indeed, we can use the commutation relations (2.7) to derive that

$$
\begin{aligned}
L_+ v_{\Lambda-2n} &= L_+ L_- v_{\Lambda-2n+2} \\
&= (L_- L_+ + L_0) v_{\Lambda-2n+2} = (r_{n-1} + \Lambda - 2n + 2) \cdot v_{\Lambda-2n+2},
\end{aligned} \qquad (2.16)
$$

where we introduced constants of proportionality r_n by

$$L_+ v_{\Lambda-2n} = r_n \cdot v_{\Lambda-2n+2}. \qquad (2.17)$$

Equation (2.16) implies that these constants satisfy the recursion relation $r_n = r_{n-1} + \Lambda - 2n + 2$. Now equation (2.11) tells us that the vector $L_+ v_\Lambda$, if non-vanishing, would have eigenvalue $\Lambda + 2 > \Lambda$, contradicting the assumption that Λ is the maximal eigenvalue; hence we need $r_0 = 0$. Using this initial condition, the recursion relation implied by (2.16) is easy to solve (see exercise 2.5), leading to

$$r_n = n(\Lambda - n + 1).$$ (2.18)

To determine the integer N in (2.14), we use the fact that the vector $L_- v_{\Lambda-2N}$ vanishes to find

$$0 = L_+ L_- v_{\Lambda-2N} = (L_- L_+ + L_0) v_{\Lambda-2N} = (r_N + \Lambda - 2N) \cdot v_{\Lambda-2N}.$$ (2.19)

Substituting $r_N = N(\Lambda - N + 1)$, this yields the quadratic equation $N^2 + (1 - \Lambda)N - \Lambda = 0$ which possesses the two solutions $N = -1$ and $N = \Lambda$. Since by construction N must be a non-negative integer, the first solution must be discarded, and thus we learn that the highest weight Λ has to be a non-negative integer.

We can now read off the dimension of the irreducible representation with highest weight Λ: it is

$$\dim = N + 1 = \Lambda + 1.$$ (2.20)

In physical applications it is common to use instead of the weights λ, i.e. the eigenvalues with respect to L_0, the eigenvalues $j_3 := \lambda/2$ with respect to the operator $L_3 = L_0/2$. The analogue of the highest weight is then called the *spin*

$$j = \tfrac{1}{2}\Lambda$$ (2.21)

of the representation. In terms of the spin, the formula for the dimension reads

$$\dim = 2j + 1.$$ (2.22)

Besides the dimension, there is another important quantity characterizing an irreducible representation, namely the eigenvalue of the operator $\vec{L}^2$ (2.9). As a consequence of the identity (see exercise 2.3)

$$[\vec{L}^2, L_i] = 0,$$ (2.23)

this eigenvalue is constant on the whole irreducible representation space. Later we will see that also for many more complicated Lie algebras there exists an analogous quantity, the so-called quadratic (or second order) *Casimir operator*. In the present case, a simple calculation (see exercise 2.6) shows that the eigenvalue of this operator is given by

$$j(j + 1) = \tfrac{1}{4}\Lambda(\Lambda + 2).$$ (2.24)

2.4 The free scalar field

Our second example also comes from quantum theory, this time in the formalism of 'second quantization'. This is one of the techniques used for the quantization of field theories, i.e. dynamical systems with an infinite number of degrees of freedom (whereas quantum mechanics deals with the quantum theory of dynamical systems with finitely many degrees of freedom). In field theory, the rôle of the position operators $q_i \equiv q_i(t)$ is taken over by the fundamental fields Φ of the theory, and the labels i corresponding to the three components of position are replaced by a continuous parameter x which takes its values in position space. The analogue of the momentum operators p_i are now the canonical momenta of the fields, which in a Lagrangian description are obtained as derivatives of the Lagrangian density $\mathcal{L}$ with respect to the time derivatives of the fields.

There is also an analogue of the Noether charge $Q = (\partial L/\partial \dot{q}) \, \delta q$ that is associated to the variation δq of q and that satisfies the conservation equation $dQ/dt = 0$. The field theory analogue is a current $j^\mu :=$ $(\partial \mathcal{L}/\partial(\partial_\mu \Phi)) \, \delta \Phi$. This current, called the *Noether current* for the variation $\delta \Phi$, depends on all coordinates $x^\mu = (x, t)$; its conservation equation reads $\sum_\mu \partial_\mu j^\mu = 0$. If the space-time can be written as the product of a real line representing a global time coordinate and a space-like hypersurface M, the integral Q of j over M does not depend on the global time coordinate; this integral Q is sometimes called Noether charge as well.

To be specific, we consider the relativistic field theory of N free bosons, i.e. of N free bosonic hermitian fields $\Phi^j(x, t)$, $j = 1, 2, \ldots, N$. The dynamics is then described by the Lagrangian density

$$\mathcal{L}(x, t) = \sum_{j=1}^{N} \left(\tfrac{1}{2} (\partial_t \Phi^j)^2 - \tfrac{1}{2} |\nabla \Phi^j|^2 \right) - V(\Phi), \qquad (2.25)$$

where $\partial_t \equiv \frac{\partial}{\partial t}$ and ∇ are the derivatives with respect to time and space, respectively, and where V is some potential function (e.g. a mass term $V_m = m^2 \Phi^2$). The canonical momenta are then

$$\pi^j(x, t) := \frac{\partial \mathcal{L}}{\partial(\partial_t \Phi^j(x, t))} = \partial_t \Phi^j(x, t). \qquad (2.26)$$

As in the example of angular momentum, again the crucial input for finding the desired Lie algebra commutators is provided by the canonical commutation relations. For the free boson field theory they read

$$[\pi^j(x, t), \Phi^k(y, t)] \equiv [\partial_t \Phi^j(x, t), \Phi^k(y, t)] = -\mathrm{i} \, \delta^{jk} \, \delta(x - y), \qquad (2.27)$$

where $\delta(x)$ is the Dirac delta function.

2.5 The Heisenberg algebra

The analysis of the free boson theory simplifies considerably when one considers the special situation that the space on which the fields Φ^j live is a one-dimensional circle of radius 1. Since the Heisenberg commutation relations (2.27) are equal-time commutators, from now on we will consider all fields at a fixed value of t and suppress the time variable in our notation. Because of the compactness of the circle, the momentum can only take discrete values $n \in \mathbb{Z}$, and we can introduce the Fourier components a_n^j, $n \in \mathbb{Z}$, of the fields:

$$\Phi^j(x) = a_0^j + \sum_{n \in \mathbb{Z} \setminus \{0\}} \frac{1}{n} a_n^j e^{2\pi i n x} . \tag{2.28}$$

As the fields on the right hand side of (2.28) are operators, the coefficients a_n^j are operators as well. The commutation relations among them are obtained by taking the Fourier transform of the canonical commutation relations (2.27). An easy calculation (see exercise 2.7) shows that

$$[a_n^i, a_m^j] = n \, \delta^{ij} \, \delta_{n+m,0} . \tag{2.29}$$

Since these commutators are derived from the canonical commutation relations which underlie Heisenberg's uncertainty relation, the Lie algebra that is generated by the operators a_n is known, in the case of one single boson ($N = 1$) as the *Heisenberg algebra*.

Note that the subscript n of a_n^j takes arbitrary integral values; thus the Heisenberg algebra is an infinite-dimensional Lie algebra, quite in contrast to the previously described Lie algebra $\mathfrak{sl}(2)$ which has three generators and hence is three-dimensional. Nevertheless there is a close similarity between certain features of these algebras. In particular, the algebra spanned by the a_n^j can be split into three subalgebras, i.e. subspaces on which the commutator closes:

▷ First, there are the N generators a_0^j for all values of j. These generators commute among each other, and provided that the quantum fields Φ^j are hermitian fields, they possess the particular property of being hermitian.

▷ There is an infinite number of so-called *creation operators*. These are the generators a_n^i with $n \geq 1$.
These operators are analogues of the raising operator L_+.

▷ Finally, there are also *annihilation operators*, namely all a_n^i with $n \leq -1$.
These are analogues of L_-.

The reason why these operators are called, respectively, creation and annihilation operators will become clear soon. Just like the angular momentum operators $L_\pm$, these operators are not hermitian, but rather we have (in case the fields Φ^j are hermitian),

$$\left(a_n^j\right)^\dagger = a_{-n}^j , \tag{2.30}$$

analogous to the relation $(L_-)^\dagger = L_+$. The operators a_0^j are somewhat similar to the single generator L_0 in the angular momentum example.

However, in contrast to L_0, according to the relations (2.29) they do not only commute among each other, but also with all other generators.

Let us remark that for $n \neq 0$ we can rescale the operators according to

$$ b_{j;n} := \tfrac{1}{\sqrt{n}}\, a^j_{-n}\,, \qquad b^+_{j;n} := (b_{j;n})^\dagger = \tfrac{1}{\sqrt{n}}\, a^j_n\,. \qquad (2.31) $$

The operators $b_{j;n}$ and $b^+_{j;n}$ that are introduced this way realize, for each value of j and each $n \in \mathbb{Z}_{>0}$ separately, the commutator algebra of the creation and annihilation operators which appear in the quantum mechanical description of the harmonic oscillator. Therefore the algebra spanned by the operators (2.31) (and sometimes also (2.29)) is called the *oscillator algebra*. The same method that is used in quantum mechanics to quantize the harmonic oscillator can then be employed to construct a representation space, the so-called (bosonic) *Fock space*, for the algebra which is defined by the relations (2.29).

We start with one distinguished non-zero vector of the representation space, which will be called the *vacuum state* or simply the *vacuum*, and be denoted by $v_\Omega \equiv |\Omega\rangle$. By definition, v_Ω is the vector which is the quantum mechanical ground state for all N harmonic oscillators that are described by the operators (2.31). This means that application of any of the operators a^j_n with $n \leq -1$ annihilates this state. This property of the vacuum v_Ω is the source of the term 'annihilation operator' that is used for the a^j_n with $n \leq -1$; it should be compared to the analogous property of the highest weight vector v_Λ in the angular momentum example.

Applying all possible combinations of creation operators to the vacuum state yields all vectors in the representation of (2.29) that we want to describe. This representation is called a *Fock space* representation. (More generally, whenever the whole space of physical states can be obtained by acting with a symmetry algebra on a single state vector v, typically a vacuum vector, the symmetry algebra is called a *spectrum generating symmetry* and the vector v a *cyclic* vector.) In the context of second quantization, the action of the raising operator a^j_n can be described in words by saying that it creates a 'particle' of type j with momentum n. Just like for the ordinary harmonic oscillator, an arbitrary state in the Fock space can then be characterized by occupation numbers. Also, since the momentum can be arbitrarily large, any Fock representation contains infinitely many independent states. Thus not only the Lie algebra, but also its representations, are infinite-dimensional.

The occupation numbers tell us how many particles of prescribed type i and momentum n are present in a specific state. Up to multiplication by a non-zero number, they characterize a state of the Fock space completely, i.e. there is no degeneracy in the quantum numbers. This is however specific to the present situation and not a generic property of Lie algebra

representations; we will encounter many examples for degeneracies later
on.

2.6 Common features

In the two situations considered above, a number of structural similarities
have shown up. These structures are in fact common to a large class of
Lie algebras, and in particular to essentially all Lie algebras that will be
treated in this book. From the two examples we have given it is also
apparent that many structural features do not depend on whether the
algebra is finite- or infinite-dimensional. As a consequence, for many
aspects the analysis of infinite-dimensional Lie algebras does not require
much special effort, so that often we need not make any assumption on
the dimensionality of the algebra. Below we will summarize the common
features of the two examples in a manner which should help to recognize
them later on when they appear in various disguises.

The main property of the algebras which we have encountered above is
that they split naturally into three parts, all of which are subalgebras:

▷   First, a maximal set of operators which all commute. (Note that
    in the case of angular momentum we do not count the operator
    $\vec{L}^2$ as belonging to this set of operators, since it is *not* an element
    of the Lie algebra $\mathfrak{sl}(2)$.) In quantum mechanical terms, they
    correspond to observables that can be measured simultaneously.
    Thus they give rise to a maximal set of quantum numbers for the
    vectors in the representations of the algebra. In the general case,
    the analogue of this subalgebra is the *maximal abelian subalgebra*,
    which for a particularly important class of Lie algebras (including
    e.g. $\mathfrak{sl}(2)$) is also known as the *Cartan subalgebra*.
    Note that there is some freedom in the choice of the maximal
    abelian subalgebra. For instance, in the angular momentum ex-
    ample we could have chosen to define

$$L_{\pm} := L_2 \pm iL_3\,, \qquad L_0 := 2L_1 \qquad\qquad (2.32)$$

instead of (2.5); then $2L_1$ rather than $2L_3$ would have been the
generator of the maximal abelian subalgebra. As will be explained
in detail later, all relevant properties of a Lie algebra and of its representations do, however, not depend on which maximal abelian
subalgebra one chooses.

▷   The remaining generators can be separated into raising (or cre-
    ation) operators and lowering (or annihilation) operators. Each of
    these two types of generators forms again a subalgebra, and they
    are related by hermitian conjugation.

A decomposition of this type will be present for almost all Lie algebras that we will analyze in this book. There are of course also important structural differences in the two examples we displayed. For instance, in the case of angular momentum the commutator of L_0 with any of the other generators $L_\pm$ is a non-vanishing multiple of $L_\pm$, whereas in the free boson example the operators a_0^j have vanishing commutator with all other generators.

In the construction of the spaces on which these algebras act, the representation spaces, there are again many similarities. In particular a 'vacuum' or 'highest weight' vector, which is annihilated by the subalgebra of annihilation operators, plays an essential role. The operators in the third subalgebra then 'create' all other vectors in the representation. In fact, one can think of the spaces on which the algebra acts simply as spaces that are *created* by the action of this subalgebra, rather than being a priori given vector spaces. Finally, the maximal abelian subalgebra provides the quantum numbers of the Hilbert space vectors. Namely, one can choose a basis in which all elements of this subalgebra act diagonally; then their eigenvalues are precisely the quantum numbers or, in mathematical terms, the weights, of the Hilbert space vectors.

Our main task in the next few chapters will be to describe all the structures we sketched above in a systematic manner and in much more detail, and to show how they generalize to a wide class of Lie algebras. However, before doing so, we will analyze in chapter 3 one more example.

To conclude this chapter, we would like to point out that with complete knowledge about $\mathfrak{sl}(2)$ one already has a lot of the information that is required when dealing with many more Lie algebras, in particular the 'simple' Lie algebras. Not all this information has been provided in the present chapter; rather, we will come back to the Lie algebra $\mathfrak{sl}(2)$ and its representations at various places in the book. In particular, in section 5.7 we treat the representation theory once more, in a somewhat different notation which is better adapted to generalization; in section 13.1 the finite-dimensional representations are summarized with emphasis on their 'highest weight' properties. Tensor product decompositions of finite-dimensional $\mathfrak{sl}(2)$-representations are described in section 15.4, and the corresponding Clebsch–Gordan coefficients in chapter 16. The relation between the Lie *groups* SU(2) and SO(3) which both possess $\mathfrak{su}(2)$ as their Lie algebra is examined in section 9.6 and section 20.5. So-called multiplier representations of the group SL(2) are described in section 21.3, and their relation with the spherical harmonics is explained in section 21.5.

Summary:

Two examples of Lie algebras are provided by $\mathfrak{su}(2)$ and by the infinite-dimensional Heisenberg algebra.
Both are well-known from elementary quantum mechanics and second quantization, respectively. This way the basic structures emerging in Lie algebra theory can be motivated naturally from basic principles of quantum theory.

Keywords:

Angular momentum, $\mathfrak{su}(2)$ and $\mathfrak{sl}(2)$, raising and lowering operators, representation;
Heisenberg algebra, creation and annihilation operators, Fock space;
maximal abelian subalgebra, Cartan subalgebra, maximal number of quantum numbers

Exercises:

Compute the Poisson brackets of the angular momentum components (2.1) in classical mechanics.

Exercise 2.1

Show that the generators L_i of angular momentum are hermitian.
Hint: Use the fact that the operators for position $\vec{q}$ and momentum $\vec{p}$ are hermitian and the canonical commutation relations (2.2).
Check the commutation relations (2.3) for angular momentum, and show that $\vec{L}^2$ commutes with each of the components L_i, $[\vec{L}^2, L_i] = 0$.

Exercise 2.2

Prove the commutation relation $[L_+, L_-] = L_0$.

Exercise 2.3

Check that the operation of forming the commutator (of matrices, or operators) satisfies the Jacobi identity (2.4).

Exercise 2.4

Show that the recursion relation $r_n = r_{n-1} + \Lambda - 2n + 2$ with initial value $r_0 = 0$ is solved by $r_n = n(\Lambda - n + 1)$.

Exercise 2.5

Check that the quadratic Casimir operator $\vec{L}^2$ satisfies (2.9), i.e. that it can be expressed as

Exercise 2.6

$$\vec{L}^2 = \tfrac{1}{4} L_0^2 + \tfrac{1}{2} \left(L_+ L_- + L_- L_+ \right). \qquad (2.33)$$

Compute its eigenvalue on an irreducible representation with highest weight Λ.

Hint: Since the value of $\vec{L}^2$ is constant on an irreducible representation, one can compute its eigenvalue by applying it on the highest weight vector v_Λ; use $L_+ v_\Lambda = 0$.

Show that the Heisenberg algebra (2.29) follows from the canonical commutation relations (2.27), and prove the relation (2.30).

Hint: Take the Fourier transform of (2.27).

Exercise 2.7

Represent the algebra (2.29) as multiplication and differential operators on the space of functions $f(t_1, t_2, ...)$ of countably many variables t_n. (For simplicity, consider only the case where the superscripts i, j in (2.29) can take only a single value.)

Exercise 2.8

3

The Lie algebra $\mathfrak{su}(3)$
and hadron symmetries

Before we start to develop the general theory of Lie algebras in the next chapter, we want to present one more example of a Lie algebra, together with some physical motivations. Those readers who want to concentrate on the mathematical theory of Lie algebras might wish to skip this chapter in a first reading.

3.1 Symmetries of hadrons

The Lie algebra $\mathfrak{su}(2)$ that was introduced in chapter 2 plays a fundamental rôle in many different areas of physics. An instructive example is provided by the mass spectrum of hadrons in elementary particle physics. The basic idea is that particles which have the same spin and parity and almost the same mass should be related by some kind of symmetry. A first step to construct the relevant symmetry is to group the hadron into irreducible representation spaces of $\mathfrak{su}(2)$; instead of $\mathfrak{su}(2)$ here one may also consider its complex version $\mathfrak{sl}(2)$, and in fact for some of the considerations below this is the more convenient algebraic structure. For instance, the proton and neutron can be regarded as the basis vectors of a two-dimensional representation space of $\mathfrak{sl}(2)$, the three pions $\pi^\pm$ and π^0 correspond to a three-dimensional representation, and the charged kaons K^+ and K^- together with the neutral kaons K^0 and $\bar{K}^0$, respectively, to two separate two-dimensional representations.

In the terminology of particle physics, the $\mathfrak{sl}(2)$-weight λ, i.e. the eigenvalue with respect to the generator L_0 of $\mathfrak{sl}(2)$, is a particular example of a *quantum number*, and the irreducible representation spaces are often referred to as *multiplets*. Also, it is conventional to use the number $\frac{1}{2}\lambda$ for the labelling of the particles; this quantum number is called (the third component of) the *isospin* and will be denoted by I. (The term isospin is short hand for 'isotopic spin'. It was introduced in nuclear physics and reflects the fact that the proton and the neutron – the two types of constituents which make up nuclei – form an isospin doublet.)

One of the major discoveries in the particle physics of the 1960s was the realization that in reactions involving electromagnetic and strong interactions another quantum number, in addition to isospin, is conserved, and that the zoo of hadrons can be organized in still larger multiplets of approximately equal mass by including this new quantum number in an appropriate way, or in other words, by enlarging sl(2) to a bigger symmetry. Rather than presenting this symmetry directly, we will try to obtain it in a more constructive manner. The basic ingredients are a few heuristic principles, the comparison with experimental data, and the representation theory of sl(2) that was sketched in chapter 2.

One should be aware of the fact that just counting the independent quantum numbers is not sufficient to describe the symmetry. Rather, one also has to know how the symmetries associated to the various quantum numbers are to be 'combined', and this determines, as we will shortly see, the structure of multiplets. In the terminology of Lie algebras, the quantum numbers correspond to the 'Cartan subalgebra', while the correct way of combining the various quantum numbers is prescribed by the commutation relations of the 'step operators'.

3.2 Combining two sl(2)-algebras

With some amount of hindsight, the generators of the isospin sl(2)-algebra will be denoted by $E^1_\pm$ and H^1 so that the sl(2) relations (2.7) read

$$[H^1, E^1_\pm] = \pm 2\,E^1_\pm\,, \qquad [E^1_+, E^1_-] = H^1\,. \qquad (3.1)$$

Since the basis vectors of the sl(2)-multiplets are labelled by the isospin $\frac{1}{2}H^1$ alone, the additional quantum number must correspond to a further independent generator of the symmetry, which will be denoted by H^2. As the two quantum numbers can be measured independently, the operators of which they are the eigenvalues must commute, i.e.

$$[H^1, H^2] = 0\,. \qquad (3.2)$$

The simplest extension of the symmetry would be obtained by requiring that H^2 commutes with E^1_+ and E^1_- as well, and that there are no further independent generators of the symmetry. But this would not enlarge the multiplets, but rather only attach one further quantum number to each isospin multiplet, and hence does not explain the observed higher (approximate) degeneracy of the hadron masses. Rather, what one also needs are additional step operators that are associated to the second quantum number analogously to the manner in which the step operators E^1_+ and E^1_- are associated to H^1. Thus we introduce operators $E^2_\pm$ which satisfy

$$[H^2, E^2_\pm] = \pm 2\,E^2_\pm\,, \qquad [E^2_+, E^2_-] = H^2\,. \qquad (3.3)$$

More precisely, the normalization of $E_\pm^2$ and H^2 can be chosen in such a way that (3.3) holds, and as seen in chapter 2, in this normalization the eigenvalues of H^2 in any finite-dimensional representation must be integers. The next step of the construction is then the specification of the commutators among $E_\pm^i$ and H^j for $i \neq j$. Requiring that the application of a step operator to any simultaneous eigenvector of H^1 and H^2 produces again such an eigenvector yields

$$[H^i, E_\pm^j] = \pm A^{ji} E_\pm^j \tag{3.4}$$

for $i, j \in \{1, 2\}$, with certain numbers A^{ji}.

We will now derive restrictions on the numbers A^{ji}. First, because of the $\mathfrak{sl}(2)$-relations (3.1) and (3.3), the diagonal elements ($i = j$) are fixed to $A^{ii} = 2$, while A^{12} and A^{21} can still be arbitrary numbers. The simplest possibility would now be to set A^{12} and A^{21} to zero, which corresponds to the situation that the application of $E_\pm^1$ does not change the H^2-eigenvalue and the application of $E_\pm^2$ does not change the H^1-eigenvalue. But this is in conflict with observation. Namely, one task of the higher symmetry is to combine the lightest mesons, i.e. the pions and the kaons, into a single multiplet; since the pions form an isospin triplet (with isospin eigenvalues 0 and ± 1) while the kaons form doublets (with eigenvalues $\pm\frac{1}{2}$), this is only possible if the step operators of the second $\mathfrak{sl}(2)$-algebra change the isospin eigenvalues: In a doublet, the isospin takes only half-integral values, while in a triplet it takes only integral values.

3.3 Multiple commutators

To proceed, it is necessary to consider also the commutators of $E_\pm^1$ with $E_\pm^2$, to which one gives names of their own,

$$E_\pm^\theta := [E_\pm^1, E_\pm^2] \,. \tag{3.5}$$

Because of the Jacobi identity

$$\begin{aligned}
[H^i, [E_\pm^1, E_\pm^2]] &= -[E_\pm^1, [E_\pm^2, H^i]] - [E_\pm^2, [E_\pm^1, H^i]] \\
&= \pm A^{2i} [E_\pm^1, E_\pm^2] \mp A^{1i} [E_\pm^2, E_\pm^1] = \pm (A^{1i} + A^{2i}) [E_\pm^1, E_\pm^2],
\end{aligned} \tag{3.6}$$

the $E_\pm^\theta$ are again step operators, and due to the similar identity

$$\begin{aligned}
[H^i, [E_-^1, E_+^2]] &= -[E_-^1, [E_+^2, H^i]] - [E_+^2, [E_-^1, H^i]] \\
&= (A^{2i} - A^{1i}) [E_-^1, E_+^2]
\end{aligned} \tag{3.7}$$

the same is true for the commutators $[E_\pm^1, E_\mp^2]$. Now a priori, the operators $E_\pm^\theta$ and / or $[E_\pm^1, E_\mp^2]$ could vanish. But in fact not both of them can, because this would be in conflict with the requirement that A^{12} and A^{21}

must not vanish. For example, the Jacobi identity implies that

$$[E^1_+, [E^1_-, E^2_+]] - [E^1_-, [E^1_+, E^2_+]] = -[E^2_+, [E^1_+, E^1_-]]$$

$$= -[E^2_+, H^1] \tag{3.8}$$

$$= A^{21} E^2_+,$$

which shows that for $A^{21} \neq 0$ not both E^θ_+ and $[E^1_-, E^2_+]$ can be zero, and analogous calculations are available for other combinations of $+$ and $-$ labels. Therefore there are at least two additional independent step operators; without loss of generality one can assume that these are E^θ_+ and E^θ_- (if they were zero with the chosen convention for $E^2_\pm$, they could be made non-zero by exchanging the rôle of E^2_+ and E^2_- which when accompanied by a redefinition $H^2 \mapsto -H^2$ would lead again to the $\mathfrak{sl}(2)$ relations (3.3)).

Thus taking commutators of step operators with $E^1_\pm$ or H^1 yields step operators again. In other words, the step operators carry a finite-dimensional representation of the $\mathfrak{sl}(2)$-subalgebra that is spanned by $E^1_\pm$ and H^1. According to the general results of chapter 2 about finite-dimensional representations of $\mathfrak{sl}(2)$, the eigenvalues of H^1 in such a representation space are integers. Thus we learn that the numbers A^{ji} appearing in the eigenvalue equation (3.4) must be integers.

The minimal possibility is then to impose the vanishing of the other two putative step operators, as well as of all further commutators among the E^1_+ and among the E^2_-, i.e. to set

$$[E^1_\pm, E^2_\mp] = 0,$$

$$[E^i_\pm, [E^1_\pm, E^2_\pm]] = 0. \tag{3.9}$$

Considering identities of the type $0 \overset{!}{=} [E^1_-, [E^1_+, E^\theta_+]] = \ldots = -(A^{11} + 2A^{21}) E^\theta_+$ shows that this corresponds to setting

$$A^{12} = A^{21} = -1. \tag{3.10}$$

With this choice the eigenvalues of H^1 change by ∓ 1 upon application of $E^2_\pm$; this is precisely the change required for relating an isospin triplet (the pions) to an isospin doublet (the kaons).

Putting these results together, the situation is described as follows. There are eight generators, the two generators H^1 and H^2 which correspond to the two distinct quantum numbers, and the six step operators $E^1_\pm$, $E^2_\pm$ and $E^\theta_\pm$. Their interrelations can be summarized in the following set of commutation relations:

$$[H^1, H^2] = 0\,, \qquad [E^1_+, E^1_-] = H^1\,, \qquad [E^2_+, E^2_-] = H^2\,,$$

$$[H^1, E^1_\pm] = \pm 2\,E^1_\pm\,, \quad [H^1, E^2_\pm] = \mp E^2_\pm\,, \quad [H^1, E^\theta_\pm] = \pm E^\theta_\pm\,,$$

$$[H^2, E^1_\pm] = \mp E^1_\pm\,, \quad [H^2, E^2_\pm] = \pm 2\,E^2_\pm\,, \quad [H^2, E^\theta_\pm] = \pm E^\theta_\pm\,, \qquad (3.11)$$

$$[E^1_\pm, E^2_\pm] = \pm E^\theta_\pm\,, \quad [E^1_\pm, E^\theta_\mp] = \mp E^2_\mp\,, \quad [E^2_\pm, E^\theta_\mp] = \pm E^1_\mp\,,$$

$$[E^1_\pm, E^2_\mp] = [E^1_\pm, E^\theta_\pm] = [E^2_\pm, E^\theta_\pm] = 0\,, \qquad [E^\theta_+, E^\theta_-] = H^1 + H^2\,.$$

Admittedly, this set of relations does not look particularly transparent. However, its structure becomes a lot clearer by introducing the notation

$$E^3_+ := E^\theta_-\,, \qquad E^3_- := E^\theta_+\,, \qquad H^3 := -H^1 - H^2\,. \qquad (3.12)$$

These operators satisfy

$$[H^3, E^3_\pm] = \pm 2\,E^3_\pm\,, \qquad [E^3_+, E^3_-] = H^3\,. \qquad (3.13)$$

Thus they generate again an $\mathfrak{sl}(2)$-algebra. (The isospin $\mathfrak{sl}(2)$-algebra given by (3.1) is sometimes also called the I-spin algebra; similarly, the $\mathfrak{sl}(2)$-algebras defined by equations (3.3) and (3.13) are then referred to as the U-spin algebra and the V-spin algebra, respectively.) Furthermore, the full set of relations (3.12) does not get changed at all if the labels $1, 2, 3$ are replaced by the cyclic permutations $2, 3, 1$ or $3, 1, 2$. This means that instead of starting with the generators with labels 1 and 2 we could have obtained the same symmetry structure in a completely analogous manner by starting either with the generators with labels 2 and 3 or with those with labels 3 and 1.

If follows in particular that the vectors of 'eigenvalues' of the eigenvalue equations $[H^i, E] \propto E$ that are obtained for $E \in \{E^1_+, E^2_+, E^\theta_+\}$ by commutation with the H^i, $i = 1, 2$, should be on an equal footing. Denoting these eigenvectors by $\alpha^{(1)}$, $\alpha^{(2)}$ and $-\theta$, respectively, one has

$$\alpha^{(1)} = (A^{11}, A^{12}) = (2, -1)\,, \qquad \alpha^{(2)} = (A^{21}, A^{22}) = (-1, 2)\,,$$

$$-\theta = -(A^{11} + A^{21}, A^{12} + A^{22}) = (-1, -1) \qquad (3.14)$$

(the reason for denoting the third eigenvector by the special symbol $-\theta$ will become clear later, see e.g. the relation (6.44)). We now require that each of the three pairs that can be formed from $\alpha^{(1)}$, $\alpha^{(2)}$ and $-\theta$ describes the same geometrical situation. It follows that the vectors $\alpha^{(1)}$ and $\alpha^{(2)}$ are not perpendicular, but rather they must form an angle of 120 degrees, so that the same angle is formed by $\alpha^{(2)}$ and $-\theta$, and by $-\theta$ and $\alpha^{(1)}$.

This is displayed in picture (3.15).

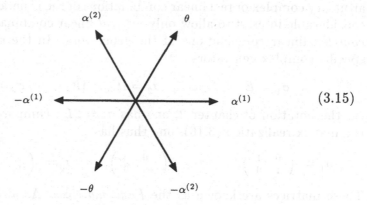

$$(3.15)$$

3.4 $\mathfrak{sl}(3)$ and $\mathfrak{su}(3)$

The algebraic structure described by the relations (3.11) is known as the Lie algebra $\mathfrak{sl}(3)$, or more precisely, as $\mathfrak{sl}(3,\mathbb{C})$ and $\mathfrak{sl}(3,\mathbb{R})$ if one allows for complex and real, respectively, linear combinations of the generators. The origin of this name is the fact that the relations can be realized through commutators of appropriate traceless complex (respectively real) 3×3-matrices. This is best understood by first treating the case of $\mathfrak{sl}(2)$ in an analogous manner. The $\mathfrak{sl}(2)$-relations (3.1) can obviously be satisfied by realizing the generators $H \equiv H^1$ and $E_\pm \equiv E_\pm^1$ as the following matrices:

$$H = \begin{pmatrix} 1 & 0 \\ 0 & -1 \end{pmatrix}, \qquad E_+ = \begin{pmatrix} 0 & 1 \\ 0 & 0 \end{pmatrix}, \qquad E_- = \begin{pmatrix} 0 & 0 \\ 1 & 0 \end{pmatrix}. \qquad (3.16)$$

These matrices are traceless, and in fact are linearly independent and span the space of all traceless 2×2-matrices. This is the reason why the Lie algebra which is characterized by the relations (3.1) is called $\mathfrak{sl}(2)$.

Similarly, the following eight traceless 3×3-matrices furnish a maximal linearly independent set of such matrices, and they satisfy the commutation relations (3.11):

$$H^1 = \begin{pmatrix} 1 & 0 & 0 \\ 0 & -1 & 0 \\ 0 & 0 & 0 \end{pmatrix}, \qquad H^2 = \begin{pmatrix} 0 & 0 & 0 \\ 0 & 1 & 0 \\ 0 & 0 & -1 \end{pmatrix},$$

$$E_+^1 = \begin{pmatrix} 0 & 1 & 0 \\ 0 & 0 & 0 \\ 0 & 0 & 0 \end{pmatrix}, \qquad E_-^1 = \begin{pmatrix} 0 & 0 & 0 \\ 1 & 0 & 0 \\ 0 & 0 & 0 \end{pmatrix},$$

$$E_+^2 = \begin{pmatrix} 0 & 0 & 0 \\ 0 & 0 & 1 \\ 0 & 0 & 0 \end{pmatrix}, \qquad E_-^2 = \begin{pmatrix} 0 & 0 & 0 \\ 0 & 0 & 0 \\ 0 & 1 & 0 \end{pmatrix}, \qquad (3.17)$$

$$E_+^\theta = \begin{pmatrix} 0 & 0 & 1 \\ 0 & 0 & 0 \\ 0 & 0 & 0 \end{pmatrix}, \qquad E_-^\theta = \begin{pmatrix} 0 & 0 & 0 \\ 0 & 0 & 0 \\ 1 & 0 & 0 \end{pmatrix}.$$

As already seen in chapter 2 for the case of $\mathfrak{sl}(2)$, instead of considering arbitrary complex or real linear combinations of the generators, it is also of considerable interest to allow only for *real* linear combinations of specific *complex* linear combinations of the generators. In the $\mathfrak{sl}(2)$ case, these specific complex generators are

$$\sigma_1 = E_+ + E_- , \qquad \sigma_2 = iE_+ - iE_- , \qquad \sigma_3 = H \qquad (3.18)$$

(in the notation of chapter 2, one has $\sigma_i = 2L_i$; compare e.g. (2.5)). In the matrix realization (3.16), one thus has

$$\sigma_1 = \begin{pmatrix} 0 & 1 \\ 1 & 0 \end{pmatrix}, \qquad \sigma_2 = \begin{pmatrix} 0 & i \\ -i & 0 \end{pmatrix}, \qquad \sigma_3 = \begin{pmatrix} 1 & 0 \\ 0 & -1 \end{pmatrix}. \qquad (3.19)$$

These matrices are known as the *Pauli matrices*. As already mentioned in section 2.2, the Lie algebra of linear combinations with real coefficients of the generators $i\sigma_j$, with σ_j as in (3.18), is called $\mathfrak{su}(2)$, and in the physics literature one commonly regards σ_j rather than $i\sigma_j$ as generators. In the matrix realization, in the former description the elements of $\mathfrak{su}(2)$ are traceless anti-hermitian 2×2-matrices, while in the latter description they are hermitian.

In the $\mathfrak{sl}(3)$-case, one considers analogously real linear combinations of H^1 and H^2, of $E_+^i + E_-^i$ and $iE_+^i - iE_-^i$ for $i = 1, 2$, and of $E_+^\theta + E_-^\theta$ and $iE_+^\theta - iE_-^\theta$ (respectively of i times these matrices). In the matrix realization (3.17), this yields a basis of the traceless hermitian (respectively anti-hermitian) 3×3-matrices, and correspondingly the real Lie algebra with these generators is called $\mathfrak{su}(3)$. In the physics literature, the matrices which realize the generators of $\mathfrak{su}(3)$ (in fact, after also replacing H^2 by the special linear combination $(H^1 + 2H^2)/\sqrt{3}$) are known as the *Gell-Mann matrices* and are conventionally denoted by λ^a, $a = 1, 2, ..., 8$.

The Lie algebra $\mathfrak{su}(2)$ can be viewed as the infinitesimal version of the Lie *group* SU(2), i.e. the group of unitary 2×2-matrices of determinant 1. In terms of the Pauli matrices (3.19), the elements of SU(2) can be written as $\exp[i(\xi_1\sigma_1 + \xi_2\sigma_2 + \xi_3\sigma_3)]$, with ξ_i three real parameters and with 'exp' the exponential power series of matrices. Analogously, the matrices $\exp(i\sum_{a=1}^8 \xi_a \lambda^a)$ with $\xi_a \in \mathbb{R}$ and λ^a the Gell-Mann matrices, form the Lie group SU(3) of unitary 3×3-matrices of unit determinant. The relation between Lie algebras and Lie groups will be studied in more detail in chapter 9.

3.5 Orthogonal basis

Instead of H^1 and H^2 let us consider the specific linear combinations

$$H_I := \tfrac{1}{2} H^1 , \qquad H_Y := \tfrac{1}{2\sqrt{3}} H^1 + \tfrac{1}{\sqrt{3}} H^2 . \qquad (3.20)$$

In terms of the matrix realization (3.17), one has e.g.

$$H_Y = \frac{1}{2\sqrt{3}} \begin{pmatrix} 1 & 0 & 0 \\ 0 & 1 & 0 \\ 0 & 0 & -2 \end{pmatrix}. \tag{3.21}$$

An advantage of the choice (3.20) shows up when one considers the trace of products of these matrices. While in the original basis one has

$$\mathrm{tr}(H^i H^j) = A^{ij} = \begin{pmatrix} 2 & -1 \\ -1 & 2 \end{pmatrix}_{i,j}, \tag{3.22}$$

the new basis elements satisfy

$$\mathrm{tr}(H_I H_I) = \mathrm{tr}(H_Y H_Y) = \tfrac{1}{2}, \qquad \mathrm{tr}(H_I H_Y) = \mathrm{tr}(H_Y H_I) = 0. \tag{3.23}$$

As we will see later, the trace operation provides a scalar product on any simple Lie algebra. The relations (3.23) tell us that the elements H_I and H_Y are orthogonal with respect to this scalar product. Correspondingly, one refers to H_I and H_Y as providing an *orthogonal basis* of the subspace spanned by H^1 and H^2. Further, writing $\vec{H} = (H_I, H_Y)$, one obtains

$$[\vec{H}, E_\pm^1] = \pm (1,0)\, E_\pm^1, \qquad [\vec{H}, E_\pm^2] = \pm \tfrac{1}{2}(-1, \sqrt{3})\, E_\pm^2,$$
$$[\vec{H}, E_\pm^\theta] = \pm \tfrac{1}{2}(1, \sqrt{3})\, E_\pm^\theta, \tag{3.24}$$

and also

$$[E_+^1, E_-^1] = 2\, H_I, \qquad [E_+^2, E_-^2] = -H_I + \sqrt{3}\, H_Y,$$
$$[E_+^\theta, E_-^\theta] = H_I + \sqrt{3}\, H_Y. \tag{3.25}$$

From (3.24) we can read off what the components of $\alpha^{(1)}$, $\alpha^{(2)}$ and θ in the orthogonal basis are:

$$\alpha^{(1)} = (1,0), \qquad \alpha^{(2)} = \tfrac{1}{2}(-1, \sqrt{3}), \qquad \theta = \tfrac{1}{2}(1, \sqrt{3}). \tag{3.26}$$

As will be seen below, the vectors orthogonal to $\alpha^{(1)}$ and $\alpha^{(2)}$ are also of interest. They are most easily obtained in the orthogonal basis, in which they are given by (in a normalization which will prove to be convenient)

$$\Lambda_{(1)} = \tfrac{1}{2}\left(1, \tfrac{1}{\sqrt{3}}\right), \qquad \Lambda_{(2)} = \tfrac{1}{2}\left(0, \tfrac{2}{\sqrt{3}}\right). \tag{3.27}$$

Expressed as linear combinations of $\alpha^{(1)}$ and $\alpha^{(2)}$, they thus read

$$\Lambda_{(1)} = \tfrac{1}{3}\left(2\alpha^{(1)} + \alpha^{(2)}\right), \qquad \Lambda_{(2)} = \tfrac{1}{3}\left(\alpha^{(1)} + 2\alpha^{(2)}\right), \tag{3.28}$$

so that in particular $\Lambda_{(1)} + \Lambda_{(2)} = \theta$. These vectors are displayed in the following picture.

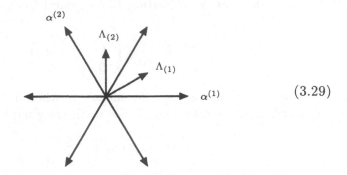

$$(3.29)$$

3.6 The eightfold way

The considerations above indicate that the mass spectrum of hadrons can be explained to a large extent by grouping the hadrons into multiplets of $su(3)$. The idea to do so has been termed the *eightfold way*. Note that at this point our description is still based on the complex Lie algebra $sl(3) \equiv sl(3, \mathbb{C})$. As we will point out later, in a quantum field theoretic explanation of the $sl(3)$ structure in terms of bound states of quarks, it is the real Lie algebra $su(3)$ that is relevant; this is not in conflict with the present description, because the representations of the real Lie algebra $su(3)$ and the complex Lie algebra $sl(3, \mathbb{C})$ are in one-to-one correspondence (whereas the representation theory of the real Lie algebra $sl(3, \mathbb{R})$ is different). Historically, the eightfold way was suggested much earlier than the interpretation of the hadrons in terms of quark fields.

The representation theory of $sl(3)$ is more complicated than that of $sl(2)$. It will follow as a special case of the representation theory of simple Lie algebras, to be described in detail in chapter 13. The main results needed here are the following. The lowest-dimensional irreducible representations have dimension 1, 3, 6, 8, 10. The one- and eight-dimensional representations are unique (up to isomorphism), whereas in dimension 3, 6 and 10 there are two inequivalent ones. In a short-hand notation, these representations are denoted by the symbols 1, 3, $\bar{3}$, 6, $\bar{6}$, 8, 10, $\overline{10}$.

The basis vectors in the representation space for the octet (8) are obtained as follows. To each of the six end points of the arrows in the diagram (3.15) we associate a basis vector. In addition we associate two distinct basis vectors to the origin of the diagram, which correspond to the two elements H^1 and H^2 that span the Cartan subalgebra. The action of the Lie algebra on the representation space can then be visualized in terms of the diagram (3.15) as transforming the basis vectors corresponding to these points into suitable linear combinations of each other.

There are altogether three pions and four kaons, and hence the minimal possibility to combine them in an irreducible representation space is to use the eight-dimensional representation. This idea indeed works, with the 8th particle provided by the η-meson. The *meson octet* so obtained looks as follows:

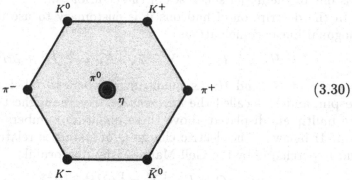

$$(3.30)$$

Similarly, the lightest baryons form another octet, and there is a decuplet (10) of heavier baryons. The octet consists of the proton and neutron (an isospin doublet) and the Λ, Σ and Ξ particles according to

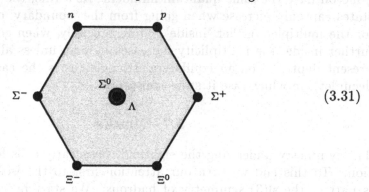

$$(3.31)$$

The decuplet

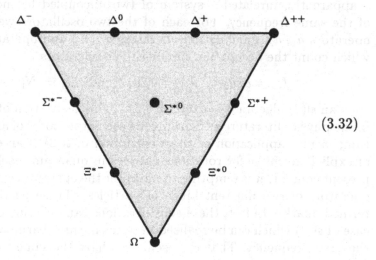

$$(3.32)$$

contains the Δ, Σ^* and Ξ^* resonances and, as an isospin singlet, the Ω^-. The prediction of the Ω^- resonance, which had not yet been detected, was one of the major successes of the eightfold way.

In the description of hadrons it is customary to use the mutually orthogonal linear combinations

$$I = H_I = \tfrac{1}{2} H^1, \qquad Y = 2\sqrt{3}\,H_Y = \tfrac{1}{3}\left(H^1 + 2H^2\right) \qquad (3.33)$$

in place of H^1 and H^2 as quantum numbers to label states. I is the isospin, and Y is called the *hypercharge*. For the highest weight states of the multiplets displayed above these quantum numbers are displayed in table II below. The electric charge Q of a state is related to the isospin and hypercharge by the Gell-Mann–Nishijima formula

$$Q = I + \tfrac{1}{2} Y = \tfrac{1}{3}\left(2H^1 + H^2\right). \qquad (3.34)$$

It is important to realize that these quantum numbers do *not* necessarily specify states within one multiplet uniquely. For example, π^0 and the η-meson have the same quantum numbers. As a rule, the multiplicity of states can only increase when going from the 'boundary' of the diagram for the multiplet further 'inside'. More precisely, when going one 'step' further inside, the multiplicity increases by one, unless all states at the present depth lie on an equilateral triangle (as is the case e.g. for the decuplet), in which case it stays constant.

3.7 Quarks

The symmetry underlying the eightfold way clearly calls for an explanation. To this end we turn our attention first to the isospin $\mathfrak{sl}(2)$ that is part of the $\mathfrak{sl}(3)$ symmetry of hadrons. We start by considering the – apparently unrelated – system of two uncoupled harmonic oscillators of the same frequency. For each of the two oscillators we have creation operators a_1, a_2, annihilation operators a_1^+, a_2^+ and operators $N_i = a_i^+ a_i$ which count the occupation number. The operators

$$J_+ := a_1^+ a_2, \quad J_- := a_2^+ a_1 \quad \text{and} \quad J_3 := N_1 - N_2 \qquad (3.35)$$

span an $\mathfrak{sl}(2)$ algebra (see exercise 3.3). (This realization of $\mathfrak{sl}(2)$ is known in the physics literature as Schwinger's oscillator model of angular momentum; a nice application of this description of $\mathfrak{sl}(2)$ is an easy derivation of explicit formulæ for rotation matrices in quantum mechanics.) In the present context, it is tempting to interpret the operators a_i as annihilation operators of two different kinds of particles. These particles have been termed quarks. In fact, these considerations can be easily extended to the case of $\mathfrak{sl}(3)$ which can be realized in terms of three harmonic oscillators of the same frequency. That the oscillators have the same frequency means that the three particles should have equal masses.

According to this *quark model,* each meson can be understood as a bound state of a quark and an anti-quark, and each baryon as a bound state of three quarks. This description of baryons allows in particular for an explanation of the mass differences which are still present within the hadron multiplets in terms of the different masses of the up (u), down (d) and strange (s) quark. (In nature the masses of the u, d and s quarks are only approximately equal. However, for many purposes, such as mass formulæ for baryon multiplets, one can first treat the quarks as being equally heavy and afterwards include the mass differences as small perturbations. The larger mass of the s quark can e.g. be incorporated by describing the system by a Hamilton operator which consists of an sl(3)-invariant main term and a perturbation which is proportional to the Gell-Mann matrix λ_8.)

The light quarks u, d, s form a triplet (3), while the anti-quarks form an anti-triplet ($\bar{3}$), according to

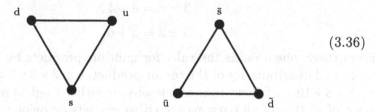

$$(3.36)$$

The highest weight states of these multiplets are u and $\bar{s}$; the quantum numbers of all states are provided in table II.

Table II. *Quantum numbers of hadrons and quarks*

particle	H^1	H^2	I	Y	Q	S
K^+	1	1	1/2	1	1	1
p	1	1	1/2	1	1	0
Δ^{++}	3	0	3/2	1	2	0
u	1	0	1/2	1/3	2/3	0
d	−1	1	−1/2	1/3	−1/3	0
s	0	−1	0	−2/3	−1/3	−1
$\bar{s}$	0	1	0	2/3	1/3	1
$\bar{d}$	1	−1	1/2	−1/3	1/3	0
$\bar{u}$	−1	0	−1/2	−1/3	−2/3	0

Note that the sl(3) symmetry that was deduced from the existence of a
second quantum number in addition to isospin, and from the combination
of several isospin multiplets into larger multiplets, requires that there are
three distinct light quark species; the quarks u and d alone cannot form
an irreducible representation of sl(3). The hypercharge $Y = -2/3$ of
the strange quark differs from the common hypercharge $Y = 1/3$ of the
up and down quarks. To stress this difference, one introduces another
(dependent) quantum number, the strangeness

$$S = Y - B,\tag{3.37}$$

as hypercharge minus baryon number B; the baryon number of the quarks
is 1/3, so that $S = 0$ for up and down quarks and $S = -1$ for the strange
quark.

   The formation of bound states of quarks (and / or anti-quarks) respects
the 'tensor product' of the corresponding sl(3)-multiplets. The basic rules
for the decomposition of tensor products into multiplets are

$$3 \times \overline{3} = 8 + 1,$$
$$3 \times 3 = \overline{3} + 6.\tag{3.38}$$

From these, one obtains the rules for multiple products by using associa-
tivity and distributivity of the tensor product, e.g. $3 \times 3 \times 3 = (\overline{3}+6) \times 3 =
1+8+8+10$. It follows that all mesons are either singlets or belong to an
octet of sl(3), and all baryons are either singlets or belong to an octet or
to a decuplet. While for the time being these rules just provide a formal
calculus described in a rather ad hoc manner, we will later learn methods
which allow for a natural and systematic treatment of arbitrary tensor
products of representations of (simple) Lie algebras (see chapter 15).

   In the quantum field theoretic description of the strong interactions,
known as quantum chromodynamics (QCD), the quarks are considered as
the elementary fields. The symmetries (which leave e.g. the Lagrangian
density invariant) then act on the quark fields by unitary 3×3-matrices
or, separating out an overall complex phase, by unitary 3×3-matrices
of determinant 1. The latter form the Lie *group* SU(3) (and the former
the Lie group U(3)). The infinitesimal transformations that correspond
to the finite SU(3) transformations are described by traceless hermitian
3×3-matrices, which can be written as real linear combinations of the
Gell-Mann matrices. In the quantum field theory framework it is therefore
the real Lie algebra su(3) which is relevant for the symmetries of quarks.

The $\mathfrak{su}(3)$ symmetry of hadrons and quarks, respectively, described here is also known as the *flavor* $\mathfrak{su}(3)$. In quantum chromodynamics, another $\mathfrak{su}(3)$ symmetry, the *color* $\mathfrak{su}(3)$ plays in fact a more fundamental rôle. This symmetry is a gauge symmetry; it governs the interactions between quarks and gluons, i.e. the gauge particles that mediate the interactions (see e.g. chapter 19 of [Nachtmann 1990]). The strong interactions among hadrons can be viewed as a kind of van der Waals-forces induced by the fundamental quark-gluon interactions.

The fact that both symmetries are described in terms of the Lie algebra $\mathfrak{su}(3)$ is purely accidental. In fact, the flavor symmetry gets still more extended (and more approximate, which limits its practical use) when one includes the heavier quark species charm, bottom and top, whereas these heavy quarks still possess the same color symmetry as the light ones.

Information

3.8 Abstract Lie algebras

Instead of considering the particular realization (3.16) of $\mathfrak{sl}(2)$ in terms of 2×2-matrices, or (2.7) in terms of angular momentum operators in quantum mechanics, one can also think of $\mathfrak{sl}(2)$ as an abstract mathematical structure, being defined by the presence of generators H and $E_\pm$ that are subject to the system

$$[H, E_\pm] = \pm 2\, E_\pm\,, \qquad [E_+, E_-] = 2\,H \tag{3.39}$$

of relations and also satisfy the Jacobi identity as well as antisymmetry for the operation '$[\,\cdot\,,\cdot\,]$'. This structure is a special example of what is called a 'simple Lie algebra'.

Instead of $\mathfrak{sl}(2)$ one also uses the symbol A_1 if one is considering the structure as an abstract Lie algebra. Similarly, the symbol A_2 is used in place of $\mathfrak{sl}(3)$ to denote the object defined abstractly through the generators H^i, $E_\pm^i$ and $E_\pm^\theta$ subject to the relations (3.11). A_2 is again a simple Lie algebra. An important feature distinguishing A_2 from A_1 is the presence of the two commuting generators H^1 and H^2. The maximal number of such generators is called the *rank* of a Lie algebra; in physical terms, the rank determines the maximal number of quantum numbers. Thus A_1 has rank one, and it is in fact the unique simple Lie algebra of rank one. In contrast, besides A_2 there are two other simple Lie algebras which have rank two. These are called B_2 and G_2 and can be constructed in a way similar to what we did in the A_2 case above; they correspond to having

$$A^{12} = -2\,, \qquad A^{21} = -1 \tag{3.40}$$

and $\qquad A^{12} = -3\,, \qquad A^{21} = -1 \tag{3.41}$

for B_2 and G_2, respectively, instead of the minimal solution (3.10). The pictures analogous to the diagram (3.15) for A_2 are:

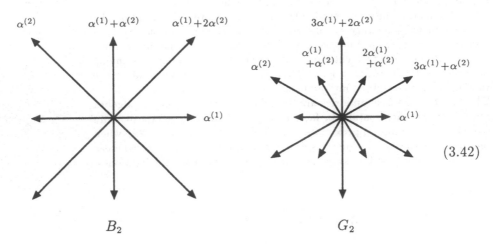

$$B_2 \qquad\qquad\qquad G_2 \tag{3.42}$$

In the theory of abstract Lie algebras, many of the quantities that were encountered in the discussion above have been given special names. Some of them are as follows.

• One can expand the Lie bracket of two elements of a basis of $\mathfrak{g}$ again in terms of this basis; the coefficients of this expansion (such as the number 2 in $[H, E_+] = 2E_+$) are known as the structure constants of the Lie algebra.

• The vectors of eigenvalues with respect to the H^i such as (3.14), which correspond to the step operators $E^j_\pm$, are called roots, and diagrams like (3.15) and (3.42) root diagrams. Arbitrary vectors of eigenvalues with respect to the H^i in some representation are called weights. The vector space spanned by the weights or roots, respectively, is called the weight respectively root space; the trace in some representation induces a natural scalar product on it.

• Roots like $\alpha^{(1)}$ and $\alpha^{(2)}$ are called simple roots, and the weights orthogonal to them (more precisely, orthonormal when some definite choice of the overall normalization is made), like $\Lambda_{(1)}$ and $\Lambda_{(2)}$, are known as fundamental weights.

• The structure constants A^{ij} appearing in the relations $[H^j, E^i_+] = A^{ij} E^i_+$ can be combined in a matrix A, which is called the Cartan matrix of the Lie algebra.

• Any subset of the algebra which is itself a Lie algebra, such as the $\mathfrak{sl}(2)$-algebras spanned by H^i and $E^i_\pm$ for some fixed $i \in \{1,2,3\}$, is called a subalgebra of $\mathfrak{g}$.

• Finally, the real Lie algebras $\mathfrak{su}(2)$ and $\mathfrak{su}(3)$ are the so-called compact real forms of the complex Lie algebras $\mathfrak{sl}(2, \mathbb{C})$ and $\mathfrak{sl}(3, \mathbb{C})$, respectively; while the representations of real and complex Lie algebras are generically rather different, the representation theories of any complex simple Lie algebra and its compact real form are isomorphic.

Information

With the help of Lie algebras one is not only able to group the hadrons into multiplets, but also many other objects which appear in physics. A large class of such objects is provided by the spectra (both for bound states and for scattering states) of a variety of Hamilton operators in quantum mechanics. This includes applications in atomic physics (see e.g. exercise 3.5 for the Coulomb potential, and [Judd 1968]), the shell structure as well as the rotational and vibrational spectra of nuclei (see e.g. section VIII of [Gruber and Samuel 1975], chapter 11 of [Hamermesh 1962], section 13.2 of [Ludwig and Falter 1988], [Iachello and Arima 1987] and [Parikh 1978]), as well as the spectra of various solvable Hamiltonians [Alhassid *et al.* 1983].

Summary:

The mass spectrum of hadrons possesses an (approximate) $\mathfrak{sl}(3)$-symmetry. The 8-dimensional Lie algebra $\mathfrak{sl}(3)$ can be obtained from two $\mathfrak{sl}(2)$-algebras interrelated in a specific way; it also contains a third $\mathfrak{sl}(2)$-subalgebra.

$A_1 = \mathfrak{sl}(2)$ is the only simple Lie algebra over the complex numbers $\mathbb{C}$ of rank one. Besides $A_2 = \mathfrak{sl}(3)$, there are two other simple Lie algebras over $\mathbb{C}$ of rank two, called B_2 and G_2.

Keywords:

Symmetries of hadrons, isospin, hypercharge, meson octet, baryon decuplet, quark triplet;
quantum number, multiplet, $\mathfrak{sl}(3)$, $\mathfrak{su}(3)$, Pauli matrix, Gell-Mann matrix, tensor product decomposition;
Lie algebra, structure constants, root diagram, Cartan matrix.

Exercises:

Write down the Gell-Mann matrices. Compute the structure constants of $\mathfrak{su}(3)$ in the basis provided by these matrices.

Exercise 3.1

Compute the traces $\text{tr}(\sigma_i \sigma_j)$ of the product of two Pauli matrices ($i, j \in \{1, 2, 3\}$). Perform the analogous calculation for the Gell-Mann matrices.

Exercise 3.2

Verify that the operators (3.35) satisfy the commutation relations of the $su(2)$ Lie algebra.

Exercise 3.3

The roots in the A_2 root diagram (3.15) (including the twofold zero 'root' corresponding to H^1 and H^2) decompose as $8 = 1 + 2 + 2 + 3$ into irreducible representation spaces (denoted by their dimension) with respect to the $sl(2)$ algebras defined by any of the roots.
Find the analogous decompositions for the root diagrams (3.42) of B_2 and G_2 with respect to the $sl(2)$ algebras defined by the two simple roots $\alpha^{(1)}$ and $\alpha^{(2)}$.
In the case of G_2, find a decomposition of the form $14 = 3 + \bar{3} + 8$ with respect to a suitable A_2 algebra.

Exercise 3.4

While when describing hadron symmetries it was not sufficient to take just two copies of $sl(2)$, there is a well known physical system where such a symmetry structure is present, namely the hydrogen problem in quantum mechanics.
We denote by $\vec{L} = (L_i)_{i=1,2,3}$ the angular momentum operators (2.1), by $H := \vec{p}^2 - 1/r$ the Hamiltonian of the hydrogen problem (with $r := |\vec{q}| \equiv (q_1^2 + q_2^2 + q_3^2)^{1/2}$, and with mass and electric charge scaled in such a way that they do not appear explicitly any more), and by

Exercise 3.5

$$\vec{R} := \vec{p} \times \vec{L} - \vec{L} \times \vec{p} - \vec{q}/r \qquad (3.43)$$

the so-called *Runge–Lenz vector*.

a) Show that both angular momentum and the Runge–Lenz vector are conserved quantities.

b) Compute the commutation relations $[R_i, L_j]$ and $[R_i, R_j]$.

c) Note that these commutation relations do *not* define a Lie algebra (but rather a structure that is sometimes, misleadingly, called a quadratic algebra).

d) Convince yourself that when the Hamiltonian is considered as a number (i.e., physically speaking, when these operators are all applied to energy eigenstates of a definite energy), the operators L_i and R_j span a six-dimensional Lie algebra.

e) Show that as long as the energy is non-zero, this Lie algebra consists of two copies of $sl(2)$, or in more mathematical terms, is the direct sum $sl(2) \oplus sl(2)$. (Hint: Look for suitable linear combinations $L_j + \xi R_j$.)

f) What happens when the energy eigenvalue is zero?

g) Show that in the hydrogen problem one has $\vec{L} \cdot \vec{R} = 0$. Discuss the restriction this imposes on the allowed weights of $sl(2) \oplus sl(2)$, and explain the degeneracies in the hydrogen spectrum.

4
Formalization: Algebras and Lie algebras

4.1 Short summary

This chapter is intended to provide the mathematical definition of what a Lie algebra is and to introduce various related algebraic concepts. It is almost inevitable that the presentation of these matters is somewhat dry and formal. Fortunately it will be possible to return to a less abstract level in the chapters that follow.

For those readers who prefer to skip the formal definitions in a first reading (and return to them when necessary), we start by briefly summarizing the main concepts: A Lie algebra $\mathfrak{g}$ is a vector space with an operation $(x, y) \mapsto [x, y]$, called the Lie bracket, which is bilinear (i.e. obeys $[\xi x + \eta y, z] = \xi[x, z] + \eta[y, z]$ and an analogous relation for $[x, \eta y + \zeta z]$ for all numbers ξ, η) and satisfies $[x, y] = -[y, x]$ and the Jacobi identity $[x, [y, z]] + [y, [z, x]] + [z, [x, y]] = 0$. The elements of a basis of $\mathfrak{g}$ are called the generators T^a of $\mathfrak{g}$, and the expansion coefficients $f^{ab}{}_c$ in their Lie brackets $[T^a, T^b] = \sum_c f^{ab}{}_c T^c$ are known as the structure constants of $\mathfrak{g}$ in this basis. A Lie algebra with $[x, y] = 0$ for all $x, y \in \mathfrak{g}$ is called *abelian*. A Lie algebra $\mathfrak{g}$ for which each element can be written as a commutator of two elements of $\mathfrak{g}$ is called a *semisimple* Lie algebra; any semisimple Lie algebra is the direct sum of (one or more) basic constituents, the *simple* Lie algebras. The Lie algebras $\mathfrak{sl}(2)$ and $\mathfrak{sl}(3)$ described in the previous two chapters are both simple.

4.2 Groups, rings, fields

We start the formal development by recalling a few very basic notions and definitions that underlie the concept of an algebra: groups, rings, fields, and vector spaces.

47

• A *group* is a set G together with a product ' $\diamond$ ', i.e. a map from $G \times G$ to G (a 'binary operation') which is associative and has a unit element and inverses.

Associativity means that no bracketing is required: $x \diamond (y \diamond z) = x \diamond y \diamond z = (x \diamond y) \diamond z$ for all $x, y, z \in G$. The unit, or identity, element $e \in G$ is an element satisfying $e \diamond x = x = x \diamond e$ for all $x \in G$; the existence of inverses means that to any element x of G there is an inverse element x^{-1} in G satisfying $x \diamond x^{-1} = e = x^{-1} \diamond x$.

If the product is commutative (i.e. $x \diamond y = y \diamond x$ for all $x, y \in G$), then the group is called *abelian*. If there is no unit element or not all elements possess an inverse, but the associativity axiom is still fulfilled, then G is called a *semigroup*. For example, the set $\mathbb{Z}$ of integers is an abelian group with respect to addition, while the set $\mathbb{Z}_{\geq 0}$ of non-negative integers is an abelian semigroup.

• A *ring* is a set $\mathcal{R}$ endowed with two binary operations ' $+$ ' and ' $\diamond$ ' that obey the usual axioms of addition and multiplication.

That is, both are associative, with respect to the addition ' $+$ ' $\mathcal{R}$ is an abelian group, and the operations are distributive (i.e. $x \diamond (y + z) = x \diamond y + x \diamond z$ etc. for all $x, y, z \in \mathcal{R}$).

The neutral element with respect to the addition is denoted by ' 0 '. If there exists a unit element for the multiplication, the ring is called unital. An important example for a unital ring is the set of all integer numbers $\mathbb{Z}$, with ordinary addition and multiplication. Also, given any ring $\mathcal{R}$, the set $\mathcal{R}[t]$ of all polynomials in some indeterminate t with coefficients in $\mathcal{R}$ is again a ring, with pointwise addition and multiplication. Note that it is not required that in a ring there exist inverses with respect to multiplication. The elements of a unital ring which do possess an inverse are often called *units* – not to be mixed up with the *unit element* which in this terminology is a rather special unit.

• A *field F* is a ring for which the ring multiplication is commutative and in which all elements except 0 possess an inverse (i.e. are units).

In chapter 2 and chapter 3 we already encountered two different fields, namely the field of complex numbers $\mathbb{C}$ and the field of real numbers $\mathbb{R}$, and in most of what follows these are the only ones we are interested in. Other fields are the rational numbers $\mathbb{Q}$ and extensions of $\mathbb{Q}$ by algebraic numbers (i.e., by solutions of algebraic equations with rational coefficients, such as $\sqrt{2}$).

4.3 Vector spaces

• A *vector space* (or *linear space*) V over some field F is a set endowed with a (commutative) addition and a multiplication by elements of F, such that distributivity laws and other compatibility properties of the two operations are valid.

The elements of V are called vectors, while the elements of the underlying field F are called scalars. Both the addition in V and that in the field F are denoted by the symbol '+', and the multiplication by scalars is denoted by the symbol '·' (or, when no confusion can arise, simply by juxtaposition). Thus the defining properties read $\xi \cdot (v + w) = \xi \cdot v + \xi \cdot w$, $(\xi + \eta) \cdot v = \xi \cdot v + \eta \cdot v$, $\xi \cdot (\eta \cdot v) = (\xi \eta) \cdot v$, $0 \cdot v = 0$, $1 \cdot v = v$ etc., for all $v, w \in V$ and all $\xi, \eta \in F$.

• Given any subset M of a vector space V, the set of all (finite) linear combinations of elements of M, called the span or linear hull of M and denoted by $\mathrm{span}_F(M) \equiv \mathrm{span}_F\{v \mid v \in M\}$, is a subspace (also called linear subspace or sub-vector space) of V, that is, a subset which has itself the structure of a vector space. A *basis* $\mathcal{B}$ of a vector space V is a subset which has the properties that it is linearly independent (i.e. no *finite* linear combination of the elements of $\mathcal{B}$ is zero) and that the subspace $\mathrm{span}_F\{v \mid v \in \mathcal{B}\}$ is already all of V. The number of basis elements is independent of the actual choice of basis $\mathcal{B}$; if this number is finite, one denotes it by

$$\mathrm{d} \equiv \dim_F V := |\mathcal{B}| \tag{4.1}$$

and calls it the *dimension* of the vector space V, while otherwise V is said to be infinite-dimensional. Note that for an infinite-dimensional vector space V the number of basis elements is infinite, but still each element of V is a *finite* linear combination of basis elements.

• The *direct sum* $V_1 \oplus V_2$ of two vector spaces V_1 and V_2 over the same field F is the set of all formal sums $v_1 \oplus v_2$ of elements of V_1 and V_2, with scalar multiplication and addition defined by $\xi (v_1 \oplus v_2) = (\xi v_1) \oplus (\xi v_2)$ and $(v_1 \oplus v_2) + (w_1 \oplus w_2) = (v_1 + w_1) \oplus (v_2 + w_2)$, respectively.

If for $j = 1, 2$ the set $\mathcal{B}_j = \{v_{(i)}^{[j]}\}$, with $i = 1, 2, \dots, \dim V_j$ is a basis of V_j, then the set

$$\mathcal{B} := \{v_{(i)}^{[1]} \oplus 0 \mid i = 1, 2, \dots, \dim V_1\} \cup \{0 \oplus v_{(i)}^{[2]} \mid i = 1, 2, \dots, \dim V_2\}$$

is a basis of the direct sum $V_1 \oplus V_2$. In particular, the dimension of $V_1 \oplus V_2$ is the sum $\dim V_1 + \dim V_2$.

• The *cartesian product* or *Kronecker product* $V_1 \times V_2$ of two vector spaces V_1 and V_2 is the set of all ordered pairs $(v_1; v_2)$ of elements $v_1 \in V_1$, $v_2 \in V_2$. The cartesian product is itself not a vector space. However, if V_1 and V_2 are vector spaces over the same field F, then one obtains a new vector space if one identifies the pairs $(\xi v_1; v_2)$ and $(v_1; \xi v_2)$ for $\xi \in F$, as well as $(v_1 + w_1; v_2 + w_2)$ and $(v_1; v_2) + (v_1; w_2) + (w_1; v_2) + (w_1; w_2)$.

• The set obtained this way from $V_1 \times V_2$ is called the *tensor product*

$V_1 \otimes V_2$ of V_1 and V_2. The element of the tensor product that is the class of the pair $(v_1; v_2)$ is denoted by $v_1 \otimes v_2$. The scalar multiplication then acts as $\xi(v_1 \otimes v_2) = (\xi v_1; v_2) = (v_1; \xi v_2)$.

Given bases $\mathcal{B}_j = \{v_{(i)}^{[j]}\}$ $(i = 1, 2, \ldots, \dim V_j)$ of V_j for $j = 1, 2$, the set

$$\mathcal{B} := \{v_{(i)}^{[1]} \otimes v_{(j)}^{[2]} \,|\, i = 1, 2, \ldots, \dim V_1,\ j = 1, 2, \ldots, \dim V_2\}$$

is a basis of the tensor product $V_1 \otimes V_2$. Thus the dimension of the tensor product $V_1 \otimes V_2$ of vector spaces of dimension d_1 and d_2 is the product $d_1 \cdot d_2$. The tensor product can also be characterized by the *universal property* that to any bilinear (i.e., linear in each argument) map from $V_1 \times V_2$ to some vector space W there is precisely one linear map φ from $V_1 \otimes V_2$ to W such that the bilinear map is the composition of φ with the mapping $(v_1; v_2) \mapsto v_1 \otimes v_2$.

- It is sometimes useful to regard a vector space over a field F as a vector space over a subfield of F, in particular a vector space over $\mathbb{C}$ as a vector space over $\mathbb{R}$. Indeed we can view any complex vector space V as a real vector space by restricting the scalar multiplication to the real numbers, and if $\mathcal{B}_{\mathbb{C}} = \{v_{(i)}\}$ is a basis over $\mathbb{C}$, then $\mathcal{B}_{\mathbb{R}} := \{v_{(i)}\} \cup \{iv_{(i)}\}$ is a basis over $\mathbb{R}$. The dimension of V over $\mathbb{R}$ is therefore twice the complex dimension. Conversely, from any real vector space $V_{\mathbb{R}}$ with basis $\mathcal{B} = \{w_{(i)}\}$ one can construct a complex vector space $V_{\mathbb{C}}$ as follows. Given a basis $\tilde{\mathcal{B}} = \{w_{(i)}\} \cup \{\tilde{w}_{(i)}\}$ of the direct sum $V_{\mathbb{R}} \oplus V_{\mathbb{R}}$, one extends scalar multiplication to $\mathbb{C}$ via $iw_{(i)} := \tilde{w}_{(i)}$ and $i\tilde{w}_{(i)} := -w_{(i)}$. This way $\mathcal{B}$ becomes the basis of a vector space over $\mathbb{C}$, which is denoted by $V_{\mathbb{C}}$ and called the *complexification* of $V_{\mathbb{R}}$. The complex dimension of $V_{\mathbb{C}}$ equals the real dimension of $V_{\mathbb{R}}$.

- The set of all linear forms on a vector space V, i.e. of linear maps $\eta: V \to F$, is again a vector space over F; it is called the vector space *dual* to V and is denoted by $V^\star$.

If $\mathcal{B} = \{v_{(i)}\}$ is a basis of V, then the *dual basis* $\mathcal{B}^\star = \{w^{(i)}\}$ of $\mathcal{B}$ is that basis of $V^\star$ which consists of the linear forms w^i which obey

$$w^{(i)}(v_{(j)}) = \delta_j^i \quad \text{for } i, j = 1, 2, \ldots, \dim_F V\,, \tag{4.2}$$

with δ the Kronecker symbol, which is defined to be $\delta_j^i = 1$ for $i = j$, and zero otherwise.

- An *inner product* or scalar product on a vector space V is a non-degenerate bilinear form, i.e. a bilinear map $\kappa: V \times V \to F$ that satisfies the non-degeneracy condition that for any $v \in V$ there exists a $w \in V$ such that $\kappa(v, w) \neq 0$.

- For $F = \mathbb{C}$, a *hermitian product* is a map $\mu: V \times V \to \mathbb{C}$ which is linear in the second factor, but 'anti-linear' in the first, i.e. satisfies $\mu(\xi v, \zeta w) = \bar{\xi} \zeta\, \mu(v, w)$.

4.4 Algebras

• An *algebra* $\mathfrak{A}$ is a vector space endowed with an additional binary operation $\circ : \mathfrak{A} \times \mathfrak{A} \to \mathfrak{A}$ on which the only requirement is that it is bilinear, i.e. $(x + y) \circ z = x \circ z + y \circ z$ and $x \circ (y + z) = x \circ y + x \circ z$, as well as $(\xi x) \circ (\eta y) = (\xi \eta) x \circ y$ for all $x, y, z \in \mathfrak{A}$ and all elements ξ, η of the underlying field F.

(Equivalently, one may define an algebra as a ring $\mathcal{R}$ together with an action of the field F on $\mathcal{R}$ which is compatible with the multiplication '$\circ$' and addition '$+$' of the ring.)

Note that this definition is still extremely general. It is therefore not surprising that in order to be of any interest, the product '$\circ$' of an algebra must have further properties. For instance, it may be required to be associative, in which case $\mathfrak{A}$ is called an *associative algebra*, or to possess a unit element, yielding a *unital algebra*. Examples of unital associative algebras are the algebras $\mathrm{GL}(n, F)$ of arbitrary $n \times n$-matrices with entries in F and $\mathrm{SL}(n, F)$ of $n \times n$ matrices of unit determinant, where the product '$\circ$' is matrix multiplication, and the algebra of bounded linear operators on a Hilbert space, where '$\circ$' is the composition of maps.

Independently of the specific form of these additional properties of the product, a large number of concepts, such as isomorphisms, derivations, subalgebras, solvability, semisimplicity and so on can be defined in full generality. However, for definiteness we will describe these concepts below directly in the setting of Lie algebras.

4.5 Lie algebras

• A *Lie algebra* $\mathfrak{g}$ is an algebra in the sense just described for which the bilinear operation is neither commutative nor associative, but rather possesses two other special properties.

The bilinear operation is then called the *Lie bracket* and commonly denoted by '$[\cdot, \cdot]$'. With this notation, the two defining properties read

$$[x, x] = 0 \qquad \text{for all } x \in \mathfrak{g} \tag{4.3}$$

and

$$[x, [y, z]] + [y, [z, x]] + [z, [x, y]] = 0 \qquad \text{for all } x, y, z \in \mathfrak{g}. \tag{4.4}$$

The second of these equalities is called the *Jacobi identity*. As a consequence of the bilinearity, the first property (4.3) of the Lie bracket implies

$$0 = [x + y, x + y] = [x, x] + [x, y] + [y, x] + [y, y] = [x, y] + [y, x], \tag{4.5}$$

or in other words,

$$[x, y] = -[y, x] \tag{4.6}$$

for all $x, y \in \mathfrak{g}$; it is therefore known as the *antisymmetry* property.

Given any associative algebra $\mathfrak{A}$ with product $\circ$, one can immediately obtain an associated Lie algebra $\mathfrak{g}$. Namely, by considering $\mathfrak{A}$ as a vector space and defining the Lie bracket as the *commutator* with respect to the original multiplication, that is

$$[x, y] := x \circ y - y \circ x, \qquad (4.7)$$

one constructs a Lie algebra structure on the same vector space $\mathfrak{A}$.

The standard example for this construction is the Lie algebra $GL(n, F)$ of $n \times n$-matrices, for which the associative product $\circ$ is just the ordinary matrix multiplication. Very often, and in particular in applications in physics, Lie algebras appear naturally as algebras of matrices; as a consequence, the term *commutator*, or also *commutation rule*, is often used as a synonym for the term Lie bracket. It must however be stressed that, while for commutators (with respect to an *associative* product) the Jacobi identity is automatic, it is an independent defining property of Lie algebras, and sometimes it requires some work to check its validity (compare e.g. the case of the Poisson bracket in exercise 1.2).

4.6 Generators

The *dimension* $\mathrm{d} = \dim \mathfrak{g}$ of a Lie algebra is the dimension of $\mathfrak{g}$ considered as a vector space. If the dimension is finite or countably infinite, one uses the notation

$$\mathcal{B} = \{T^a \mid a = 1, 2, ..., \mathrm{d}\} \qquad (4.8)$$

for any basis $\mathcal{B}$ of $\mathfrak{g}$ and one refers to the elements T^a of $\mathcal{B}$ as the *generators* of the Lie algebra.

Thus the generators T^a span (linearly) the Lie algebra $\mathfrak{g}$. There is also a rather different meaning of the term generator, namely in the sense of elements from which the whole algebra can be generated algebraically, that is, not only by the use of linear combinations, but also by performing products, i.e. Lie brackets. (Furthermore, neither of these notions must be confused with the concept of a generator of the symmetry of a physical system. If that symmetry is described by the Lie algebra $\mathfrak{g}$, then *any* element of $\mathfrak{g}$ realizes an infinitesimal symmetry transformation and hence is a generator of the symmetry.)

Because of the bilinearity, the Lie bracket is already determined uniquely if it is known on a basis $\mathcal{B}$. Therefore one can define the Lie bracket, and hence the Lie algebra $\mathfrak{g}$, abstractly through the expansion

$$[T^a, T^b] = \sum_{c=1}^{\mathrm{d}} f^{ab}{}_c \, T^c \qquad (4.9)$$

of the bracket of two generators with respect to the basis; the expansion coefficients $f^{ab}_{c} \in F$ are called the *structure constants* of the Lie algebra $\mathfrak{g}$. Of course, the explicit values of the structure constants depend on the chosen basis. Expressed through the structure constants, the Lie algebra property (4.3) reads $f^{aa}_{b} = 0$, while (4.4) becomes

$$\sum_{c=1}^{d} (f^{ab}_{c} f^{cd}_{e} + f^{da}_{c} f^{cb}_{e} + f^{bd}_{c} f^{ca}_{e}) = 0 \, ; \tag{4.10}$$

the version (4.6) of the antisymmetry property amounts to

$$f^{ab}_{c} = -f^{ba}_{c} \tag{4.11}$$

i.e. to antisymmetry of the structure constants in their two upper indices.

As already mentioned, many concepts appearing in Lie algebra theory can be introduced for arbitrary algebras. The structure constants are among these. For any algebra $\mathfrak{A}$ with generators T^a, they are defined by $T^a \circ T^b = \sum_{c=1}^{d} C^{ab}_{c} T^c$. For an associative algebra, the associativity of $\circ$ translates into the property

$$\sum_{c=1}^{d} C^{ab}_{c} C^{cd}_{e} = \sum_{c=1}^{d} C^{bd}_{c} C^{ac}_{e} \, . \tag{4.12}$$

It is often convenient to suppress the summation symbols in formulæ like (4.10). Doing so, it is implicit that a summation has to be performed over any symbol that appears once as an upper and once as a lower index. We will use this summation convention only when we are sure that it does not create confusion, but write out the summations otherwise. Note that if this convention is applied consistently, it provides quite a useful check of many manipulations: In case that at some stage of a calculation one is summing over an index which appears twice as an upper (or twice as a lower) index, it is likely that one made a mistake. (The formula $f^{aa}_{b} = 0$ does make sense, however, because it does not involve a summation on the index a.) Note that there is a clear distinction between upper and lower indices, unless one can define some (invariant) inner product on the Lie algebra which can be used to raise and lower indices. A prominent example of a Lie algebra for which such a scalar product does not exist is the Virasoro algebra which will be analyzed in section 12.12.

4.7 Homomorphisms, isomorphisms, derivations

A *homomorphism* from the Lie algebra $\mathfrak{g}$ to the Lie algebra $\mathfrak{h}$ is a map $\varphi \colon \mathfrak{g} \to \mathfrak{h}$ that preserves the algebraic structures, i.e. that, besides being linear, carries Lie brackets to Lie brackets, that is

$$[x, y] \mapsto \varphi([x, y]) = [\varphi(x), \varphi(y)] \tag{4.13}$$

for all $x, y \in \mathfrak{g}$. A homomorphism which has trivial kernel (i.e. sends only the zero element 0 of $\mathfrak{g}$ to the zero of $\mathfrak{h}$) and hence is an injective map is called a *monomorphism* or embedding, and a homomorphism $\varphi \colon \mathfrak{g} \to \mathfrak{h}$ whose image $\varphi(\mathfrak{g})$ is all of $\mathfrak{h}$ (i.e. which is a surjective map) is called an *epimorphism*. If φ possesses both of these properties, it is called an *isomorphism* from $\mathfrak{g}$ to $\mathfrak{h}$. One of the fundamental goals in Lie algebra theory is the classification of Lie algebras up to isomorphism. If there exists an isomorphism (epimorphism) from $\mathfrak{g}$ to $\mathfrak{h}$, then $\mathfrak{h}$ is said to be *isomorphic* to (a *homomorphic image* of) $\mathfrak{g}$; this situation is denoted by $\mathfrak{h} \cong \mathfrak{g}$ ($\mathfrak{h} \sim \mathfrak{g}$). A homomorphism $\varphi\colon \mathfrak{g} \to \mathfrak{g}$ of a Lie algebra to itself is called an *endomorphism*, and an isomorphism of a Lie algebra onto itself is called an *automorphism*. The specific properties of these mappings are often indicated by using special symbols in place of ' $\to$ '; this is summarized in the following table.

homomorphisms		symbol
*mono*morphism	injective = one-to-one = embedding	$\hookrightarrow$
*epi*morphism	surjective = onto	$\longrightarrow\!\!\!\!\to$
*iso*morphism	injective and surjective	$\stackrel{\cong}{\longrightarrow}$
*endo*morphism	to the same space	
*auto*morphism	isomorphism to the same space	(4.14)

There also exists an important class of maps from $\mathfrak{g}$ to itself which are not endomorphisms, namely the derivations. A *derivation* δ of the Lie algebra $\mathfrak{g}$ is a linear map which satisfies a condition analogous to the product rule of differentiation, the so-called *Leibniz rule*

$$[x, y] \;\mapsto\; \delta([x, y]) = [x, \delta(y)] + [\delta(x), y] \tag{4.15}$$

for all $x, y \in \mathfrak{g}$. The notion of a derivation can be defined analogously for any product, not only for the Lie bracket; the Leibniz rule then takes the form $\delta(x \circ y) = \delta x \circ y + x \circ \delta y$. An important example in geometry is the algebraic definition of the tangent space of a manifold as the vector space of derivations on the germs of smooth functions. In the case of Lie algebras, an example for a derivation is provided by the 'multiplication' (i.e., forming a Lie bracket) with an arbitrary, but fixed, element of the algebra from the left: For any $x \in \mathfrak{g}$, the map $\mathrm{ad}_x\colon \mathfrak{g} \to \mathfrak{g}$ defined by

$$y \;\mapsto\; \mathrm{ad}_x(y) := [x, y] \tag{4.16}$$

for all $y \in \mathfrak{g}$, is a derivation (see exercise 4.7). The derivation (4.16) will often be referred to as the *adjoint map* associated to $x \in \mathfrak{g}$.

4.8 Subalgebras and ideals

A subspace $\mathfrak{h} \subseteq \mathfrak{g}$ of a Lie algebra $\mathfrak{g}$ which itself is a Lie algebra (with respect to the Lie bracket of $\mathfrak{g}$) is called a Lie *subalgebra* of $\mathfrak{g}$; in this context, the Lie algebra $\mathfrak{g}$ itself is sometimes called the *ambient* Lie algebra. Now for any two subsets $\mathfrak{h}$, $\mathfrak{k}$ of a Lie algebra we may introduce the notation

$$[\mathfrak{h}, \mathfrak{k}] \equiv \mathrm{span}_F\{[x, y] \mid x \in \mathfrak{h}, \ y \in \mathfrak{k}\}. \tag{4.17}$$

With this abbreviation one can write various relations in a more convenient form. For instance, the property of a subspace $\mathfrak{h} \subseteq \mathfrak{g}$ to be a subalgebra is in this notation expressed as

$$[\mathfrak{h}, \mathfrak{h}] \subseteq \mathfrak{h}. \tag{4.18}$$

Any Lie algebra $\mathfrak{g}$ has two subalgebras, namely $\mathfrak{g}$ itself and the one-element subspace $\{0\}$. These subalgebras are called *trivial* subalgebras; any other subalgebra of $\mathfrak{g}$ is called a *proper* subalgebra of $\mathfrak{g}$.

Above we have introduced subalgebras of $\mathfrak{g}$ as specific subsets of $\mathfrak{g}$, but the same situation can also be described from a slightly different perspective. Namely, one considers a pair of Lie algebras $\mathfrak{g}$ and $\mathfrak{h}$ for which there exists an injective Lie algebra homomorphism $\mathfrak{h} \hookrightarrow \mathfrak{g}$, called the *embedding* of $\mathfrak{h}$ in $\mathfrak{g}$. The first description is recovered by identifying $\mathfrak{h}$ and the image of $\mathfrak{h}$ via the embedding homomorphism. Conversely, the inclusion map $\imath\colon \mathfrak{h} \hookrightarrow \mathfrak{g}$ of a Lie subalgebra $\mathfrak{h}$ into the Lie algebra $\mathfrak{g}$ is a special example of an embedding.

If not only (4.18), but also the stronger property

$$[\mathfrak{h}, \mathfrak{g}] \subseteq \mathfrak{h} \tag{4.19}$$

(that is, $[x, y] \in \mathfrak{h}$ for all $x \in \mathfrak{h}$ and *all* $y \in \mathfrak{g}$) is satisfied, then the subalgebra $\mathfrak{h} \hookrightarrow \mathfrak{g}$ is called an *invariant* subalgebra or *ideal* of the Lie algebra $\mathfrak{g}$. A *proper* ideal is an ideal that is neither equal to $\{0\}$ nor to $\mathfrak{g}$ itself, which are two obvious ideals of $\mathfrak{g}$.

It is important to realize that a subalgebra $\mathfrak{h}$ of a Lie algebra $\mathfrak{g}$ must not just be thought of as a pair $(\mathfrak{g}, \mathfrak{h})$ (the latter mis-interpretation is unfortunately suggested by the notation used in many applications). Rather, as we will see in chapter 18, generically a Lie algebra $\mathfrak{h}$ can be embedded in $\mathfrak{g}$ in various inequivalent ways. An example that we already encountered in chapter 3 is the embedding $\mathfrak{sl}(2) \hookrightarrow \mathfrak{sl}(3)$; as we saw there (see the equations (3.1), (3.3) and (3.13)), there are three obvious possibilities to embed $\mathfrak{sl}(2)$, corresponding to taking the generators denoted by H^i and $E^i_\pm$ with fixed $i \in \{1, 2, 3\}$ of $\mathfrak{sl}(3)$. But there also exists another embedding which is of a different type; it is obtained by setting $H = 2\,(H^1 + H^2)$ and $E_\pm = \sqrt{2}\,(E^1_\pm + E^2_\pm)$.

The *center* $\mathcal{Z}(\mathfrak{g})$ of $\mathfrak{g}$, i.e. the set

$$\mathcal{Z}(\mathfrak{g}) := \{x \in \mathfrak{g} \mid [x, \mathfrak{g}] = 0\} \tag{4.20}$$

of all elements of $\mathfrak{g}$ which possess zero bracket with all of $\mathfrak{g}$, provides an example for an ideal of $\mathfrak{g}$; $\mathcal{Z}(\mathfrak{g})$ may or may not be trivial. (The center of a Lie algebra must not be confused with the center of the associated Lie *group*, in case such a group exists; for simple Lie groups the latter is a finite group, see chapter 9.) Also, if $\mathfrak{h}$ and $\mathfrak{k}$ are ideals of $\mathfrak{g}$, then so are $[\mathfrak{h}, \mathfrak{k}]$, $\mathfrak{h} \cap \mathfrak{k}$, and

$$\mathfrak{h} + \mathfrak{k} := \{x \in \mathfrak{g} \mid x = y + z, \ y \in \mathfrak{h}, \ z \in \mathfrak{k}\}. \tag{4.21}$$

A class of subalgebras of $\mathfrak{g}$ that are obtained similarly to the center are the *central-izers* $C_{\mathfrak{g}}(\mathfrak{k})$ of subsets $\mathfrak{k}$ of $\mathfrak{g}$: $C_{\mathfrak{g}}(\mathfrak{k})$ is the set

$$C_{\mathfrak{g}}(\mathfrak{k}) := \{x \in \mathfrak{g} \mid [x, \mathfrak{k}] = 0\} \tag{4.22}$$

of all elements of $\mathfrak{g}$ which possess zero bracket with all of $\mathfrak{k}$. Note that $C_{\mathfrak{g}}(\mathfrak{g}) = \mathcal{Z}(\mathfrak{g})$. Similarly, the *normalizer* $\mathcal{N}_{\mathfrak{g}}(\mathfrak{h})$ of a subalgebra $\mathfrak{h} \subseteq \mathfrak{g}$ is the subalgebra

$$\mathcal{N}_{\mathfrak{g}}(\mathfrak{h}) := \{x \in \mathfrak{g} \mid [x, \mathfrak{h}] \subseteq \mathfrak{h}\} \tag{4.23}$$

consisting of all elements of $\mathfrak{g}$ whose brackets with $\mathfrak{h}$ lie in $\mathfrak{h}$. $\mathcal{N}_{\mathfrak{g}}(\mathfrak{h})$ is the largest subalgebra of $\mathfrak{g}$ which contains $\mathfrak{h}$ as an ideal. In particular one has $\mathcal{N}_{\mathfrak{g}}(\mathfrak{h}) = \mathfrak{g}$ if $\mathfrak{h}$ is an ideal of $\mathfrak{g}$.

Further on, we will use the symbol ' $\oplus$ ', i.e. write

$$\mathfrak{g} = \bigoplus_{i=1}^{n} \mathfrak{g}_i \equiv \mathfrak{g}_1 \oplus \mathfrak{g}_2 \oplus \cdots \oplus \mathfrak{g}_n \tag{4.24}$$

to indicate that $\mathfrak{g}$ is, as a vector space, the direct sum of the vector spaces $\mathfrak{g}_i$. If each of the subspaces $\mathfrak{g}_i$ is actually an ideal of $\mathfrak{g}$, we will rather use the notation

$$\mathfrak{g} = \bigoplus_{i=1}^{n} \mathfrak{g}_i \equiv \mathfrak{g}_1 \oplus \mathfrak{g}_2 \oplus \cdots \oplus \mathfrak{g}_n \tag{4.25}$$

and call $\mathfrak{g}$ the *direct sum* of the Lie algebras $\mathfrak{g}_i$, $i = 1, 2, \ldots, n$. If the Lie brackets of the algebras $\mathfrak{g}_i$ are denoted by $[\cdot\,,\cdot]_i$, then the bracket $[\cdot\,,\cdot]$ of the direct sum $\mathfrak{g}$ satisfies

$$\begin{aligned} [x, y] = [x, y]_i \quad &\text{for} \quad x, y \in \mathfrak{g}_i \ (i = 1, 2, \ldots, n) \quad \text{and} \\ [\mathfrak{g}_i, \mathfrak{g}_j] = 0 \quad &\text{for} \quad i \neq j \ (i, j = 1, 2, \ldots, n) \,. \end{aligned} \tag{4.26}$$

If in the case of two summands, $\mathfrak{g} = \mathfrak{g}_1 \oplus \mathfrak{g}_2$, the second of these relations is replaced by (say)

$$[\mathfrak{g}_1, \mathfrak{g}_2] \subseteq \mathfrak{g}_1 \,, \tag{4.27}$$

but still $\mathfrak{g}_1$ and $\mathfrak{g}_2$ are Lie subalgebras of $\mathfrak{g}$, then $\mathfrak{g}$ is called the *semidirect sum* of the subalgebras $\mathfrak{g}_1$ and $\mathfrak{g}_2$, denoted by $\mathfrak{g}_1 \uplus \mathfrak{g}_2$.

Trivially, if one is given an isomorphism of each of the summands $\mathfrak{g}_i$ in (4.25), one can combine them to an isomorphism of their direct sum $\mathfrak{g}$. As a consequence, the analysis of direct sums of Lie algebras – in particular their classification up to isomorphism – reduces to the analysis of the summands. More generally, any Lie algebra is to a large extent characterized by the structure of its proper ideals.

*4.9 Solvable and nilpotent Lie algebras

In this section we describe a few notions which are helpful to deal with more general Lie algebras. For any Lie algebra $\mathfrak{g}$ one obtains an ideal by considering its *derived* algebra $\mathfrak{g}'$, which is defined as the set

$$\mathfrak{g}' := [\mathfrak{g}, \mathfrak{g}] \tag{4.28}$$

of all linear combinations of brackets of $\mathfrak{g}$. Because of $[\mathfrak{g}, [\mathfrak{g}, \mathfrak{g}]] \subseteq [\mathfrak{g}, \mathfrak{g}]$, this is indeed an ideal of $\mathfrak{g}$. One can think of two distinct ways of iterating the procedure leading to $\mathfrak{g}'$. First, one may define the sequence $\{\mathfrak{g}^{\{i\}}\}$ by considering first the elements that can be written as Lie brackets, then those that can be written as brackets among brackets, etc.; this amounts to setting $\mathfrak{g}^{\{1\}} := \mathfrak{g}'$ and

$$\mathfrak{g}^{\{i\}} := [\mathfrak{g}^{\{i-1\}}, \mathfrak{g}^{\{i-1\}}] \tag{4.29}$$

for $i \geq 2$. Second, there is the sequence $\{\mathfrak{g}_{\{i\}}\}$, where again $\mathfrak{g}_{\{1\}} := \mathfrak{g}'$, but at each step one allows for brackets between the subspace and *all* elements of $\mathfrak{g}$, so that

$$\mathfrak{g}_{\{i\}} := [\mathfrak{g}, \mathfrak{g}_{\{i-1\}}] \tag{4.30}$$

for $i \geq 2$. The sequence of subspaces labelled by superscripts that is defined by (4.29) is known as the upper central series or *derived series* of $\mathfrak{g}$, while the one with subscripts defined by (4.30) is called the *lower central series*. Each of the subspaces $\mathfrak{g}^{\{i\}}$ and $\mathfrak{g}_{\{i\}}$ is an ideal of $\mathfrak{g}$.

The derived algebra $\mathfrak{g}'$ of $\mathfrak{g}$ is not necessarily a proper subalgebra, but can also be $\{0\}$ (and also all of $\mathfrak{g}$, of course). More generally, the procedures defining the derived and lower central series can, after some finite number of steps, end up with zero. These properties provide a convenient criterion for the classification of Lie algebras. If the derived series of a Lie algebra $\mathfrak{g}$ ends up with $\{0\}$, then one calls $\mathfrak{g}$ *solvable*, and if the lower central series of $\mathfrak{g}$ ends up with $\{0\}$ one calls it *nilpotent*. One has $\mathfrak{g}^{\{j\}} \subseteq \mathfrak{g}_{\{j\}}$ for all $j \in \mathbb{Z}_{>0}$, so that nilpotency implies solvability. If $\mathfrak{g}$ is solvable (nilpotent), then so are all subalgebras $\mathfrak{h}$ of $\mathfrak{g}$, since $\mathfrak{h}^{\{j\}} \subseteq \mathfrak{g}^{\{j\}}$ ($\mathfrak{h}_{\{j\}} \subseteq \mathfrak{g}_{\{j\}}$) for all $j \in \mathbb{Z}_{>0}$. All homomorphic images of a solvable (nilpotent) Lie algebra are solvable (nilpotent) as well.

A *maximal* solvable ideal is one that is not contained in any larger solvable ideal. The maximal solvable ideal of a Lie algebra $\mathfrak{g}$ is unique; it is called the *radical* of $\mathfrak{g}$. Thus $\mathfrak{g}$ is solvable if and only if it equals its own radical.

4.10 Semisimple and abelian Lie algebras

The Lie algebras which are the most important ones for applications in physics are the abelian Lie algebras and simple Lie algebras, and their direct sums. An *abelian* Lie algebra is a Lie algebra which satisfies $[\mathfrak{g}, \mathfrak{g}] = 0$. A *simple* Lie algebra is a Lie algebra which contains no proper ideals and which is not abelian. A direct sum of simple Lie algebras is called *semisimple*, and a direct sum of simple and abelian Lie algebras is referred to as *reductive*. Semisimple Lie algebras can also be characterized by the fact that the elements $[x, y]$ with arbitrary $x, y \in \mathfrak{g}$ already exhaust all of $\mathfrak{g}$.

An abelian Lie algebra equals its own center, $\mathcal{Z}(\mathfrak{g}) = \mathfrak{g}$. Also, the derived algebra of an abelian algebra is zero (thus abelianness can be viewed as a very strong version of solvability or nilpotency), whereas semisimple Lie algebras are equal to their derived algebras, $\mathfrak{g}' \equiv [\mathfrak{g}, \mathfrak{g}] = \mathfrak{g}$, and their center vanishes, $\mathcal{Z}(\mathfrak{g}) = \{0\}$.

Further, a Lie algebra is semisimple if and only if it does not possess a non-zero solvable ideal, i.e. if and only if its radical is zero (thus, in a sense, semisimplicity is the opposite of solvability). Similarly, a Lie algebra is reductive if and only if its radical equals its center. More generally, any Lie algebra $\mathfrak{g}$ can be written as a (semidirect, but in general non-direct) sum

$$\mathfrak{g} = \mathfrak{s} \oplus \mathfrak{g}_{\mathrm{rad}} \tag{4.31}$$

where $\mathfrak{s}$ is semisimple and $\mathfrak{g}_{\mathrm{rad}}$ is the radical of $\mathfrak{g}$. The splitting (4.31) is called a Levi decomposition of $\mathfrak{g}$.

If $\mathfrak{g}$ is one-dimensional, then because of the antisymmetry property (4.3) the single generator T has bracket $[T, T] = 0$, and hence any one-dimensional Lie algebra is abelian. (Also, just as simple Lie algebras, it does not contain any proper ideal; this is the reason why in the definition of simple Lie algebras one requires that $\mathfrak{g}$ must not be abelian.) Up to isomorphism, there is thus a unique one-dimensional Lie algebra. This is isomorphic to the base field as a vector space, so it would be most natural to denote it simply by F. However, in the case of $F = \mathbb{R}$ or $\mathbb{C}$ we are mainly interested in, the common notation for the one-dimensional Lie algebra is $\mathfrak{u}(1)$. Moreover, any d-dimensional abelian Lie algebra $\mathfrak{g}$ is isomorphic to the d-fold direct sum of one-dimensional abelian Lie algebras,

$$\mathfrak{g} = \bigoplus_{i=1}^{\mathrm{d}} \mathfrak{u}(1) \, . \tag{4.32}$$

As a consequence, the non-trivial part of the classification of reductive Lie algebras is the classification of simple Lie algebras; this classification will be the subject of chapter 7.

We have already encountered a simple Lie algebra in chapter 2, namely the three-dimensional Lie algebra $\mathfrak{sl}(2)$, which is spanned by generators E_+, E_- and H with brackets

$$[E_+, E_-] = H \, , \qquad [H, E_\pm] = \pm\, 2\, E_\pm \tag{4.33}$$

(compare (2.7)). That this Lie algebra is simple can be checked as follows. Applying $\mathrm{ad}_{E_\pm}$ twice to an arbitrary non-zero element $x = \xi_+ E_+ + \xi_- E_- + \zeta H$ of the algebra yields $-2\xi_\mp E_\pm$. Thus for $\xi_+ \neq 0$ ($\xi_- \neq 0$) any subalgebra containing x also contains E_+ (E_-), while for $\xi_+ = \xi_- = 0$ it contains of course H. This observation implies that any ideal containing x is in fact equal to the algebra itself. Since x was arbitrary, it follows that $\mathfrak{sl}(2)$ does not possess a proper ideal and hence is simple. In fact, for $F = \mathbb{C}$, $\mathfrak{sl}(2)$ is the unique (up to isomorphism) three-dimensional simple Lie algebra, and there are no lower-dimensional simple

Lie algebras (any one-dimensional Lie algebra is isomorphic to the abelian algebra $\mathfrak{u}(1)$, and any two-dimensional Lie algebra is either abelian or isomorphic to the solvable algebra with non-trivial bracket $[T^1, T^2] = T^2$).

*4.11 Gradations and Lie superalgebras

A *gradation* of a Lie algebra $\mathfrak{g}$ by an abelian semigroup G (with group operation denoted by ' + ') is by definition a decomposition

$$\mathfrak{g} = \bigoplus_{p \in G} \mathfrak{g}_{(p)} \tag{4.34}$$

of $\mathfrak{g}$ into a direct sum of vector spaces such that

$$[\mathfrak{g}_{(p)}, \mathfrak{g}_{(q)}] \subseteq \mathfrak{g}_{(p+q)} \tag{4.35}$$

for all $p, q \in G$. $\mathfrak{g}$ is said to be *graded* by the semigroup G. Also, if $x \in \mathfrak{g}_{(p)}$ for some $p \in G$, then x is called a *homogeneous element* (with respect to the grading) of $\mathfrak{g}$, and the label p is referred to as the grade or *degree* of x and denoted by $p = \deg x$.

In many applications G is the group $\mathbb{Z}$ of all integer numbers. As an example, consider the eight-dimensional Lie algebra $\mathfrak{sl}(3)$ characterized by the brackets (3.11) and set

$$\deg H^i = 0, \quad \deg E^i_\pm = \pm 1, \quad \deg E^\theta_\pm = \pm 2. \tag{4.36}$$

This provides a $\mathbb{Z}$-gradation for $\mathfrak{g} = \mathfrak{sl}(3)$ in which $\mathfrak{g}_{(p)} = 0$ for all p with $|p| > 2$.

The concept of gradation can again be defined analogously for arbitrary algebras. As an example, consider the algebra of polynomials in some indeterminate t; a $\mathbb{Z}_{\geq 0}$-gradation of this algebra is obtained by defining the degree of a polynomial to be n if the leading power in t is t^n.

In the special case of the two-element group $G = \mathbb{Z}_2$, the gradation can be employed to construct a 'supersymmetric' generalization of Lie algebras. Namely, one defines a *Lie superalgebra* $\mathfrak{g}_s$ as a $\mathbb{Z}_2$-graded algebra whose product, denoted by $[\cdot, \cdot]_s$, satisfies $\mathbb{Z}_2$-graded analogues of the Jacobi identity (4.4) and the antisymmetry property of the Lie bracket, namely the *super Jacobi identity*

$$(-1)^{|x||z|} [x, [y, z]_s]_s + (-1)^{|x||y|} [y, [z, x]_s]_s + (-1)^{|y||z|} [z, [x, y]_s]_s = 0 \tag{4.37}$$

for all homogeneous $x, y, z \in \mathfrak{g}_s$, as well as

$$[x, y]_s = (-1)^{1+|x||y|} [y, x]_s . \tag{4.38}$$

Here the notation $|x| \equiv \deg x$ is used, and $\mathbb{Z}_2$ is understood as the set $\{0, 1\}$ with addition modulo 2.

Similarly as of Lie algebras, Lie superalgebras can be realized via an underlying associative algebra, namely by expressing $[x, y]_s$ as

$$[x, y]_s := x \circ y - (-1)^{|x||y|} y \circ x = \begin{cases} x \circ y + y \circ x & \text{for } |x| = |y| = 1, \\ x \circ y - y \circ x & \text{else}, \end{cases} \tag{4.39}$$

through the associative product $\circ$. One then often uses the notation $\{x, y\}$ for the former and $[x, y]$ for the latter case and refers to the bracket symbol as the anti-commutator and commutator of x and y, respectively.

Another example for Lie superalgebras can be obtained from the vector space $\wedge M$ of all differential forms of arbitrary degree on some manifold M. To any form we assign its degree d; this turns $\wedge M$ into a $\mathbb{Z}$-graded vector space. The exterior (wedge) product respects this grading. However, we can consider on the vector space $\wedge M$ also the coarser grading by $\mathbb{Z}_2$, where the grade is given by d mod 2. For any two elements η_1, η_2 in $\wedge M$ we define

$$[\eta_1, \eta_2]_s = \eta_1 \eta_2 - (-1)^{d_1 d_2} \eta_2 \eta_1 \, . \tag{4.40}$$

It is easy to check that this is a super Lie bracket; hence $\wedge M$ has the natural structure of a Lie superalgebra. Note, however, that for forms that take values in the complex numbers, or more generally in some abelian (Lie) algebra, the super Lie bracket vanishes. This is not true any more for forms that take their values in some nonabelian (Lie) algebra. We also remark that $\wedge M$ carries in fact a much richer structure since the exterior derivative is compatible with the Lie superalgebra structure; this leads to the notion of a differential (super) Lie algebra.

As another specific Lie superalgebra let us consider the algebra with four generators Q_α, $\bar{Q}_\alpha$ ($\alpha = 1, 2$) of degree 1 (*odd* generators) and four generators P_m ($m = 0, 1, 2, 3$) of degree 0 (*even* generators) which are subject to the relation

$$[Q_\alpha, \bar{Q}_\beta]_s = 2 \sum_{m=0}^{3} \sigma_{\alpha\beta}^m P_m \, , \tag{4.41}$$

with all other brackets among generators vanishing. The matrices σ^m are the Pauli matrices (3.19) (for $m = 1, 2, 3$) together with the 2×2 unit matrix. This Lie superalgebra is known as the *supersymmetry algebra*; the generators Q_α and $\bar{Q}_\alpha$ are then called the supercharges, and P_μ is the energy-momentum four-vector. For more information on this algebra see section 20.9.

*4.12 Open and non-linear algebras

We have already mentioned that one way Lie algebras arise is by considering commutators of the elements of some associative algebra. The Jacobi identity and the antisymmetry property are then fulfilled automatically. In this situation one often starts with a set $\{T^i\}$ that generates the associative algebra algebraically, i.e. upon taking both linear combinations and products. Usually, one can then express the commutator $[T^i, T^j]$ in the associative algebra as a sum over terms $f^{ij}_k(T^l) T^k$, where the f^{ij}_k are polynomials in the generators T^l:

$$[T^i, T^j] = \sum_k f^{ij}_k(T^l) T^k \, . \tag{4.42}$$

Superficially this looks like the relations of a Lie algebra, if it were not for the fact the $f^{ij}_k(T^l)$ are functions of T^l rather than constants. Such a structure is therefore sometimes referred to as an *open* or *soft* (Lie) algebra, or as a *non-linear* (Lie) algebra (e.g. as quadratic algebra when the polynomials $f^{ij}_k(T^l)$ are linear in the generators T^l). Another terminology that is sometimes used is to speak of 'field dependent structure constants'. None of these misleading terms will be used in the rest of this book.

Note that to be able to write down relations like (4.42) at all, one implicitly has to assume that there is more structure than just the Lie algebraic one, since already the product of operators, rather than only their commutator, must make sense. In order to forget about this additional structure, and hence to turn an open algebra into a decent Lie algebra, one has to declare all new *linearly* independent terms on the right hand side of the commutators of the original generators as additional generators of the Lie algebra. Since the commutators involving these additional generators may again give rise to new terms, this procedure must usually be iterated until one finally ends up with a basis of generators which close among themselves under taking Lie brackets. While this recipe involves only simple manipulations, very often the basis so obtained will contain infinitely many elements even if one started with a finite number of generators T^l.

A rather special example in which this prescription is easily implemented is the situation that the commutator of two elements is simply a 'number' c, i.e. an element of the base field F of the algebra. In this case one has to interpret the result of the commutator as a new generator C of the Lie algebra which with respect to commutation behaves analogously to the unit element of an associative algebra, that is, C satisfies $[C, T^a] = 0$ for all generators T^a. Such a generator thus belongs to the center of the Lie algebra, and correspondingly is called a central element or central charge. We have actually already encountered this situation in chapter 2: On the right hand side of the Heisenberg algebra relations (2.29) only numbers appear; in order to interpret these relations as defining a proper Lie algebra, they should therefore be written as $[a_n^i, a_m^j] = n\delta^{ij}\delta_{n+m,0}\, C$ with a central charge C.

Summary:

This chapter explained the basic notions which allow to deal with the mathematical structure of Lie algebras. A Lie algebra is an algebra for which the product, called the Lie bracket, is antisymmetric and satisfies the Jacobi identity.

Special subclasses are the abelian and simple Lie algebras, which are the building blocks of semisimple and reductive Lie algebras.

Keywords:

Group, ring, field, vector space, basis, direct sum, tensor product space, complexification, dual space, inner product, hermitian product; algebra, associative algebra, unital algebra, Lie algebra, Lie bracket, commutator, Jacobi identity, generators, structure constants; homomorphism, isomorphism, embedding, derivation, adjoint map, subalgebra, ambient algebra, ideal, center, (semi-) direct sum; abelian Lie algebra, (semi-) simple Lie algebra, reductive Lie algebra; derived algebra, nilpotent Lie algebra, solvable Lie algebra, radical, gradation, Lie superalgebra, open algebra, non-linear algebra.

Exercises:

Consider the set $\{a + b\sqrt{2} \mid a, b \in \mathbb{Z}\}$. Show that this is a unital associative commutative ring. Determine its units.

Exercise 4.1

Express the addition and scalar multiplication for the direct sum $V_1 \oplus V_2$ of two vector spaces explicitly in terms of the corresponding operations of V_1 and V_2.
Do the same for the tensor product $V_1 \otimes V_2$.

Exercise 4.2

By the *characteristic* of a field F one means the smallest positive integer n such that $\underbrace{\xi + \xi + ... + \xi}_{n \text{ summands}} = 0$ for some non-zero $\xi \in F$.

If no such integer exists, F is said to be of characteristic zero.

Show that for a Lie algebra over an arbitrary field F, the properties (4.3) and (4.6) of the Lie bracket are equivalent provided that the characteristic of F is different from 2.

Exercise 4.3

Give an example for two non-isomorphic associative algebras for which the commutator yields isomorphic Lie algebras.
Hint: Find two non-isomorphic commutative associative algebras having the same dimension.

Exercise 4.4

Is the commutator with respect to a non-associative product a Lie bracket?

Exercise 4.5

Show that the cross product

$$[\vec{x}, \vec{y}] := \vec{x} \times \vec{y} \qquad (4.43)$$

of vectors in $\mathbb{R}^3$ turns $\mathbb{R}^3$ into a Lie algebra. Compute the structure constants in the canonical basis of $\mathbb{R}^3$.
Is this Lie algebra isomorphic to the Lie algebra structure that is induced on $\mathbb{R}^3$ by the group of translations of three-dimensional euclidean space?

Exercise 4.6

Show that for any $x \in \mathfrak{g}$ the map ad_x defined by (4.16) is a derivation.
Show that the map

$$x \mapsto \mathrm{ad}_x \qquad (4.44)$$

is a Lie algebra homomorphism from $\mathfrak{g}$ to the space of linear maps on $\mathfrak{g}$.

Exercise 4.7

Check that the commutator of two derivations is again a derivation. Is the product of two derivations a derivation?

Exercise 4.8

Show that for any element x of a Lie algebra over $\mathbb{C}$ the subspace of $\mathfrak{g}$ that is spanned by the eigenvectors of the adjoint map ad_x is a Lie subalgebra of $\mathfrak{g}$.
Hint: Formulate the problem in such a way that the statement follows with the help of the Jacobi identity.

Exercise 4.9

a) Verify that if $\mathfrak{h}$ and $\mathfrak{k}$ are ideals of $\mathfrak{g}$, then $[\mathfrak{h}, \mathfrak{k}]$, $\mathfrak{h} \cap \mathfrak{k}$, and $\mathfrak{h} + \mathfrak{k}$ as defined by (4.21) are ideals of $\mathfrak{g}$, too.
b) Check that $C_{\mathfrak{g}}(\mathfrak{g}) = \mathcal{Z}(\mathfrak{g})$.
c) Prove that for any subset $\mathfrak{k} \subset \mathfrak{g}$, and for any subalgebra $\mathfrak{h}$ of $\mathfrak{g}$, $C_{\mathfrak{g}}(\mathfrak{k})$ and $\mathcal{N}_{\mathfrak{g}}(\mathfrak{h})$ are subalgebras of $\mathfrak{g}$, and that $\mathcal{N}_{\mathfrak{g}}(\mathfrak{h}) = \mathfrak{g}$ if $\mathfrak{h}$ is an ideal of $\mathfrak{g}$.

Exercise 4.10

Show that the subspaces $\mathfrak{g}^{\{i\}}$ and $\mathfrak{g}_{\{i\}}$ defined by (4.29) and (4.30), respectively, are ideals of $\mathfrak{g}$.

Exercise 4.11

Let $\mathfrak{h}$ be a maximal solvable ideal of a Lie algebra. Show that if $\mathfrak{k}$ is any other solvable ideal, then so is $\mathfrak{h} + \mathfrak{k}$, so that the maximality requirement implies $\mathfrak{h} + \mathfrak{k} = \mathfrak{h}$ and hence $\mathfrak{k} \subseteq \mathfrak{h}$. Conclude that the maximal solvable ideal is unique.

Exercise 4.12

Consider the two-dimensional Lie algebra with generators T and U and Lie bracket $[T, U] = U$, and the three-dimensional Lie algebra with generators X, Y and Z and brackets

$$[X, Y] = Z, \qquad [X, Z] = [Y, Z] = 0. \qquad (4.45)$$

Are these Lie algebras solvable? Are they nilpotent?

Exercise 4.13

a) Classify all two-dimensional Lie algebras.
b) Classify all three-dimensional Lie algebras.

Exercise 4.14

Classify all Lie algebras of rank 1 and dimension 4. More specifically, show that they are all direct sums of lower-dimensional Lie algebras.

Exercise 4.15

5

Representations

5.1 Representations and representation matrices

Lie algebras were defined in the previous chapter quite abstractly, namely as a vector space endowed with a Lie bracket $[\cdot\,,\cdot]$. To make contact to physical systems, these abstract objects have to 'act' in a much more concrete manner on some space, typically on a 'space of physical states' as sketched in chapter 1. The investigation of such spaces will provide us with indispensable tools for the study of Lie algebras. The question on what spaces a Lie algebra $\mathfrak{g}$ can act is therefore of great importance.

Before being able to answer this question, we have to make it more precise: By 'act' on some space V we mean that to any element x of $\mathfrak{g}$ there is associated a map

$$R(x): \quad V \to V\,, \tag{5.1}$$

such that for any two elements x, y of $\mathfrak{g}$ the commutator of $R(x)$ and $R(y)$ exists and reproduces the Lie bracket of x and y in $\mathfrak{g}$:

$$R(x)\circ R(y) - R(y)\circ R(x) = R([x,y])\,. \tag{5.2}$$

Here by '$\circ$' we denote the composition of maps; this composition endows the set of maps from V to V with more than just a Lie algebra bracket, namely even with an associative product from which the Lie bracket is derived. If one is given a set V and mappings $R(x)$ which obey the relation (5.2), one says that V is a representation space of $\mathfrak{g}$ and calls R a representation of $\mathfrak{g}$.

The reader might have noticed that we have been very unspecific concerning the spaces V and the mappings $R(x)$. In principle one can consider for V just any set, and for $R(x)$ arbitrary mappings. However, of particular interest is the case when V is a vector space and the $R(x)$ are linear mappings. This situation arises naturally in quantum mechanics, where the space of physical states is a Hilbert space and therefore carries a linear structure. Nonetheless, there are situations in physics when it is useful to allow for more general mappings, e.g. affine mappings.

An affine mapping of a vector space V is a map $\Phi\colon V \to V$ which acts as $v \mapsto \Phi(v) = Bv + w$, where w is an arbitrary fixed element of V and $v \mapsto Bv$ is a linear mapping of V. The prototypical example for an affine mapping is the action of the group of translations on euclidean space.

Another important example is the action of the gauge group of electrodynamics or Yang–Mills theories on the gauge potentials. In electrodynamics a gauge transformation acts on the four-potential $A_\mu(x)$ by adding the gradient of some smooth function $\phi(x)$: $A_\mu \mapsto A_\mu + \partial_\mu \phi$ (compare equation (1.34)).

It is also important to realize that no restriction on the dimensionality of V has been made. In particular, V may well be an infinite-dimensional vector space even if the Lie algebra $\mathfrak{g}$ which acts on V by some representation R is finite-dimensional.

To gain further structural insight, it is convenient to formulate the definition of a representation in a slightly more abstract way. To this end we first present a few simple general considerations which apply to any vector space V over an arbitrary field F – in practice F will be either the real numbers $\mathbb{R}$ or the complex numbers $\mathbb{C}$. The space of all linear mappings from V to V is itself a vector space over F, which will be denoted by $\mathfrak{gl}(V)$. Since the composition '$\circ$' of linear mappings is again a linear mapping, the operation '$\circ$' is an associative product, so that the vector space $\mathfrak{gl}(V)$ becomes an associative algebra, or, when considering commutators with respect to this product, also a Lie algebra over F. This construction works for any vector space V; the Lie algebra $\mathfrak{gl}(V)$ is called the *general linear algebra* of V. If V has finite dimension n, then after fixing a basis of V, any element of $\mathfrak{gl}(V)$ can be described by an $n \times n$-matrix with entries in F, and the composition of maps is implemented by matrix multiplication. The set of all such matrices is denoted by $\mathfrak{gl}(n)$; thus $\mathfrak{gl}(n)$ is a Lie algebra of dimension n^2.

According to the formula (5.2), in a representation R of a Lie algebra $\mathfrak{g}$ we want to reproduce the Lie bracket by the commutator in $\mathfrak{gl}(V)$ for some vector space V, with $\mathfrak{g}$ and V defined over the same field F. Moreover, the construction should be compatible with the underlying vector space structures. This is achieved by requiring that

$$R\colon \quad \mathfrak{g} \to \mathfrak{gl}(V) \tag{5.3}$$

is a homomorphism of Lie algebras. If this requirement is satisfied, then (5.2) indeed holds, and we say that the vector space V *carries* a (linear) representation R of $\mathfrak{g}$. If the homomorphism is injective, i.e. a monomorphism, then R is called a *faithful* representation of $\mathfrak{g}$. The vector space V on which the representation matrices act is called a *representation space* of $\mathfrak{g}$, or more briefly, a $\mathfrak{g}$-*module*. By a slight abuse of terminology, in the physics literature the term 'representation' instead of 'representation space' is also often used for the space V. The *dimension* of a $\mathfrak{g}$-module is its dimension as a vector space.

Representations and representation spaces (modules) always come in pairs. Both notions describe the same situation, but from a somewhat different point of view. A module of the Lie algebra $\mathfrak{g}$ is a vector space V (over the same field F as $\mathfrak{g}$) together with an action of $\mathfrak{g}$ on V, i.e. an operation

$$\begin{aligned} \bullet : \quad & \mathfrak{g} \times V \to V \\ & (x; w) \mapsto x \bullet w = w' \in V \,, \end{aligned} \tag{5.4}$$

which possesses the properties

$$\begin{aligned} (\xi x + \zeta y) \bullet w &= \xi \, (x \bullet w) + \zeta \, (y \bullet w) \,, \\ x \bullet (\xi v + \zeta w) &= \xi \, (x \bullet v) + \zeta \, (x \bullet w) \,, \\ [x, y] \bullet w &= x \bullet (y \bullet w) - y \bullet (x \bullet w) \end{aligned} \tag{5.5}$$

for all $\xi, \zeta \in F$, all $x, y \in \mathfrak{g}$ and all $v, w \in V$. Given any representation $R \colon \mathfrak{g} \to \mathfrak{gl}(V)$, the vector space V becomes a module of $\mathfrak{g}$ via the definition

$$x \bullet w := (R(x))(w) \,. \tag{5.6}$$

Conversely, given any $\mathfrak{g}$-module V, a representation R of $\mathfrak{g}$ is obtained by reading this equation 'backwards':

$$(R(x))(w) := x \bullet w \,. \tag{5.7}$$

The notation $(R(x))(w)$ used here is well adapted to the fact that $R(x)$ is an endomorphism acting on $w \in V$, but is a bit clumsy because it requires writing several brackets. We therefore prefer in the following the shorter notation $R(x)w$ in place of $(R(x))(w)$, which also fits well with the fact that $R(x)$ can be identified with a matrix.

5.2 The adjoint representation

Obviously, there always exists at least one representation for any Lie algebra, namely the one which maps each element of $\mathfrak{g}$ to the zero map, i.e. the map that maps any vector on the zero vector. This representation, which is clearly not faithful, is called the trivial representation or *singlet* representation of $\mathfrak{g}$. In contrast, it is a priori not clear at all whether for an arbitrary Lie algebra non-trivial representations do exist. However, since any Lie algebra $\mathfrak{g}$ is itself a vector space, it is possible to represent $\mathfrak{g}$ on itself; thereby one obtains the *adjoint representation* or *regular representation*. This representation exists for any Lie algebra, independent of its dimensionality or structure. The adjoint representation is defined by

$$R_{\mathrm{ad}} : \quad \mathfrak{g} \to \mathfrak{gl}(\mathfrak{g}), \qquad x \mapsto \mathrm{ad}_x \,, \tag{5.8}$$

with the map ad_x as defined in (4.16), i.e.

$$\mathrm{ad}_x(y) = [x, y] \,. \tag{5.9}$$

The kernel of the mapping (5.8) consists of all elements x of $\mathfrak{g}$ for which $R_{\mathrm{ad}}(x)$ is the zero map, i.e. of those elements of $\mathfrak{g}$ that commute with all of $\mathfrak{g}$. Hence the kernel of the adjoint map is precisely the center, which is an ideal of $\mathfrak{g}$. If $\mathfrak{g}$ is simple, this means that the kernel is either all of $\mathfrak{g}$ or zero, since by definition simple Lie algebras do not possess non-trivial ideals. If the kernel were all of $\mathfrak{g}$, all Lie brackets would vanish and the algebra would be abelian. Hence the kernel is zero, which means that the adjoint representation of a simple Lie algebra is faithful. In contrast, for any abelian Lie algebra the adjoint representation is not faithful.

The dimension of the adjoint matrix representation is equal to the dimension of the algebra. Also, in R_{ad} the Lie brackets are the commutators $[\mathrm{ad}_x, \mathrm{ad}_y] \equiv \mathrm{ad}_x \circ \mathrm{ad}_y - \mathrm{ad}_y \circ \mathrm{ad}_x$. Thus one has

$$([\mathrm{ad}_x, \mathrm{ad}_y])(z) = [x, [y, z]] - [y, [x, z]] \,, \tag{5.10}$$

so that the antisymmetry and Jacobi identity for the Lie bracket of $\mathfrak{g}$ imply that $([\mathrm{ad}_x, \mathrm{ad}_y])(z) = [[x, y], z] \equiv \mathrm{ad}_{[x,y]}(z)$; this verifies the homomorphism property of R_{ad}. The fact that any Lie algebra has at least one non-trivial representation provides another motivation for the study of representations: By examining representations we can learn more about the structure of the Lie algebra itself.

As an example, we display the matrices describing the adjoint representations of $\mathfrak{sl}(2)$ and $\mathfrak{su}(2)$. For $\mathfrak{sl}(2)$, H is represented by the diagonal matrix $R_{\mathrm{ad}}(H) = \mathrm{diag}(2, 0, -2)$, and

$$R_{\mathrm{ad}}(E_+) = \begin{pmatrix} 0 & -2 & 0 \\ 0 & 0 & 1 \\ 0 & 0 & 0 \end{pmatrix}, \qquad R_{\mathrm{ad}}(E_-) = \begin{pmatrix} 0 & 0 & 0 \\ -1 & 0 & 0 \\ 0 & 2 & 0 \end{pmatrix}. \tag{5.11}$$

Here the first row and column refer to E_+, the second to H, and the third to E_-. For $\mathfrak{su}(2)$ the matrices are

$$R_{\mathrm{ad}}(\mathrm{i}L_1) = \begin{pmatrix} 0 & 0 & 0 \\ 0 & 0 & 1 \\ 0 & -1 & 0 \end{pmatrix}, \qquad R_{\mathrm{ad}}(\mathrm{i}L_2) = \begin{pmatrix} 0 & 0 & -1 \\ 0 & 0 & 0 \\ 1 & 0 & 0 \end{pmatrix},$$

$$\tag{5.12}$$

$$R_{\mathrm{ad}}(\mathrm{i}L_3) = \begin{pmatrix} 0 & 1 & 0 \\ -1 & 0 & 0 \\ 0 & 0 & 0 \end{pmatrix}.$$

In terms of the generators, the adjoint representation is just given by the structure constants,

$$\mathrm{ad}_{T^a}(T^b) = [T^a, T^b] = \sum_c f^{ab}{}_c T^c, \qquad (5.13)$$

so that the entries of the matrices $R_{\mathrm{ad}}(T^a)$ read

$$(R_{\mathrm{ad}}(T^a))^b{}_c = f^{ab}{}_c. \qquad (5.14)$$

In the physics literature the adjoint representation is therefore sometimes called the representation of a Lie algebra 'on its structure constants'.

Let us also recall the following elementary fact from linear algebra. Any endomorphism φ of a finite-dimensional complex vector space can be uniquely written as the sum

$$\varphi = \varphi_{\mathrm{s}} + \varphi_{\mathrm{n}} \qquad (5.15)$$

of a diagonalizable endomorphism φ_{s} and a nilpotent endomorphism φ_{n} which commute, $\varphi_{\mathrm{s}} \circ \varphi_{\mathrm{n}} = \varphi_{\mathrm{n}} \circ \varphi_{\mathrm{s}}$ (an endomorphism is called nilpotent if some finite power of it vanishes; this means that its matrix realization with respect to a suitable basis is strictly upper triangular). The endomorphisms φ_{s} and φ_{n} are called the *semisimple* and the *nilpotent* part of φ, respectively, and (5.15) is called the *Jordan decomposition* of φ. Moreover, both φ_{s} and φ_{n} can be written as polynomials in φ, and φ_{s} has the same eigenvalues as φ, while φ_{n} has only a single degenerate eigenvalue, namely zero.

Now the adjoint representation of a semisimple Lie algebra $\mathfrak{g}$ is faithful and hence we can use it to identify the elements x of $\mathfrak{g}$ with their images ad_x in the adjoint representation. We can then analogously decompose any $x \in \mathfrak{g}$ as $x = x_{\mathrm{s}} + x_{\mathrm{n}}$ into its semisimple and nilpotent parts x_{s} and x_{n}. A rather deep theorem states that for semisimple $\mathfrak{g}$ the decomposition $x = x_{\mathrm{s}} + x_{\mathrm{n}}$ induces the Jordan decomposition of $R(x)$ for each finite-dimensional $\mathfrak{g}$-representation R and each $x \in \mathfrak{g}$, i.e. $R(x_{\mathrm{s}})$ and $R(x_{\mathrm{n}})$ are the semisimple and nilpotent parts of $R(x)$, respectively. In contrast, for non-semisimple $\mathfrak{g}$ the semisimple and nilpotent parts of $R(x)$ in different representations R are completely unrelated.

5.3 Constructing more representations

Given one representation of $\mathfrak{g}$, one can construct more of them using elementary machinery from linear algebra. Suppose we are given a matrix representation R on some vector space V. Then we can define a new set of matrices by

$$R^+(x) := -(R(x))^{\mathrm{t}} \quad \text{for all } x \in \mathfrak{g}, \qquad (5.16)$$

where ' ^{t} ' stands for the transposition of matrices. These matrices realize again a representation of $\mathfrak{g}$ (see exercise 5.1), denoted by R^+ and called the representation *conjugate* (or *contragredient*) to R. The vector space on which this representation acts has the same dimension as V; however, since we have transposed the matrices, it is natural to interpret it as the dual vector space $V^\star$. (Recall that the vector space $V^\star$ dual to V is the space of linear forms on V, i.e. of linear maps from V to the base field F.)

Given two $\mathfrak{g}$-modules V and W, one can represent $\mathfrak{g}$ also on the vector space direct sum $V \oplus W$, namely by

$$R_V \oplus R_W: \quad ((R_V \oplus R_W)(x))(v \oplus w) := R_V(x)\,v + R_W(x)\,w, \quad (5.17)$$

and on the tensor product $V \otimes W$ by

$$R_V \otimes R_W: \quad ((R_V \otimes R_W)(x))(v \otimes w) :=$$
$$(R_V(x)\,v) \otimes w \,+\, v \otimes (R_W(x)\,w). \quad (5.18)$$

(The reader can check in exercise 5.1 that these prescriptions indeed lead to representations.) These representations are called the *direct sum* and the *tensor product* of the representations R_V and R_W, respectively. An immediate consequence of the definition of the tensor product is that it is associative (up to isomorphism), i.e. that

$$(R \otimes R') \otimes R'' \cong R \otimes (R' \otimes R'') \quad (5.19)$$

for all representations R, R', R'' of $\mathfrak{g}$.

Because of the summation on the right hand side of (5.18), quantum numbers (or, in mathematical terms, weights) add up when forming the tensor product representation. In the physics literature the quantum numbers coming from infinitesimal symmetries, i.e. from symmetries described by Lie algebras, are therefore sometimes called *additive* quantum numbers. Another important property of the tensor product of representations of *simple* Lie algebras is that the tensor product of two irreducible representations R and R' contains the singlet, i.e. the trivial one-dimensional representation, at most once as a submodule; it does contain the singlet precisely in the case when R' is the representation R^+ conjugate to R.

Yet another possibility to obtain new representations of $\mathfrak{g}$ is by means of endomorphisms of $\mathfrak{g}$. Namely, if R is a $\mathfrak{g}$-representation and φ an endomorphism of $\mathfrak{g}$, then $R^\varphi := R \circ \varphi$, acting as

$$R^\varphi(x) := R(\varphi(x)), \quad (5.20)$$

is a representation of $\mathfrak{g}$, too.

Representations obtained with the help of endomorphisms play a prominent rôle in the algebraic theory of superselection sectors, which describes relativistic quantum field theory in terms of von Neumann algebras of bounded operators on a Hilbert space. The representation theory of these

Information

associative algebras is extremely complicated. But of most direct physical interest is the class of those representations which are of the form $R_o \circ \varphi$, with φ an endomorphism of the algebra and R_o the vacuum representation (whose representation space contains the Lorentz invariant vacuum state of the theory). These representations are comparatively easy to access. A distinctive feature of the relevant maps φ are their localization properties, which are implied by relativistic causality. For details, see chapter IV.2 of [Haag 1992].

We also would like to introduce maps which relate two modules of one and the same Lie algebra $\mathfrak{g}$. Such a map must certainly be linear. Furthermore, it should be compatible with the action of $\mathfrak{g}$. More precisely, we are interested in linear maps $f: V \to W$ which commute with the action of $\mathfrak{g}$ in the sense that

$$R_W(x) \circ f = f \circ R_V(x) \quad \text{for all } x \in \mathfrak{g}. \tag{5.21}$$

One says that f *intertwines* the action of $\mathfrak{g}$, and calls f an *intertwiner* or $\mathfrak{g}$-morphism. The intertwining property (5.21) can also be expressed by stipulating that for each $x \in \mathfrak{g}$ the diagram

$$
\begin{array}{ccc}
V & \xrightarrow{R_V(x)} & V \\
f \downarrow & & \downarrow f \\
W & \xrightarrow{R_W(x)} & W
\end{array}
\tag{5.22}
$$

is a so-called *commutative diagram*. A diagram of mappings is said to be commutative if different paths in the diagram which correspond to different compositions of mappings between one and the same pair of objects describe in fact identical maps. In the case of the diagram (5.22), starting at the top left corner, going down and then turning to the right reproduces the left hand side of the relation (5.21), while first going right and then down yields the right hand side. The requirement that both paths give rise to identical maps therefore indeed coincides with the defining property (5.21) of an intertwiner.

In case there exists a surjective intertwiner from some $\mathfrak{g}$-module V to a $\mathfrak{g}$-module W, the module W is called a *homomorphic image* of V. If an intertwiner f is a vector space *iso*morphism, the two modules V and W are said to be *isomorphic* (as modules). Such modules are not only indistinguishable as abstract vector spaces, but also cannot be distinguished by using the action of the Lie algebra $\mathfrak{g}$.

We remark that the concepts of representations and modules can also be defined in an analogous manner for algebras which are not Lie algebras. A (linear) representation of the algebra $\mathfrak{A}$ is a linear map to the general linear algebra of some vector space V which preserves the product of $\mathfrak{A}$. However, unlike in the case of Lie algebras, there is generically no natural way of defining tensor products of representations. This problem, which will be described in more detail in section 15.2, eventually leads to the concept of Hopf algebras that will be introduced in chapter 22.

5.4 Schur's Lemma

The various constructions sketched in the previous section show that typically a Lie algebra will have many different, but related, representations. To get an overview over all possible representations (up to isomorphism), one therefore would like to identify the fundamental building blocks into which other representations can be decomposed, but which cannot be further decomposed themselves. This leads to the notion of *irreducible* representations. To explain what these are, it is necessary to present a few further concepts.

First, a *submodule*, or invariant subspace, of a module V is a sub-vector space U of V which is mapped into itself by all $R(x)$, i.e. $(R(x))(U) \subseteq U$ for all $x \in \mathfrak{g}$. By restriction, the representation R on V induces a representation on any submodule U of V. Any module possesses two trivial submodules, the module itself and the linear subspace consisting of the zero vector. Also, the kernel and the image of any intertwiner (defined as in equation (5.21)) $f \colon V \to W$ provide examples for submodules (see exercise 5.3). Now an *irreducible module* of a Lie algebra is by definition a module which does not contain a non-trivial submodule. Any other, i.e. non-irreducible, module is called *reducible*.

Irreducible modules are distinguished by a number of properties which are a consequence of *Schur's Lemma*, which we are going to present now. Consider two irreducible representation spaces V and W of $\mathfrak{g}$. As just noted, the kernel of an intertwiner $f \colon V \to W$ is a submodule; since V is irreducible and hence does not possess any non-trivial submodule, the kernel must be either all of V, in which case the map is zero, or else consist of only the zero vector, in which case f is injective. Similarly, since the image of f is a submodule of W, it has to be either W or the zero vector, which shows that f is also surjective, unless it is the zero map. These findings are summarized by

Schur's Lemma: Any intertwiner between two irreducible modules V and W is either an isomorphism or zero.

It is important that this assertion is valid for any Lie algebra over an arbitrary field F; in particular it holds for both real and complex Lie algebras. This result can be employed to describe the set $\mathcal{O}$ of those endomorphisms of an irreducible representation V over some number field F which intertwine the representation with itself, i.e. which commute with all endomorphisms that represent elements of the algebra. $\mathcal{O}$ is an algebra of maps and hence an associative algebra over F; also, as just shown, any element of $\mathcal{O}$ except for the zero map is an isomorphism and therefore has an inverse with respect to multiplication. Since this inverse is an isomorphism as well, all elements of $\mathcal{O}$ except for the zero possess inverses. An algebra $\mathcal{O}$ with this property is called a *division algebra* over the field F.

The situations of main practical interest of Schur's Lemma are the cases where $F = \mathbb{R}$ or $F = \mathbb{C}$, and where in addition the relevant irreducible modules are finite-dimensional. For simplicity let us first assume that the Lie algebra $\mathfrak{g}$ is defined over the complex numbers. Consider a *self-intertwiner* f of a finite-dimensional irreducible $\mathfrak{g}$-module V. Since V is a finite-dimensional complex vector space, f has at least one eigenvector v to some eigenvalue ξ. As a consequence, the kernel of the map $f - \xi\,id$ (which is an intertwiner as well) does not consist of only the zero vector. Thus Schur's Lemma immediately implies that this map is zero, or in other words, that any self-intertwiner of V must be a multiple $f = \xi\,id$ of the identity map id, i.e. it must be of the form $v \mapsto \xi v$ for some complex number ξ. This statement about the spaces of self-intertwiners of irreducible *complex* representations is again referred to as Schur's Lemma.

The contents of Schur's Lemma can also be rephrased as follows. Any endomorphism of an n-dimensional vector space V can be described by an $n{\times}n$-matrix. The endomorphisms of $\mathfrak{g}$ are thus precisely the representation matrices, and the self-intertwiners are the matrices that commute with all representation matrices. Schur's Lemma then tells us that if a matrix commutes with all representation matrices of an irreducible representation, it must be a multiple of the unit matrix. Conversely, it also allows us to test whether a given representation is irreducible: In case one can find a matrix which commutes with all representation matrices, but which is not a multiple of the unit matrix, the representation cannot be irreducible.

As an illustration consider the anti-unitary operator T that implements time reversal in a quantum mechanical system. Its square T^2 is a unitary operator, and one can show that it commutes with all observables of the system, in particular with the generators J_i of the total angular momentum. Hence T^2 acts as a constant on any irreducible module of the $\mathfrak{sl}(2)$ algebra spanned by the J_i. This constant is $+1$ if the highest weight Λ of the irreducible module is even, and -1 else (see chapter XV, §19 of [Messiah 1986]).

Information

For finite-dimensional modules Schur's Lemma has another important consequence: In the case of two finite-dimensional complex irreducible modules V and W the dimension, over $\mathbb{C}$, of the space of intertwiners between V and W is either 0 or 1; in the latter case the two modules are isomorphic. This follows because assuming that there are two different intertwiners f_1 and f_2 from V to W, the map $f_2^{-1}f_1$ is an isomorphism of V, so that by Schur's Lemma $f_2^{-1}f_1 = \xi\,id$, or in other words, f_2 is a scalar multiple of f_1.

For the case of $F = \mathbb{C}$, Schur's Lemma can also be derived by employing a theorem from algebra, which states that the only division algebras over algebraically closed fields (such as the complex numbers $\mathbb{C}$) are the fields themselves. To deal with algebras over the real numbers (which are not algebraically closed) one has to invoke in addition the so-called Frobenius theorem, which asserts that up to isomorphism the only finite-dimensional associative division algebras over the field $\mathbb{R}$ are the real numbers themselves, the complex numbers and the quaternions. (The quaternion algebra $\mathbb{H}$, which has dimension 4 over $\mathbb{R}$, is non-commutative. For more information on the quaternions, see section 8.3). If also the requirement of associativity is dropped, one obtains one more division algebra, the octonions or Cayley-Dickson algebra $\mathbb{O}$. Since in the present context the algebras can be written as matrix algebras and are therefore associative, the octonions do not appear.) Thus for algebras over $\mathbb{R}$ the space of intertwiners between two finite-dimensional irreducible modules, if not zero, is isomorphic either to $\mathbb{R}$, or to $\mathbb{C}$, or to the quaternions, with real dimension 1, 2 and 4, respectively.

5.5 Reducible modules

As seen above, one can build more modules out of irreducible ones by forming the direct sum. Correspondingly one introduces the notion of *fully reducible* or *completely reducible* modules of a Lie algebra, i.e. of modules which are the direct sum of irreducible modules. This notion can of course also be translated to the language of representations (see exercise 5.4). It is important to note that a reducible (i.e. non-irreducible) module needs not necessarily be fully reducible.

A representation over $\mathbb{C}$ is fully reducible if and only if there is a basis of the underlying vector space V such that all representation matrices $R(x)$ are simultaneously of a block-diagonal form,

$$R(x) = \begin{pmatrix} R_1(x) & 0 & \cdots & 0 \\ 0 & R_2(x) & \cdots & 0 \\ \vdots & \vdots & \ddots & \vdots \\ 0 & 0 & \cdots & R_n(x) \end{pmatrix}. \tag{5.23}$$

Here the R_i are square matrices of appropriate dimension, and '0' stands for matrices for which all entries are zero. To ensure that the R_i describe irreducible representations, we have to impose the condition that the only matrices that commute with $R_i(x)$ for a fixed value of i and all $x \in \mathfrak{g}$ are multiples of the unit matrix. Similarly, for representations which are reducible, but not fully reducible, there is a basis of V such that all representation matrices have the block form

$$R(x) = \begin{pmatrix} R_1(x) & Q_1(x) \\ 0 & Q_2(x) \end{pmatrix}, \qquad (5.24)$$

where $R_1(x)$ are the representation matrices for some (possibly reducible) representation R_1. (For more details, see exercise 5.5.)

If $\mathfrak{g}$ is a *semisimple* Lie algebra, then the tensor product of finite-dimensional irreducible modules possesses the crucial property of being fully reducible. Even more, all finite-dimensional modules can be obtained as irreducible components of tensorial powers of a small set of basic modules (these will be listed in section 13.6). Thus, for semisimple algebras, it is possible to construct the whole representation theory of finite-dimensional modules using only these basic modules and the concept of the tensor product. This result proves to be particularly useful for the simple Lie algebras $\mathfrak{sl}(n)$; it will be used extensively in chapter 19 to construct a diagrammatic calculus for dealing with representations of these (and other) simple Lie algebras.

Finally, we remark that the set of finite-dimensional irreducible modules of a semisimple Lie algebra and their tensor product may be viewed as describing the generators and the multiplication, respectively, of an associative ring over the integers $\mathbb{Z}$. This was in fact already implicitly used in chapter 3 (compare equation (3.38)) to describe tensor products of $\mathfrak{sl}(3)$-representations in a formal manner. The analysis of this ring structure leads to the concept of character rings and fusion rings, which we will describe in chapter 22.

5.6 Matrix Lie algebras

Above we started from an abstract Lie algebra $\mathfrak{g}$ and investigated matrix representations of $\mathfrak{g}$. Now we will turn this procedure around, thereby obtaining many simple Lie algebras by studying certain matrix algebras. The Lie bracket is then given by the commutator of matrices. These Lie algebras, which are frequently called *classical* Lie algebras, come in infinite series, as opposed to the exceptional simple Lie algebras of which there is only a small finite number and which are not introduced via a defining matrix representation.

So let V be a finite-dimensional vector space over some field F – in practice again the real or complex numbers. We have already seen that the space $\mathfrak{gl}(V)$ of all endomorphisms of V forms a Lie algebra. This Lie algebra is not semisimple; it contains an abelian ideal, which consists of all scalar multiples of the identity matrix. By imposing the condition that the matrices should be traceless, one eliminates this ideal and arrives at a complex Lie algebra which is simple (see exercise 5.7). If V has complex dimension n, this algebra is denoted by $\mathfrak{sl}(n)$ or A_{n-1}. The second of these names emerges in the classification of simple Lie algebras (see chapter 7), while the first name refers to the special properties of the matrix realization. Namely, to $\mathfrak{sl}(n)$ there corresponds (see chapter 9) the Lie *group* $\mathrm{SL}(n)$ of $n \times n$-matrices of determinant 1, which is known as the *special linear group*; similarly, the Lie group associated to $\mathfrak{gl}(n)$ is the *general linear group* $\mathrm{GL}(n)$ of invertible $n \times n$-matrices.

There also exist a few further natural simple subalgebras of $\mathfrak{gl}(V)$. First, there are the *symplectic* Lie algebras, a series denoted by $\mathfrak{sp}(n)$ or C_n. The Lie algebra $C_n \subset \mathfrak{gl}(2n)$ consists of those matrices $M \in \mathfrak{gl}(2n)$ which obey the relation $M^{\mathrm{t}} = JMJ$, where J is the $2n \times 2n$-matrix

$$J := \begin{pmatrix} 0_n & \mathbb{1}_n \\ -\mathbb{1}_n & 0_n \end{pmatrix}, \tag{5.25}$$

with $\mathbb{1}_n$ the $n \times n$ unit matrix and 0_n the $n \times n$-matrix whose entries are all zero. Another simple subalgebra of $\mathfrak{gl}(2n)$ consists of those matrices M which fulfill the relation $M^{\mathrm{t}} K + KM = 0$, where K is the $2n \times 2n$-matrix

$$K := \begin{pmatrix} 0_n & \mathbb{1}_n \\ \mathbb{1}_n & 0_n \end{pmatrix}. \tag{5.26}$$

This algebra is called $\mathfrak{so}(2n)$ or D_n. Finally there is also a simple subalgebra of $\mathfrak{gl}(2n+1)$, called $\mathfrak{so}(2n+1)$ or B_n, which is given by all matrices M which obey $M^{\mathrm{t}} K' + K'M = 0$, where K' is the $(2n+1) \times (2n+1)$-matrix

$$K' := \begin{pmatrix} 1 & 0 & 0 \\ 0 & 0_n & \mathbb{1}_n \\ 0 & \mathbb{1}_n & 0_n \end{pmatrix}. \tag{5.27}$$

The Lie algebras in the two series B_n and D_n are also known as the *orthogonal* Lie algebras. Note that, just like $\mathfrak{sl}(n)$, here the symbols $\mathfrak{sp}(n)$ and $\mathfrak{so}(n)$ refer to Lie algebras over the *complex* numbers. Both symplectic and orthogonal algebras over the *real* numbers appear very naturally in physics: Orthogonal Lie algebras describe infinitesimal rotations, while infinitesimal symplectic mappings provide infinitesimal canonical transformations in classical mechanics which leave the form of the Hamilton equations invariant.

For each of the four types of classical Lie algebras, there exists a convenient basis whose elements are matrices which have only a few non-zero entries. To describe these bases, one introduces for $i, j = 1, 2, \ldots, n$ the following $n \times n$-matrices $\mathcal{E}_{i,j}$, which are called matrix units. By definition, the entry in the ith row and jth column of $\mathcal{E}_{i,j}$ is equal to 1 while all other entries are zero:

$$(\mathcal{E}_{i,j})_{kl} = \delta_{ik}\delta_{jl}. \tag{5.28}$$

The matrix units $\mathcal{E}_{i,j}$ with $i,j \in \{1,2,\dots,n\}$ provide a basis for the vector space of $n \times n$-matrices. Furthermore, the product of two matrix units is again a matrix unit:

$$\mathcal{E}_{i,j}\mathcal{E}_{k,l} = \delta_{jk}\,\mathcal{E}_{i,l}\,. \tag{5.29}$$

A basis of $\mathfrak{sl}(n) \cong A_{n-1}$ is then provided by the union

$$\mathcal{B}_+ \cup \mathcal{B}_\circ \cup \mathcal{B}_- \tag{5.30}$$

of three sets $\mathcal{B}_\circ \equiv \mathcal{B}_\circ(A_{n-1})$ and $\mathcal{B}_\pm \equiv \mathcal{B}_\pm(A_{n-1})$, which are defined as

$$\mathcal{B}_\circ(A_{n-1}) = \{\mathcal{E}_{i,i} - \mathcal{E}_{i+1,i+1} \mid i = 1,2,\dots,n-1\}\,,$$
$$\mathcal{B}_+(A_{n-1}) = \{\mathcal{E}_{i,j} \mid 1 \le i < j \le n\}\,, \tag{5.31}$$
$$\mathcal{B}_-(A_{n-1}) = \{\mathcal{E}_{i,j} \mid 1 \le j < i \le n\}\,.$$

Together there are thus $(n-1) + 2 \cdot \frac{n(n-1)}{2} = n^2 - 1$ generators, so that the dimension of A_{n-1} is $n^2 - 1$.

For the other classical algebras, i.e. $\mathfrak{sp}(n) \cong C_n$, $\mathfrak{so}(2n) \cong D_n$ and $\mathfrak{so}(2n+1) \cong B_n$, the bases are again conveniently described by the union (5.30) of three sets $\mathcal{B}_\circ$ and $\mathcal{B}_\pm$. In the formulæ below, the basis elements are again expressed in terms of matrix units $\mathcal{E}_{i,j}$, which are now, however, $2n \times 2n$-matrices (for C_n and D_n) and $(2n+1) \times (2n+1)$-matrices (for B_n), respectively, and in the latter case it proves to be convenient to choose the labelling such that $i,j \in \{0,1,\dots,2n\}$. The subsets then look as follows, with the integers i and j all taking values in the set $\{1,2,\dots,n\}$:

$$C_n: \quad \mathcal{B}_\circ = \{\mathcal{E}_{i,i} - \mathcal{E}_{i+n,i+n}\}\,,$$
$$\mathcal{B}_+ = \{\mathcal{E}_{i,j+n} + \mathcal{E}_{j,i+n} \mid i \le j\} \cup \{\mathcal{E}_{i,j} - \mathcal{E}_{j+n,i+n} \mid i < j\}\,,$$
$$\mathcal{B}_- = \{\mathcal{E}_{j+n,i} + \mathcal{E}_{i+n,j} \mid i \le j\} \cup \{\mathcal{E}_{i,j} - \mathcal{E}_{j+n,i+n} \mid j < i\}\,.$$

$$D_n: \quad \mathcal{B}_\circ = \{\mathcal{E}_{i,i} - \mathcal{E}_{i+n,i+n}\}\,,$$
$$\mathcal{B}_+ = \{\mathcal{E}_{i,j+n} - \mathcal{E}_{j,i+n} \mid i < j\} \cup \{\mathcal{E}_{i,j} - \mathcal{E}_{j+n,i+n} \mid i < j\}\,,$$
$$\mathcal{B}_- = \{-\mathcal{E}_{j+n,i} + \mathcal{E}_{i+n,j} \mid i < j\} \cup \{\mathcal{E}_{i,j} - \mathcal{E}_{j+n,i+n} \mid j < i\}\,.$$

$$B_n: \quad \mathcal{B}_\circ = \{\mathcal{E}_{i,i} - \mathcal{E}_{i+n,i+n}\}\,,$$
$$\mathcal{B}_+ = \{\mathcal{E}_{i,j+n} + \mathcal{E}_{j,i+n} \mid i \le j\} \cup \{\mathcal{E}_{i,j} - \mathcal{E}_{j+n,i+n} \mid i < j\}$$
$$\cup \{\mathcal{E}_{i,0} - \mathcal{E}_{0,i+n}\}\,,$$
$$\mathcal{B}_- = \{-\mathcal{E}_{j+n,i} + \mathcal{E}_{i+n,j} \mid i \le j\} \cup \{\mathcal{E}_{i,j} - \mathcal{E}_{j+n,i+n} \mid j < i\}$$
$$\cup \{\mathcal{E}_{0,i} - \mathcal{E}_{i+n,0}\}\,. \tag{5.32}$$

We have deliberately described these bases as the union (5.30) of three specific subsets. As is easily checked (see exercise 5.7), for all four series the subsets $\mathcal{B}_o$ and $\mathcal{B}_\pm$ possess the following properties.

• First, the generators in $\mathcal{B}_o$ all commute among each other and therefore span an abelian subalgebra; in fact this algebra is even a maximal abelian subalgebra of $\mathfrak{g}$. The number n appearing as a subscript in the A-B-C-D-notation of the algebras is the dimension of this subalgebra.

• Second, each of the subsets $\mathcal{B}_+$ and $\mathcal{B}_-$ spans a nilpotent subalgebra.

• And third, the basis consists only of matrices which are eigenvectors under the adjoint map ad_x for all elements x of $\mathcal{B}_o$.

Thus we encounter the same type of decomposition which we already have noticed in the basic examples considered in chapter 2; a splitting of this type is called a triangular decomposition of these algebras. As will be explained in detail in chapter 6, this structure is actually common to all semisimple Lie algebras (see equation (6.32)) and to many other Lie algebras as well.

5.7 The representation theory of $\mathfrak{sl}(2)$ revisited

The representation theory of abelian Lie algebras is most easy to understand. Any abelian Lie algebra is a direct sum of $\mathfrak{u}(1)$ algebras, so that its representation theory directly follows from the one of $\mathfrak{u}(1)$. As a consequence, any fully reducible module is the direct sum of one-dimensional modules. Also, any finite-dimensional *irreducible* representation R is one-dimensional, and the generators T^a act multiplicatively, $R(T^a)\,v = q^a \cdot v$, on any irreducible module. The complex numbers q^a are called the *charges* of the module. (For a few more details, see exercise 5.6.)

Any irreducible representation of a direct sum of Lie algebras $\mathfrak{g}_i$ can be considered as the tensor product of irreducible representations of the individual summands $\mathfrak{g}_i$. Recalling that each finite-dimensional representation of a semisimple Lie algebra is the direct sum of irreducible representations, it therefore follows that the hard part in working out the representation theory of any semisimple Lie algebra $\mathfrak{g}$ is to determine the irreducible representations of the simple direct summands of $\mathfrak{g}$. (In view of the easy description of representations of abelian Lie algebras, this remark applies equally to direct sums of semisimple and abelian algebras, i.e. to all reductive Lie algebras.) As will be demonstrated in chapter 13, the latter can in turn be reduced to a large extent to the representation theory of $\mathfrak{sl}(2)$. The representation theory of $\mathfrak{sl}(2)$ therefore constitutes one of the central aspects of the theory of reductive Lie algebras. Therefore the results on $\mathfrak{sl}(2)$-representations which have already been obtained in chapter 2 will be displayed here once more, in a notation which will prove to be convenient for the generalization to arbitrary simple Lie algebras.

According to chapter 2, any finite-dimensional irreducible module of

$\mathfrak{sl}(2)$ contains an element v_Λ which, first, is annihilated by the action of the raising operator E_+ and second, is an eigenvector with eigenvalue Λ for the action of H. The number Λ is called the *highest weight* of the module. We denote the irreducible module and representation with these properties by V_Λ and R_Λ, respectively. The defining properties then read

$$R_\Lambda(E_+)\, v_\Lambda = 0\,, \qquad R_\Lambda(H)_, v_\Lambda = \Lambda \cdot v_\Lambda\,. \tag{5.33}$$

Up to multiplication with a non-zero number, the element $v_\Lambda \in V_\Lambda$ is unique; it is called the *highest weight vector* of the module V_Λ.

All other elements of V_Λ which are eigenvectors of H can be obtained, up to scalar factors, by multiple application of the lowering operator E_- on the highest weight vector (compare equation (2.13)). Each application of this generator changes the weight, i.e. H-eigenvalue, by 2. A basis of V_Λ is then given by elements v_λ obeying

$$v_\lambda \propto (R_\Lambda(E_-))^{(\Lambda-\lambda)/2}\, v_\Lambda\,, \qquad R_\Lambda(H)\, v_\lambda = \lambda \cdot v_\lambda\,, \tag{5.34}$$

with $\Lambda - \lambda$ a non-negative even integer. Furthermore, the basis element $v_{-\Lambda}$ is annihilated by E_- (it is therefore called the *lowest weight* vector of V_Λ), so that the allowed values of the weight λ for the irreducible representation R_Λ are

$$\lambda \in \{-\Lambda,\, -\Lambda+2,\, \ldots,\, \Lambda-2,\, \Lambda\}\,. \tag{5.35}$$

Each of these weights 'occurs' precisely once, i.e. v_λ is again unique up to normalization. As a vector space, V_Λ is thus the direct sum

$$V_\Lambda = \bigoplus_{\substack{\lambda=-\Lambda \\ \Lambda-\lambda\in 2\mathbb{Z}}}^{\Lambda} V_{(\lambda)} \tag{5.36}$$

of one-dimensional subspaces $V_{(\lambda)}$.

From the Lie brackets of $\mathfrak{sl}(2)$ it also follows immediately (compare equations (2.13) and (2.16)) that

$$R_\Lambda(E_\pm)R_\Lambda(E_\mp)\, v_\lambda = \left(\tfrac{1}{2}\,(\Lambda \pm \lambda) + \tfrac{1}{4}\,(\Lambda^2 - \lambda^2)\right) \cdot v_\lambda \tag{5.37}$$

for all weights λ in the range (5.35). While this result is independent of the chosen normalization of the basis elements v_λ, the action of the individual step operators $E_\pm$ does depend on this choice. A convenient normalization is

$$v_\lambda := \left[(\tfrac{1}{2}\,(\Lambda+\lambda)-1)!\,/\,(\tfrac{1}{2}\,(\Lambda-\lambda))!\,\Lambda!\right]^{1/2} (R_\Lambda(E_-))^{(\Lambda-\lambda)/2}\, v_\Lambda\,. \tag{5.38}$$

It is, however, not necessary at all to express the vectors v_λ in the closed form (5.38) through the vectors $(R_\Lambda(E_-))^{(\Lambda-\lambda)/2} v_\Lambda$. Rather, the recurrence relations

$$R_\Lambda(E_+)\, v_\lambda = \tfrac{1}{2}\,\sqrt{2\,(\Lambda-\lambda)+\Lambda^2-\lambda^2}\cdot v_{\lambda+2} \tag{5.39}$$

and

$$R_\Lambda(E_-)\, v_\lambda = \tfrac{1}{2}\,\sqrt{2\,(\Lambda+\lambda)+\Lambda^2-\lambda^2}\cdot v_{\lambda-2} \tag{5.40}$$

are already sufficient for all purposes. (Here it is understood that $v_{\Lambda+2} = 0 = v_{-\Lambda-2}$.) This is a phenomenon one frequently encounters when performing explicit calculations: While closed expressions may look 'nicer', recurrence relations are by far easier to handle and to evaluate, especially on a computer.

As will become clear later on, for general simple Lie algebras it proves to be extremely convenient to have only integers as weights. However, as long as one is dealing with $\mathfrak{sl}(2)$, one may equally well work with the spin $j = \Lambda/2$ and its 'third component' $m = \lambda/2$, which take half integral values; indeed this is what is commonly done in the treatment of angular momentum in non-relativistic quantum mechanics. In terms of the Cartan subalgebra of $\mathfrak{sl}(2)$, using the spin quantum numbers amounts to rescaling the generator L_0 by $1/2$. Writing $J_0 := \frac{1}{2}L_0$ as well as $J_\pm := L_\pm$, the Lie brackets of $\mathfrak{sl}(2)$ read

$$[J_0, J_\pm] = \pm J_\pm \,,$$
$$[J_+, J_-] = 2J_0 \,. \tag{5.41}$$

The eigenvalues m of J_0 are half-integers, and the eigenvalue j of the highest weight vector of an irreducible module is called the spin of the module.

In quantum mechanics one also frequently uses the 'bra-ket' notation for the vectors of $\mathfrak{su}(2)$-modules and their dual vector spaces, and suppresses the symbols for the representations in which the elements of the Lie algebra are to be taken. The basis of J_0-eigenvectors of the irreducible module with spin j that is provided by the vectors (5.38) is then denoted by $\{\,|j,m\rangle\,|\,m = -j, -j + 1, \ldots, j\,\}$, and the action of the generators on the basis vectors reads

$$J_0 \,|j,m\rangle = m \cdot |j,m\rangle \,,$$
$$J_\pm |j,m\rangle = \sqrt{j(j+1) - m(m \pm 1)} \cdot |j, m \pm 1\rangle \,. \tag{5.42}$$

Summary:

A (linear) representation of a Lie algebra $\mathfrak{g}$ is a homomorphism from $\mathfrak{g}$ to the general linear algebra of some vector space V. Any Lie algebra possesses a faithful representation acting on $\mathfrak{g}$ itself, the adjoint representation. An irreducible module does not contain any non-trivial submodule. According to Schur's Lemma, any non-zero intertwiner between irreducible modules is an isomorphism.

Keywords:

(Matrix) representation, general linear algebra $\mathfrak{gl}(V)$, $\mathfrak{gl}(n)$, module, representation space, singlet, adjoint representation;
conjugate representation, direct sum and tensor product of representations, submodule;
intertwiner, (ir)reducible representation, Schur's Lemma, fully reducible representation;
matrix Lie algebra, classical Lie algebra, orthogonal and symplectic algebras.

Exercises:

Check that the contragredient representation as defined in (5.16) is indeed a representation.
Show that also the direct sum (5.17) and the tensor product (5.18) of two representations are again representations.
Check that the tensor product obeys the associativity property (5.19), and that the prescription (5.20) defines a representation.

Exercise 5.1

Show that the prescription $x \mapsto R(x) := R_1(x) \otimes R_2(x)$, with R_1 and R_2 linear representations of some algebra $\mathfrak{A}$, does not yield a map linear in x, and hence in particular does not define a tensor product representation.

Exercise 5.2

Verify that the kernel $\mathrm{Ker}(f) \subseteq V$ and the image $\mathrm{Im}(f) \subseteq W$ of any intertwiner $f \colon V \to W$ of $\mathfrak{g}$-modules is a $\mathfrak{g}$-submodule of V respectively W.

Exercise 5.3

Show that a representation R of a Lie algebra $\mathfrak{g}$ is irreducible (respectively fully reducible, not fully reducible) if and only if the image $R(\mathfrak{g})$ is a simple (respectively semisimple, non-semisimple) Lie algebra.
As an application, prove that the adjoint representation of a Lie algebra $\mathfrak{g}$ is irreducible (fully reducible, not fully reducible) if and only if $\mathfrak{g}$ is simple (semisimple, non-semisimple).

Exercise 5.4

Prove that a matrix representation R is fully reducible if and only if it can be transformed by a change of basis to the block diagonal form (5.23).
Show that a matrix representation R on a module V is reducible, but not fully reducible, if it can be transformed to the block form (5.24).
Hint: identify the submodule V_1 which corresponds to the representation R_1, and analyze the action of R on the vector space direct sum $V = V_1 \oplus V_2$.

Exercise 5.5

Work out the representation theory of abelian Lie algebras $\mathfrak{g}$:

a) Show that the representation matrices $R(x)$ for a fully reducible representation of $\mathfrak{g}$ are diagonal matrices.

Hint: First show that the representation matrices are normal, i.e. commute with their transpose. Normal matrices that commute can be simultaneously diagonalized. Conclude that any fully reducible module is the direct sum of one-dimensional modules.

b) Prove that any finite-dimensional irreducible representation of an abelian Lie algebra is one-dimensional. (Actually this is even true for an arbitrary solvable Lie algebra.) Conclude that on any irreducible module of $\mathfrak{g} \cong \mathfrak{u}(1)$ the single generator T acts on each element of the module as multiplication

$$R(T) \cdot v = q\,v \qquad (5.43)$$

by the charge q of the module.

c) What is the module conjugate to a $\mathfrak{u}(1)$-module of charge q? Determine its $\mathfrak{u}(1)$-charge.

Exercise 5.6

a) Verify that the matrices $\mathcal{E}_{i,j}$ with entries $(\mathcal{E}_{i,j})_{kl} = \delta_{ik}\delta_{jl}$ obey the commutation relation (5.29).

b) Check that the generators in $\mathcal{B}_0$ as defined by (5.31) all commute among each other, and that the linear span of both $\mathcal{B}_+$ and $\mathcal{B}_-$ is a subalgebra. Compute the eigenvalues of the elements of $\mathcal{B}_+$ under the adjoint map for elements of $\mathcal{B}_0$.

c) Use these results to show that the Lie algebras A_n are semisimple.

Hint: As a characterization of semisimplicity, use the property that $[\mathfrak{g}, \mathfrak{g}] = \mathfrak{g}$. Use the fact that the elements of $\mathcal{B}_\pm$ are eigenvectors of ad_x with $x \in \mathcal{B}_0$; find generators x, y such that $[x, y]$ is an element of $\mathcal{B}_0$.

Exercise 5.7

Use the matrices (5.11) to compute the structure constants of $\mathfrak{sl}(2)$ in the basis $\{L_i\}$ (2.3).

Exercise 5.8

Determine the dimension of the Lie algebras B_n, C_n and D_n. Perform the calculations analogous to exercise 5.7 b) and c) for these algebras.

Exercise 5.9

6

The Cartan–Weyl basis

The present chapter is somewhat longer than the previous ones. This is unavoidable, because many central concepts are introduced. Having mastered this chapter, the reader will have acquired all the basic knowledge that is necessary to understand the chapters that follow and will be able to read many of those chapters independently according to personal interest or taste.

The key concept introduced below is the Cartan–Weyl basis. The choice of a Cartan–Weyl basis, besides being extremely convenient for explicit calculations, reveals a lot of structural information about semisimple Lie algebras. Similar structures can also be found for more general classes of Lie algebras.

6.1 Cartan subalgebras

The structure constants of a Lie algebra depend of course on the choice of basis. Just like in other areas of modern algebra, one may therefore try to avoid the explicit use of bases as much as possible. But as we will see now and in the chapters that follow, on the contrary for Lie algebras and their representation theory it is most convenient to choose a specific type of basis, which allows us in particular to write down the structure constants in a standard form. Among other applications, this will provide a basic tool for the classification of finite-dimensional semisimple Lie algebras. In fact, for (finite-dimensional) semisimple Lie algebras there is a completely canonical (type of) basis, the so-called Cartan–Weyl basis. In contrast, for more general Lie algebras, and in particular for solvable Lie algebras, a fully canonical form of the structure constants is not available. Unless stated otherwise, in the following it will therefore be assumed that $\mathfrak{g}$ is finite-dimensional and semisimple; some remarks on non-semisimple algebras are made at the end of the chapter.

The starting point in the construction of Cartan–Weyl bases is the identification of a certain abelian subalgebra of $\mathfrak{g}$. More precisely, one considers elements $x \in \mathfrak{g}$ with the property that the map ad_x is diagonalizable, which means that there is a choice of basis $\tilde{\mathcal{B}} = \{\tilde{T}^a\}$ such that $[x, \tilde{T}^a]$ is proportional to $\tilde{T}^a$ for any element of $\tilde{\mathcal{B}}$. Such elements called *ad-diagonalizable* or *semisimple* elements of $\mathfrak{g}$. We now restrict to the case that the base field F is algebraically closed, which means that any algebraic equation with coefficients in F has a solution in F; this is true for the field $\mathbb{C}$ of complex numbers, whereas the field $\mathbb{R}$ of real numbers is *not* algebraically closed (consider e.g. the equation $x^2 + 1 = 0$). Thus from now on we will assume that $F = \mathbb{C}$. The analysis of Lie algebras over $\mathbb{R}$, which will be sketched in chapter 8, is considerably harder than that of complex Lie algebras, and several problems are still open for real Lie algebras which in the complex case can be solved easily.

It can be shown that each semisimple Lie algebra $\mathfrak{g}$ over an algebraically closed field F contains indeed ad-diagonalizable elements x. When comparing the eigenvalue equation $[x, \tilde{T}^a] = \zeta \tilde{T}^a$ for a semisimple element to the general expansion $[x, T^a] = \sum_b \xi_b^a T^b$, it follows that the eigenvalue ζ obeys the equation $\det(\xi - \zeta \mathbb{1}) = 0$, which is called the *characteristic* or *secular equation*. In order that ad_x is diagonalizable, all roots of the secular equation must lie in the number field F; for this it is sufficient that F is algebraically closed.

Now among the ad-diagonalizable elements of $\mathfrak{g}$, one chooses a maximal set of linearly independent elements, denoted by H^i, which possess zero Lie brackets among themselves,

$$[H^i, H^j] = 0 \qquad \text{for } i, j = 1, 2, \ldots, \mathrm{r}. \tag{6.1}$$

The linear hull

$$\mathfrak{g}_\circ := \mathrm{span}_\mathbb{C}\{H^i \mid i = 1, 2, \ldots, \mathrm{r}\} \tag{6.2}$$

of these elements is called a *Cartan subalgebra* of $\mathfrak{g}$. The following properties of Cartan subalgebras are important:

• A Cartan subalgebra of $\mathfrak{g}$ is a maximal abelian subalgebra consisting entirely of semisimple elements. (This is just the definition of a Cartan subalgebra.)

• A semisimple Lie algebra can possess many different Cartan subalgebras. Fortunately, they are all related by automorphisms of $\mathfrak{g}$ (for more information, see section 11.3), so that the freedom in choosing a Cartan subalgebra does not lead to any arbitrariness in the description of semisimple Lie algebras.

• All Cartan subalgebras possess the same dimension r. This common dimension is thus a property of the Lie algebra $\mathfrak{g}$ and not of the specific choice of a Cartan subalgebra; it is called the *rank* of $\mathfrak{g}$:

$$r \equiv \operatorname{rank}\mathfrak{g} = \dim\mathfrak{g}_\circ. \qquad (6.3)$$

In particular, two Lie algebras can only be isomorphic if they have the same rank.

• The rank r is the minimal dimension of all subalgebras $\mathfrak{g}_h := \{x \in \mathfrak{g} \mid [x,h] = 0\}$ of $\mathfrak{g}$, where h is any semisimple element of $\mathfrak{g}$. Whenever for a semisimple element h of $\mathfrak{g}$ the dimension of $\mathfrak{g}_h$ is equal to r, $\mathfrak{g}_h$ is a Cartan subalgebra of $\mathfrak{g}$ that contains h. In this case h is called a *regular* element of $\mathfrak{g}$. The subset of regular elements of a given Cartan subalgebra $\mathfrak{g}_\circ$ is an open and dense subset of $\mathfrak{g}_\circ$ (in the natural topology of $\mathfrak{g}_\circ$ that follows from the isomorphism $\mathfrak{g}_\circ \equiv \mathbb{C}^r$). We will find a convenient characterization of regular elements in section 10.6.

In physical terms, the rank provides the maximal number of quantum numbers which can be used to label (at least partially) the states of a physical system that has $\mathfrak{g}$ as its symmetry algebra (compare chapter 3 for the case $\mathfrak{g} = \mathfrak{sl}(3)$, which has rank 2).

6.2 Roots

Because the generators H^i of a Cartan subalgebra $\mathfrak{g}_\circ$ have zero bracket among themselves, the adjoint maps ad_{H^i} for $i = 1, 2, \ldots, r$, and thus in fact ad_h for all Cartan subalgebra elements h, are *simultaneously* diagonalizable. As a consequence, $\mathfrak{g}$ is spanned by such elements y which are simultaneous eigenvectors of all the maps ad_h, $h \in \mathfrak{g}_\circ$, i.e. satisfy

$$[h, y] \equiv \operatorname{ad}_h(y) = \alpha_y(h)\, y. \qquad (6.4)$$

For any fixed element $y \in \mathfrak{g}$ of this type, the eigenvalue $\alpha_y(h)$ of y is some complex number which depends linearly on $h \in \mathfrak{g}_\circ$, and hence α_y is a linear function $\mathfrak{g}_\circ \to \mathbb{C}$, i.e. an element of the vector space $\mathfrak{g}_\circ^\star$ dual to $\mathfrak{g}_\circ$. (Recall from chapter 4 that the dual vector space $V^\star$ of a vector space V is the space of linear maps from V to the base field F.) The eigenvalues of ad_h are the roots of the characteristic equation for h. Such a function α is therefore called a *root* of the Lie algebra $\mathfrak{g}$ (relative to the chosen Cartan subalgebra $\mathfrak{g}_\circ$). For later convenience, however, one declares the function α to be a root only if it is non-zero; in addition to the roots, there are always r = $\operatorname{rank}\mathfrak{g}$ solutions which are identically zero (compare exercise 6.1).

As $\mathfrak{g}$ is spanned by elements satisfying (6.4), it can be written as a direct sum of vector spaces $\mathfrak{g}_\alpha$ according to

$$\mathfrak{g} = \bigoplus_\alpha \mathfrak{g}_\alpha, \qquad \mathfrak{g}_\alpha = \{x \in \mathfrak{g} \mid [h, x] = \alpha(h)\cdot x \text{ for all } h \in \mathfrak{g}_\circ\}. \qquad (6.5)$$

Separating the elements of the Cartan subalgebra $\mathfrak{g}_\circ$, this decomposition of $\mathfrak{g}$ can be refined to

$$\mathfrak{g} = \mathfrak{g}_\circ \oplus \bigoplus_{\alpha \neq 0} \mathfrak{g}_\alpha. \tag{6.6}$$

The splitting (6.6) is called the *root space decomposition* of $\mathfrak{g}$ relative to the Cartan subalgebra $\mathfrak{g}_\circ$. This decomposition means in particular that there is a basis $\mathcal{B}$ of $\mathfrak{g}$ which apart from a basis $\{H^i\}$ of the chosen Cartan subalgebra $\mathfrak{g}_\circ$ consists entirely of elements E^α which satisfy

$$[H^i, E^\alpha] = \alpha^i \, E^\alpha \qquad \text{for } i = 1, 2, \ldots, \mathrm{r}, \tag{6.7}$$

where for each α the eigenvalue $\alpha^i := \alpha(H^i)$ is non-vanishing for at least one value of i. The r-dimensional vector $(\alpha^i)_{i=1,\ldots r}$ of eigenvalues of E^α with respect to ad_{H^i} is called a *root vector* or, shortly, a *root* of $\mathfrak{g}$ (relative to the basis $\{H^i\}$ of $\mathfrak{g}_\circ$). Root vectors are naturally interpreted as elements of the vector space dual to the Cartan subalgebra $\mathfrak{g}_\circ$.

The set of all roots of $\mathfrak{g}$ will be denoted by $\Phi \equiv \Phi(\mathfrak{g})$ and be called the *root system* of $\mathfrak{g}$. A fundamental property of the root system of a semisimple Lie algebra, which we will deduce in section 6.4, is that the roots are not degenerate, i.e. the eigenspaces $\mathfrak{g}_\alpha$ introduced in (6.5) are all one-dimensional. Thus in particular the generators E^α are uniquely specified, up to normalization, by the label α, and the root spaces $\mathfrak{g}_\alpha$ can be written as

$$\mathfrak{g}_\alpha = \mathrm{span}_{\mathbb{C}}\{E^\alpha\}. \tag{6.8}$$

Therefore, the basis $\mathcal{B}$ of $\mathfrak{g}$ described above can be summarized as

$$\mathcal{B} = \{H^i \mid i = 1, \ldots, \mathrm{r}\} \cup \{E^\alpha \mid \alpha \in \Phi\}. \tag{6.9}$$

A basis of this form, with H^i and E^α obeying (6.1) and (6.7), is called a *Cartan–Weyl basis* of $\mathfrak{g}$. It is also sometimes called a *canonical* or *standard basis* of $\mathfrak{g}$, and its elements H^i and E^α the canonical generators of $\mathfrak{g}$. We prefer the term Cartan–Weyl basis because this basis is not completely canonical (several arbitrary choices are involved: One has to choose a specific Cartan subalgebra $\mathfrak{g}_\circ$, a basis of $\mathfrak{g}_\circ$, and then specific non-zero elements E^α in the one-dimensional spaces $\mathfrak{g}_\alpha$). Also, the generators E^α are often called the *step operators* or *ladder operators* associated to the roots α. The reason for these terms is that, because of (6.7), the application of E^α to an element of a representation space having eigenvalues λ^i with respect to H^i yields an element with eigenvalues $\lambda^i + \alpha^i$, in a manner completely analogous to the special case of $\mathfrak{g} = \mathfrak{sl}(2)$ that was considered for instance in formula (5.39).

The root space decomposition (6.7) has important implications in gauge $\quad$ Information
theories. The gauge bosons of a non-abelian Yang–Mills theory carry
the adjoint representation of some simple Lie algebra; the quantum numbers of the gauge bosons are zero if they correspond to generators of the
Cartan subalgebra, and otherwise they are given by the roots (see e.g.
[O'Raifeartaigh 1986]). Since the roots are non-zero, the gauge bosons of
a non-abelian gauge theory are charged and therefore self-interact. For
example, in the theory of electro-weak interactions the $W^{\pm}$ gauge bosons
correspond to the positive and negative root, respectively, of an $\mathfrak{sl}(2)$ algebra and hence are charged, while a linear combination of the photon
and the Z boson, which are uncharged, corresponds to the Cartan subalgebra. This is in sharp contrast to the situation in abelian gauge theories
like electrodynamics, in which there are no roots at all so that the gauge
bosons do not self-interact.

6.3 The Killing form

For the further analysis of root systems, it will be necessary to define
an inner product on the space $\mathfrak{g}_\circ^\star$ of roots. This can be derived from
an analogous structure on its dual space $\mathfrak{g}_\circ$, which in turn is obtained
as the restriction of an inner product on the whole Lie algebra $\mathfrak{g}$. This
inner product on $\mathfrak{g}$ is called the *Killing form* (or Cartan–Killing form) κ
of $\mathfrak{g}$; for finite-dimensional $\mathfrak{g}$ it is defined by taking traces in the adjoint
representation, i.e.

$$\kappa(x,y) := \mathrm{tr}(\mathrm{ad}_x \circ \mathrm{ad}_y), \qquad (6.10)$$

where ' $\circ$ ' denotes the composition of maps, and ' tr ' the trace of linear
maps.

The linearity of the adjoint maps implies that $\kappa \colon \mathfrak{g} \times \mathfrak{g} \to \mathbb{C}$ is bilinear, and the cyclic invariance of the trace implies that it is symmetric,
$\kappa(x,y) = \kappa(y,x)$. Further important properties of the Killing form are its
invariance

$$\kappa([x,y],z) = \kappa(x,[y,z]), \qquad (6.11)$$

which follows with the help of $\mathrm{ad}_{[x,y]} = \mathrm{ad}_x \circ \mathrm{ad}_y - \mathrm{ad}_y \circ \mathrm{ad}_x$ (see exercise
6.2), and the fact that κ is preserved by any automorphism of $\mathfrak{g}$. As will
be shown in chapter 8, for simple $\mathfrak{g}$ these properties already define κ up
to an overall multiplicative constant.

Given a basis $\{T^a \,|\, a = 1, 2, \ldots, \mathrm{d}\}$ of $\mathfrak{g}$, the Killing form is represented
by the matrix

$$\kappa^{ab} := \tfrac{1}{I_{\mathrm{ad}}}\,\kappa(T^a, T^b) = \tfrac{1}{I_{\mathrm{ad}}} \cdot \mathrm{tr}\,(\mathrm{ad}_{T^a} \circ \mathrm{ad}_{T^b}), \qquad (6.12)$$

where for convenience a normalization constant I_{ad} has been introduced
(I_{ad} will be fixed later, see equation (14.32)). When acting with $\mathrm{ad}_{T^a} \circ \mathrm{ad}_{T^b}$

on another basis element T^c, one has

$$\mathrm{ad}_{T^a} \circ \mathrm{ad}_{T^b}(T^c) = [T^a, [T^b, T^c]] = \sum_{e,f=1}^{d} f^{bc}{}_e f^{ae}{}_f T^f \,. \tag{6.13}$$

Taking the trace amounts to keeping in the expression (6.13) the coefficient of T^c and summing over c, and hence one can express κ^{ab} through the structure constants as

$$\kappa^{ab} = \frac{1}{I_{\mathrm{ad}}} \sum_{c,e=1}^{d} f^{bc}{}_e f^{ae}{}_c \,. \tag{6.14}$$

While the Killing form is bilinear and symmetric, for an arbitrary Lie algebra it is degenerate, i.e. there may exist elements $x \neq 0$ of $\mathfrak{g}$ obeying $\kappa(x, y) = 0$ for all $y \in \mathfrak{g}$; in this case κ does not provide a proper inner product on $\mathfrak{g}$. It is an important result (obtained by Cartan) that a finite-dimensional Lie algebra $\mathfrak{g}$ is semisimple if and only if κ is non-degenerate (compare exercise 6.3), and hence is a proper inner product on $\mathfrak{g}$.

Consider now a Cartan–Weyl basis of a semisimple Lie algebra. Actually not only is κ non-degenerate, but its restriction to the Cartan subalgebra $\mathfrak{g}_\circ$ is non-degenerate as well (see exercise 6.6). Now any non-degenerate bilinear form on a vector space can be used to identify the vector space and its dual space. Hence we are lead to associate to any root α an element H^α of $\mathfrak{g}_\circ$, which up to normalization is unique, such that

$$\alpha(h) = c_\alpha \, \kappa(H^\alpha, h) \tag{6.15}$$

for all $h \in \mathfrak{g}_\circ$; here c_α is a normalization constant that will be chosen conveniently later on in (6.53). With the help of the elements H^α, one can then define a (non-degenerate) inner product on $\mathfrak{g}_\circ^\star$; namely, one sets

$$(\alpha, \beta) := c_\alpha c_\beta \, \kappa(H^\alpha, H^\beta) = c_\beta \, \alpha(H^\beta) \,, \tag{6.16}$$

for all roots α, β, and this extends by bilinearity to all of $\mathfrak{g}_\circ^\star \times \mathfrak{g}_\circ^\star$.

6.4 Some properties of roots and the root system

In this section we derive a few properties of roots and the root system of finite-dimensional simple Lie algebras. While the arguments are not complicated, they are somewhat lengthy. For the convenience of the reader we therefore list the most important results here:

- The roots span all of $\mathfrak{g}_\circ^\star$: $\mathrm{span}_{\mathbb{C}}(\Phi) = \mathfrak{g}_\circ^\star$.
- The root spaces $\mathfrak{g}_\alpha$ are one-dimensional.
- The only multiples of $\alpha \in \Phi$ which are roots are $\pm\alpha$.
- There is a basis $\{H^i\}$, $i = 1, 2, \ldots, \mathrm{r}$, of the Cartan subalgebra $\mathfrak{g}_\circ$ such

that $\beta(H^i)$ is a real number (in fact, an integer) for all i and for each root $\beta \in \Phi$.

Let us first show that the roots span all of $\mathfrak{g}_\circ^*$,

$$\mathrm{span}_{\mathbb{C}}(\Phi) = \mathfrak{g}_\circ^* . \tag{6.17}$$

Otherwise there would exist a non-zero Cartan subalgebra element h with $\alpha(h) = 0$ for all $\alpha \in \Phi$. But then due to (6.4) we would have $[h, \mathfrak{g}] = 0$, i.e. h would be an element of the center of $\mathfrak{g}$. Now the center is an abelian ideal, whereas by definition any ideal of a semisimple Lie algebra is simple.

Next we explain why the roots of a finite-dimensional semisimple Lie algebra are non-degenerate. To this end, consider two roots α, β which obey $\alpha + \beta \neq 0$ (for the following argument we temporarily also allow for zero as a root). If $x_\alpha \in \mathfrak{g}_\alpha$ and $x_\beta \in \mathfrak{g}_\beta$ are elements of the corresponding root spaces, then the endomorphism $\mathrm{ad}_{x_\alpha} \circ \mathrm{ad}_{x_\beta}$ maps the root space $\mathfrak{g}_\gamma$ to $\mathfrak{g}_{\alpha+\beta+\gamma}$. Due to $\alpha + \beta \neq 0$ this differs from $\mathfrak{g}_\gamma$, and accordingly the endomorphism $\mathrm{ad}_{x_\alpha} \circ \mathrm{ad}_{x_\beta}$ has zero trace. Thus the Killing form on these elements is zero: $\kappa(x_\alpha, x_\beta) = 0$. On the other hand, the Killing form is non-degenerate, and therefore for fixed x_α the number $\kappa(x_\alpha, x_\beta)$ cannot be zero for *all* root spaces $\mathfrak{g}_\beta$. The only root space for which $\kappa(x_\alpha, x_\beta)$ can be non-zero is however $\mathfrak{g}_{-\alpha}$. Thus there exists a non-zero element $x_{-\alpha} \in \mathfrak{g}_{-\alpha}$ such that $\kappa(x_\alpha, x_{-\alpha}) \neq 0$. In particular, $\mathfrak{g}_{-\alpha}$ contains non-zero elements, and we conclude that if α is a root, then so is $-\alpha$.

From now on, we again assume that $\alpha \neq 0$; by the invariance (6.11) of κ, for $h \in \mathfrak{g}_\circ$ we have

$$\kappa(h, [x_\alpha, x_{-\alpha}]) = \kappa([h, x_\alpha], x_{-\alpha}) = \alpha(h)\, \kappa(x_\alpha, x_{-\alpha}) . \tag{6.18}$$

Since for each root α there is a $h \in \mathfrak{g}_\circ$ with $\alpha(h) \neq 0$, it follows that $[x_\alpha, x_{-\alpha}] \neq 0$. Also, combining the equations (6.18) and (6.15) shows that $\kappa(h, [x_\alpha, x_{-\alpha}]) = c_\alpha \kappa(x_\alpha, x_{-\alpha})\kappa(h, H^\alpha)$ for all $h \in \mathfrak{g}_\circ$, which implies that $[x_\alpha, x_{-\alpha}] = c_\alpha \kappa(x_\alpha, x_{-\alpha}) \cdot H^\alpha$.

Together with (6.7) it follows in particular that the three generators x_α, $x_{-\alpha}$ and H^α span a subalgebra $\mathfrak{h}$ of $\mathfrak{g}$. To determine whether the structure constant $\alpha(H^\alpha)$ of this Lie algebra can be zero, we consider the trace of the restriction of the endomorphism ad_{H^α} to the subspace $\bigoplus_{n \in \mathbb{N}} \mathfrak{g}_{\beta+n\alpha}$ for some fixed root β. Because of $\mathrm{ad}_{H^\alpha} = \mathrm{ad}_{x_\alpha} \circ \mathrm{ad}_{x_{-\alpha}} - \mathrm{ad}_{x_{-\alpha}} \circ \mathrm{ad}_{x_\alpha}$ and the cyclic invariance of the trace this trace is zero; on the other hand, it can be directly evaluated as $\sum_{n \in \mathbb{N}} (\beta + n\alpha)(H^\alpha)\dim(\mathfrak{g}_{\beta+n\alpha})$, and hence we have $\alpha(H^\alpha) \sum_{n \in \mathbb{N}} n \dim(\mathfrak{g}_{\beta+n\alpha}) = -\beta(H^\alpha) \sum_{n \in \mathbb{N}} \dim(\mathfrak{g}_{\beta+n\alpha})$. Since there is at least one root β with $\beta(H^\alpha) \neq 0$, it follows that also $\alpha(H^\alpha)$ is non-vanishing. This implies that the subalgebra $\mathfrak{h}$ is isomorphic to $\mathfrak{sl}(2)$.

Now consider, for fixed root α, the adjoint action of $\mathfrak{h}$ on the subspace $\mathfrak{k}$ of $\mathfrak{g}$ that is defined by $\mathfrak{k} := \mathfrak{g}_\circ \oplus \bigoplus_{\zeta \in \mathbb{C}} \mathfrak{g}_{\zeta\alpha}$. Using the representation theory of $\mathfrak{sl}(2)$ (compare section 5.7) one can show that $\mathfrak{h}$ acts irreducibly on $\mathfrak{k}$. Since $\mathfrak{h}$ already acts irreducibly on itself and $\mathfrak{h}$ is a subset of $\mathfrak{k}$, this implies that in fact $\mathfrak{k} = \mathfrak{h}$. This means that only the values $\zeta = \pm 1$ occur, and that each of the corresponding root spaces is one-dimensional. Hence we have shown that the root spaces $\mathfrak{g}_\alpha$ are one-dimensional and that the only multiples of $\alpha \in \Phi$ which are roots are $\pm\alpha$.

We now assume that we have normalized the elements $E^{\pm\alpha}$ that span $\mathfrak{g}_{\pm\alpha}$ and H^α in such a way that they obey the standard relations of $\mathfrak{sl}(2)$ as in equation (2.7). Upon taking commutators we have

$$[H^\alpha, E^\beta] = \beta(H^\alpha)E^\beta , \tag{6.19}$$

which shows that $\mathfrak{g}$ decomposes into finite-dimensional irreducible modules of this $\mathfrak{sl}(2)$. Now from the results of section 2.3 concerning the representation theory of $\mathfrak{sl}(2)$ we infer that for any root β the number $\beta(H^\alpha)$ has to be an integer; hence it is in particular a real number.

Now we denote by $\mathfrak{h}$ the span of all H^α over the real numbers $\mathbb{R}$. This is a real subspace of the Cartan subalgebra $\mathfrak{g}_\circ$, and we claim that its complexification is precisely the Cartan subalgebra. Indeed, it is not difficult to see that

$$\mathfrak{h}_{\mathbb{C}} \oplus \underset{\alpha}{\bigoplus} \mathbb{C} E^\alpha \tag{6.20}$$

is an ideal of $\mathfrak{g}$ (the derived ideal). But a semisimple Lie algebra does not possess any non-trivial ideals, and hence this has to be the whole algebra. Thus $\mathfrak{h}_{\mathbb{C}}$ has to be identical to the Cartan subalgebra $\mathfrak{g}_\circ$. Let us choose a basis $\{H^i\}$ which generates $\mathfrak{h}$ over the real numbers. Since the complexification of $\mathfrak{h}$ is just $\mathfrak{g}_\circ$, this is also a basis of $\mathfrak{g}_\circ$ over $\mathbb{C}$, and we have

$$\beta(H^i) \in \mathbb{R} \tag{6.21}$$

for all roots β. This proves the last property announced above.

6.5 Structure constants of the Cartan–Weyl basis

So far only part of the bracket relations in a Cartan–Weyl basis has been established, namely (6.1) and (6.7). We have also learned that $[E^\alpha, E^{-\alpha}]$ is a non-vanishing element H^α of the Cartan subalgebra. The brackets $[E^\alpha, E^\beta]$ can be obtained by imposing the Jacobi identity for H^i, E^α and E^β, analogously as was already done in (3.6) for $\mathfrak{g} = \mathfrak{sl}(3)$. When combined with the known bracket relations, the Jacobi identity implies that $[H^i, [E^\alpha, E^\beta]] = (\alpha + \beta)^i [E^\alpha, E^\beta]$. For $\alpha + \beta \neq 0$, conformity with (6.7) then requires (since $\alpha + \beta$ is non-degenerate) that $[E^\alpha, E^\beta]$ is a multiple of the element $E^{\alpha+\beta}$ generating the root space $\mathfrak{g}_{\alpha+\beta}$:

$$[E^\alpha, E^\beta] = e_{\alpha,\beta}\, E^{\alpha+\beta} \qquad \text{for } \alpha + \beta \in \Phi \tag{6.22}$$

with $e_{\alpha,\beta} \in \mathbb{C}$; we will see below that if $\alpha + \beta$ is a root of $\mathfrak{g}$, $e_{\alpha,\beta}$ is non-zero. Similarly, if $\alpha + \beta = 0$, then

$$[E^\alpha, E^{-\alpha}] = \sum_{i=1}^{r} \tilde{\alpha}_i\, H^i\,, \tag{6.23}$$

and finally

$$[E^\alpha, E^\beta] = 0 \qquad \text{if } \alpha + \beta \neq 0 \text{ and } \alpha + \beta \notin \Phi\,. \tag{6.24}$$

The structure constants in Cartan–Weyl basis thus read as follows:

$$f^{ij}{}_k = f^{ij}{}_\alpha = 0\,, \qquad f^{i\alpha}{}_\beta = \alpha^i\, \delta_{\alpha,\beta}\,,$$

$$f^{\alpha\beta}{}_i = \tilde{\alpha}_i\, \delta_{\alpha,-\beta}\,, \qquad f^{\alpha\beta}{}_\gamma = \begin{cases} e_{\alpha,\beta}\, \delta_{\alpha+\beta,\gamma} & \text{if } \alpha + \beta \in \Phi\,, \\ 0 & \text{else}\,. \end{cases} \tag{6.25}$$

Note that, as we have not yet specified the normalization of the generators E^α nor the choice of the basis $\{H^i\}$ of the Cartan subalgebra $\mathfrak{g}_\circ$, the structure constants $e_{\alpha,\beta}$ and $\tilde{\alpha}_i$ are not yet determined uniquely.

Inserting (6.25) into the expression (6.14) of the Killing form in terms of the structure constants yields

$$
I_{\mathrm{ad}}\,\kappa^{ab} = \begin{cases} \sum_{\alpha\in\Phi} \alpha^i \alpha^j & \text{for } a=i,\ b=j\,, \\ \{2(\alpha,\tilde\alpha) + \sum_\gamma e_{\alpha,\gamma} e_{-\alpha,\alpha+\gamma}\}\, \delta_{\alpha,-\beta} & \text{for } a=\alpha,\ b=\beta\,, \\ 0 & \text{else}\,. \end{cases} \tag{6.26}
$$

This result has far reaching consequences. First note that because of the first line, the inner product of two elements λ and μ of $\mathfrak{g}_\circ^*$ can be computed as

$$
(\lambda,\mu) = \sum_{\alpha\in\Phi} (\alpha,\lambda)\,(\alpha,\mu)\,. \tag{6.27}
$$

In particular, the norm squared of a weight is a sum of squares:

$$
(\lambda,\lambda) = \sum_{\alpha\in\Phi} (\alpha,\lambda)^2\,. \tag{6.28}
$$

We have seen that the generators H^i of $\mathfrak{g}_\circ$ can be chosen in such a way that the components $\alpha^i = \alpha(H^i)$ of all roots are real. Therefore it makes sense to consider the dual vector space of the span of the H^i over the *real* numbers $\mathbb{R}$. Due to our specific choice of the H^i, this real vector space of dimension r contains all roots and is in fact spanned by the roots $\alpha\in\Phi$ over $\mathbb{R}$. It is therefore sometimes also called the *root space* of $\mathfrak{g}$ (this must not be confused with the spaces $\mathfrak{g}_\alpha$, which are also called root spaces, which are one-dimensional subspaces of the Lie algebra $\mathfrak{g}$; it will always be clear from the context in which sense the term root space is used).

On $\mathrm{span}_{\mathbb{R}}(\Phi)$, the *real* span of the roots, the inner product induced by the Killing form is also real; it follows that for any real linear combination λ of the roots one has $(\lambda,\lambda)\geq 0$, and that the norm is zero if and only if λ is zero. Thus the Killing form provides in fact a *euclidean* metric on the root space, so that the root space is isomorphic to $\mathbb{R}^r$. This highly non-trivial fact will be one key ingredient in the classification of all finite-dimensional simple Lie algebras.

6.6 Positive roots

The presence of a euclidean inner product on the root system enables us to develop a geometrical picture of the roots. This geometrical interpretation is the source of the term root 'vector' for the linear functions α, for which otherwise the terms 'linear form' or 'one-form' would be more appropriate. Finite-dimensional Lie algebras have only finitely many roots, and hence it is possible to find a hyperplane in root space which does not contain any root. This hyperplane divides the root space into two disjoint half-spaces $V_{\pm}$. This observation can be used to divide also the set of roots into two subsets, the so-called positive and the negative roots. Namely, one declares $\alpha \in \Phi$ to be a *positive root* of $\mathfrak{g}$ if and only if it is contained in (say) the half-space V_+; otherwise α is said to be a *negative root*. One writes $\alpha > 0$ and $\alpha < 0$ to denote that α is positive and negative, respectively. The sets of positive respectively negative roots are denoted by

$$\Phi_+ := \{\alpha \in \Phi \mid \alpha > 0\}, \qquad \Phi_- := \Phi \setminus \Phi_+. \qquad (6.29)$$

Since except for $-\alpha$ no other multiple of $\alpha \in \Phi$ is a root, one has $\alpha \in \Phi_+$ precisely if $(-\alpha) \in \Phi_-$, and hence we can write

$$\{E^\alpha \mid \alpha \in \Phi\} = \{E^\alpha \mid \alpha > 0\} \cup \{E^{-\alpha} \mid \alpha > 0\}. \qquad (6.30)$$

The step operator E^α associated to a positive root α is also called a *raising* operator, while $E^{-\alpha}$ with $\alpha > 0$ is called a *lowering* operator. Note that it depends on the convention that was chosen for dividing the root system into positive and negative roots whether a step operator is to be considered as a lowering or as a raising operator. It also follows that the number of elements of Φ_+ is $|\Phi_+| = \frac{1}{2}(d - r)$; in particular the difference $d - r$ between the dimension and the rank of $\mathfrak{g}$ is always even. Given a Cartan subalgebra $\mathfrak{g}_\circ$ of $\mathfrak{g}$, the subspaces of $\mathfrak{g}$ that are spanned by the step operators for positive and negative roots, respectively, are in fact sub*algebras*; they will be denoted by $\mathfrak{g}_+$ and $\mathfrak{g}_-$, respectively:

$$\mathfrak{g}_\pm := \mathrm{span}_{\mathbb{C}}\{E^{\pm\alpha} \mid \alpha > 0\}. \qquad (6.31)$$

According to the decompositions (6.4) and (6.30), $\mathfrak{g}$ can be written as the sum (direct sum of vector spaces)

$$\mathfrak{g} = \mathfrak{g}_+ \oplus \mathfrak{g}_\circ \oplus \mathfrak{g}_-. \qquad (6.32)$$

This is called the *triangular* or *Gauss decomposition* (sometimes also the Cartan decomposition) of $\mathfrak{g}$. We have already encountered such a splitting in the examples of chapter 2, and for the classical Lie algebras in equations (5.31) and (5.32).

6.7 Simple roots and the Cartan matrix

Given the set of positive roots with respect to some chosen basis, a *simple root* of $\mathfrak{g}$ is by definition a positive root which cannot be obtained as a linear combination of other positive roots with positive coefficients. Simple roots also possess the following properties:

• The simple roots are those positive roots which are closest to the hyperplane that was used to separate positive and negative roots. (Distances are measured in the norm that is induced by the Killing form.)

• Independently of the choice of that hyperplane, there are exactly $r =$ rank $\mathfrak{g}$ simple roots. They will be denoted by $\alpha^{(i)}$, so that the set of simple roots is written as

$$\Phi_s := \{\alpha^{(i)} \mid i = 1, \dots, r\}. \tag{6.33}$$

• The simple roots provide a basis for the root space, i.e. they are linearly independent and they span the whole root space, $\mathrm{span}_\mathbb{R}(\Phi_s) = \mathrm{span}_\mathbb{R}(\Phi)$.

• The linear combination $\alpha^{(i)} - \alpha^{(j)}$ of simple roots is never a root. (If it were a positive root, then the simple root $\alpha^{(i)}$ could be written as the sum $(\alpha^{(i)} - \alpha^{(j)}) + \alpha^{(j)}$ of two positive roots; if it were a negative root, the same argument would go through with the rôle of i and j interchanged.)

• Any positive root is a linear combination of simple roots with non-negative *integral* coefficients. This can be deduced from the properties of root strings (6.64) which will be discussed below.

Generically the basis of simple roots is *not* orthonormal, as we already established for $\mathfrak{g} = \mathfrak{sl}(3)$ in chapter 3. The non-orthonormality is encoded in the *Cartan matrix A* of $\mathfrak{g}$, which is defined as the $r \times r$-matrix with entries

$$A^{ij} := 2 \frac{(\alpha^{(i)}, \alpha^{(j)})}{(\alpha^{(j)}, \alpha^{(j)})}. \tag{6.34}$$

(Note that this definition is asymmetric in i and j; also, in the literature the Cartan matrix is occasionally defined as the transpose of the matrix (6.34).)

The Cartan matrix summarizes in fact the structure of a semisimple Lie algebra completely. To see how this comes about requires, however, still a lot more understanding of this structure than is possible at this stage.

6.8 Root and weight lattices. The Dynkin basis

It is convenient to introduce for any root α another element

$$\alpha^\vee := \frac{2\alpha}{(\alpha, \alpha)} \tag{6.35}$$

of $\mathfrak{g}_\circ^\star$. Using the vectors $\alpha^\vee$ simplifies various formulæ. For example, the defining equation (6.34) for the Cartan matrix becomes

$$A^{ij} := (\alpha^{(i)}, \alpha^{(j)\vee}). \tag{6.36}$$

The vector $\alpha^\vee$ is called the *dual root* or *coroot* of $\alpha \in \Phi$, and for a simple root $\alpha^{(i)}$ the vector $\alpha^{(i)\vee}$ is referred to as a *simple coroot*.

This terminology stems from the fact that for some purposes it is convenient to regard $\alpha^\vee$ not as an element of $\mathfrak{g}_\circ^\star$, but as an element of the vector space that is dual to $\mathfrak{g}_\circ^\star$, i.e. the bi-dual space $(\mathfrak{g}_\circ^\star)^\star$ of $\mathfrak{g}_\circ$. Now the bi-dual $(V^\star)^\star$ of a vector space V is isomorphic to V whenever V is finite-dimensional. Thus the dual of $\mathfrak{g}_\circ^\star$ can be identified with the Cartan subalgebra $\mathfrak{g}_\circ$. In the interpretation as elements of the bi-dual space, the coroots $\alpha^\vee$ just correspond to the Cartan subalgebra elements H^α; in particular, the simple coroots $\alpha^{(i)\vee}$ correspond to the Cartan subalgebra generators H^i. For the purposes of this book, however, we consider $\alpha^\vee$ just as the scalar multiple (6.35) of α, and hence as an element of $\mathfrak{g}_\circ^\star$.

When one considers the root space as consisting of real linear combinations of the roots, then the dual space of the root space is called the *weight space*, and its elements the *weights* of $\mathfrak{g}$. (Previously we referred by the term 'weight' to a vector of eigenvalues with respect to the Cartan subalgebra generators. That both usages of this term are equivalent will be seen in section 13.2.) It is often convenient to choose as a basis $\mathcal{B}$ of the root space the simple coroots,

$$\mathcal{B} := \{\alpha^{(i)\vee} \mid i = 1, 2, ..., r\}. \tag{6.37}$$

According to (4.2), the basis of the weight space which is dual to $\mathcal{B}$ then consists of those weights, denoted by $\Lambda_{(i)}$, which obey

$$\Lambda_{(i)}(\alpha^{(j)\vee}) = \delta_i^j \quad \text{for } i, j = 1, 2, ..., r. \tag{6.38}$$

These r weights $\Lambda_{(i)}$ are called the *fundamental weights* of the semisimple Lie algebra $\mathfrak{g}$, and the basis

$$\mathcal{B}^\star := \{\Lambda_{(i)} \mid i = 1, 2, ..., r\} \tag{6.39}$$

is referred to as the *Dynkin basis* of the weight space. (The fundamental weights of $\mathfrak{sl}(3)$ have already been displayed in figure (3.29).) The components of a weight in the Dynkin basis are called *Dynkin labels*. The Dynkin basis is very convenient because, as follows from the representation theory of $\mathfrak{sl}(2)$ (and as we will see in detail in chapter 13), the Dynkin labels of the weights that appear most frequently in applications are integers; this is also one of the reasons why already in chapter 2 we preferred to use weights λ rather than the spin components $j_3 = \lambda/2$ to label vectors in representation spaces.

As a consequence of the representation theory of $\mathfrak{sl}(2)$, the root system has the highly non-trivial property that all roots are not just arbitrary linear combinations of the simple roots, but even linear combinations with integral coefficients. This motivates us to consider a special subset of root space, namely the set of all linear combinations of simple roots with integral coefficients. These vectors form a lattice. (The *lattice* associated to some discrete subset V_0 (without accumulation points) of a vector space is the set of all linear combinations of elements of V_0 with *integral* coefficients.) The integer span $L(\mathfrak{g}) := \mathrm{span}_{\mathbb{Z}}(\Phi_\mathrm{s}) = \mathrm{span}_{\mathbb{Z}}(\Phi)$ of the simple roots is called the *root lattice*, the integer span of the simple coroots the *coroot lattice* $L^\vee(\mathfrak{g})$, and the integer span of the fundamental weights the *weight lattice* $L_\mathrm{w}(\mathfrak{g})$ of $\mathfrak{g}$. The weight lattice is the lattice dual (over $\mathbb{Z}$) to the coroot lattice, i.e.

$$L_\mathrm{w}(\mathfrak{g}) = (L^\vee(\mathfrak{g}))^* \equiv \{\lambda \mid \lambda(\alpha^\vee) \in \mathbb{Z} \text{ for all } \alpha \in \Phi\}. \qquad (6.40)$$

As an illustration, the root and weight lattices of $A_2 \cong \mathfrak{sl}(3)$ are indicated by solid and dashed lines, respectively, in the following figure:

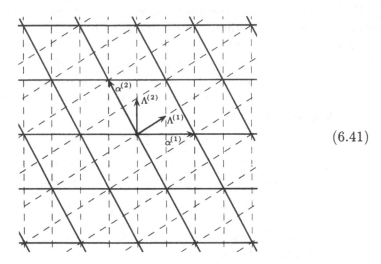

$$(6.41)$$

The integer that is obtained as the sum $\sum_{i=1}^{r} b_i$ of the components of a root (or more generally, of a root lattice vector) $\beta = \sum_{i=1}^{r} b_i \alpha^{(i)}$ in the basis of simple roots is called the *height* of β. As can be read off the bracket relations (6.22) – (6.24), the subspaces $\mathfrak{g}_\alpha$ of $\mathfrak{g}$ satisfy $[\mathfrak{g}_\alpha, \mathfrak{g}_\beta] \subseteq \mathfrak{g}_{\alpha+\beta}$, with the convention that $\mathfrak{g}_\alpha$ is interpreted as $\mathfrak{g}_\circ$ if $\alpha = 0$. As a consequence, the height concept allows to define a natural $\mathbb{Z}$-gradation of $\mathfrak{g}$, the so-called *root space gradation*, according to

$$\begin{aligned}\mathfrak{g}_{(0)} &= \mathfrak{g}_\circ, \\ \mathfrak{g}_{(j)} &= \mathrm{span}_{\mathbb{C}}\{E^\alpha \mid \mathrm{height}(\alpha) = j\} \qquad \text{for } j \in \mathbb{Z}\backslash\{0\}.\end{aligned} \qquad (6.42)$$

In the special case of $\mathfrak{g} = \mathfrak{sl}(3)$, this gradation is the one that was already presented in the example (4.36).

For simple $\mathfrak{g}$, there is a unique root, called the *highest root* of $\mathfrak{g}$ and denoted by θ, such that the height of θ is larger than that of any other root. θ also satisfies

$$(\theta, \theta) \geq (\alpha, \alpha) \qquad \text{for all } \alpha \in \Phi. \qquad (6.43)$$

Furthermore, for any $\beta \in \Phi_+ \setminus \{\theta\}$ the vector $\theta - \beta$ is a linear combination of simple roots in which all coefficients are positive integers; thus, writing $\beta = \sum_{i=1}^r b_i \alpha^{(i)}$ as well as $\theta =: \sum_{i=1}^r a_i \alpha^{(i)}$, one has

$$a_i \geq b_i(\beta) \quad \text{for all } i = 1, 2, ..., r \text{ and all } \beta \in \Phi \setminus \{\theta\}. \tag{6.44}$$

In the mathematical literature, it is conventional to fix the (up to now still arbitrary) normalization of the inner product in such a way that the highest root has length squared 2, $(\theta, \theta) = 2$. Occasionally, other conventions are however used as well; therefore in this book the normalization dependence will always be made explicit.

6.9 The metric on weight space

Any vector space endowed with a (non-degenerate) inner product, and hence also the root space which is isomorphic to $\mathbb{R}^r$, can be identified with its dual space via the inner product. Accordingly we will from now on identify the root and weight spaces of $\mathfrak{g}$ and treat roots and weights on an equal footing. For any element β of the root space, the element $\beta^\star$ in the weight space with which it is identified is characterized by the property that $\beta^\star(\gamma) = (\beta, \gamma)$ for all γ in the root space. In particular, the roots themselves will from now on be considered as weights; in fact, they are the weights of the adjoint representation.

In the description of representations, the Dynkin components of a weight λ play the rôle of eigenvalues with respect to the generators H^i of the Cartan subalgebra. Therefore it is natural to use superscripts for these components, and correspondingly subscripts for the components in the basis of simple coroots; this is in accordance with the previous use of upper indices, as e.g. in the bracket relation (6.7) for $[H^i, E^\alpha]$.

As we will see later, all entries of the Cartan matrix are integers, and this implies that the Dynkin labels of any root are integral, too. In order to describe the inner products on the root and weight spaces explicitly, we express roots and weights through their components with respect to the basis of simple coroots and the Dynkin basis, respectively. Thus we write

$$\lambda = \sum_{i=1}^r \lambda_i \, \alpha^{(i)\vee} = \sum_{j=1}^r \lambda^j \Lambda_{(j)} \tag{6.45}$$

with

$$\lambda_i = (\lambda, \Lambda_{(i)}), \qquad \lambda^i = (\lambda, \alpha^{(i)\vee}). \tag{6.46}$$

In particular, as a consequence of $\Lambda_{(i)}(\alpha^{(j)\vee}) = \delta_i^j$ one has

$$(\alpha^{(j)\vee})_i = \delta_i^j = (\Lambda_{(i)})^j, \qquad (\alpha^{(j)})_i = \tfrac{1}{2} (\alpha^{(j)}, \alpha^{(j)}) \, \delta_i^j. \tag{6.47}$$

It is conventional to use one and the same symbol 'G' for the two metrics which raise and lower indices, respectively:

$$\lambda_i = \sum_{j=1}^{r} G_{ij}\lambda^j, \qquad \lambda^i = \sum_{j=1}^{r} G^{ij}\lambda_j. \qquad (6.48)$$

(Thus the fact that they are inverse to each other is expressed by the identity $\sum_{j=1}^{r} G_{ij}G^{jk} = \delta_i^k$.) These $r \times r$-matrices are expressed in terms of the basis elements as

$$G_{ij} = (\Lambda_{(i)}, \Lambda_{(j)}), \qquad G^{ij} = (\alpha^{(i)\vee}, \alpha^{(j)\vee}) = \frac{2}{(\alpha^{(i)}, \alpha^{(i)})} A^{ij}. \qquad (6.49)$$

The matrix G^{ij} provides an inner product on the weight space; the reasoning presented after equation (6.26) shows that for simple Lie algebras this product is in fact euclidean, i.e. its signature is $(r, 0)$. (It is this product we implicitly used in section 3.3 when we talked about angles between vectors in weight space.) The inner product of two weights λ and μ may be expressed through the components in various ways:

$$(\lambda, \mu) = \sum_{i=1}^{r} \lambda_i \mu^i = \sum_{i,j=1}^{r} G_{ij}\lambda^i \mu^j = \sum_{i,j=1}^{r} G^{ij}\lambda_i \mu_j. \qquad (6.50)$$

The metric $G = (G_{ij})$ with lower indices will be referred to as the *quadratic form matrix* of $\mathfrak{g}$; the explicit form of this matrix will be listed in table VI for all simple Lie algebras. Also, G with upper indices is often called the symmetrized Cartan matrix; as we will see in (6.61) below, the symmetrized Cartan matrix coincides with the restriction of the Killing form to the Cartan subalgebra; thus in short, the quadratic form matrix is dual to the Killing form.

Combining (6.47) and (6.49), it follows that

$$(\alpha^{(i)})^j \equiv \sum_{k=1}^{r}(\alpha^{(i)})_k G^{kj} = A^{ij}. \qquad (6.51)$$

In words:

> The components of the simple roots in the Dynkin basis coincide with the rows of the Cartan matrix.

In the special case of simple roots $\alpha = \alpha^{(i)}$, the formula (6.7) for the Lie bracket of H^i and E^α thus yields

$$[H^i, E^{\alpha^{(j)}}] = (\alpha^{(j)})^i E^{\alpha^{(j)}} = A^{ji} E^{\alpha^{(j)}}. \qquad (6.52)$$

6.10 The Chevalley basis

So far the normalization of the step operators E^α has not been fully specified. To fix them in a convenient manner, we have to fix also the constants of proportionality c_α that were introduced in (6.15), and we will do this first. To this end we set $h = H^\beta$ in (6.15) and use the definition $\alpha^i = \alpha(H^i)$ to learn that $\alpha(H^\beta) = c_\beta^{-1}(\alpha, \beta) = c_\beta^{-1} \sum_{i=1}^r \beta_i \alpha^i = c_\beta^{-1} \sum_{i=1}^r \beta_i \, \alpha(H^i)$ for all roots α. Since the roots span $\mathfrak{g}_\circ^*$ (see (6.17)), by the linearity of the functions α it follows that $H^\beta = c_\beta^{-1} \sum_{i=1}^r \beta_i H^i$. Let us now fix the normalization to

$$c_\beta = \tfrac{1}{2} (\beta, \beta) \,. \tag{6.53}$$

We then have

$$H^\beta = \sum_{i=1}^r (\beta^\vee)_i \, H^i \,. \tag{6.54}$$

It follows in particular that

$$\alpha(H^\beta) = c_\beta^{-1}(\alpha, \beta) = (\alpha, \beta^\vee) \,, \tag{6.55}$$

which means that the eigenvalues of H^β in the adjoint representation are given by (in addition to the r-fold eigenvalue zero) the inner products $(\alpha, \beta^\vee)$ with $\alpha \in \Phi$. Also note that $\alpha(H^\alpha) = (\alpha, \alpha^\vee) = 2$. Thus upon the identification of the Cartan subalgebra $\mathfrak{g}_\circ$ and its dual space, H^α corresponds to $\alpha^\vee$. For this reason, some authors also refer to the Cartan subalgebra elements H^α as coroots.

The results of section 6.4 tell us that $[E^\alpha, E^{-\alpha}] = c_\alpha \kappa(E^\alpha, E^{-\alpha}) \cdot H^\alpha$. With the choice (6.53), this reads $[E^\alpha, E^{-\alpha}] = \kappa(E^\alpha, E^{-\alpha}) \sum_{i=1}^r \alpha_i H^i$. Now by normalizing the generators E^α appropriately, the constant of proportionality can be set to any desired value for each root α. A convenient choice is to set $\kappa(E^\alpha, E^{-\alpha}) = 2/(\alpha, \alpha)$ for any root α. (In terms of the structure constants $\tilde{\alpha}_i$ used in (6.23), this amounts to setting $\tilde{\alpha} = \alpha^\vee$.) With this choice, the Lie bracket between E^α and $E^{-\alpha}$ takes the canonical form

$$[E^\alpha, E^{-\alpha}] = H^\alpha \,, \tag{6.56}$$

while the bracket relations (6.7) imply $[H^\alpha, E^{\pm\alpha}] = \sum_{i=1}^r (\alpha^\vee)_i \, [H^i, E^{\pm\alpha}] = \pm(\alpha, \alpha^\vee) \, E^{\pm\alpha}$, i.e.

$$[H^\alpha, E^{\pm\alpha}] = \pm 2 \, E^{\pm\alpha} \,. \tag{6.57}$$

A Cartan–Weyl basis with normalizations chosen such that (6.56) and (6.57) hold is called a *Chevalley basis* of $\mathfrak{g}$. As is suggested by the notation H^α, in the case of simple roots $\alpha^{(i)}$ the Chevalley generators H^α should be just the original Cartan subalgebra generators H^i; this is indeed the case:

$$H^{\alpha^{(i)}} = \sum_{j=1}^r (\alpha^{(i)\vee})_j H^j = \sum_{j=1}^r \delta_j^{\,i} \, H^j = H^i \,. \tag{6.58}$$

It is then natural to introduce also a special notation for the step operators associated to simple roots,

$$E^i_\pm := E^{\pm\alpha^{(i)}} \,. \tag{6.59}$$

These generators obey $[E^i_+, E^j_-] = 0$ for $i \neq j$ because the difference $\alpha^{(i)} - \alpha^{(j)}$ of two simple roots is by definition never a root, while according to (6.56) we have $[E^i_+, E^i_-] = H^{\alpha^{(i)}} = H^i$; thus

$$[E^i_+, E^j_-] = \delta_{ij} \, H^i \,. \tag{6.60}$$

With the identification (6.58) it follows that the metric G with upper indices is nothing but the restriction of the Killing form to the Cartan subalgebra. Namely, the choice (6.53) of the normalization constants c_α means that $G^{ij} \equiv (\alpha^{(i)\vee}, \alpha^{(j)\vee}) = c^{-1}_{\alpha^{(i)}} c^{-1}_{\alpha^{(j)}} (\alpha^{(i)}, \alpha^{(j)})$, which because of the relation (6.16) between the inner product of roots and the Killing form can be written as

$$
\begin{aligned}
G^{ij} &= \kappa(H^{\alpha^{(i)}}, H^{\alpha^{(j)}}) \\
&\equiv \kappa(H^i, H^j) = I_{\mathrm{ad}}\, \kappa^{ij} \,.
\end{aligned}
\tag{6.61}
$$

Moreover, according to the results of section 6.4, on the step operators corresponding to the roots the Killing form is given by

$$\kappa(E^\alpha, E^\beta) = \frac{2}{(\alpha, \alpha)} \delta_{\alpha, -\beta} \tag{6.62}$$

in the Chevalley basis.

We summarize the bracket relations found above in table III. The left column of table III contains the bracket relations between the generators of a Chevalley–Serre basis (for the last relation, compare section 7.1), while in the right column we have written the analogous relations for an arbitrary Cartan–Weyl basis.

Table III. *Lie bracket relations in Chevalley–Serre and Cartan–Weyl bases*

$[H^i, H^j] = 0$	$[H^\alpha, H^\beta] = 0$
$[H^i, E^j_\pm] = \pm A^{ji} \, E^j_\pm$	$[H^\alpha, E^\beta] = (\alpha^\vee, \beta) \, E^\beta$
$[E^i_+, E^j_-] = \delta_{ij} \, H^j$	$[E^\alpha, E^{-\alpha}] = H^\alpha$
$(\mathrm{ad}_{E^i_\pm})^{1 - A^{ji}} E^j_\pm = 0$	$[E^\alpha, E^\beta] = e_{\alpha, \beta}\, E^{\alpha+\beta} \quad$ for $\alpha \neq -\beta$

6.11 Root strings

The crucial advantage of the normalization of the generators of $\mathfrak{g}$ we have chosen in the previous section is that for each root α the step operators E^α and $E^{-\alpha}$ together with the Cartan subalgebra element H^α satisfy the same relations as the generators E_+, E_- and H of the simple Lie algebra $\mathfrak{sl}(2)$ (compare (3.39)). Thus these elements of $\mathfrak{g}$ generate an $\mathfrak{sl}(2)$-*subalgebra* of $\mathfrak{g}$; this subalgebra will be denoted by $\mathfrak{sl}(2)_\alpha$. Because of this connection, many of the properties of simple Lie algebras can be analyzed to a large extent by only making judicious use of the properties of $\mathfrak{sl}(2)$. As a first application, we note that, since $\mathfrak{g}$ is finite-dimensional, the eigenvalues $(\alpha, \beta^\vee)$ of H^β in the adjoint representation are weights of finite-dimensional representations of the subalgebra $\mathfrak{sl}(2)_\beta$; hence the representation theory of $\mathfrak{sl}(2)$ tells us that they are integers:

$$(\alpha, \beta^\vee) \in \mathbb{Z} \quad \text{for all } \alpha, \beta \in \Phi. \tag{6.63}$$

In particular, as already announced, all entries $A^{ij} = (\alpha^{(i)}, \alpha^{(j)\vee})$ of the Cartan matrix are integral.

Thus the structure constants in the bracket relations (6.52) and (6.57) of the Chevalley basis are integers. In fact this holds for *all* structure constants in the Chevalley basis. To see this, recall first that for any root α the only multiples of α which are roots are $\pm\alpha$. A more complicated question is whether, given another root $\beta \neq \alpha$, the combination $\alpha + m\beta$ with $m \in \mathbb{Z}$ is again a root. This problem is most easily investigated in terms of the representation theory of $\mathfrak{sl}(2)$. The basic observation is that, because of $[E^\alpha, E^{m\alpha+\beta}] \propto E^{(m+1)\alpha+\beta}$, the subspace $\bigoplus_m \mathfrak{g}_{m\alpha+\beta}$ of $\mathfrak{g}$ can be regarded as a module of $\mathfrak{sl}(2)_\alpha$. Moreover, since the root spaces $\mathfrak{g}_\alpha$ are non-degenerate, this module is even irreducible, and the roots of the form $\beta + m\alpha$ are the weights of this irreducible module. From this observation it follows, first, that the 'string' of roots of this form does not have any 'holes', i.e. has the structure

$$S_{\alpha;\beta} = \{\beta + m\alpha \mid m = -n_-, -n_-+1, \dots, n_+-1, n_+\} \tag{6.64}$$

for some non-negative integers $n_\pm$; the dimension of this module is then $n_+ + n_- + 1$, so that the highest $\mathfrak{sl}(2)_\alpha$-weight is $\Lambda = n_+ + n_-$. $S_{\alpha;\beta}$ is called a *root string*, or more specifically, the α-*string through* β. Further, since the lowest weight of an irreducible module of $\mathfrak{sl}(2)$ is minus the highest weight, it also follows that $(\beta - n_-\alpha)(H^\alpha) = -(\beta + n_+\alpha)(H^\alpha)$, which implies that $(\beta, \alpha^\vee) = \beta(H^\alpha) = \frac{1}{2}(n_- - n_+)\alpha(H^\alpha)$, i.e.

$$(\beta, \alpha^\vee) = n_- - n_+. \tag{6.65}$$

This shows once more that the inner product $(\beta, \alpha^\vee)$ of roots is an integer. It also implies that $n_- > n_+ \geq 0$ if $(\alpha, \beta) > 0$, while $n_+ > n_- \geq 0$ if

$(\alpha, \beta) < 0$, and hence

$$\alpha - \beta \in \Phi \quad \text{if } (\alpha, \beta) > 0,$$
$$\alpha + \beta \in \Phi \quad \text{if } (\alpha, \beta) < 0. \tag{6.66}$$

It is also not difficult to obtain the following relations among the structure constants $e_{\alpha,\beta}$ of the Chevalley basis (see exercise 6.5):

$$\frac{e_{\beta,-\alpha-\beta}}{e_{-\alpha-\beta,\alpha}} = \frac{(\alpha,\alpha)}{(\beta,\beta)}, \qquad \frac{e_{-\alpha-\beta,\alpha}}{e_{\alpha,\beta}} = \frac{(\beta,\beta)}{(\alpha+\beta,\alpha+\beta)}, \tag{6.67}$$

provided that $\alpha + \beta$ is a root. Similarly, if each of $\alpha + \beta$, $\alpha + \gamma$, $\beta + \gamma$ and $\alpha + \beta + \gamma$ is a root, the Jacobi identity applied to E^α, E^β and E^γ gives

$$e_{\alpha,\beta} e_{\alpha+\beta,\gamma} + e_{\beta,\gamma} e_{\beta+\gamma,\alpha} + e_{\gamma,\alpha} e_{\alpha+\gamma,\beta} = 0. \tag{6.68}$$

The highest weight vector v_Λ of the $\mathfrak{sl}(2)_\alpha$-module under consideration is the step operator $E^{\beta+n+\alpha}$. Analogously, E^β is the vector $v_{\Lambda-2n_+}$ with weight $\lambda = \Lambda - 2n_+ = n_- - n_+$, so that according to (5.39) application of E^α yields a factor of $1 + \frac{1}{2}(\Lambda + \lambda) = n_- + 1$, i.e. in terms of commutators we have $[E^\alpha, E^\beta] = (n_- + 1)E^{\alpha+\beta}$. A similar reasoning (see exercise 6.8) shows that

$$[E^{-\alpha}, [E^\alpha, E^\beta]] = n_+(n_- + 1)\, E^\beta. \tag{6.69}$$

Note that because of the bracket relation (6.56) one cannot change the normalization of E^α independently of that of $E^{-\alpha}$ (thus, while the result for $[E^\alpha, E^\beta]$ just obtained depends in fact on the specific normalization of E^α, (6.69) is normalization independent). In particular, the relative sign is fixed; one can then show (with the help of the so-called Chevalley involution which will be described in section 11.9) that

$$e_{-\alpha,-\beta} = -e_{\alpha,\beta}. \tag{6.70}$$

Combining the identities (6.69) and (6.70), one arrives at the explicit formula

$$(e_{\alpha,\beta})^2 = n_+(n_- + 1)\frac{(\alpha+\beta,\alpha+\beta)}{(\beta,\beta)} \tag{6.71}$$

(see exercise 6.9). With some further effort (exercise 6.10), one can show that

$$\frac{n_- + 1}{n_+} = \frac{(\alpha+\beta,\alpha+\beta)}{(\beta,\beta)} \tag{6.72}$$

if $\alpha + \beta$ is a root (i.e. if $n_+ \neq 0$, so that the left hand side makes sense). When combined with (6.71), it follows that $e_{\alpha,\beta} = \pm (n_- + 1)$, i.e.

$$[E^\alpha, E^\beta] = \pm (n_- + 1)\, E^{\alpha+\beta} \tag{6.73}$$

for any pair α, β of roots. This finally not only proves that $e_{\alpha,\beta} \neq 0$, as announced after (6.22), but also completes our argument that all structure constants in a Chevalley basis are integers. Furthermore, once a definite convention about which roots are the positive ones is adopted, then up to the sign freedom in (6.73) (which is still restricted by the identities (6.70) and (6.67)), and of course up to the choice of a basis of simple roots, for a given choice of the Cartan subalgebra the Chevalley basis of a semisimple Lie algebra is unique.

6.12 Examples

Let us now illustrate the above analysis with a few examples.

• First, for $\mathfrak{g} = A_1 \cong \mathfrak{sl}(2)$, the description is very simple. Choosing H as the generator of the (one-dimensional) Cartan subalgebra, there is a unique (up to normalization) Cartan–Weyl basis, namely $\{H, E_\pm\}$ as used in (3.39), and the Chevalley basis is precisely this Cartan–Weyl basis. There is a single positive root, which is hence the highest root θ; it has Dynkin label equal to 2. The Cartan 'matrix' is the number 2, and the quadratic form 'matrix' equals $\frac{1}{4}(\theta, \theta)$, i.e. $\frac{1}{2}$ in the conventional normalization $(\theta, \theta) = 2$. The single fundamental weight is $\Lambda = \Lambda_{(1)} = 1$ (i.e. in terms of the spin which equals one half of the weight, the fundamental weight corresponds to the spin-$\frac{1}{2}$ representation). Thus the weight lattice is isomorphic to $\mathbb{Z}$ and embedded into the root lattice which is isomorphic to $2\mathbb{Z}$.

• Next consider $\mathfrak{g} = A_2 \cong \mathfrak{sl}(3)$. Most of the relevant formulæ have already been obtained in chapter 3. The distinction between positive and negative roots can be imposed in such a way that the two simple roots are $\alpha^{(1)}$ and $\alpha^{(2)}$, and there is one other positive root, the highest root $\theta = \alpha^{(1)} + \alpha^{(2)}$. The root system of A_2 can then be summarized as in figure (3.15). The corresponding triangular decomposition reads $\mathfrak{g} = \mathfrak{g}_+ \oplus \mathfrak{g}_\circ \oplus \mathfrak{g}_-$ with $\mathfrak{g}_+ = \text{span}_{\mathbb{C}}\{E^\alpha, E^\beta, E^\theta\}$, $\mathfrak{g}_- = \text{span}_{\mathbb{C}}\{E^{-\alpha}, E^{-\beta}, E^{-\theta}\}$, and $\mathfrak{g}_\circ = \text{span}_{\mathbb{C}}\{H^1, H^2\}$. Further, the scalar products of the roots are given by $(\alpha^{(1)}, \alpha^{(1)}) = (\alpha^{(2)}, \alpha^{(2)}) = (\theta, \theta)$ and $(-\alpha^{(1)}, \alpha^{(2)}) = (\alpha^{(1)}, \theta) = (\alpha^{(2)}, \theta) = \frac{1}{2}(\theta, \theta)$ (in (3.26) the overall normalization was chosen such that $(\theta, \theta) = 1$). Thus in particular the Cartan matrix and the quadratic form matrix read

$$A(\mathfrak{sl}(3)) = \begin{pmatrix} 2 & -1 \\ -1 & 2 \end{pmatrix}, \qquad G(\mathfrak{sl}(3)) = \tfrac{1}{6}(\theta, \theta) \begin{pmatrix} 2 & 1 \\ 1 & 2 \end{pmatrix}. \qquad (6.74)$$

Further, the fundamental weights are given up to the normalization factor $2/(\theta, \theta)$ by the formula (3.27), in agreement with the general expression $G_{ij} = (\Lambda_{(i)}, \Lambda_{(j)})$ for the quadratic form matrix. The root and weight lattice thus look as already displayed in figure (6.41).

Finally, the generators presented in the triangular decomposition above in fact form a Chevalley basis for A_2. The bracket relations of this Chevalley basis have already been displayed in (3.11); for completeness we list the non-zero brackets once again:

$$[E^{\alpha^{(1)}}, E^{-\alpha^{(1)}}] = H^{\alpha^{(1)}}, \quad [E^{\alpha^{(2)}}, E^{-\alpha^{(2)}}] = H^{\alpha^{(2)}}, \quad [E^{\theta}, E^{-\theta}] = H^{\theta},$$

$$[H^{\alpha^{(1)}}, E^{\pm\alpha^{(1)}}] = \pm 2\, E^{\pm\alpha^{(1)}}, \quad [H^{\alpha^{(1)}}, E^{\pm\alpha^{(2)}}] = \mp\, E^{\pm\alpha^{(2)}},$$

$$[H^{\alpha^{(2)}}, E^{\pm\alpha^{(2)}}] = \pm 2\, E^{\pm\alpha^{(2)}}, \quad [H^{\alpha^{(2)}}, E^{\pm\alpha^{(1)}}] = \mp\, E^{\pm\alpha^{(1)}}, \tag{6.75}$$

$$[E^{\pm\alpha^{(1)}}, E^{\pm\alpha^{(2)}}] = \pm\, E^{\pm\theta}, \quad [H^{\alpha^{(1)}}, E^{\theta}] = [H^{\alpha^{(2)}}, E^{\theta}] = \pm\, E^{\theta},$$

$$[E^{\pm\alpha^{(1)}}, E^{\mp\theta}] = \mp\, E^{\mp\alpha^{(2)}}, \quad [E^{\pm\alpha^{(2)}}, E^{\mp\theta}] = \mp\, E^{\mp\alpha^{(1)}}.$$

- Let us also present some relations for the case of arbitrary classical Lie algebras. First consider $\mathfrak{g} = A_{n-1}$. In terms of the matrix units $\mathcal{E}_{i,j}$ (5.28) with entries $(\mathcal{E}_{i,j})_{kl} = \delta_{ik}\delta_{jl}$, a Cartan-Weyl basis is provided by the basis already displayed in (5.31). The generators of a Cartan subalgebra are given by the set $\mathcal{B}_\circ$ defined there, i.e. by

$$H^i = \mathcal{E}_{i,i} - \mathcal{E}_{i+1,i+1} \tag{6.76}$$

for $i = 1, 2, \ldots, n-1$; in particular, the rank is $n - 1$. The step operators are the elements of $\mathcal{B}_\pm$, i.e. $\mathcal{E}_{i,j}$ with $i \neq j$; for $i < j$ ($\mathcal{B}_+$) they correspond to positive roots, and for $i > j$ ($\mathcal{B}_-$) to negative roots. The generators for the simple roots are

$$E^i_+ = \mathcal{E}_{i,i+1} \tag{6.77}$$

for $i = 1, 2, \ldots, n-1$, and the generator corresponding to the highest root is $E^\theta = \mathcal{E}_{1,n}$. (The root space gradation of A_{n-1} is then the gradation over $\mathbb{Z}$ in which the grade of $\mathcal{E}_{i,j}$ is defined to be $j - i$.)

Similarly, for the simple Lie algebras B_n, C_n and D_n a Cartan-Weyl basis is again given by the sets introduced in section 5.6, see the list (5.32). For all three series B_n, C_n and D_n the step operators for the first $n - 1$ simple roots are $E^i_+ = \mathcal{E}_{i,i+1} - \mathcal{E}_{i+n+1,i+n}$ $(i = 1, 2, \ldots, n-1)$. The generator for the remaining simple root is $E^n_+ = 2\mathcal{E}_{n,2n}$ for C_n, $E^n_+ = \mathcal{E}_{n,0} - \mathcal{E}_{0,2n}$ for B_n, and $E^n_+ = \mathcal{E}_{n-1,2n} - \mathcal{E}_{n,2n-1}$ for D_n. The step operator for the highest root θ reads $E^\theta = 2\mathcal{E}_{1,n+1}$ for C_n and $E^\theta = \mathcal{E}_{1,n+2} - \mathcal{E}_{2,n+1}$ for both B_n and D_n.

*6.13 Non-semisimple Lie algebras

For finite-dimensional semisimple Lie algebras, we have constructed the Cartan sub-algebra $\mathfrak{g}_\circ$ as a maximal abelian subalgebra of $\mathfrak{g}$ consisting of elements x which are ad-diagonalizable. (Also, it was assumed that the base field F is algebraically closed. If this is not the case, the qualification 'diagonalizable' must be replaced by 'semisim-ple', where by a semisimple endomorphism one now means an endomorphism for which the roots of its minimal polynomial over F are all distinct.) Another characterization of the Cartan subalgebra is that it is a maximal abelian subalgebra containing a regular element, i.e. an element x for which the number of zero eigenvalues of the map ad_x is as small as possible (namely equal to rank $\mathfrak{g}$). The concept of a Cartan subalgebra is however available also for arbitrary Lie algebras: A Cartan subalgebra $\mathfrak{g}_\circ$ of the Lie algebra $\mathfrak{g}$ is by definition a nilpotent subalgebra which equals its normalizer, i.e. for which the relation $[x, \mathfrak{g}_\circ] \subseteq \mathfrak{g}_\circ$ for some $x \in \mathfrak{g}$ implies that in fact $x \in \mathfrak{g}_\circ$.

In general, however, a triangular decomposition relative to $\mathfrak{g}_\circ$ no longer exists. If $\mathfrak{g}$ is solvable, the remaining generators can still be arranged in a special manner: One can order the generators in such a way that each $x \in \mathfrak{g}$ is represented by upper triangular matrices, i.e. one has $[x, T^a] = \sum_b \xi^a_b T^b$ with $\xi^a_b = 0$ for all $b > a$ (and also for $b = a$ if $\mathfrak{g}$ is nilpotent).

Let us also recall the Levi decomposition (4.31) of a Lie algebra $\mathfrak{g}$ into its maximal solvable ideal, the radical $\mathfrak{g}_{\mathrm{rad}}$, and a semisimple part $\mathfrak{s}$. Each element of $\mathfrak{g}$ can be written uniquely as the sum of an element in the radical and an element of $\mathfrak{s}$. Also, while the radical is unique, the semisimple part $\mathfrak{s}$ is unique only in the trivial case that it is an ideal of $\mathfrak{g}$; however, if $\mathfrak{s}_1$ and $\mathfrak{s}_2$ are two different possible semisimple parts, then there exists an automorphism ω of $\mathfrak{g}$ such that $\omega(\mathfrak{s}_1) = \mathfrak{s}_2$. As a consequence, the investigation of general (finite-dimensional) Lie algebras can be reduced to the study of semisimple and solvable Lie algebras.

Also, as we have learned above, a finite-dimensional Lie algebra $\mathfrak{g}$ is simple if and only if the Killing form is non-degenerate; similarly, a finite-dimensional Lie algebra $\mathfrak{g}$ is solvable if and only if $\kappa(x, y) = 0$ for all $x \in \mathfrak{g}$ and all y in the derived algebra $\mathfrak{g}' = [\mathfrak{g}, \mathfrak{g}]$. (The 'if' parts of these statements are known as the *Cartan criteria* for semisimplicity and solvability, respectively.) As a consequence, the Levi decomposition of $\mathfrak{g}$ can be accomplished as follows. One first determines the subspace of $\mathfrak{g}$ on which the Killing form vanishes identically, i.e. the space of all $x \in \mathfrak{g}$ for which $\kappa(x, x) = 0$; if this space is zero, then $\mathfrak{g}$ is semisimple. Otherwise the subspace is a maximal nilpotent ideal of $\mathfrak{g}$. In a second step one divides out this ideal, and then again analyzes the Killing form of the resulting algebra; the subspace on which the Killing form vanishes identically then forms, together with the previous ideal, the radical of $\mathfrak{g}$; the semisimple part can then be obtained as the quotient (see section 14.2) of $\mathfrak{g}$ by this ideal. In short, if $\mathfrak{g} = \mathfrak{s} \oplus \mathfrak{g}_{\mathrm{rad}}$ is a Levi decomposition of $\mathfrak{g}$ into its radical and semisimple part, and if $\mathfrak{k}$ is the maximal nilpotent subalgebra of $\mathfrak{g}_{\mathrm{rad}}$, then κ is non-degenerate on $\mathfrak{s}$ and on $\mathfrak{g}_{\mathrm{rad}}$ mod $\mathfrak{k}$, and is identically zero on $\mathfrak{k}$.

Another concept which can be made concrete in the case of semisimple Lie algebras, and which also plays an important rôle in the general case, is the notion of Borel subalgebras. For semisimple Lie algebras, the subalgebras $\mathfrak{g}_+$ and $\mathfrak{g}_-$ associated to the positive and negative roots, respectively, are nilpotent, and according to (6.7) the sums

$$\mathfrak{b}_\pm := \mathfrak{g}_\circ \oplus \mathfrak{g}_\pm = \mathfrak{g}_\circ \uplus \mathfrak{g}_\pm \qquad (6.78)$$

are semidirect sums of Lie algebras. These algebras $\mathfrak{b}_\pm$ are called the *Borel subalgebras* of $\mathfrak{g}$. Their derived algebras are $\mathfrak{g}_\pm$; since $\mathfrak{g}_\pm$ are nilpotent, it follows that the algebras $\mathfrak{b}_\pm$ are solvable. In fact, they are maximal solvable subalgebras of $\mathfrak{g}$. (Namely, let $\mathfrak{h} \subseteq \mathfrak{g}$ be any subalgebra *properly* containing, say, $\mathfrak{b}_+$. Then it must contain some linear combination of step operators for negative roots, and hence, as can be seen by taking Lie brackets of this element with the Cartan subalgebra, at least one step operator $E^{-\alpha} \in \mathfrak{g}_-$ for some positive root α. As a consequence, $\mathfrak{h}$ contains the semisimple Lie algebra $\mathfrak{sl}(2)_\alpha$. Thus $\mathfrak{h}$ is not solvable, and therefore $\mathfrak{b}_\pm$ is maximal among the solvable subalgebras of $\mathfrak{g}$.) Now for general Lie algebras, the latter property is used for the definition, i.e. a Borel subalgebra $\mathfrak{b}$ of $\mathfrak{g}$ is defined as a maximal solvable subalgebra of $\mathfrak{g}$. Also, any subalgebra of $\mathfrak{g}$ which contains a Borel subalgebra is called a parabolic subalgebra of $\mathfrak{g}$.

While a general Lie algebra $\mathfrak{g}$ does not possess a triangular decomposition like (6.32), i.e. no distinction between positive and negative roots can be made, there is still a root space decomposition with respect to each Cartan subalgebra $\mathfrak{g}_\circ \subseteq \mathfrak{g}$. It reads

$$\mathfrak{g} = \mathcal{C}_\mathfrak{g}(\mathfrak{g}_\circ) \oplus \bigoplus_{\alpha \neq 0} \mathfrak{g}_\alpha \,, \tag{6.79}$$

where the first summand is the centralizer of $\mathfrak{g}_\circ$. However, the root spaces $\mathfrak{g}_\alpha$ can be very complicated. But many non-semisimple Lie algebras that are relevant in physics do possess a decomposition analogous to (6.32), which is then again called a *triangular decomposition* of the algebra. For some of these even the notion of simple roots makes sense. An important class of such Lie algebras is provided by the so-called *Kac–Moody algebras* for which one can define a non-zero inner product of the simple roots so that there is an associated Cartan matrix. A particular subclass of these is given by the *affine* Lie algebras, which will be described in chapter 7. These Kac–Moody algebras play an important rôle in both mathematics and physics. Mathematically, they are of interest because there are numerous connections to other branches of mathematics such as number theory, topology, singularity theory, or the theory of finite simple groups. Moreover, their representation theory is fully under control and they can be classified completely, with both the classification and the representation theory (to be presented in section 7.9 and sections 8 – 10 of chapter 13, respectively) being very similar to those of simple Lie algebras. In physics, there are a variety of applications of Kac–Moody algebras, the most important ones being in two-dimensional conformal field theory and in the theory of completely integrable systems (for more details, see the end of section 12.4 and section 12.12).

Among the simple Lie algebras, there are in particular finite-dimensional algebras; these will be classified in chapter 7 (but there exist also simple Lie algebras that are infinite-dimensional). In contrast, all non-semisimple Kac–Moody algebras are infinite-dimensional. Note that many interesting physical systems possess infinitely many independent symmetries. In particular, the gauge symmetries which are a basic ingredient of the standard model of elementary particle physics form infinite-dimensional Lie groups; infinite-dimensional symmetries also arise in conformal field theory (and hence in string theory and in the description of phase transitions in condensed matter physics), and in integrable systems which possess an infinite number of conservation laws. In physics infinite-dimensional Lie algebras are therefore just as important as finite-dimensional ones. Even systems with a finite number of degrees of freedom can possess infinite-dimensional symmetries (see exercise 12.7).

There are essentially three classes of infinite-dimensional Lie algebras whose structure is known in some detail, namely the Lie algebras of vector fields which generate the diffeomorphisms of some manifold, the Lie algebras of operators in some Hilbert or Banach space, and the Lie algebras of smooth mappings from some manifold to a finite-dimensional Lie algebra. Provided that one allows for a so-called central extension of the algebra, the affine Kac–Moody algebras constitute examples of the latter class, namely with the manifold being the circle S^1.

Summary:

The structure of a semisimple Lie algebra is most transparent in the Cartan–Weyl basis. A Cartan–Weyl basis consists of the generators of the Cartan subalgebra, which is abelian, and of step operators which are labelled by the roots.

For semisimple Lie algebras, the root space can be taken to be a real vector space, on which the Killing form provides a euclidean inner product. The roots are not degenerate, and for any root α the only multiples which are roots are $\pm\alpha$; each root is an integral linear combination of simple roots, with all coefficients of the same sign. Root strings correspond to irreducible modules of the subalgebras $\mathfrak{sl}(2)_\alpha$ of the semisimple Lie algebra.

The basis of the weight space which is dual to the basis of simple coroots is called the Dynkin basis; its elements are the fundamental weights. Via the inner product one can identify the root and weight spaces.

A Chevalley–Serre basis is a special Cartan–Weyl basis in which the normalizations are chosen such that all structure constants are integral.

(The bracket relations in these special bases are summarized in table III.)

Keywords:

Semisimple (ad-diagonalizable) element, rank, Cartan subalgebra, root, Cartan–Weyl basis, Killing form, positive root, triangular decomposition, simple root, coroot, fundamental weight, Dynkin basis; root lattice, coroot lattice, weight lattice, quadratic form matrix, Chevalley basis, root string;
regular element, Levi decomposition, Borel subalgebra, triangular decomposition, Kac–Moody algebra.

Exercises:

Explain why the secular equation for elements of the Cartan subalgebra has precisely rank $\mathfrak{g}$ 'roots' which are identically zero.

Exercise 6.1

Prove the invariance property $\kappa([x,y],z) = \kappa(x,[y,z])$ (6.11) of the Killing form.

Exercise 6.2

Show that if the Killing form of a finite-dimensional Lie algebra $\mathfrak{g}$ is non-degenerate, then $\mathfrak{g}$ is semisimple.

Hint: Show that if $\mathcal{I}$ is an ideal, then also its orthogonal complement with respect to the Killing form is an ideal. Use this result to decompose $\mathfrak{g}$ into mutually orthogonal ideals that cannot be further decomposed. (Why must $\mathfrak{g}$ be finite-dimensional?)

Exercise 6.3

Show that the Lie algebra $\mathfrak{su}(3)$ possesses abelian subalgebras of dimension two which are not Cartan subalgebras, i.e. whose elements are generically not ad-diagonalizable.

Hint: Consider step operators associated to positive or negative simple roots. To show that the action cannot be diagonalized, choose a matrix representation.

Exercise 6.4

Verify the relations (6.67) for the structure constants in a Chevalley basis.

Hint: Write down the Jacobi identity for the generators E^{α}, E^{β} and $E^{-(\alpha+\beta)}$. To evaluate the various contributions, use the definition of the Chevalley generators H^{α} to show that $H^{-\alpha} = -H^{\alpha}$, and that if α, β and $\alpha + \beta$ are roots, then

$$H^{\alpha+\beta} = [(\alpha + \beta, \alpha + \beta)]^{-1} \left((\alpha,\alpha) H^{\alpha} + (\beta,\beta) H^{\beta} \right). \qquad (6.80)$$

Exercise 6.5

Prove that the restriction of the Killing form of a semisimple Lie algebra to the Cartan subalgebra is non-degenerate.

Hint: Use (6.26) to compute $\kappa(h,h)$, with h any real linear combination of the generators H^{i}.

Exercise 6.6

Draw pictures analogous to (6.41) for the root and weight lattices of $A_1 \oplus A_1$, B_2, and G_2.

Exercise 6.7

Prove the bracket relation (6.69).

Hint: Observe that in terms of the $\mathfrak{sl}(2)_{\alpha}$-representation one has to consider the action of $E^{-\alpha}$ on $v_{n_- - n_+ + 2}$ (analogously as in the computation of $[E^{\alpha}, E^{\beta}]$ via the action of E^{α} on $v_{n_- - n_+}$).

Exercise 6.8

Derive the formula (6.71) for the structure constants $e_{\alpha,\beta}$.

Hint: Deduce from (6.70) that

$$- [[E^{\alpha}, E^{\beta}], [E^{-\alpha}, E^{-\beta}]] = e_{\alpha,\beta}^2 [E^{\alpha+\beta}, E^{-\alpha-\beta}] = e_{\alpha,\beta}^2 H^{\alpha+\beta}. \qquad (6.81)$$

Then employ the Jacobi identity and the relation (6.69) to evaluate the left hand side of this equation as $m_+(m_- + 1)H^{\alpha} + n_+(n_- + 1)H^{\beta}$, with $n_\pm \equiv n_\pm^{(\alpha;\beta)}$ and $m_\pm \equiv n_\pm^{(\beta;\alpha)}$. On the right hand side, use (6.80). The formula then follows by comparing the coefficients of H^{β}.

Exercise 6.9

Prove the relation (6.72).
Hint: Consider the quantity

$$M := n_- - \frac{(\alpha+\beta,\alpha+\beta)}{(\beta,\beta)}\, n_+ + 1\,. \tag{6.82}$$

Eliminate n_- with the help of (6.65) to obtain

$$M = [(\beta,\alpha^\vee) + 1] \cdot [1 - \frac{(\alpha,\alpha)}{(\beta,\beta)}\, n_+]\,. \tag{6.83}$$

Then show that one of the two factors in square brackets vanishes; to this end, consider the cases $(\alpha,\alpha) \geq (\beta,\beta)$ and $(\alpha,\alpha) < (\beta,\beta)$ separately.

Given two roots α and β of a semisimple Lie algebra, set $m := (\beta,\alpha^\vee)$ and $n := (\alpha,\beta^\vee)$. Conclude that

$$mn \leq 4 \quad \text{and} \quad m/n = (\beta,\beta)/(\alpha,\alpha)\,. \tag{6.84}$$

Use this result to constrain the allowed angles between α and β and the ratio of their lengths.

Compute the structure constants of A_{n-1} in the basis described by (6.76) and (6.77).

The *dual representation* R^* of a $\mathfrak{g}$-representation R is a representation of $\mathfrak{g}$ on the dual space V^* of the representation space V of R. Now recall that the adjoint representation R_{ad} of a Lie algebra $\mathfrak{g}$ is defined on the vector space $\mathfrak{g}$. Its dual representation R_{ad}^* is then defined on the dual space $\mathfrak{g}^*$ of $\mathfrak{g}$; it acts as

$$\mathrm{ad}_x^* b(y) = b(-[x,y]) \tag{6.85}$$

for all $x, y \in \mathfrak{g}$ and all $b \in \mathfrak{g}^*$. R_{ad}^* is called the *co-adjoint representation* of $\mathfrak{g}$.
Show that the existence of a non-degenerate invariant bilinear form on a Lie algebra $\mathfrak{g}$ implies that the adjoint representation and the co-adjoint representation are isomorphic.

Assuming that $\mathfrak{g}$ is simple, show that Schur's Lemma implies that any $\mathbb{C}$-bilinear $\mathfrak{g}$-invariant form

$$B: \quad \mathfrak{g} \times \mathfrak{g} \to \mathbb{C} \tag{6.86}$$

is symmetric. Conclude that such form is proportional to the Killing form.
Hint: Since $\mathfrak{g}$ is simple, the adjoint representation is irreducible. Show that B induces an intertwiner between the adjoint and the co-adjoint representation.

7

Simple and affine Lie algebras

7.1 Chevalley–Serre relations

The main purpose of this chapter is the classification of the finite-dimensional simple Lie algebras over the complex numbers. This then yields immediately also the classification of the reductive Lie algebras over $\mathbb{C}$, because reductive algebras are direct sums of simple and abelian algebras, and because any abelian Lie algebra is just the direct sum of $\mathfrak{u}(1)$ algebras. More surprisingly, it also turns out that this classification can easily be extended to obtain also the classification of an even larger class of Lie algebras, the so-called affine Kac–Moody algebras, which also play an important rôle in physics and mathematics. In this context (and also in many other situations) the fact that the latter are infinite-dimensional whereas simple Lie algebras are finite-dimensional is not of any particular relevance. It is worth emphasizing that the classification only covers *finite-dimensional* simple Lie algebras. In fact, as we will note in section 7.8, many infinite-dimensional Lie algebras are simple as well.

In order to classify finite-dimensional simple Lie algebras up to isomorphism, we only have to classify the possible Cartan subalgebras and root systems. As Cartan subalgebras are abelian, the first part of this task is trivial. Further, it turns out that the whole information about the root system Φ is already contained in the set Φ_s of *simple* roots. Indeed, it is possible to characterize a simple Lie algebra completely by specifying only the Lie brackets which involve the step and Cartan subalgebra operators associated to the simple roots in a Chevalley basis, together with a few further relations that contain the information about the full root system.

Namely, the semisimple Lie algebra $\mathfrak{g}(\Phi_s)$ associated to a set $\Phi_s = \{\alpha^{(i)} \mid i = 1, 2, ..., r\}$ of r simple roots $\alpha^{(i)}$ is uniquely determined as the Lie algebra that is algebraically generated by

- 3r generators $\{E^i_\pm, H^i \mid i = 1, 2, \ldots, r\}$

which are subjected to the following Lie bracket relations.

- First,

$$[H^i, H^j] = 0.\tag{7.1}$$

- Second,

$$[H^i, E^j_\pm] = \pm A^{ji} E^j_\pm, \qquad [E^i_+, E^j_-] = \delta_{ij} H^i.\tag{7.2}$$

- And third,

$$(\mathrm{ad}_{E^i_\pm})^{1 - A^{ji}} E^j_\pm = 0\tag{7.3}$$

for $i, j = 1, 2, \ldots, r$, $i \neq j$. Here $E^i_\pm \equiv E^{\pm\alpha^{(i)}}$, and $(\mathrm{ad}_x)^n$ is used as a shorthand notation for $\underbrace{\mathrm{ad}_x \circ \mathrm{ad}_x \circ \ldots \circ \mathrm{ad}_x}_{n \text{ times}}$, so that e.g. $(\mathrm{ad}_x)^2(y) \equiv [x, [x, y]]$.

This last set of relations replaces (6.22), which we had to postulate for all roots, not only the simple roots.

By 'algebraically generated' we mean that the elements of $\mathfrak{g}$ are obtained as arbitrary linear combinations of arbitrary (multiple) Lie brackets of the basic generators. A characterization of a Lie algebra $\mathfrak{g}$ of the type above, which must not be mixed up with the description of $\mathfrak{g}$ as the linear span of a basis, is called a *presentation* of $\mathfrak{g}$ by *generators modulo relations*. (An analogous presentation is often also used for other structures, e.g. for finite groups. If there are no relations at all, the resulting structure is called the *free* Lie algebra (respectively free group, etc.) with the given generators.) It is worth stressing that the only non-universal numbers in the presentation of $\mathfrak{g}$ above are the integers A^{ij}, i.e. the entries of the r×r Cartan matrix $A(\Phi_s)$ that is associated to the simple roots Φ_s according to (6.34). In particular the *same* numbers appear in the two different types of relations (7.2) and (7.3).

The relations (7.3) determine which multiple brackets of step operators must vanish and hence fix the length of the $\alpha^{(i)}$-string through $\alpha^{(j)}$; they are called the *Serre relations*, and the full set of relations listed above is known as the *Chevalley–Serre relations*. Note that while these relations summarize the structure of $\mathfrak{g}$ in a very compact form, it is by no means obvious that they indeed characterize $\mathfrak{g}$ uniquely. (It is even difficult to see that $\mathfrak{g}$ is finite-dimensional.)

To prove that the whole Lie algebra $\mathfrak{g}$ can be reconstructed uniquely from the given information is provided by an algorithm for reconstructing $\mathfrak{g}$ which is known as the *Serre construction*. With this algorithm one obtains all step operators E^α as multiple Lie brackets of the $E^i_\pm$. This is achieved by induction on the height of the roots: One starts with the simple roots, i.e. the roots of height 1; from the Serre relations one gets the $\alpha^{(i)}$-string through $\alpha^{(j)}$ for $i \neq j$ and hence in particular finds all roots of height

2, and thus all ladder operators which can be written in the form $[E_{\pm}^{i_1}, E_{\pm}^{i_2}]$ for some $i_1, i_2 \in \{1, 2, ..., r\}$. Afterwards the Serre relations can be used once again to find the $\alpha^{(i)}$-strings through the roots of height 2, in particular all roots of height 3, and so on.

Let us consider as an example the case of $\mathfrak{g} = A_2 \cong \mathfrak{sl}(3)$. According to the result (6.74) for the Cartan matrix of A_2, the Serre relations for this algebra read

$$[E_{\pm}^1, [E_{\pm}^1, E_{\pm}^2]] = 0, \qquad [E_{\pm}^2, [E_{\pm}^2, E_{\pm}^1]] = 0. \tag{7.4}$$

From these constraints one learns that $\pm(\alpha^{(1)} + \alpha^{(2)})$ are roots of height 2, but that the possible combinations of height three, namely $\pm(2\alpha^{(1)} + \alpha^{(2)})$ and $\pm(\alpha^{(1)} + 2\alpha^{(2)})$, are not roots, and more generally, that there are no roots of height larger than two at all. Thus the roots of A_2 are given by the linear combinations $\pm\alpha^{(1)}$, $\pm\alpha^{(2)}$, and $\pm(\alpha^{(1)} + \alpha^{(2)})$ of simple roots, in agreement with the results of chapter 3.

7.2 The Cartan matrix

The only free parameters occurring in (7.2) and (7.3) are the entries of the Cartan matrix. The classification of simple Lie algebras therefore amounts to the classification of their Cartan matrices. (In particular, the Cartan matrix of a simple Lie algebra is independent of the choice of the basis of simple roots, up to the numbering of rows and columns. This follows from the results of chapter 6, and can be deduced most directly by using the concept of the Weyl group of a simple Lie algebra, see chapter 10.) Therefore we now list several properties of Cartan matrices of simple Lie algebras.

• First, by setting $i = j$ in the definition (6.34) of the matrix elements A^{ij}, one obtains $A^{ii} = 2$ for all $i = 1, 2, ..., r$.

• Next, the symmetry of the scalar product in root space implies that $A^{ij} = 0$ is equivalent to $A^{ji} = 0$.

• Further, $A^{ij} \equiv (\alpha^{(i)}, \alpha^{(j)\vee}) \in \mathbb{Z}$ for $i, j = 1, 2, ..., r$ (more generally, we have already seen in chapter 6 that the inner product $(\alpha^\vee, \beta)$ is an integer for any pair of roots $\alpha, \beta \in \Phi$).

• The difference of any two simple roots is never a root; together with the fact (compare equation (6.66)) that the sign of the inner product (α, β) determines whether $\alpha \pm \beta$ is a root, it follows that $(\alpha^{(i)}, \alpha^{(j)}) \leq 0$ for $i \neq j$, and hence that $A^{ij} \in \mathbb{Z}_{\leq 0}$ for $i, j = 1, 2, ..., r$.

• Finally, according to the relations (6.49) the symmetrization of the Cartan matrix is just the weight space metric G (with upper indices). Therefore the Cartan matrix is non-degenerate: $\det A \neq 0$.

In terms of the root system Φ, the splitting of a semisimple Lie algebra $\mathfrak{g} = \mathfrak{g}_{(1)} \oplus \cdots \oplus \mathfrak{g}_{(n)}$ into its simple summands $\mathfrak{g}_{(p)}$ is expressed through the fact that Φ can be written as the sum of 'irreducible' subsystems $\Phi_{(p)}$ – the root systems of the direct summands $\mathfrak{g}_{(p)}$ – such that $(\alpha, \beta) = 0$ for all $\alpha \in \Phi_{(p)}$ and all $\beta \in \Phi_{(q)}$ with $p \neq q$ (and such that none of the

$\Phi_{(p)}$ can itself be decomposed in the same way further). The restriction from semisimple to simple Lie algebras thus means that the Cartan matrix must be *indecomposable* (or *irreducible*) in the sense that there is no renumbering of the simple roots which would bring A to the block diagonal form

$$A = \begin{pmatrix} A_{(1)} & 0 \\ 0 & A_{(2)} \end{pmatrix}. \tag{7.5}$$

We have already seen (cf. the discussion after equation (6.26)) that the root space $\text{span}_{\mathbb{R}}(\Phi)$ of a finite-dimensional simple Lie algebra is a real euclidean vector space. This implies that the property $\det A \neq 0$ can be strengthened to

$$\det A > 0. \tag{7.6}$$

In summary, the classification of simple Lie algebras amounts to a classification of all square matrices A which possess the following properties:

$$
\begin{aligned}
&\text{a)} && A^{ii} = 2, \\
&\text{b)} && A^{ij} = 0 \Leftrightarrow A^{ji} = 0, \\
&\text{c)} && A^{ij} \in \mathbb{Z}_{\leq 0} \ \text{for} \ i \neq j, \\
&\text{d)} && \det A > 0, \\
&\text{e)} && A \ \text{not equivalent to the form (7.5)}.
\end{aligned}
\tag{7.7}
$$

By employing the Serre construction, one finds that Cartan matrices which cannot be transformed into each other by a relabelling of the rows and columns correspond to non-isomorphic Lie algebras. (Inspecting the table V below, for finite-dimensional simple Lie algebras this also follows more directly from the fact that isomorphic Lie algebras must have both the same rank and the same dimension.)

The enumeration of all matrices which satisfy these constraints is not too difficult. First write $(\alpha, \beta) = \cos \vartheta \, [(\alpha, \alpha)(\beta, \beta)]^{1/2}$, where ϑ is the angle between α and β; this implies that $(\alpha, \beta^{\vee}) = 2 \cos \vartheta \, [(\alpha, \alpha)/(\beta, \beta)]^{1/2}$, and hence (compare exercise 6.11)

$$(\alpha, \beta^{\vee})(\alpha^{\vee}, \beta) = 4 \cos^2 \vartheta \leq 4. \tag{7.8}$$

Further, this inequality is saturated only if $\alpha = \pm \beta$. Applied to the Cartan matrix, this implies $A^{ij} A^{ji} \leq 4$ with equality only for $i = j$, i.e.

$$A^{ij} A^{ji} \in \{0, 1, 2, 3\} \quad \text{for} \ i \neq j. \tag{7.9}$$

Taking into account the specific requirements on the off-diagonal entries of A, this leaves us only with the possibilities

$$
\begin{aligned}
& A^{ij} = A^{ji} = 0 && \text{or} && A^{ij} = A^{ji} = -1 && \text{or} \\
& A^{ij} = -1, \ A^{ji} = -2 && \text{or} && A^{ij} = -1, \ A^{ji} = -3.
\end{aligned}
\tag{7.10}
$$

The rest of the work consists of finding further restrictions, thereby implementing in particular also the requirement $\det A > 0$. Actually, with the help of this stronger property one can also derive the restrictions (7.10) on $A^{ij} A^{ji}$ without using the definition of the Cartan matrix elements as inner products of simple roots, i.e. without any reference to a root space with positive definite metric at all.

7.3 Dynkin diagrams

The enumeration of all possible solutions to the conditions (7.7) is a somewhat lengthy procedure, but it is purely combinatorial. It does not lead to any further insight in the intrinsic structure of Lie algebras, nor does it require any new techniques. Therefore we will state only the final result (some intermediate steps can be worked out in exercise 7.2). To do so, it is most convenient to introduce the notion of a *Dynkin diagram*: To each Cartan matrix one associates a diagram consisting of vertices and lines connecting them. Each vertex of the diagram represents a simple root; the vertices for $\alpha^{(i)}$ and $\alpha^{(j)}$ ($i \neq j$) are connected by $\max\{|A^{ij}|, |A^{ji}|\}$ lines. (In particular, when the weight space is euclidean, which is precisely the case if $\mathfrak{g}$ is finite-dimensional and semisimple, then vertices connected by a single, double or triple bond correspond to simple roots spanning an angle of $\frac{2}{3}\pi$, $\frac{3}{4}\pi$ and $\frac{5}{6}\pi$, respectively, while simple roots not connected by a line are mutually orthogonal.) Furthermore, an arrowhead ' > ' is added to the lines from the ith to the jth node if $A^{ij} \neq 0$ and $|A^{ij}| > |A^{ji}|$ (which is equivalent to $(\alpha^{(i)}, \alpha^{(i)}) > (\alpha^{(j)}, \alpha^{(j)})$). Alternatively, one may specify 'long' roots by open dots ' o ' and 'short' roots by filled dots ' • '; this qualification makes sense because the analysis shows that at most two different lengths are allowed for the roots of any given finite-dimensional simple Lie algebra. For Cartan matrices which differ only by renumbering the simple roots (and, as a consequence, describe the same Lie algebra), one obtains identical Dynkin diagrams.

Let us illustrate this with the Lie algebras of rank two. The Dynkin diagram of each of these has two nodes. From the root diagrams (3.15) and (3.42) we read off that the two simple roots form an angle of $\frac{2}{3}\pi$ for A_2, of $\frac{3}{4}\pi$ for B_2, and of $\frac{5}{6}\pi$ for G_2. Correspondingly, we obtain the rank two Dynkin diagrams listed in table IV.

The Dynkin diagrams are in fact multi-purpose graphs. They are extremely useful to visualize many different features of a Lie algebra. The ith node can be regarded as representing e.g. the following objects:

• simply the row index of the Cartan matrix (A^{ij});

• the ith simple root $\alpha^{(i)}$;

• the ith fundamental weight $\Lambda_{(i)}$;

• the Cartan subalgebra generator H^i;

• yet another possibility is the ith generator $w_{(i)}$ of the Weyl group, see chapter 10.

Furthermore, as we will see in chapter 11, an important subclass of automorphisms of the Lie algebra can be interpreted as symmetries of the Dynkin diagram.

7.4 Simple Lie algebras

The arguments indicated at the end of section 7.2 yield the following classification of finite-dimensional simple Lie algebras. There are four infinite series, which are denoted by

$$A_r \ (r \geq 1), \quad B_r \ (r \geq 3), \quad C_r \ (r \geq 2), \quad D_r \ (r \geq 4), \qquad (7.11)$$

and in addition five isolated cases, which are called

$$E_6, \quad E_7, \quad E_8, \quad G_2, \quad F_4. \qquad (7.12)$$

In all cases the subscript denotes the rank of the algebra. This classification is displayed in table IV in terms of Dynkin diagrams.

The algebras in the infinite series of simple Lie algebras are called the *classical* (Lie) algebras; they are isomorphic to the matrix algebras that were described in section 5.6:

$$A_r \cong \mathfrak{sl}(r+1), \qquad B_r \cong \mathfrak{so}(2r+1),$$

$$C_r \cong \mathfrak{sp}(r), \qquad D_r \cong \mathfrak{so}(2r); \qquad (7.13)$$

the five isolated cases are referred to as the *exceptional* Lie algebras. The restrictions on the rank r of the classical algebras are imposed to avoid double-counting; if one includes all values $r \geq 1$ in all four series, then one has to take into account the isomorphisms

$$A_1 \cong B_1 \cong C_1 \cong D_1, \qquad B_2 \cong C_2,$$

$$D_2 \cong A_1 \oplus A_1, \qquad D_3 \cong A_3. \qquad (7.14)$$

For the simple Lie algebras of types A_r, D_r, E_6, E_7 and E_8, all roots have the same length, and any two nodes of the Dynkin diagram are connected by at most one line; these algebras are therefore called *simply laced*. In the other cases there are roots of two different lengths, the length of the long roots being $\sqrt{2}$ times the length of the short roots for B_r, C_r and F_4, and $\sqrt{3}$ times that length for G_2, respectively. The dual root system $\Phi^\vee(\mathfrak{g}) := \{\alpha^\vee \mid \alpha \in \Phi(\mathfrak{g})\}$ of $\mathfrak{g}$ is isomorphic to the root system of another simple Lie algebra, which is called the *dual Lie algebra* $\mathfrak{g}^\vee$ of $\mathfrak{g}$, $\Phi^\vee(\mathfrak{g}) \cong \Phi(\mathfrak{g}^\vee)$. The simple roots of $\mathfrak{g}^\vee$ are the simple coroots of $\mathfrak{g}$. Of course, any simply laced algebra is self-dual, i.e. satisfies $\mathfrak{g} = \mathfrak{g}^\vee$, but this is also true for C_2, G_2 and F_4, while $(B_r)^\vee = C_r$ and vice versa.

The classification of finite-dimensional simple Lie algebras implies in particular the following statements concerning root strings. When $\mathfrak{g}$ is simply laced, the sum $\alpha + \beta$ of

Table IV.　*Dynkin diagrams of the finite-dimensional simple Lie algebras*

name	numbering of the nodes	dual Coxeter labels *Coxeter labels*
A_r		
B_r		
C_r		
D_r		
E_6		
E_7		
E_8		
F_4		
G_2		

any two positive roots α and β is again a root if and only if $(\alpha^\vee, \beta) = -1$. Because of $(\alpha^\vee, \beta) = n_- - n_+$ (see equation (6.65)) it follows that $n_+ = n_- + 1$; in addition, one can show (see exercise 7.4) that the only solution to the requirement that $e_{\alpha, \beta + (n_+ - 1)\alpha}$ is integral is that $n_+ = 1$, in which case the value is ± 1. Thus for simply laced $\mathfrak{g}$, one has, for any pair of positive roots, $n_+ = 1$, $n_- = 0$, and $|e_{\alpha, \beta}| = 1$; in particular, any root string consists of at most two elements. For non-simply laced algebras one has $(\alpha^\vee, \beta) \geq -3$ for any pair of positive roots, and from this it follows that $n_+ + n_- \leq 3$, i.e. the root strings have at most four elements.

Some characteristic numbers of the simple Lie algebras are listed in table V. These numbers have the following meaning. First, d is the dimension of $\mathfrak{g}$, $|\Phi_+| = \frac{1}{2}(d - r)$ is the number of positive roots, and $|\Phi_+^{(<)}|$ the number of short positive roots. $g^\vee$ and g are the dual Coxeter number and Coxeter number which will be introduced shortly. $I_c := |L_w/L|$, called the index of connection, is the maximal number of weights of finite-dimensional $\mathfrak{g}$-modules that do not differ pairwise by an element of the root lattice (such weights form a conjugacy class of $\mathfrak{g}$-weights, see section 15.5). By inspection one finds that I_c is equal to the determinant of the Cartan matrix. Finally, the *exponents* are integers which are related to the eigenvalues of the Cartan matrix A (see equation (7.21) below). From the table one can also read off the relation $g + 1 = d/r$; in particular that the rank of $\mathfrak{g}$ is always a divisor of the dimension.

Table V. *Some characteristic numbers of simple Lie algebras*

$\mathfrak{g}$	d	$|\Phi_+|$	$|\Phi_+^{(<)}|$	$g^\vee$	g	I_c	exponents + 1
A_r	$r^2 + 2r$	$\frac{1}{2}(r^2+r)$	–	$r+1$		$r+1$	$2, 3, \ldots, r+1$
B_r	$2r^2 + r$	r^2	r	$2r-1$	$2r$	2	$2, 4, \ldots, 2r$
C_r	$2r^2 + r$	r^2	$r^2 - r$	$r+1$	$2r$	2	$2, 4, \ldots, 2r$
D_r	$2r^2 - r$	$r^2 - r$	–	$2r-2$		4	$2, 4, 6, \ldots, 2r-2, r$
E_6	78	36	–	12		3	$2, 5, 6, 8, 9, 12$
E_7	133	63	–	18		2	$2, 6, 8, 10, 12, 14, 18$
E_8	248	120	–	30		1	$2, 8, 12, 14, 18, 20, 24, 30$
F_4	52	24	12	9	12	1	$2, 6, 8, 12$
G_2	14	6	3	4	6	1	$2, 6$

7.5 Quadratic form matrices

As described in chapter 6, the metric on the weight space or quadratic form matrix is the inverse of the symmetrized Cartan matrix. Closed formulæ for the entries of the quadratic form matrices of the classical simple Lie algebras look as follows.

$$A_r: \quad G_{ij} = \frac{1}{r+1} \min(i,j) \cdot (r+1 - \max(i,j)),$$

$$B_r: \quad G_{ij} = \begin{cases} \min(i,j) & \text{for } i,j \neq r, \\ \frac{1}{2}i & \text{for } i \neq r,\, j = r, \\ \frac{1}{2}j & \text{for } i = r,\, j \neq r, \\ \frac{1}{4}r & \text{for } i = j = r, \end{cases}$$

$$C_r: \quad G_{ij} = \frac{1}{2} \min(i,j), \tag{7.15}$$

$$D_r: \quad G_{ij} = \begin{cases} \min(i,j) & \text{for } i,j \notin \{r, r-1\}, \\ \frac{1}{2}i & \text{for } i \notin \{r, r-1\},\, j \in \{r, r-1\}, \\ \frac{1}{2}j & \text{for } i \in \{r, r-1\},\, j \notin \{r, r-1\}, \\ \frac{1}{4}r & \text{for } i = j = r \text{ or } i = j = r-1, \\ \frac{1}{4}(r-2) & \text{for } i = r,\, j = r-1 \text{ or vice versa}. \end{cases}$$

For convenience, in table VI the quadratic form matrices of the simple Lie algebras, including the exceptional ones, are also displayed explicitly, together with the Cartan matrices which can directly be read off the Dynkin diagrams.

7.6 The highest root and the Weyl vector

In table IV we have displayed each Dynkin diagram twice. In the left column the numbers adjoined to the nodes define the numbering of the nodes (or, equivalently, of simple roots, or of fundamental weights). In the right column, the numbers are the so-called *Coxeter labels* a_i and *dual Coxeter labels* $a_i^\vee$ (the Coxeter labels are written in italics, and they are only displayed if they are different from the dual Coxeter labels). These numbers are defined as the components of the highest root θ in its expansions

$$\theta =: \sum_{i=1}^{r} a_i\, \alpha^{(i)}, \qquad \frac{2}{(\theta,\theta)}\, \theta =: \sum_{i=1}^{r} a_i^\vee\, \alpha^{(i)\vee}. \tag{7.16}$$

with respect to the basis of simple roots and simple coroots, respectively. As expressed by the inequality (6.44), the expansion coefficients of θ in the basis of simple roots, i.e. the Coxeter labels, are strictly positive integers (not just non-negative as for any positive root); the

Table VI. *Cartan matrices and quadratic form matrices of simple Lie algebras.*
(*For better readability, zero entries are written in smaller font.*)

$\mathfrak{g}$	Cartan matrix	quadratic form matrix
A_r	$$\begin{pmatrix} 2 & -1 & 0 & \cdots & 0 & 0 & 0 & 0 \\ -1 & 2 & -1 & \cdots & 0 & 0 & 0 & 0 \\ 0 & -1 & 2 & & 0 & 0 & 0 & 0 \\ \vdots & \vdots & \vdots & \ddots & \vdots & \vdots & \vdots & \vdots \\ 0 & 0 & 0 & \cdots & 2 & -1 & 0 & 0 \\ 0 & 0 & 0 & \cdots & -1 & 2 & -1 & 0 \\ 0 & 0 & 0 & \cdots & 0 & -1 & 2 & -1 \\ 0 & 0 & 0 & \cdots & 0 & 0 & -1 & 2 \end{pmatrix}$$	$\dfrac{1}{r+1}\times$ $$\begin{pmatrix} r & r-1 & r-2 & \cdots & 3 & 2 & 1 \\ r-1 & 2(r-1) & 2(r-2) & \cdots & 6 & 4 & 2 \\ r-2 & 2(r-1) & 3(r-2) & \cdots & 9 & 6 & 3 \\ \vdots & \vdots & \vdots & \ddots & \vdots & \vdots & \vdots \\ 4 & 8 & 12 & \cdots & 3(r-3) & 2(r-3) & r-3 \\ 3 & 6 & 9 & \cdots & 3(r-2) & 2(r-2) & r-2 \\ 2 & 4 & 6 & \cdots & 2(r-2) & 2(r-1) & r-1 \\ 1 & 2 & 2 & \cdots & r-2 & r-1 & r \end{pmatrix}$$
B_r	$$\begin{pmatrix} 2 & -1 & 0 & \cdots & 0 & 0 & 0 & 0 \\ -1 & 2 & -1 & \cdots & 0 & 0 & 0 & 0 \\ 0 & -1 & 2 & \cdots & 0 & 0 & 0 & 0 \\ \vdots & \vdots & \vdots & \ddots & \vdots & \vdots & \vdots & \vdots \\ 0 & 0 & 0 & \cdots & 2 & -1 & 0 & 0 \\ 0 & 0 & 0 & \cdots & -1 & 2 & -1 & 0 \\ 0 & 0 & 0 & \cdots & 0 & -1 & 2 & -2 \\ 0 & 0 & 0 & \cdots & 0 & 0 & -1 & 2 \end{pmatrix}$$	$$\begin{pmatrix} 1 & 1 & 1 & \cdots & 1 & 1 & 1 & \frac{1}{2} \\ 1 & 2 & 2 & \cdots & 2 & 2 & 2 & 1 \\ 1 & 2 & 3 & \cdots & 3 & 3 & 3 & \frac{3}{2} \\ \vdots & \vdots & \vdots & \ddots & \vdots & \vdots & \vdots & \vdots \\ 1 & 2 & 3 & \cdots & r-3 & r-3 & r-3 & \frac{r-3}{2} \\ 1 & 2 & 3 & \cdots & r-3 & r-2 & r-2 & \frac{r-2}{2} \\ 1 & 2 & 3 & \cdots & r-3 & r-2 & r-1 & \frac{r-1}{2} \\ \frac{1}{2} & 1 & \frac{3}{2} & \cdots & \frac{r-3}{2} & \frac{r-2}{2} & \frac{r-1}{2} & \frac{r}{4} \end{pmatrix}$$
C_r	$$\begin{pmatrix} 2 & -1 & 0 & \cdots & 0 & 0 & 0 & 0 \\ -1 & 2 & -1 & \cdots & 0 & 0 & 0 & 0 \\ 0 & -1 & 2 & \cdots & 0 & 0 & 0 & 0 \\ \vdots & \vdots & \vdots & \ddots & \vdots & \vdots & \vdots & \vdots \\ 0 & 0 & 0 & \cdots & 2 & -1 & 0 & 0 \\ 0 & 0 & 0 & \cdots & -1 & 2 & -1 & 0 \\ 0 & 0 & 0 & \cdots & 0 & -1 & 2 & -1 \\ 0 & 0 & 0 & \cdots & 0 & 0 & -2 & 2 \end{pmatrix}$$	$\dfrac{1}{2}$ $$\begin{pmatrix} 1 & 1 & 1 & \cdots & 1 & 1 & 1 & 1 \\ 1 & 2 & 2 & \cdots & 2 & 2 & 2 & 2 \\ 1 & 2 & 3 & \cdots & 3 & 3 & 3 & 3 \\ \vdots & \vdots & \vdots & \ddots & \vdots & \vdots & \vdots & \vdots \\ 1 & 2 & 3 & \cdots & r-3 & r-3 & r-3 & r-3 \\ 1 & 2 & 3 & \cdots & r-3 & r-2 & r-2 & r-2 \\ 1 & 2 & 3 & \cdots & r-3 & r-2 & r-1 & r-1 \\ 1 & 2 & 3 & \cdots & r-3 & r-2 & r-1 & r \end{pmatrix}$$
D_r	$$\begin{pmatrix} 2 & -1 & 0 & \cdots & 0 & 0 & 0 & 0 \\ -1 & 2 & -1 & \cdots & 0 & 0 & 0 & 0 \\ 0 & -1 & 2 & \cdots & 0 & 0 & 0 & 0 \\ \vdots & \vdots & \vdots & \ddots & \vdots & \vdots & \vdots & \vdots \\ 0 & 0 & 0 & \cdots & 2 & -1 & 0 & 0 \\ 0 & 0 & 0 & \cdots & -1 & 2 & -1 & -1 \\ 0 & 0 & 0 & \cdots & 0 & -1 & 2 & 0 \\ 0 & 0 & 0 & \cdots & 0 & -1 & 0 & 2 \end{pmatrix}$$	$$\begin{pmatrix} 1 & 1 & 1 & \cdots & 1 & 1 & \frac{1}{2} & \frac{1}{2} \\ 1 & 2 & 2 & \cdots & 2 & 2 & 1 & 1 \\ 1 & 2 & 3 & \cdots & 3 & 3 & \frac{3}{2} & \frac{3}{2} \\ \vdots & \vdots & \vdots & \ddots & \vdots & \vdots & \vdots & \vdots \\ 1 & 2 & 3 & \cdots & r-3 & r-3 & \frac{r-3}{2} & \frac{r-3}{2} \\ 1 & 2 & 3 & \cdots & r-3 & r-2 & \frac{r-2}{2} & \frac{r-2}{2} \\ \frac{1}{2} & 1 & \frac{3}{2} & \cdots & \frac{r-3}{2} & \frac{r-2}{2} & \frac{r}{4} & \frac{r-2}{4} \\ \frac{1}{2} & 1 & \frac{3}{2} & \cdots & \frac{r-3}{2} & \frac{r-2}{2} & \frac{r-2}{4} & \frac{r}{4} \end{pmatrix}$$

Table VI   (*continued*).

$\mathfrak{g}$	Cartan matrix	quadratic form matrix
E_6	$\begin{pmatrix} 2 & -1 & 0 & 0 & 0 & 0 \\ -1 & 2 & -1 & 0 & 0 & 0 \\ 0 & -1 & 2 & -1 & 0 & -1 \\ 0 & 0 & -1 & 2 & -1 & 0 \\ 0 & 0 & 0 & -1 & 2 & 0 \\ 0 & 0 & -1 & 0 & 0 & 2 \end{pmatrix}$	$\begin{pmatrix} \frac{4}{3} & \frac{5}{3} & 2 & \frac{4}{3} & \frac{2}{3} & 1 \\ \frac{5}{3} & \frac{10}{3} & 4 & \frac{8}{3} & \frac{4}{3} & 2 \\ 2 & 4 & 6 & 4 & 2 & 3 \\ \frac{4}{3} & \frac{8}{3} & 4 & \frac{10}{3} & \frac{5}{3} & 2 \\ \frac{2}{3} & \frac{4}{3} & 2 & \frac{5}{3} & \frac{4}{3} & 1 \\ 1 & 2 & 3 & 2 & 1 & 2 \end{pmatrix}$
E_7	$\begin{pmatrix} 2 & -1 & 0 & 0 & 0 & 0 & 0 \\ -1 & 2 & -1 & 0 & 0 & 0 & 0 \\ 0 & -1 & 2 & -1 & 0 & 0 & -1 \\ 0 & 0 & -1 & 2 & -1 & 0 & 0 \\ 0 & 0 & 0 & -1 & 2 & -1 & 0 \\ 0 & 0 & 0 & 0 & -1 & 2 & 0 \\ 0 & 0 & -1 & 0 & 0 & 0 & 2 \end{pmatrix}$	$\begin{pmatrix} 2 & 3 & 4 & 3 & 2 & 1 & 2 \\ 3 & 6 & 8 & 6 & 4 & 2 & 4 \\ 4 & 8 & 12 & 9 & 6 & 3 & 6 \\ 3 & 6 & 9 & \frac{15}{2} & 5 & \frac{5}{2} & \frac{9}{2} \\ 2 & 4 & 6 & 5 & 4 & 3 & 2 \\ 1 & 2 & 3 & \frac{5}{2} & 2 & \frac{3}{2} & \frac{3}{2} \\ 2 & 4 & 6 & \frac{9}{2} & 3 & \frac{3}{2} & \frac{7}{2} \end{pmatrix}$
E_8	$\begin{pmatrix} 2 & -1 & 0 & 0 & 0 & 0 & 0 & 0 \\ -1 & 2 & -1 & 0 & 0 & 0 & 0 & 0 \\ 0 & -1 & 2 & -1 & 0 & 0 & 0 & 0 \\ 0 & 0 & -1 & 2 & -1 & 0 & 0 & 0 \\ 0 & 0 & 0 & -1 & 2 & -1 & 0 & -1 \\ 0 & 0 & 0 & 0 & -1 & 2 & -1 & 0 \\ 0 & 0 & 0 & 0 & 0 & -1 & 2 & 0 \\ 0 & 0 & 0 & 0 & -1 & 0 & 0 & 2 \end{pmatrix}$	$\begin{pmatrix} 2 & 3 & 4 & 5 & 6 & 4 & 2 & 3 \\ 3 & 6 & 8 & 10 & 12 & 8 & 4 & 6 \\ 4 & 8 & 12 & 15 & 18 & 12 & 6 & 9 \\ 5 & 10 & 15 & 20 & 24 & 16 & 8 & 12 \\ 6 & 12 & 18 & 24 & 30 & 20 & 10 & 15 \\ 4 & 8 & 2 & 16 & 20 & 14 & 7 & 10 \\ 2 & 4 & 6 & 8 & 10 & 7 & 4 & 5 \\ 3 & 6 & 9 & 12 & 15 & 10 & 5 & 8 \end{pmatrix}$
F_4	$\begin{pmatrix} 2 & -1 & 0 & 0 \\ -1 & 2 & -2 & 0 \\ 0 & -1 & 2 & -1 \\ 0 & 0 & -1 & 2 \end{pmatrix}$	$\begin{pmatrix} 2 & 3 & 2 & 1 \\ 3 & 6 & 4 & 2 \\ 2 & 4 & 3 & \frac{3}{2} \\ 1 & 2 & \frac{3}{2} & 1 \end{pmatrix}$
G_2	$\begin{pmatrix} 2 & -3 \\ -1 & 2 \end{pmatrix}$	$\begin{pmatrix} 2 & 1 \\ 1 & \frac{2}{3} \end{pmatrix}$

same is then true for the components $a_i^\vee \equiv (\theta^\vee)_i$. According to the equations (7.16) we also have $\theta = \sum_{i=1}^{r} a_i \alpha^{(i)} = \frac{1}{2}(\theta,\theta) \sum_{i=1}^{r} a_i^\vee \alpha^{(i)\vee} = \sum_{i=1}^{r} [(\theta,\theta)/(\alpha^{(i)},\alpha^{(i)})] \, a_i^\vee \alpha^{(i)}$. Since the simple roots are linearly independent, this implies that

$$\frac{a_i}{a_i^\vee} = \frac{(\theta,\theta)}{(\alpha^{(i)},\alpha^{(i)})}. \tag{7.17}$$

As the highest root θ is always a long root, this shows that for simple Lie algebras one has $a_i^\vee \le a_i$ for all $i = 1, 2, \dots, \mathrm{r}$.

The *Coxeter number* and *dual Coxeter number* of a simple Lie algebra are by definition the sum of all (dual) Coxeter labels augmented by 1,

$$g := 1 + \sum_{i=1}^{r} a_i, \qquad g^\vee := 1 + \sum_{i=1}^{r} a_i^\vee. \tag{7.18}$$

Both the Coxeter and the dual Coxeter number appear in a variety of circumstances. For instance, consider the *Weyl vector* ρ of a semisimple Lie algebra $\mathfrak{g}$ which is defined as half the sum of positive $\mathfrak{g}$-roots,

$$\rho := \tfrac{1}{2} \sum_{\alpha>0} \alpha. \tag{7.19}$$

The length of the Weyl vector can be expressed through the dimension and the dual Coxeter number of $\mathfrak{g}$ via the so-called *strange formula*

$$(\rho, \rho) = \tfrac{1}{24} g^\vee (\theta, \theta) \dim \mathfrak{g}. \tag{7.20}$$

Another example is provided by the eigenvalues of the Cartan matrices of simply laced Lie algebras. They are given by

$$4 \sin^2\left(\tfrac{\pi m_i}{2g}\right), \tag{7.21}$$

where m_i are the exponents; also, the maximal exponent is equal to $g - 1$. Equivalently, the eigenvalues of the matrix $2 \cdot \mathbb{1} - A$ (this matrix is known as the adjacency matrix of the Dynkin diagram because its entries equal the number of lines between the nodes) are $2 \cos(\tfrac{\pi m_i}{g})$.

As will be seen in section 10.5, the Weyl vector (7.19) also obeys $\rho^i \equiv (\rho, \alpha^{(i)\vee}) = 1$ for all i, i.e. the expansion of ρ with respect to the fundamental weights reads

$$\rho = \sum_{i=1}^{r} \Lambda_{(i)}. \tag{7.22}$$

As a consequence, the strange formula can be rewritten as the identity $\sum_{i,j=1}^{r} G_{ij} = \tfrac{1}{24} \mathrm{d} g^\vee (\theta, \theta)$ for the quadratic form matrix.

7.7 The orthonormal basis of weight space

As already mentioned, for the study of representations of simple Lie algebras the most convenient basis of the weight space is usually the Dynkin basis. Occasionally, however, the use of a suitable orthonormal basis is more convenient than the Dynkin basis. The existence of an orthonormal basis is an immediate consequence of the fact that a (positive definite) inner product exists on the weight space.

We recall that in the Dynkin basis (6.39) the components of the simple roots are just given by the rows of Cartan matrix, and hence are in particular integers (they can be read off table VI). Now for all simple Lie algebras except A_r, E_6, E_7, and G_2 there actually exists an orthonormal basis $\{e_i\}$ such that the coordinates of any root are also very simple numbers, namely they are all half integers between -2 and 2. In contrast, for the latter algebras the expressions in any orthonormal basis of the weight space involve square roots (for A_2 this is already apparent from the results (3.26) for the simple roots). However, in these cases it is possible to choose an embedding of the weight space in $\mathbb{R}^D$, with $D = r + 1$ (for A_r, E_7, and G_2) or $D = r + 2$ (for E_6), together with an orthonormal basis in the larger space with respect to which the coordinates of any root are again half integers between -2 and 2.

The overall normalization of the roots which makes this basis an ortho*normal* one is such that for C_r and G_2 the short roots, and for all other cases the long roots, have length squared equal to two. The resulting expressions for the simple roots $\alpha^{(i)} \in \Phi_s$ and arbitrary roots $\alpha \in \Phi$ are listed in table VII. For the classical algebras, these orthonormal bases of Φ correspond precisely to the Lie algebra bases that were already introduced in the formulæ (5.31) and (5.32) in section 5.6 (see exercise 7.5).

For the cases in which the dimension D is strictly larger than r, the relevant subspace of $\mathbb{R}^D$ which is isomorphic to the weight space is identified as the space orthogonal to the vector $\sum_{i=1}^{r+1} e_i$ for A_r and G_2, orthogonal to $e_7 + e_8$ for E_7, and orthogonal to $e_6 - e_7$ and to $e_7 + e_8$ for E_6. Also note that according to table VII we can obtain D_r from A_{r-1} by including $\pm(e_i + e_j)$ as additional roots, and B_r and C_r from D_r by including $\pm e_i$, respectively $\pm 2e_i$, as roots.

By imposing the orthonormality condition $(\alpha^{(i)}, \Lambda_{(j)}) = \delta_i^j$, one easily obtains the expressions for the fundamental weights in the orthonormal basis of $\mathfrak{g}_0^*$. We present only the results for the classical Lie algebras; for A_r one finds that

$$\Lambda_{(j)} = \sum_{i=1}^{j} e_i - \frac{j}{r+1} \sum_{i=1}^{r+1} e_i \qquad (7.23)$$

for all $j = 1, 2, \ldots, r$. For the remaining classical algebras the result is

$$\Lambda_{(j)} = \sum_{i=1}^{j} e_i \equiv (\underbrace{1, 1, \ldots, 1}_{j \text{ times}}, 0, 0, \ldots, 0) \qquad (7.24)$$

for all $j = 1, 2, \ldots, r$, except for the following special cases:

$$\Lambda_{(r)} = \tfrac{1}{2}(1, 1, \ldots, 1, 1, 1) \qquad \text{for } B_r \text{ and } D_r,$$
$$\Lambda_{(r-1)} = \tfrac{1}{2}(1, 1, \ldots, 1, 1, -1) \quad \text{for } D_r. \qquad (7.25)$$

From these formulæ one then deduces the following relations between the components

Table VII. *Roots in the orthonormal basis*

$\mathfrak{g}$	D	simple roots		positive roots	
A_r	r+1	$e_i - e_{i+1},$	$1 \le i \le r$	$e_i - e_j,$	$1 \le i < j \le r+1$
B_r	r	$e_i - e_{i+1},$ e_r	$1 \le i \le r-1;$	$e_i \pm e_j,$ $e_i,$	$1 \le i < j \le r;$ $1 \le i \le r$
C_r	r	$e_i - e_{i+1},$ $2e_r$	$1 \le i \le r-1;$	$e_i \pm e_j,$ $2e_i,$	$1 \le i < j \le r;$ $1 \le i \le r$
D_r	r	$e_i - e_{i+1},$ $e_{r-1} + e_r$	$1 \le i \le r-1;$	$e_i \pm e_j,$	$1 \le i < j \le r$
E_6	8	$e_{i+1} - e_i,$ $\frac{1}{2}\left(e_8 + e_1 - \sum\limits_{j=2}^{7} e_j\right),$ $e_1 + e_2$	$1 \le i \le 4;$	$e_i \pm e_j,$ $\frac{1}{2}\left(e_8 - e_7 - e_6 + \sum\limits_{j=1}^{5}(\pm)e_j\right),$ even number of minus signs	$1 \le j < i \le 5;$
E_7	8	$e_{i+1} - e_i,$ $\frac{1}{2}\left(e_8 + e_1 - \sum\limits_{j=2}^{7} e_j\right),$ $e_1 + e_2$	$1 \le i \le 5;$	$e_i \pm e_j,$ $\frac{1}{2}\left(e_8 - e_7 + \sum\limits_{j=1}^{6}(\pm)e_j\right),$ odd number of minus signs; $e_8 - e_7$	$1 \le j < i \le 6;$
E_8	8	$e_{i+1} - e_i,$ $\frac{1}{2}\left(e_8 + e_1 - \sum\limits_{j=2}^{7} e_j\right),$ $e_1 + e_2$	$1 \le i \le 6;$	$e_i \pm e_j,$ $\frac{1}{2}\left(e_8 + \sum\limits_{j=1}^{7}(\pm)e_j\right),$ even number of minus signs	$1 \le j < i \le 8;$
F_4	4	$e_2 - e_3,\ e_3 - e_4,$ $e_4,$ $\frac{1}{2}(e_1 - e_2 - e_3 - e_4)$		$e_i \pm e_j,$ $e_i,$ $\frac{1}{2}(e_1 \pm e_2 \pm e_3 \pm e_4)$	$1 \le i < j \le 4;$ $1 \le i \le 4;$
G_2	3	$e_1 - e_2,$ $e_2 + e_3 - 2e_1$		$e_1 - e_2,\ e_3 - e_1,\ e_3 - e_2,$ $e_2 + e_3 - 2e_1,\ e_1 + e_3 - 2e_2,$ $-e_1 - e_2 + 2e_3$	

λ^i of a weight λ in the Dynkin basis and its components ℓ^i in the orthogonal basis:

$$A_r: \quad \ell^j(\lambda) = \sum_{i=j}^{r} \lambda^i - \frac{1}{r+1} \sum_{i=1}^{r} i\,\lambda^i,$$

$$B_r: \quad \ell^j(\lambda) = \sum_{i=j}^{r-1} \lambda^i + \frac{1}{2}\lambda^r, \qquad\qquad C_r: \quad \ell^j(\lambda) = \sum_{i=j}^{r} \lambda^i, \qquad (7.26)$$

$$D_r: \quad \ell^j(\lambda) = \begin{cases} \sum_{i=j}^{r-2} \lambda^i + \frac{1}{2}\left(\lambda^{r-1} + \lambda^r\right) & \text{for } i = 1, 2, \dots, r-1, \\ \frac{1}{2}\left(\lambda^{r-1} - \lambda^r\right) & \text{for } i = r. \end{cases}$$

The inverse transformation is given by

$$\lambda^j = \ell^j(\lambda) - \ell^{j+1}(\lambda) \tag{7.27}$$

for all classical Lie algebras and all $i = 1, 2, \ldots, r$, except for

$$\lambda^r = 2\,\ell^r(\lambda) \qquad \text{for } \mathfrak{g} = B_r \,,$$
$$\lambda^r = \ell^r(\lambda) \qquad \text{for } \mathfrak{g} = C_r \,, \tag{7.28}$$
$$\lambda^{r-1} = \ell^{r-1} + \ell^r \,, \quad \lambda^r = \ell^{r-1} - \ell^r \quad \text{for } \mathfrak{g} = D_r \,.$$

In particular, the components of the Weyl vector ρ in the orthogonal basis read

$$\ell^j(\rho) = \tilde{\ell}(\mathfrak{g}) - j \qquad \text{with}$$
$$\tilde{\ell}(A_r) = \tfrac{r}{2} + 1 \,, \quad \tilde{\ell}(B_r) = r + \tfrac{1}{2} \,, \quad \tilde{\ell}(C_r) = r + 1 \,, \quad \tilde{\ell}(D_r) = r \,. \tag{7.29}$$

7.8 Kac–Moody algebras

We now show how the results of the previous sections generalize to a larger class of Lie algebras. Those readers who are mainly interested in finite-dimensional (semi-)simple Lie algebras might wish to skip the next three sections in a first reading.

As described above, the simple Lie algebras are obtained by requiring that the $r \times r$-matrix A appearing in (7.2) and (7.3) is an (irreducible) Cartan matrix, i.e. is indecomposable in the sense of (7.5) and obeys $A^{ii} = 2$, $A^{ij} = 0 \Leftrightarrow A^{ji} = 0$, $A^{ij} \in \mathbb{Z}_{\leq 0}$, and $\det A > 0$. In particular the rank of A is equal to r. The *Kac–Moody* Lie algebras are obtained by requiring again that they are algebraically generated by Chevalley generators $E_\pm^i$, H^i modulo the relations (7.1), (7.2) and (7.3), but by also relaxing the conditions on the matrix A, which is then called a *generalized* Cartan matrix. In addition, one must usually enlarge the Cartan subalgebra by including so-called derivations, as will be described in the affine case in chapter 12.

Let us remark that when one weakens the axioms of some given mathematical structure, one destroys most probably much of the power and beauty of the theory. In the present situation, however, the new theory is in fact equally interesting as the original one, provided that one only relaxes the condition $\det A > 0$ (7.6). When one removes this condition completely, one obtains the general class of Kac–Moody algebras. By analysing the root system of these algebras one can show that all Kac–Moody algebras, except for those already listed in table IV, are infinite-dimensional (for the case of affine Kac–Moody algebras, see sections 12.7 and 12.9). The class of general Kac–Moody algebras is in fact extremely big. To obtain a nice theory, it is necessary to restrict oneself to a (still very comprehensive) subclass, the so-called *symmetrizable*

Kac–Moody algebras. A Kac–Moody algebra is called symmetrizable if there exists a non-degenerate diagonal matrix D such that the matrix DA, with A the generalized Cartan matrix, is symmetric. Clearly, any finite-dimensional simple Lie algebra is symmetrizable; the entries of the matrix D are $D_{ij} = \delta_{ij} D_i$ with $D_i := a_i / a_i^\vee$. From equation (7.17) it is already apparent that the existence of the matrix D is intimately linked to the existence of a norm on weight space. In fact, for any symmetrizable Kac–Moody algebra there exists a bilinear symmetric invariant form. One can show that a symmetrizable Kac–Moody algebra is simple, i.e. does not possess any non-trivial ideal, if an only if its Cartan matrix is indecomposable and has non-zero determinant (see exercise 7.7).

The most important subclass of symmetrizable Kac–Moody algebras is obtained if (7.6) is replaced by the requirement

$$\det A_{\{i\}} > 0 \quad \text{for all } i = 0, 1, \ldots, r. \tag{7.30}$$

Here $A_{\{i\}}$ denote those matrices which are obtained from A by deleting the ith row and ith column (the determinants $\det A_{\{i\}}$ are called the principal minors of A). In (7.30) we also changed the labelling convention for the rows and columns of the Cartan matrix: We allow for $i \in \{0, 1, \ldots, r\}$, so that A is now a $(r+1) \times (r+1)$-matrix.

For general Kac–Moody algebras the rank of A can be any integer between 0 and $r+1$, while the validity of (7.30) implies that the rank is at least r. A matrix obeying the inequalities (7.30) is called degenerate positive semidefinite, while a matrix for which the requirement on the determinant in (7.7) is completely dropped is called a *generalized Cartan matrix*. A generalized irreducible Cartan matrix which is degenerate positive semidefinite is called an (irreducible) *affine Cartan matrix*. The Lie algebras defined by generators and relations as in (7.1), (7.2) and (7.3), with A an affine Cartan matrix, are called *affine Lie algebras*. Instead of requiring (7.30), one can characterize an affine Cartan matrix also by the requirement that there exists an invertible diagonal matrix D such that DA is symmetric and positive semidefinite, but not positive definite. (If DA is positive definite, then A is the Cartan matrix of a finite-dimensional simple Lie algebra.) This is so because this requirement already implies that DA can have at most one zero eigenvalue.

The relaxed requirement (7.30) proves to be almost as restrictive as the original condition $\det A > 0$. In particular, as we will see shortly, all affine Cartan matrices can be classified. The classification is also known for the *hyperbolic* Kac–Moody algebras. These correspond to those generalized Cartan matrices A for which there exists a non-singular diagonal matrix D such that DA is symmetric and of so-called indefinite type, i.e. which is not positive semi-definite, and for which any proper connected subdiagram is the one of a finite-dimensional simple or of an affine Lie algebra. It can be shown that these Cartan matrices have signature $(r, 1)$, i.e. they possess precisely one negative eigenvalue in addition to r positive eigenvalues. There are no hyperbolic algebras of rank larger than 10, and the number of hyperbolic algebras with rank larger than 2 is finite. In contrast, for Kac–Moody algebras which are not simple, affine, or hyperbolic, one is still far from having completed the classification.

7.9 Affine Lie algebras

The rank of the affine $(r + 1) \times (r + 1)$ Cartan matrices is

$$\operatorname{rank} A = r. \tag{7.31}$$

Once the classification of simple Lie algebras up to some given rank has been obtained, the classification of affine Lie algebras with that rank is straightforward. For $r = 1$ one only has to determine the possible off-diagonal entries of A for which $\det A \equiv A^{11}A^{22} - A^{21}A^{12} = 4 - A^{21}A^{12}$ is zero. Just like requiring $\det A > 0$ immediately leads to the three simple rank-two Cartan matrices

$$A(A_2) = \begin{pmatrix} 2 & -1 \\ -1 & 2 \end{pmatrix}, \quad A(B_2) = \begin{pmatrix} 2 & -2 \\ -1 & 2 \end{pmatrix}, \quad A(G_2) = \begin{pmatrix} 2 & -3 \\ -1 & 2 \end{pmatrix},$$

the requirement $\det A = 0$ yields the Cartan matrices of two rank-one affine algebras, which are denoted by $A_1^{(1)}$ and $A_1^{(2)}$:

$$A(A_1^{(1)}) = \begin{pmatrix} 2 & -2 \\ -2 & 2 \end{pmatrix}, \qquad A(A_1^{(2)}) = \begin{pmatrix} 2 & -4 \\ -1 & 2 \end{pmatrix}. \tag{7.32}$$

For $r > 1$, we can use as a starting point the observation that by deleting the ith row and ith column ($i \in \{0, 1, \dots, r\}$ arbitrary) from an affine Cartan matrix, one must produce the Cartan matrix of a finite-dimensional semisimple Lie algebra. Furthermore, inspection of table IV shows that when deleting further rows and columns one stays within this class of Lie algebras. Hence in particular any 2×2-matrix obtained from A by deleting $r - 1$ rows and the corresponding columns must be a rank-two Cartan matrix, i.e. be one of the Cartan matrices of A_2, B_2 or G_2 just listed. Therefore for $i \neq j$, one has either $A^{ij} = A^{ji} = 0$, or else

$$\min\{|A^{ij}|, |A^{ji}|\} = 1, \qquad \max\{|A^{ij}|, |A^{ji}|\} \leq 3. \tag{7.33}$$

In particular, these submatrices must obey $A^{ij}A^{ji} \leq 3$; thus the rank-1 algebras $A_1^{(1)}$ and $A_1^{(2)}$ which have $A^{12}A^{21} = 4$ are somewhat exceptional. Implementing the constraints listed above, it becomes a matter of straightforward combinatorics to enumerate the affine Cartan matrices for any given value of r (for an example of the combinatorics involved see exercise 7.2). The results are listed in table VIII, which provides the names of the algebras and their Dynkin diagrams. The numbers which are adjoined to the nodes are the Coxeter and dual Coxeter labels; their precise definition will be presented below.

The nomenclature chosen for these algebras requires some explanation. For the algebras in table VIII A, a unique nomenclature is accepted in

Table VIII. *Dynkin diagrams of the affine Lie algebras*

the literature; we will explain its origin below. In contrast, for the affine Lie algebras in table VIII B, several different conventions can be found in the literature. We will use the one displayed in the first column of table B, which is also adopted in [Fuchs 1992] (for a similar notation, see [Kass *et al.* 1990]); the one in the second column is used in [Kac 1990]. We also warn the reader that sometimes different conventions for drawing the Dynkin diagram are used. For example, as already mentioned in section 7.3, the distinction between nodes for long and short simple roots is often not indicated by arrows, but by some convention for coloring the nodes. Also, sometimes the Dynkin diagram for $A_1^{(1)}$ is depicted as ⬤⇉⬤ .

According to table VIII there are seven infinite series

$$A_r^{(1)}\,(r \geq 2)\,, \quad B_r^{(1)}\,(r \geq 3)\,, \quad C_r^{(1)}\,(r \geq 2)\,, \quad D_r^{(1)}\,(r \geq 4)\,,$$
$$B_r^{(2)}\,(r \geq 3)\,, \quad C_r^{(2)}\,(r \geq 2)\,, \quad \tilde{B}_r^{(2)}(r \geq 2) \tag{7.34}$$

of affine Lie algebras, and in addition nine exceptional affine algebras, which are denoted by

$$A_1^{(1)}\,, \quad E_6^{(1)}\,, \quad E_7^{(1)}\,, \quad E_8^{(1)}\,, \quad F_4^{(1)}\,, \quad G_2^{(1)}\,, \quad A_1^{(2)}\,, \quad F_4^{(2)}\,, \quad G_2^{(3)}\,. \tag{7.35}$$

Also, for many purposes, it is natural to regard $A_1^{(1)}$ as the first element of the series $A_r^{(1)}$, and $A_1^{(2)}$ as the first element of the series $\tilde{B}_r^{(2)}$.

The affine algebras are thus all denoted by symbols $X_r^{(\ell)}$, where X_r is the symbol for a simple Lie algebra and $\ell \in \{1, 2, 3\}$. The algebras with $\ell = 1$ whose Dynkin diagrams are listed in table VIII A are called *untwisted* affine algebras, while those with $\ell = 2, 3$, whose Dynkin diagrams appear in table VIII B, are the *twisted* affine algebras. For the time being, the distinction between these two types of algebras is only a matter of nomenclature. The reasons for this choice of terminology will become clear in chapter 12 where the affine Lie algebras are constructed by starting from certain underlying simple Lie algebras.

As an illustration, let us write down the bracket relations (7.1), (7.2) and (7.3) of the Chevalley–Serre basis explicitly for the affine Lie algebra $A_1^{(1)}$. One has

$$[H^0, H^1] = 0\,, \qquad [E_+^0, E_-^0] = H^0\,, \qquad [E_+^1, E_-^1] = H^1\,,$$
$$[H^0, E_\pm^0] = \pm 2\, E_\pm^0\,, \quad [H^1, E_\pm^1] = \pm 2\, E_\pm^1\,, \tag{7.36}$$
$$[H^0, E_\pm^1] = \mp 2\, E_\pm^1\,, \quad [H^1, E_\pm^0] = \mp 2\, E_\pm^0\,, \quad [E_\pm^0, E_\mp^1] = 0\,,$$

while the Serre relations are given by

$$0 = (\mathrm{ad}_{E_\pm^1})^3 E_\pm^0 = [E_\pm^1, [E_\pm^1, [E_\pm^1, E_\pm^0]]] \tag{7.37}$$

and by the same equation with the labels 0 and 1 interchanged.

From (7.37) we learn the following. Denoting, with foresight, the simple roots by $\alpha^{(1)} = \bar{\alpha}$ and $\alpha^{(0)} = \delta - \bar{\alpha}$, the vanishing of the bracket $[E_\pm^0, E_\mp^1] \equiv [E^{\pm(\delta-\bar{\alpha})}, E^{\mp\bar{\alpha}}]$

tells us that $\delta - 2\bar{\alpha}$ is not a root. Similarly, from the first of the Serre relations we learn that δ and $\delta + \bar{\alpha}$ are roots, but $\delta + 2\bar{\alpha}$ is not a root, and from the second that δ and $2\delta - \bar{\alpha}$ are roots, but not $3\delta - 2\bar{\alpha}$. Next one can verify that $[E_{\pm}^1, [E_{\pm}^1, [E_{\pm}^0, E_{\pm}^1]]]$ and $[E_{\pm}^1, [E_{\pm}^1, [E_{\pm}^0, [E_{\pm}^0, E_{\pm}^1]]]]$ are not constrained to vanish so that 2δ and $2\delta + \bar{\alpha}$ are roots, but, using again the Serre relations, not $3\delta + 2\bar{\alpha}$. Iterating this procedure, one obtains the characterization

$$\Phi(A_1^{(1)}) = \{\pm\bar{\alpha} + n\delta \mid n \in \mathbb{Z}\} \cup \{n\delta \mid n \in \mathbb{Z}\setminus\{0\}\} \qquad (7.38)$$

of the root system of $A_1^{(1)}$.

Kac–Moody algebras arise e.g. in dimensional reduction of gravity and supergravity theories. As was first observed by Geroch and Julia, there is an infinite-dimensional symmetry which transforms different solutions of the (vacuum) Einstein equations into each other, and the corresponding Lie algebra of infinitesimal symmetries is the affine Kac–Moody algebra $A_1^{(1)}$. The algebra $A_1^{(1)}$ has a central element, which in this context acts as a scaling operator on the conformal factor of the space-time metric. The presence of this symmetry can be understood in terms of a dimensional reduction of the gravity theory from four to two dimensions. When matter is included into the theory or when it is extended to supergravity, dimensional reduction again generically leads to infinite-dimensional symmetries. In particular, reduction of supergravity from four to one dimension gives rise to a hyperbolic Kac–Moody algebra. For more information, see [Nicolai 1991, 1992].

<u>Information</u>

Affine and hyperbolic Kac–Moody algebras also play the rôle of dynamical symmetries in string theory [Gebert and Nicolai 1995]. More specifically, affine Lie algebras can be realized in terms of tachyon and photon emission operators, while hyperbolic algebras correspond to higher-mass states of the string.

There is also a different way to look at Cartan matrices which we will sketch now. A generalized Cartan matrix is a matrix A with integral entries defined over some finite index set I, such that $A^{ii} \leq 2$ for all $i \in I$, $A^{ij} \leq 0$ for all $i, j \in I$ with $i \neq j$ and $A^{ij} = 0$ if and only if $A^{ji} = 0$. To any generalized Cartan matrix one can associate a graph in the same way as we associated a Dynkin diagram to an ordinary Cartan matrix, with the additional convention that there are $2 - A^{ii}$ loops beginning and ending at the ith node (thus an ordinary Cartan matrix leads to a graph without such loops). By an *additive function* for a generalized Cartan matrix one means an $|I|$-dimensional vector d with positive integral components such that $d^t A = 0$; similarly, if such a vector obeys the weaker condition $d^t A \geq 0$, then it is called a *subadditive* function. One can show that the only Cartan matrices which admit an additive function are the ones of affine Lie algebras, and that the only Cartan matrices which admit a subadditive function that is not additive are those of simple Lie algebras. If one allows for generalized Cartan matrices, one obtains four more infinite series that admit additive functions; these correspond to the following generalized

Dynkin diagrams:

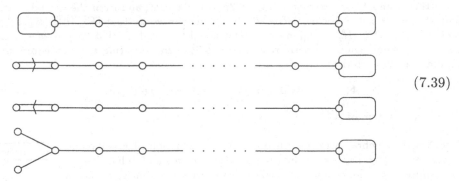

$$(7.39)$$

When considering subadditive functions that are not additive, one obtains one more infinite series, with diagrams

$$(7.40)$$

All additive functions of a given Cartan matrix are multiples of each other. In the case of affine Lie algebras the components of a normalized additive function are the so-called Coxeter labels to be discussed below. The connection to subadditive functions is one reason why affine Lie algebras make their appearance in many rather different contexts.

7.10 Coxeter labels and dual Coxeter labels

As in the case of the third column of table IV, the numbers attached to the vertices in table VIII are called the Coxeter labels a_i and dual Coxeter labels $a_i^\vee$ (if a_i differs from $a_i^\vee$, then a_i is listed in italics below $a_i^\vee$). These are defined by requiring that

$$\sum_{j=0}^{r} a_j A^{ji} = 0 = \sum_{j=0}^{r} A^{ij} a_j^\vee,$$

$$(7.41)$$

i.e. that $(a_i)_{i=0,\dots,r}$ and $(a_i^\vee)_{i=0,\dots,r}$ are, respectively, left and right eigenvectors with eigenvalue zero of the affine Cartan matrix, and in addition that the normalization condition

$$\min\{a_i \mid i = 0, 1, \dots, r\} = 1 = \min\{a_i^\vee \mid i = 0, 1, \dots, r\}$$

$$(7.42)$$

is obeyed. Note that with the basic properties of the Cartan matrix it follows immediately from the formula (7.41) that the dual Coxeter labels obey $a_i^\vee = \frac{1}{2} \sum_{j \neq i} |A^{ij}| a_j^\vee$.

According to the analysis of the requirement (7.30), by removing any node from the Dynkin diagram of an affine Lie algebra one gets the Dynkin diagram of some simple Lie algebra (this can be checked by inspecting table VIII). The notations for the affine algebras that were used in the lists (7.34) and (7.35) have been chosen such that for $\mathfrak{g} = X_r^{(\ell)}$, with X standing for one of the letters $A, B, \ldots, G$, there is one node in the Dynkin diagram of $\mathfrak{g}$ whose removal leads to the Dynkin diagram of the simple Lie algebra X_r, with X denoting the same letter as before. The numbering of the nodes of the affine Dynkin diagram is then the one inherited from the Dynkin diagram of $\bar{\mathfrak{g}} := X_r$, with the additional node (the removal of which reduces the Dynkin diagram of $\mathfrak{g}$ to that of $\bar{\mathfrak{g}}$) labelled by '0'. In table VIII we have distinguished this *zeroth node* in an affine Dynkin diagram by writing its (dual) Coxeter label in boldface; the table also shows that in all cases except $A_1^{(2)}$ and $\tilde{B}_r^{(2)}$ both the Coxeter label and the dual one are equal to one,

$$a_0 = 1 = a_0^\vee. \tag{7.43}$$

Inspecting also the other Coxeter labels of the Dynkin diagrams, one sees that removal of the zeroth node from the Dynkin diagram of an untwisted algebra produces both the Dynkin diagram of the subalgebra $\bar{\mathfrak{g}} = X_r$ and at the same time also the Coxeter and dual Coxeter labels of X_r that were defined in terms of the highest root of X_r in equation (7.16). In contrast, in the twisted case the Coxeter labels of X_r are not produced correctly (nor would they if any other node with $a_i = 1$ were removed from the affine Dynkin diagram).

The Dynkin diagram of $\bar{\mathfrak{g}}$ can be obtained from that of $\mathfrak{g}$ not only by removing the zeroth node, but also by removing instead any other node having both its Coxeter label and its dual Coxeter label equal to one. Moreover, all nodes of the Dynkin diagram of $\mathfrak{g}$ with these properties are related by a symmetry of the Dynkin diagram.

The sums of the Coxeter and dual Coxeter labels are known as the *Coxeter number* g and *dual Coxeter number* $g^\vee$ of the Lie algebra $\mathfrak{g}$, respectively:

$$g := \sum_{i=0}^{r} a_i, \qquad g^\vee := \sum_{i=0}^{r} a_i^\vee. \tag{7.44}$$

From the relation between the (dual) Coxeter labels of simple and affine Lie algebras described above, it follows immediately that both the Coxeter number and the dual Coxeter number of an untwisted affine algebra $\mathfrak{g}$ coincide with those of its simple subalgebra $\bar{\mathfrak{g}}$, which are defined by (7.18).

Information

If the dual Coxeter label has value $a_i^\vee = 1$, then the corresponding fundamental weight $\Lambda_{(i)}$ is called a *minimal* fundamental weight, and if the Coxeter label is $a_i = 1$, then $\Lambda_{(i)}$ is called *cominimal*. The index of connection $|L_w/L|$ of a simple Lie algebra $\mathfrak{g}$ is one plus the number of cominimal fundamental weights. The cominimal fundamental weights of a simple Lie algebra $\mathfrak{g}$ are in one-to-one correspondence to the non-trivial elements of the center of the universal covering group G whose Lie algebra is the compact real form of $\mathfrak{g}$.

In two-dimensional conformal field theory (compare section 12.12), any integrable weight corresponds to a so-called primary quantum field. If the weight is an integral multiple of a cominimal fundamental weights, then the primary field is called a simple current. Simple currents play an important rôle in conformal field theory, e.g. they are essential in the description of coset conformal field theories and the classification of modular invariant partition functions [Schellekens and Yankielowicz 1990].

Minimal fundamental weights appear naturally in the description of stable magnetic monopoles and of symmetry breaking in grand unified theories [Goddard and Olive 1981] and in relations among Toda field theories [Olive and Turok 1983].

Summary:

Finite-dimensional simple Lie algebras can be classified with the help of a Chevalley–Serre basis, in which the problem is reduced to the one of classifying all Cartan matrices. Cartan matrices and various other characteristic quantities of finite-dimensional simple Lie algebras can be pictorially represented by Dynkin diagrams.

The inverse of the symmetrized Cartan matrix, called quadratic form matrix, provides a scalar product on weight space. Other important data for finite-dimensional simple Lie algebras are the highest root, the Weyl vector and the (dual) Coxeter labels and numbers.

The generalization of the structure present in finite-dimensional simple Lie algebras leads to Kac–Moody algebras, in particular to twisted and untwisted affine Lie algebras and to hyperbolic Lie algebras.

Keywords:

Serre relations, Chevalley–Serre presentation, irreducible Cartan matrix, Dynkin diagram, simply laced Lie algebra, (dual) Coxeter labels and number;

Weyl vector, highest root, dual Lie algebra, quadratic form matrix; generalized Cartan matrix, (symmetrizable) Kac–Moody algebra, affine Cartan matrix, zeroth node, (un)twisted affine Lie algebra; hyperbolic Lie algebra.

Exercises:

Use the conditions (7.7) to derive the following facts about the Dynkin diagrams of finite-dimensional simple Lie algebras. (These are crucial steps in the classification of these algebras.)

Exercise 7.1

a) The Dynkin diagram of G_2 cannot be a subdiagram of any larger diagram.

b) A diagram can contain the Dynkin diagram of B_2 at most once as a subdiagram.

c) The allowed Dynkin diagrams have at most three end points.

Hint: For any diagram of forbidden type, construct a vector with negative length squared as a suitable linear combination of simple roots. For example, in the case of G_2 consider an additional node labelled '3' attached to the second node of the G_2 Dynkin diagram, and find an upper bound on the length squared (β, β) of $\beta = \sqrt{3}\alpha^{(1)} + 2\alpha^{(2)} + \alpha^{(3)}$.

As an illustration of the classification of affine Lie algebras, take $r = 2$ and consider the case where a rank two 2×2 submatrix of the Cartan matrix A is equal to the Cartan matrix of G_2. Thus, possibly up to a renumbering of rows and columns, A reads

Exercise 7.2

$$A = \begin{pmatrix} 2 & p & q \\ r & 2 & -3 \\ s & -1 & 2 \end{pmatrix}. \qquad (7.45)$$

Show that there are only two allowed combinations of the integers p, q, r, s for which the determinant of A is zero, and use the table VIII to identify the associated affine Lie algebras.

By computing the inner product of the simple roots given in table VII, check that they provide indeed a system of simple roots for the respective simple Lie algebras.

Exercise 7.3

a) Derive the recursion relation

Exercise 7.4

$$(e_{\alpha,\beta+(j-1)\alpha})^2 - (e_{\alpha,\beta+j\alpha})^2 = (\beta + j\alpha, \alpha^\vee) \qquad (7.46)$$

for the structure constants $e_{\alpha,\beta}$.

Hint: Evaluate the Jacobi identity for E^α, $E^{-\alpha}$ and $E^{\beta+j\alpha}$ and employ the relations (6.67).

b) Solve the recursion relation with the initial condition $e_{\alpha,\beta+n_+\alpha} = 0$. Deduce that $-n_+ \leq (\beta, \alpha^\vee) \leq n_-$ from the fact that $\mathfrak{g}$ is finite-dimensional.

c) For $\mathfrak{g}$ simply laced, use the identity (7.46) to show that the structure constant $e_{\alpha,\beta+(n_+-1)\alpha}$ can be an integer only if $n_+ = 1$.

Show that the roots in the orthonormal basis of a classical simple Lie algebra $\mathfrak{g}$ as given in table VII correspond to the Cartan-Weyl basis of $\mathfrak{g}$ that was defined in equations (5.31) and (5.32).

Exercise 7.5

Check that the Kac-Moody algebra associated to the Dynkin diagram

$$(7.47)$$

is hyperbolic.

Exercise 7.6

As a comparison of this Dynkin diagram to those of the E series shows, it is natural to denote this algebra by E_{10}. Likewise, one could write E_9 for $E_8^{(1)}$, and define E_r with r > 10 in the obvious way by joining consecutively further nodes to the rightmost one of the Dynkin diagram, and also $E_5 \equiv D_5$ and $E_4 \equiv A_4$.

Show that the determinant of the Cartan matrices for this extended E series is given by the formula $\det A(E_r) = 9 - r$. Conclude that E_{10} is the only hyperbolic algebra in the E series, and E_9 the only affine one.

Show that a Kac-Moody algebra is simple, i.e. does not contain any non-trivial ideal, if and only if its Cartan matrix A is indecomposable and satisfies $\det A \neq 0$. (This implies in particular that hyperbolic Kac-Moody algebras are simple.)

Exercise 7.7

Hint: Take the fact for granted that any ideal of a Kac-Moody algebra has non-zero intersection with the Cartan subalgebra. Also use that $\det A \neq 0$ implies that the center of the Kac-Moody algebra is trivial.

8

Real Lie algebras and real forms

8.1 More about the Killing form

In chapter 6 we have introduced a bilinear form, the Killing form κ, on a Lie algebra. The definition of κ through (6.10) involves taking the trace over adjoint maps ad_x. In order for κ to be well-defined one must therefore in fact assume that the adjoint representation of $\mathfrak{g}$, and hence $\mathfrak{g}$ itself, is finite-dimensional. In terms of the representation matrices $R_{\mathrm{ad}}(x)$ in the adjoint representation, the formula (6.10) for Killing form reads

$$\kappa(x, y) = \mathrm{tr}\left(R_{\mathrm{ad}}(x) R_{\mathrm{ad}}(y)\right), \tag{8.1}$$

with the symbol 'tr' now denoting the trace of matrices, which is well-defined for finite-dimensional matrices.

Several important properties of κ follow from the cyclicity property $\mathrm{tr}(MN) = \mathrm{tr}(NM)$ of the trace: It is symmetric in x and y, and the invariance property (6.11) holds, i.e.

$$\kappa(x, [y, z]) = \kappa([x, y], z). \tag{8.2}$$

If $\mathfrak{g}$ is a simple Lie algebra, κ is, up to normalization, the unique bilinear form with these properties.

To show this, we use a Chevalley basis as introduced in chapter 6. (For a shorter and more conceptual proof, see exercise 6.14). First we note that any form $\tilde{\kappa}$ with these properties obeys

$$0 = \tilde{\kappa}([H^i, H^j], E^\alpha) = \tilde{\kappa}(H^i, [H^j, E^\alpha]) = \alpha^j\, \tilde{\kappa}(H^i, E^\alpha). \tag{8.3}$$

Since the roots α are non-zero functions, one can choose $j \in \{1, 2, \dots, \mathrm{r}\}$ such that α^j does not vanish. Hence (8.3) states that $\tilde{\kappa}(H^i, E^\alpha) = 0$ for any generator H^i of the Cartan subalgebra and any root α. Next, suppose that α and β are two roots with

$\alpha \neq -\beta$. The calculation

$$0 = e_{\alpha,\beta}\,\tilde{\kappa}(H^i, E^{\alpha+\beta}) = \tilde{\kappa}(H^i, [E^\alpha, E^\beta]) = \tilde{\kappa}([H^i, E^\alpha], E^\beta) = \alpha^i\,\tilde{\kappa}(E^\alpha, E^\beta)$$

shows (again by choosing i suitably) that $\tilde{\kappa}(E^\alpha, E^\beta)$ is zero for any two roots, except possibly for $\alpha = -\beta$. Hence

$$\tilde{\kappa}(E^\alpha, E^\beta) = n_\alpha \delta_{\alpha+\beta,0}\,. \tag{8.4}$$

Next we show that the number n_α depends only on the length of α. To this end we calculate

$$n_\alpha\, e_{\beta,\gamma}\delta_{\alpha+\beta+\gamma,0} = \tilde{\kappa}(E^\alpha, [E^\beta, E^\gamma]) = \tilde{\kappa}([E^\alpha, E^\beta], E^\gamma) = n_\gamma\, e_{\alpha,\beta}\delta_{\alpha+\beta+\gamma,0}\,,$$

which when combined with the relation (6.67) for the structure constants $e_{\alpha,\beta}$ in the Chevalley basis shows that

$$n_{-\alpha-\beta}\, e_{\alpha,\beta} = n_\alpha\, e_{\beta,-\alpha-\beta} = n_\alpha\, e_{\alpha,\beta} \cdot \frac{(\alpha,\alpha)}{(\alpha+\beta,\alpha+\beta)}\,. \tag{8.5}$$

Now the root system of a simple Lie algebra is 'connected' in the sense that starting from any root one can obtain all others by moving along root strings. Therefore the result (8.5) in fact implies that all n_α are of the form $n_\alpha = c/(\alpha,\alpha)$, with c a fixed number, and hence by comparison with (6.62) we learn that on the subspace spanned by the step operators $E_\pm^\alpha$, $\tilde{\kappa}$ is a multiple of the Killing form. Writing the elements of the Cartan subalgebra as commutators of step operators and using the invariance (8.2) once again, we see that on $\mathfrak{g}_\circ$ (and hence on all of $\mathfrak{g}$) $\tilde{\kappa}$ must be a multiple of the Killing form as well.

The invariance (8.2) is in fact a key property of the Killing form. For certain infinite-dimensional Lie algebras (which only allow for infinite-dimensional non-trivial representations), e.g. affine Lie algebras, one cannot define a Killing form using traces over representation spaces. Nevertheless, in some cases there is still a unique invariant bilinear form. This form, which is again called the Killing form, can be used for the same purposes as in the simple case; for affine Lie algebras it will be described in chapter 12. However, we warn the reader that for an arbitrary Lie algebra $\mathfrak{g}$ such a form need not exist. In exercise 8.2 the reader can show that e.g. the so-called Virasoro algebra does not admit such a form.

Another important consequence of the uniqueness is that instead of the adjoint representation, for simple Lie algebras one can in fact use any other non-trivial finite-dimensional irreducible representation R to define the Killing form. Namely, if $\mathfrak{g}$ is simple, then the trace

$$\kappa_R(x,y) := \mathrm{tr}\,(R(x)\,R(y))\,, \tag{8.6}$$

in this irreducible representation satisfies

$$\kappa_R(x,y) \propto \kappa(x,y)\,, \tag{8.7}$$

with a constant of proportionality that depends on the chosen representation; the value I_R of this constant will be obtained later on, in equation (14.34). (For non-simple Lie algebras the relation (8.7) no longer holds.

For instance, the Killing form of $u(1)$ is zero, but nevertheless evaluation of the trace (8.6) in a representation of non-zero charge q yields the non-vanishing result q^2.)

As already mentioned in chapter 6, finite-dimensional semisimple and solvable Lie algebras are conveniently characterized by the (non-) degeneracy properties of the Killing form (the Cartan criteria). A few more properties of κ are the following. The ideals in a direct sum $\mathfrak{g} \oplus \mathfrak{h}$ of Lie algebras are mutually orthogonal with respect to the Killing form, i.e. $\kappa(x, y) = 0$ for all $x \in \mathfrak{g}$, $y \in \mathfrak{h}$. As a consequence, the Killing form of a semisimple Lie algebra is already determined by the Killing forms of its simple ideals. The radical of $\mathfrak{g}$ is the orthogonal complement with respect to κ of the derived algebra $\mathfrak{g}' = [\mathfrak{g}, \mathfrak{g}]$. Furthermore, if $\mathfrak{g}$ is nilpotent, then $\kappa \equiv 0$.

8.2 The Killing form of real Lie algebras

Since in the semisimple case the Killing form is non-degenerate, it possesses an inverse, and one can use it as a metric on weight space. Given this 'metric', it is tempting to ask what its signature is, i.e. what the signs of its eigenvalues are. Now κ is bilinear (rather than a hermitian form; hermitian forms are antilinear in one and linear in the other argument), so that by replacing the arguments x, y of κ by ix, iy one changes the sign of the Killing form, $\kappa(ix, iy) = i^2 \kappa(x, y) = -\kappa(x, y)$. Hence for complex Lie algebras this question doesn't make sense.

In contrast, for Lie algebras over the *real* numbers $\mathbb{R}$ the question does make sense, and actually turns out to be quite interesting. Since also for any simple Lie algebra over the real numbers $\mathbb{R}$, κ is non-degenerate, there exists an orthogonal basis, i.e. by a suitable choice of basis one can bring κ^{ab} to the canonical diagonal form

$$\kappa = \begin{pmatrix} \mathbb{1}_p & 0 \\ 0 & -\mathbb{1}_{d-p} \end{pmatrix}. \tag{8.8}$$

Here $\mathbb{1}_p$ is the $p \times p$ unit matrix, and 0 stands for matrices of the appropriate size which have all entries equal to zero.

As metrics, the Killing form and its inverse can be used to raise and lower indices, respectively. For example, structure constants with only upper indices are defined as

$$f^{abc} := \sum_{e=1}^{d} f^{ab}{}_e \, \kappa^{ec}. \tag{8.9}$$

One can show (see exercise 8.3) that in a basis in which the Killing form is of the form (8.8), these structure constants are completely antisymmetric in their three upper indices. Such a basis is sometimes useful in explicit calculations. For example, in the case of the complex Lie algebra $\mathfrak{sl}(2, \mathbb{C})$ for some purposes it can be preferable to work in the basis spanned by the L^a, $a = 1, 2, 3$, in which the structure constants are $f^{abc} = \epsilon^{abc}$ (cf. equation (2.1)), instead of using L_0 and $L_\pm$. For the real Lie algebra $\mathfrak{su}(2)$ $\{iL^a\}$ is anyway the natural choice of basis, while for $\mathfrak{sl}(2, \mathbb{R})$ besides L_0 one must take the linear combinations $L_+ \pm L_-$ as generators, which then give rise to one positive and one negative diagonal entry for κ.

If $\mathfrak{g}$ is a finite-dimensional simple Lie algebra over $\mathbb{C}$, one can ask what the real Lie algebras are such that $\mathfrak{g}$ is the 'complexification' of it. We will see below that one can associate with $\mathfrak{g}$ several distinct Lie algebras over $\mathbb{R}$, called *real forms* of $\mathfrak{g}$, and that there is always a unique (up to isomorphism) real Lie algebra (compare table (8.14)) called the *compact real form* of $\mathfrak{g}$, for which $p = 0$ in the formula (8.8), so that in an appropriate basis the Killing form is negative definite:

$$\kappa^{ab} = -\delta^{ab} \,. \tag{8.10}$$

It turns out that for an arbitrary real form $\mathfrak{h}$ the $(\dim \mathfrak{h} - p)$-dimensional subspace of $\mathfrak{h}$ on which $\kappa < 0$ is a sub*algebra* of $\mathfrak{h}$. This subalgebra is called the maximal compact subalgebra of $\mathfrak{h}$. On the other hand, the p-dimensional subspace on which $\kappa > 0$ is not a subalgebra.

In Lagrangian quantum field theory, Goldstone bosons arising from spontaneous symmetry breaking can be described as scalar quantum fields of a non-linear sigma model. In supersymmetric non-linear sigma models they take their values in a coset space G/H, where G is a non-compact Lie group and H its maximal compact subgroup. These models possess a hidden gauge symmetry whose structure group is H [Buchmüller and Lerche 1987].

Information

A very similar coset structure occurs in extended supergravity theories with at least five supersymmetries. The largest on-shell invariances of these theories are of the form G×H, with H the largest compact subgroup of the non-compact group G, and again H plays the rôle of the structure group of a gauge symmetry [de Wit and Nicolai 1984].

8.3 The compact and the normal real form

In the classification of simple Lie algebras in chapter 7 we have assumed that the base field of $\mathfrak{g}$ is $\mathbb{C}$. The fact that the field $\mathbb{C}$ of complex numbers is algebraically closed simplified the analysis considerably. (Recall that a field is called algebraically closed if any polynomial over the field has at least one zero.) In contrast, the field $\mathbb{R}$ of real numbers is not algebraically closed, so that in particular the eigenvalue equation (6.7) determining the roots in general need not have a solution within $\mathbb{R}$. (If the roots of the characteristic polynomial of each $x \in \mathfrak{g}$ all lie in the base field, then $\mathfrak{g}$ is called a *split* Lie algebra.)

Accordingly, the theory of real Lie algebras is more involved. For instance, it is not in general true that all Cartan subalgebras are conjugate. In fact, also rather little is known about the representation theory of real Lie algebras which are not compact forms. (One notable exception is the Lorentz algebra which is a non-compact form of $\mathfrak{so}(4)$; we will treat this algebra and some aspects of its representation theory in more detail in chapter 20.)

Fortunately there are intimate connections between a complex semisimple Lie algebra and its various real forms. First, as we will learn in section 11.9, the real forms can all be constructed with the help of the involutive automorphisms of $\mathfrak{g}$. Having established the classification of semisimple Lie algebras over $\mathbb{C}$, one therefore obtains the classification of the semisimple Lie algebras over $\mathbb{R}$ by classifying the involutive automorphisms of the complex Lie algebras.

Second, analogously as for vector spaces, to any complex Lie algebra $\mathfrak{g}$ one can associate a real Lie algebra, denoted by $\mathfrak{g}_{\mathbb{R}}$, by splitting each element of the base field $\mathbb{C}$ of $\mathfrak{g}$ into its real and imaginary part; the real dimension of $\mathfrak{g}_{\mathbb{R}}$ is thus twice the complex dimension of $\mathfrak{g}$. Conversely, again as in the case of vector spaces, one defines the *complexification* $\mathfrak{h}_{\mathbb{C}}$ of a real Lie algebra $\mathfrak{h}$ as

$$\mathfrak{h}_{\mathbb{C}} := \mathfrak{h} \otimes_{\mathbb{R}} \mathbb{C} \equiv \mathfrak{h} \oplus i\,\mathfrak{h} \qquad (8.11)$$

(direct sum of vector spaces), with the Lie bracket of $\mathfrak{h}$ extended by linearity with respect to $\mathbb{C}$. The real dimension of $\mathfrak{h}_{\mathbb{C}}$ is twice the real dimension of $\mathfrak{h}$; as a real vector space the complexification is isomorphic to $\mathfrak{h} \oplus \mathfrak{h}$. A real Lie algebra $\mathfrak{h}$ is semisimple if and only if its complexification is semisimple (because, relative to a fixed basis of $\mathfrak{h}$, the matrix for the Killing form is the same for $\mathfrak{h}_{\mathbb{C}}$ as for $\mathfrak{h}$), while if $\mathfrak{h}$ is simple then $\mathfrak{h}_{\mathbb{C}}$ is either simple or is the direct sum of two isomorphic simple ideals.

The real forms of a complex simple Lie algebra $\mathfrak{g}$ are those real subalgebras $\mathfrak{h}$ of $\mathfrak{g}_{\mathbb{R}}$ whose complexification $\mathfrak{h}_{\mathbb{C}}$ is isomorphic to $\mathfrak{g}$. (In other words, each $z \in \mathfrak{g}$ can be written uniquely as $z = x + iy$ with $x, y \in \mathfrak{h}$.) Several methods are available to construct such subalgebras $\mathfrak{h}$. One method uses the quaternion field.

The field $\mathbb{H}$ of *quaternions* is defined on a real vector space of dimension four as follows. A basis of $\mathbb{H}$ is provided by the multiplicative unit $\mathbf{1}$ and three 'imaginary units' τ_j, $j = 1, 2, 3$. Thus when writing $\mathbf{1} =: \tau_0$, the quaternions $\mathbb{H}$ are the set of all real linear combinations $\sum_{j=0}^{3} \xi^j \tau_j$. The product on $\mathbb{H}$ is defined on the basis by $(\tau_j)^2 = -\mathbf{1}$ as well as $\tau_j \tau_k = \sum_l \epsilon_{jkl} \tau_l$ for $j \neq k$, and is extended linearly (over $\mathbb{R}$) to the rest of $\mathbb{H}$.

The quaternions are a non-commutative field which contains $\mathbb{R}$ as a subfield; $\mathbb{H}$ also contains several subfields isomorphic to the complex numbers (see exercise 8.4). The formula

$$\Big(\sum_{k=0}^{3} \xi^k \tau_k\Big)^* := \xi^0 \tau_0 - \sum_{i=1}^{3} \xi^i \tau_i \,, \tag{8.12}$$

defines a conjugation on the quaternions, which gives rise to a norm $|\xi|^2 := \xi^* \xi$ on $\mathbb{H}$.

A convenient realization of the quaternions is obtained in terms of 2×2-matrices, namely by setting $\tau_i := \mathrm{i}\sigma_i$ for $i = 1, 2, 3$, where σ_i are the Pauli matrices, as well as $\tau_0 := \mathbf{1}_{2\times2}$.

To construct real forms, one may restrict the field over which $\mathfrak{g}$ is defined to a subfield (typically restrict $\mathbb{C}$ to $\mathbb{R}$) or embed $\mathbb{C}$ into the quaternions $\mathbb{H}$. Furthermore, for matrix Lie algebras one may decompose the elements of the algebra into block matrices and multiply the off-diagonal blocks by the imaginary unit i. For the classical Lie algebras, these three methods are sufficient to obtain all real forms.

A simple Lie algebra possesses several non-isomorphic real forms. We have already encountered one example in chapter 2: The complex Lie algebra $\mathfrak{sl}(2, \mathbb{C})$ has two real forms, the compact $\mathfrak{su}(2)$, and non-compact $\mathfrak{sl}(2, \mathbb{R})$, with generators (2.3) and (2.5), respectively. As real Lie algebras, these are not isomorphic because their Killing forms have different signatures. Also note that, although $\mathfrak{su}(2)$ is a real Lie algebra, the entries of the elements of its 2×2-matrix realization are generically complex.

There are two standard real forms that can be constructed for any complex simple Lie algebra. Their construction is based on the observation that in a Cartan–Weyl basis all structure constants are real. As a consequence, the *real* vector space spanned by all real linear combinations of the form $\sum_i \xi_i H^i + \sum_{\alpha \in \Phi} \xi_\alpha E^\alpha$ is a real Lie algebra. The Killing form of this algebra is of block diagonal form; according to (6.62), on the subspace spanned by E^α and $E^{-\alpha}$ it is given by

$$\frac{2}{(\alpha,\alpha)} \begin{pmatrix} 0 & 1 \\ 1 & 0 \end{pmatrix}. \tag{8.13}$$

This real Lie algebra is called the *normal* or *split* real form; it is the 'least compact' real form. Note that when one takes the bases of section 5.6 for the complex classical Lie algebras and restricts the base field from $\mathbb{C}$ to $\mathbb{R}$, then one obtains the normal real form (see also the list (8.14) below).

Starting from the normal form, one can perform a basis transformation with *complex* coefficients, analogously to the transition from $\mathfrak{sl}(2, \mathbb{R})$ to $\mathfrak{su}(2)$, so as to arrive at the basis $\{\mathrm{i}H^i\}$ for the Cartan subalgebra and $\{(\mathrm{i}\sqrt{(\alpha,\alpha)}/2)\cdot(E^\alpha + E^{-\alpha})\} \cup \{(\sqrt{(\alpha,\alpha)}/2)\cdot(E^\alpha - E^{-\alpha})\}$ for its orthogonal complement. It is straightforward to check (see exercise 8.5) that the span of these elements over the *real* numbers is again a subalgebra $\mathfrak{h}$ of $\mathfrak{g}$, and that its Killing form is $\kappa^{ab} = -\delta^{ab}$; thus $\mathfrak{h}$ is in fact the *compact* real form of $\mathfrak{g}$. (That the Killing form is *minus* the Kronecker symbol is a consequence of the factors of the imaginary unit i which are needed for the compact real form.) All other real forms can be obtained from the compact one by multiplying suitable generators by a factor of the imaginary unit i.

Let us also note that by construction the normal real form has a root space decomposition completely analogous to the complex Lie algebra. In contrast, for the compact real form $\mathfrak{h}$ all the roots with respect to the Cartan subalgebra $\mathfrak{h}_o$ are purely imaginary, so that one rather has a decomposition $\mathfrak{h} = \mathfrak{h}_o \oplus \bigoplus_\alpha \mathfrak{h}_{(\alpha)}$, where $\mathfrak{h}_{(\alpha)}$ are two-dimensional subspaces on which $\mathfrak{h}_o$ acts by rotation.

8.4 The real forms of simple Lie algebras

Let us now list the real forms of the classical simple Lie algebras.
The real forms of $A_{n-1} = \mathfrak{sl}(n) \equiv \mathfrak{sl}(n, \mathbb{C})$ are:

- $\mathfrak{sl}(n, \mathbb{R})$, which is obtained by subfield restriction;
- $\mathfrak{su}(n)$, the real Lie algebra of traceless anti-hermitian complex $n \times n$-matrices;
- $\mathfrak{su}(p, n - p)$ for $p = 1, 2, \dots, [\frac{n}{2}]$ (here the square bracket stands for the integer part of a rational number), obtained from $\mathfrak{su}(n)$ by analytic continuation;
- and, for even n, $\mathfrak{su}^*(n) := \mathfrak{sl}(\frac{n}{2} + 1, \mathbb{H})$, obtained by embedding $\mathbb{C}$ into the quaternions.

The real forms of $\mathfrak{so}(n, \mathbb{C})$, i.e. of $B_{(n-1)/2}$ (n odd) respectively $D_{n/2}$ (n even), are:

- $\mathfrak{so}(n)$, the algebra of antisymmetric real $n \times n$-matrices;
- $\mathfrak{so}(p, n - p)$ for $p = 1, 2, \dots, [\frac{n}{2}]$;
- and, for even n, $\mathfrak{so}^*(n)$, obtained by embedding $\mathbb{R}$ into $\mathbb{C}$ (this algebra consists of those elements x of $\mathfrak{so}(n, \mathbb{C})$ which satisfy $xJ + J\bar{x}^t = 0$, where J is the symplectic matrix (5.25)).

The real forms of $C_n = \mathfrak{sp}(n, \mathbb{C})$ are:

- $\mathfrak{sp}(n, 0)$, the algebra of those anti-hermitian complex $2n \times 2n$-matrices which lie in the intersection of $\mathfrak{sp}(n, \mathbb{C})$ and $\mathfrak{su}(2n)$
(this algebra is often denoted simply by $\mathfrak{sp}(n)$; we use a different notation because the latter might be mixed up with the complex Lie algebra $\mathfrak{sp}(n) \equiv \mathfrak{sp}(n, \mathbb{C})$);
- $\mathfrak{sp}(n, \mathbb{R})$, the algebra of real $2n \times 2n$-matrices that is obtained from $\mathfrak{sp}(n, \mathbb{C})$ by restricting $\mathbb{C}$ to $\mathbb{R}$;
- $\mathfrak{sp}(p, 2n - p)$ for $p = 1, 2, \dots, n$, obtained from $\mathfrak{sp}(n, \mathbb{R})$ by analytic continuation.

The compact and normal real forms of the classical Lie algebras are as given in the following list:

complex algebra	compact form	normal form	
$\mathfrak{sl}(n,\mathbb{C})$	$\mathfrak{su}(n)$	$\mathfrak{sl}(n,\mathbb{R})$	
$\mathfrak{so}(2\ell+1,\mathbb{C})$	$\mathfrak{so}(2\ell+1)$	$\mathfrak{so}(\ell+1,\ell)$	(8.14)
$\mathfrak{sp}(n,\mathbb{C})$	$\mathfrak{sp}(n,0)$	$\mathfrak{sp}(n,\mathbb{R})$	
$\mathfrak{so}(2\ell,\mathbb{C})$	$\mathfrak{so}(2\ell)$	$\mathfrak{so}(\ell,\ell)$	

The algebra $\mathfrak{su}(n)$ may be characterized by the requirement that its elements leave a positive definite symmetric bilinear form invariant, and for $\mathfrak{su}(p,q)$ an analogous statement holds with the bilinear form being non-positive definite, but rather of the signature $(\underbrace{++...+}_{p \text{ times}}\underbrace{--...-}_{q \text{ times}})$.

The algebras $\mathfrak{so}(n)$ and $\mathfrak{so}(p,q)$ are related in an analogous manner. The real forms $\mathfrak{su}^*(n)$ and $\mathfrak{so}^*(n)$ can be constructed as follows. Write the imaginary unit as the 2×2-matrix $\begin{pmatrix} 0 & 1 \\ -1 & 0 \end{pmatrix}$, i.e. represent any complex number $\xi+i\eta$ with $\xi,\eta\in\mathbb{R}$ as the 2×2-matrix $\begin{pmatrix} \xi & \eta \\ -\eta & \xi \end{pmatrix}$. Then the matrix realization of $\mathfrak{u}(n)$ becomes a realization through real $2n\times 2n$-matrices, denoted by $\mathfrak{ou}(2n)$; this is contained as a subalgebra in $\mathfrak{so}(2n)$. Analytically continuing $\mathfrak{so}(2n)$ in such a way that $\mathfrak{ou}(2n)$ is left unchanged while its orthogonal complement (with respect to the Killing form) is multiplied by a factor of i, one arrives at the real Lie algebra $\mathfrak{so}^*(2n)$. Similarly, by expressing quaternions through complex 2×2-matrices, one obtains from the matrix Lie algebra $\mathfrak{sp}(n,\mathbb{H})$ a realization in terms of complex $2n\times2n$-matrices, called $\mathfrak{usp}(2n)$, which is a subalgebra of $\mathfrak{su}(2n)$. By an analytic continuation of $\mathfrak{su}(2n)$ which leaves $\mathfrak{usp}(2n)$ unchanged one obtains $\mathfrak{su}^*(2n)$.

The real forms of the exceptional simple Lie algebras are less easy to describe. For completeness we list them, too, in the notation $X_{r|\delta}$, where δ is the difference between the numbers of positive and negative eigenvalues of the Killing form:

$$E_{6|-78}\,,\quad E_{6|-26}\,,\quad E_{6|-14}\,,\quad E_{6|2}\,,\quad E_{6|6}\,;$$

$$E_{7|-133}\,,\quad E_{7|-25}\,,\quad E_{7|-5}\,,\quad E_{7|7}\,;$$

$$E_{8|-248}\,,\quad E_{8|-24}\,,\quad E_{8|8}\,;$$ (8.15)

$$F_{4|-52}\,,\quad F_{4|-20}\,,\quad F_{4|4}\,;$$

$$G_{2|-14}\,,\quad G_{2|2}\,.$$

The first algebra $\mathfrak{h}$ in each line has $\delta=-\dim\mathfrak{h}$ and hence is the compact real form, and the last algebra in each line is the normal real form.

The isomorphisms (7.14) between complex Lie algebras of low rank induce similar isomorphisms between their real forms. This is displayed in table IX.

Finally we recall that the concept of a Lie algebra was introduced for any field F, i.e. F needs not necessarily be the field of real or complex numbers. Since in a Chevalley–Serre basis all structure constants are integers, we can also consider the Lie algebra generated modulo those relations over the field $\mathbb{Q}$ of rational numbers, or more

Table IX. *Isomorphisms between low-dimensional semisimple Lie algebras*

$A_1 \cong B_1 \cong C_1$	$\mathfrak{su}(2) \cong \mathfrak{so}(3) \cong \mathfrak{sp}(1,0) \cong \mathfrak{usp}(2)$
	$\mathfrak{su}(1,1) \cong \mathfrak{sl}(2,\mathbb{R}) \cong \mathfrak{so}(2,1) \cong \mathfrak{sp}(1,\mathbb{R})$
$B_2 \cong C_2$	$\mathfrak{so}(5) \cong \mathfrak{sp}(2,0) \cong \mathfrak{usp}(4)$
	$\mathfrak{so}(4,1) \cong \mathfrak{sp}(1,1) \cong \mathfrak{usp}(2,2)$
	$\mathfrak{so}(3,2) \cong \mathfrak{sp}(2,\mathbb{R})$
$D_2 \cong A_1 \oplus A_1$	$\mathfrak{so}(4) \cong \mathfrak{so}(3) \oplus \mathfrak{so}(3)$
	$\mathfrak{so}^*(4) \cong \mathfrak{so}(3) \oplus \mathfrak{so}(2,1)$
	$\mathfrak{so}(2,2) \cong \mathfrak{so}(2,1) \oplus \mathfrak{so}(2,1)$
	$\mathfrak{so}(3,1) \cong \mathfrak{sl}(2,\mathbb{C})$
$D_3 \cong A_3$	$\mathfrak{so}(6) \cong \mathfrak{su}(4)$
	$\mathfrak{so}^*(6) \cong \mathfrak{su}(3,1)$
	$\mathfrak{so}(5,1) \cong \mathfrak{su}^*(4)$
	$\mathfrak{so}(4,2) \cong \mathfrak{su}(2,2)$
	$\mathfrak{so}(3,3) \cong \mathfrak{sl}(4,\mathbb{R})$

generally over any other field. Fields extending the rational numbers, like e.g. algebraic number fields, are of particular interest. The reader interested in the structure theory of these Lie algebras is referred to volume VIII of [Bourbaki 1982].

Summary:

Many aspects of Lie algebras over the field $\mathbb{R}$ of the real numbers can be described by considering their complexification. Various real Lie algebras can have the same complexification.
Distinguished real forms of a complex simple Lie algebra are the compact real form and the normal real form. The Killing form of the compact real form is negative definite.

Keywords:

Real forms of a complex Lie algebra, compact real form, normal real form, complexification, signature of the Killing form.

Exercises:

According to the results of exercise 3.5, the hydrogen problem in quantum mechanics gives rise to a six-dimensional symmetry algebra which is spanned by the components of angular momentum $\vec{L}$ and of the Runge–Lenz vector $\vec{R}$. This algebra acts on the energy eigenstates of fixed energy. In exercise 3.5 the algebra was regarded as a complex Lie algebra, but in the application to the hydrogen problem one deals with hermitian operators so that the relevant structure is that of a Lie algebra over $\mathbb{R}$.

Exercise 8.1

a) At zero energy the complex symmetry algebra is the semidirect sum of an abelian Lie algebra with $\mathfrak{sl}(2,\mathbb{C})$. When considering real Lie algebras, which is the relevant real form of $\mathfrak{sl}(2,\mathbb{C})$?

b) At non-zero energy the complex symmetry algebra is the direct sum $\mathfrak{sl}(2,\mathbb{C}) \oplus \mathfrak{sl}(2,\mathbb{C})$. Show that two different real forms of this Lie algebra appear, depending on whether the energy is positive (scattering states) or negative (bound states). Are these real forms simple as real Lie algebras?

c) The symmetry algebra acting on the bound states, i.e. the states with negative energy, can even be extended further. To see this, compute the commutation relations among $\vec{L}$ and the seven operators

$$\vec{S}^{(\pm)} := \vec{q}\,(\vec{p}\cdot\vec{p}) - 2\,\vec{p}\,(\vec{q}\cdot\vec{p}) \pm \vec{q}\,, \qquad S^{(0)} := \vec{q}\cdot\vec{p} - \mathrm{i} \qquad (8.16)$$

Conclude that together these operators span a real form of B_2; which one?

d) Analyze the five additional operators

$$\vec{T}^{(0)} := r\,\vec{p}\,, \qquad T^{(\pm)} := r\,(\vec{p}\cdot\vec{p} \pm 1) \qquad (8.17)$$

($r \equiv |\vec{q}\,|$) in the same manner. Which real form of $\mathfrak{sl}(4)$ does one obtain? What is the real Lie algebra spanned by $S^{(0)}$ and $T^{(\pm)}$?

Consider the infinite-dimensional Lie algebra spanned by generators L_n with $n \in \mathbb{Z}$ and by a generator K, and having the bracket relations

Exercise 8.2

$$[L_n, L_m] = (n-m)L_{n+m} + \tfrac{1}{12}\,n(n^2-1)\delta_{n+m,0}\,K\,,$$
$$[L_n, K] = 0\,. \qquad (8.18)$$

This Lie algebra is called the Virasoro algebra. Show that the Virasoro algebra does not admit for a symmetric invariant form.

Hints: Apply the invariance property (8.2) of a symmetric invariant form to the expression $\kappa([K, L_n], L_m)$ and conclude that $\kappa(K, \cdot) = 0$. Then use the invariance for expressions of the type $\kappa([L_m, L_n], L_p)$ with $p = 0$ to conclude that κ is of the form $\kappa_{nm} \propto \delta_{n+m,0}$. Show that for a non-zero constant of proportionality, this ansatz leads to a contradiction.

Show that in any basis in which the Killing form is of the form (8.8) the structure constants with only upper indices are totally antisymmetric.

Exercise 8.3

Hint: Use the invariance property (8.2) of the Killing form. Be careful to keep track of the signs of the metric.

Verify that the quaternions $\mathbb{H}$ form a field.

Exercise 8.4

Hint: Use that

$$\tau^i \tau^j = -\delta^{ij} \mathbf{1} + \sum_{k=1}^{3} \epsilon^{ijk} \tau^k \qquad (8.19)$$

for $i, j = 1, 2, 3$ and $\tau^0 \tau^i = \tau^i \tau^0 = \tau^i$ for any i.

Show that $\mathbb{H}$ contains both $\mathbb{R}$ and $\mathbb{C}$ as a subfield. How many subfields isomorphic to $\mathbb{C}$ are there?

Compute the norm $\xi^* \xi$ of a quaternion $\xi \in \mathbb{H}$.

Show that the vector space spanned over $\mathbb{R}$ by the elements of a Chevalley–Serre basis of a complex simple Lie algebra is a real Lie algebra, and that the real vector space spanned by the generators

Exercise 8.5

$$\{i H^j \mid j = 1, 2, \dots, r\}$$
$$\cup \{ \tfrac{i}{2} \sqrt{(\alpha, \alpha)} \, (E^\alpha + E^{-\alpha}), \tfrac{1}{2} \sqrt{(\alpha, \alpha)} \, (E^\alpha - E^{-\alpha}) \mid \alpha \in \Phi \} \qquad (8.20)$$

is a real Lie algebra as well. Compute the Killing form to conclude that the latter is the compact real form.

(In the case of $\mathfrak{sl}(2)$, these bases are precisely the bases (2.5) and (2.3) of $\mathfrak{sl}(2, \mathbb{R})$ and $\mathfrak{su}(2)$, respectively.)

Show that the Lie algebra $\mathfrak{gl}(V)$ is not simple.

Exercise 8.6

Hint: Identify an abelian ideal.

9

Lie groups

9.1 Lie group manifolds

Already in our introductory discussion in chapter 1 we have distinguished between symmetry groups and local one-parameter groups of symmetries. So far we have concentrated on the structures describing the latter, which has led us to the study of Lie algebras and their representations. In the present chapter we analyze the relation between symmetry groups and Lie algebras more carefully. More precisely, we consider Lie groups over the real numbers $\mathbb{R}$. A (real) Lie group G of dimension d is characterized by the property that it possesses two rather different structures which must nevertheless be compatible. Namely, it carries the algebraic structure of a *group* and is at the same time a finite-dimensional differentiable *manifold*. Combining these two extremely different structures has highly non-trivial consequences. In particular, both as groups and as manifolds Lie groups are quite special. To a large extent this can be traced back to the fact that the manifold structure allows us to make contact to linear spaces, and in fact to Lie algebras.

A standard example of a Lie group is the group SO(3) of all rotations in a three-dimensional space. This can be described as the group of all 3×3-matrices M which are orthogonal (i.e. satisfy $M^{-1} = M^t$) and have determinant one. To understand the manipulations with arbitrary Lie groups which we will perform in this chapter, it is convenient if the reader has some familiarity with basic notions from differential topology, e.g. that of a manifold – i.e. a topological space which looks locally like $\mathbb{R}^d$, or that of the tangent space and the fundamental group of a manifold. For an account of these notions we refer the reader e.g. to [Gilmore 1974] and [Nakahara 1990]. Those readers who are not at all acquainted with such concepts should simply imagine Lie groups as generalizations of the rotation group SO(3) and recall the description of *infinitesimal* rotations in terms of the Lie *algebra* $\mathfrak{su}(2)$. (They may also prefer to skip parts of this chapter in a first reading; fortunately, many of the results presented here will not be needed in the sequel.)

Compatibility of the manifold and group structures of a Lie group means that they are linked by the requirement that the group operations of multiplication,

$$G \times G \to G, \qquad (\gamma, \gamma') \mapsto \gamma \cdot \gamma', \qquad (9.1)$$

and inverse,

$$G \to G, \qquad \gamma \mapsto \gamma^{-1}, \qquad (9.2)$$

are differentiable maps. Actually differentiability of the product map (9.1) already implies the differentiability (9.2) of the inverse (see exercise 9.1).

To describe the implications of combining the group and manifold structures, we first realize that as a consequence of the group multiplication there are families of differentiable mappings of the Lie group manifold which act transitively, i.e. such that for any two group elements γ and γ' there is a member of the family which maps γ to γ'. These are the so-called left-translations

$$L_\beta: \quad \gamma \mapsto \beta\gamma, \qquad (9.3)$$

and the right-translations $R_\beta: \gamma \mapsto \gamma\beta$, respectively, where β is an arbitrary element of G. The translations which map a given group element $\gamma \in G$ to a prescribed $\gamma' \in G$ are $L_{\gamma'\gamma^{-1}}$ and $R_{\gamma^{-1}\gamma'}$, respectively. The translations can in particular be used to map any point of the manifold to the unit element of group multiplication, which in the sequel we will denote by e. (The name translation is motivated by the fact that the group of all translations of a vector space V over $\mathbb{R}$ is a (non-compact) Lie group, which can be identified with V. In this special case left- and right-translations coincide.)

As a consequence, we can perform the following construction. Consider the tangent space $T_\gamma G$ at some fixed group element γ. By left- or right-translation one can transport a basis of the vector space $T_\gamma G$ to any other point of the manifold. Since these translations are invertible, this yields a basis in the tangent space of each point of the group manifold. Thereby one obtains a global 'moving frame', and hence the tangent bundle of any Lie group is trivial; a manifold with this property is called *parallelizable*. As a parallelizable manifold a Lie group manifold is in particular orientable.

9.2 Global vector fields

An immediate consequence of this construction is that on a Lie group manifold there exist global vector fields that vanish nowhere. To construct them, one just fixes a vector v_γ in some arbitrary tangent space $T_\gamma G$ and uses left- or right-translation to transport it to the tangent space of any other element β of G. This way one obtains global vector fields v_β which possess the specific property that they are invariant under left- and right-translation, respectively. They are called left- or right- *invariant vector fields*. By construction, the vector space of left- or right-invariant vector fields are finite-dimensional and are naturally isomorphic to any tangent space $T_\gamma G$. As a matter of convenience, we will from now on take the unit element e of G as the fixed group element γ and work with left-invariant vector fields.

To obtain more structure on the space of invariant vector fields, let us first have a look at the space of all vector fields on an arbitrary differentiable manifold $\mathcal{M}$. For any two vector fields A, B we define a bracket $[A, B]$ (which is sometimes also called the Lie derivative) by the following expression in local coordinates ξ_a:

$$([A, B])^a := \sum_{b=1}^{d} B^b \frac{\partial A^a}{\partial \xi^b} - \sum_{b=1}^{d} A^b \frac{\partial B^a}{\partial \xi^b} . \tag{9.4}$$

(One may think of vector fields as differential operators acting on the functions on the group manifold: $A = \sum_a A^a \, \partial/\partial\xi^a$. In this description, the Lie derivative is just the commutator of these operators.) It can be checked that this definition is covariant under change of the coordinate system, and hence the bracket $[A, B]$ is again a vector field. (Note, however, that no covariant derivatives enter, and hence the definition of this bracket does not require the existence of a connection on the manifold). Moreover, this bracket is manifestly bilinear and antisymmetric and can be shown to fulfill the Jacobi identity (see exercise 9.2). In other words, when endowed with the bracket operation (9.4), the vector space of all vector fields on an arbitrary differentiable manifold $\mathcal{M}$ becomes a (real) Lie algebra. In the special case where $\mathcal{M} = G$ is a Lie group one can show that the *invariant* vector fields form a Lie subalgebra of the Lie algebra of all vector fields. In short, the space of invariant vector fields on a finite-dimensional Lie group manifold G carries a natural structure of a finite-dimensional real Lie algebra.

Via the identification of this space with any tangent space, in particular also the tangent space $T_e G$ at the unit element e carries the structure of a finite-dimensional real Lie algebra. We will denote the abstract Lie algebra that is associated to a Lie group G in this manner by $\mathfrak{g}$, i.e. write $\mathfrak{g} \cong T_e G$, and refer to $\mathfrak{g}$ simply as the Lie algebra of G. From the identification of $\mathfrak{g}$ with $T_e G$ it follows that the real dimension of $\mathfrak{g}$ is equal to the dimension d of the manifold G. Whenever this is convenient, we will identify $\mathfrak{g}$ with the space of left- or right-invariant vector fields or with $T_e G$.

A closer investigation shows that the Lie algebra carries an amazing amount of information on the Lie group manifold. The only information about a finite-dimensional Lie group G which is not already contained in its Lie algebra are properties that depend either on the set $\pi_0(G)$ of different connected components or on the fundamental group $\pi_1(G)$. In particular, for any simple compact real form $\mathfrak{g}$, there is a unique compact simple Lie group $\tilde{G}$, for which the Lie algebra of invariant vector fields is isomorphic to $\mathfrak{g}$, and such that $\tilde{G}$ is connected and simply connected, i.e. satisfies $\pi_0(\tilde{G}) = 0 = \pi_1(\tilde{G})$. Moreover, it can be shown that for any connected Lie group G with Lie algebra $\mathfrak{g}$ there is a surjective Lie group homomorphism φ from $\tilde{G}$ to G such that the kernel of φ is a subgroup of the center of $\tilde{G}$ which is isomorphic to the fundamental group π_1 of G. (The center of a group G is the subgroup consisting of all those elements γ of G which commute with all $\gamma' \in G$ in the sense that $\gamma\gamma' = \gamma'\gamma$.) For this reason $\tilde{G}$ is also called the *universal covering group* associated with $\mathfrak{g}$.

A Lie group is called reductive or (semi-) simple if its Lie algebra is reductive or (semi-) simple as a real Lie algebra. It is worth mentioning that a simple *Lie* group is not necessarily a simple group in the sense that it does not possess any non-trivial normal subgroup. (A subgroup H of a group G is called an *invariant* or *normal subgroup* of G if it possesses the property that $\gamma\gamma'\gamma^{-1} \in H$ for all $\gamma' \in H$ and all $\gamma \in G$.) However, all normal subgroups of a simple Lie group are discrete groups.

9.3 Compactness

Let us now study some consequences of the results of chapter 8 about real Lie algebras for Lie groups. The following important result explains the qualification *compact* of the compact real form:

> If G is a compact manifold, then the underlying real Lie algebra $\mathfrak{g}$ is precisely the compact real form.

One can also show that if a Lie algebra is both semisimple and compact, then its universal covering group is a compact manifold. The theory of real Lie groups is best developed for those Lie groups which are compact manifolds. As for non-compact groups, we note that two Lie groups are isomorphic if and only if their respective largest compact subgroups are isomorphic.

The Lorentz group $SO(n,1)$ is not compact (see exercise 9.3). This fact has important consequences. In particular, it implies that some global features of space-time, i.e. global Lorentzian geometry, are rather different from the global features of a Riemannian manifold. For example, there exist Lorentzian manifolds for which all curvature invariants coincide and which are nonetheless not isometric. Another crucial difference to the

Information

Riemannian case is that for the topology of these manifolds there is no neighborhood basis which consists entirely of SO(n, 1)-invariant sets. As a consequence, there exist connected and geodesically complete space-times in which not any two points can be joined by a geodesic, unlike in the Riemannian case (see e.g. section 5.2 of [Hawking and Ellis 1980]).

In chapter 8 we have seen that the Killing form κ turns the real Lie algebra $\mathfrak{g}$ into a (pseudo-) euclidean vector space. Using the identification of $\mathfrak{g}$ and the tangent space T_eG we can use $-\kappa$ to endow also T_eG with the structure of a (pseudo-) euclidean space. Using left-translation, we can transport this structure to any other tangent space $T_\gamma G$; this way we obtain an invariant (pseudo-) Riemannian metric on the group manifold. Hence any finite-dimensional simple Lie group has in a natural way also the structure of a (pseudo-) Riemannian manifold. One can show that the geodesics with respect to this metric are precisely the one-parameter subgroups of G. Also, the group manifold of G is a genuine Riemannian manifold, i.e. the metric is positive definite, if and only if $\mathfrak{g}$ is the compact real form, or in other words, precisely if the manifold G is compact.

So far we have been concerned with *real* Lie groups, i.e. Lie groups that are real manifolds. Now as seen in section 8.3, for a real Lie *algebra* there always exists a complexification, which is a complex Lie algebra. In contrast, an analogous statement does not hold for Lie *groups* (compare exercise 9.11 for a real Lie group which does not possess a complexification). It can be shown, however, that each *compact* real algebraic Lie group does possess a complexification. (An algebraic group is a subgroup of a general linear group GL(m) which can be characterized by the property that certain polynomials in the group elements vanish. The complexification is then obtained by considering these algebraic equations over $\mathbb{C}$ rather than over $\mathbb{R}$.)

The compactness is required because it implies the existence of an invariant measure with respect to which the group has a finite volume (see section 21.8). By averaging over the group with respect to this measure, one obtains on any finite-dimensional G-representation an invariant scalar product (see equation (21.48)), so that the representation is unitary. It follows that G has a faithful unitary representation R of finite dimension, say $\dim R = n$; as a consequence, G is a subgroup of U(n). In short, each compact real Lie group can be viewed as a subgroup of a unitary group. Furthermore, for unitary groups U(n), a complexification manifestly exists; the complexification of U(n) is the group GL($n, \mathbb{C}$). In the special case of $n = 1$ the complexification of U(1) is $\mathbb{C}^\times$, i.e. the multiplicative group of non-zero complex numbers. For more details about the complexification of Lie groups we refer to chapter III, §6.10 of [Bourbaki 1989].

9.4 The exponential map

We now present a mapping which provides us with natural coordinates on the Lie group manifold. We first recall that the Lie group is a Riemannian manifold and that the geodesics with respect to its metric are precisely the one-parameter subgroups of G. As for any Riemannian manifold, we can define an exponential mapping from the tangent space $T_\gamma G$ at some point γ to G, which provides coordinates of a neighborhood of γ. We describe this construction for the case that $\gamma = e$; we can then use group multiplication to translate the result to the neighborhood of any other point.

Any geodesic through the point e is uniquely characterized by its derivative (i.e., its velocity) at e, which is an element of $T_e G$. Hence to any $x \in T_e G$ we can associate the geodesic $s(t)$ which is defined by $\frac{ds}{dt}\big|_{t=0} = x$ and the initial condition $s(0) = e$. We now define the exponential mapping from $T_e G$ to G by

$$x \mapsto \text{EXP}(x) := s(1) \in G . \tag{9.5}$$

This mapping also gives the relation between vectors $x \in T_e G$ – which we identify with elements of the Lie algebra $\mathfrak{g}$ – and local one-parameter subgroups of G:

$$t \in [-\epsilon, \epsilon] \mapsto x_t := \text{EXP}(tx) \in G . \tag{9.6}$$

It turns out that when G is a matrix Lie group, the exponential mapping EXP is just given by the usual exponential power series of matrices.

In the case of finite-dimensional Lie groups it can be shown that *locally*, i.e. in a suitable neighborhood of the identity element, one can write any element $\gamma \in G$ as $\gamma = \text{EXP}(x)$ for some $x \in \mathfrak{g}$. In terms of the generators T^a of $\mathfrak{g}$, this exponential relation reads

$$\gamma = \text{EXP}\Big(\sum_{a=1}^{d} \xi_a T^a\Big) \tag{9.7}$$

for a suitable choice of coefficients $\xi_a \in F$, $a = 1, 2, ..., d = \dim \mathfrak{g}$. Hence the numbers $\xi_a = \xi_a(\gamma)$ can be used as coordinates on the Lie group manifold in the neighborhood of the identity element. The inverse of the exponential mapping EXP is a coordinate mapping. While this map keeps a substantial part of the information about the structure of G, it avoids problems with non-linearities and hence allows for simple explicit calculations.

The exponential map of a *compact* connected Lie group is surjective, i.e. not only locally, but also *globally* any element of the Lie group can be written as the exponential of some element of the Lie algebra. If the group is compact, but not connected, this is true for any element in the connected component of the unit element e. This statement is not necessarily true any more for non-compact Lie groups G, i.e. exponentiation does not always yield the full group (compare exercise 9.11). However, as long as G is still simple, two successive exponential mappings are sufficient. (Also, in an explicit parametrization of G by coordinates, an infinite range of parameter values is needed for non-compact groups, while for compact groups a finite parameter range is sufficient; compare exercise 9.4.)

Another important application of the exponential map is that it relates the representations of a Lie group with the ones of its Lie algebra. It can be shown that the finite-dimensional representations of a compact Lie algebra $\mathfrak{g}$ and of its universal covering group are in one-to-one correspondence and act on the same representation spaces. In contrast, whenever $\pi_1(G) \neq 0$, a truncation takes place, i.e. only part of the finite-dimensional $\mathfrak{g}$-representations are representations of the group G. We will encounter this truncation in a different guise in section 22.12.

At this point we introduce some terminology which is convenient for describing representations R of a group G on some space M. For any $v \in M$ we call the set $\{w \in G \,|\, w = (R(\gamma))(v) \text{ for some } \gamma \in G\}$ the G-*orbit* $\mathcal{O}_v$ of v. The space M can be decomposed as a set as the disjoint union of all G-orbits. If M consists of a single orbit, the representation R of G is said to be *transitive*. R is called *faithful* or *effective* if for any two distinct elements $\gamma, \beta \in G$ the maps $R(\gamma)$ and $R(\beta)$ are different. To any point v of M one associates a subgroup of G, the *isotropy subgroup* or *stabilizer* S_v of v, defined by $S_v := \{\gamma \in G \,|\, (R(\gamma))(v) = v\}$. If the stabilizer is non-trivial, v is called a *fixed point* of R. The stabilizers of elements on the same orbit are conjugate subgroups: for $w \in \mathcal{O}_v$ one has $w = \gamma v$, and then the stabilizers are related by $S_w = \gamma S_v \gamma^{-1}$ for some $\gamma \in G$. A representation is called *free* if only the map $R(e)$ representing the identity has fixed points. For example, the action of a (Lie) group G on itself by left or right multiplication is transitive and free (for the adjoint action of G, see exercise 9.1).

We would also like to make another remark: As emphasized at several places in this book, in many respects infinite-dimensional Lie *algebras* can be treated in a manner rather parallel to finite-dimensional ones. It is therefore appropriate to point out that the relation between infinite-dimensional Lie *groups* and infinite-dimensional Lie algebras is considerably more involved than the corresponding relation in the finite-dimensional case. For example, there are Lie algebras which do not correspond to any Lie group, i.e. naive exponentiation does not work in this case. Conversely, there are Lie groups for which the exponential map is not locally bijective.

*9.5 Maurer–Cartan theory

In the previous sections we have employed invariant vector fields on the Lie group manifold to identify the structure of a Lie algebra. Alternatively, one can also use differential forms on this manifold to capture information on the Lie algebra; this approach is known as Maurer–Cartan theory.

Given d linearly independent invariant vector fields A^a, $a = 1, 2, ..., d = \dim G$, on G, the dual one-forms ζ_a defined by

$$\zeta_a(A^b) = \delta_a^b,\tag{9.8}$$

are called the *Maurer–Cartan forms* of G. The exterior derivative $d\zeta_a$ of a Maurer–Cartan form ζ_a can be written as a linear combination of the exterior (wedge) products of all Maurer–Cartan forms. More precisely, if the Lie bracket of the invariant vector fields A^a reads

$$[A^a, A^b] = \sum_c f^{ab}{}_c A^c,\tag{9.9}$$

then the so-called *Maurer–Cartan structure equation*

$$d\zeta_a = -\tfrac{1}{2} \sum_{b,c} f^{bc}{}_a \zeta_b \wedge \zeta_c\tag{9.10}$$

holds. This equation contains in fact the full information on the Lie algebra structure. Moreover, using the antisymmetry of the wedge product, the nilpotency property $d^2\zeta_a = 0$ of the exterior derivative is equivalent to the Jacobi identity (see exercise 9.7).

The structure equation can also be written in a basis independent form. To this end we introduce a one-form ζ on G with values in the Lie algebra $\mathfrak{g}$, called again the Maurer–Cartan form. By definition, ζ acts on the space $\mathfrak{g}$ of left-invariant vector fields as the identity map: $\zeta|_\mathfrak{g} = id_\mathfrak{g}$. After fixing a basis of $\mathfrak{g}$, ζ can be expressed as $\zeta = \sum_a \zeta_a A^a$ through the Maurer–Cartan forms ζ_a. In this formulation, the Maurer–Cartan structure equation takes the form

$$d\zeta = -\tfrac{1}{2}[\zeta, \zeta],\tag{9.11}$$

where the associative product with respect to which the commutator is taken consists of the wedge product of one-forms combined with the matrix product of the A^a. For more information on Maurer–Cartan theory we refer to chapter III, §3.14 of [Bourbaki 1989].

9.6 Spin

As an illustration of our discussion of Lie groups we now present an example for two different Lie groups which share the same Lie algebra and hence only differ in global topological aspects. As already pointed out in section 2.1, the group SO(3) of rotations of a three-dimensional euclidean space has a Lie algebra which is spanned by three generators iL_i with commutation relations

$$[iL_i, iL_j] = -\sum_k \epsilon_{ijk}\, iL_k.\tag{9.12}$$

The group manifold SO(3) plays an important rôle in classical mechanics. | Information
It is the configuration space of the rigid body with one point fixed, or what
is the same, the configuration space of the rigid body with the center of
mass degrees of freedom split off. A treatment of the rigid body which is
based on this observation can be found in section 28 of [Arnold 1978] and
in chapter 15 of [Marsden and Ratiu 1994]. For an application, see also
exercise 9.9.

On the other hand let us have a look at the group SU(2) of complex
unitary 2×2-matrices with determinant 1. The Lie algebra of this group
is spanned by the three traceless and anti-hermitian matrices $i\sigma_i$, $i \in$
$\{1, 2, 3\}$, where σ_i denote the Pauli matrices (3.19) (see exercise 9.8).
These matrices span a Lie algebra which is isomorphic to the Lie algebra
defined by (9.12). One might therefore wonder whether also both Lie
groups are isomorphic. As we will see, this is not the case. First we note
that any matrix in the group SU(2) is of the form

$$U = \begin{pmatrix} a + ib & c + id \\ -c + id & a - ib \end{pmatrix} \tag{9.13}$$

with four real parameters a, b, c, d. The condition that this matrix has
determinant 1 translates into $\det U \equiv a^2 + b^2 + c^2 + d^2 = 1$. Thus the
group manifold of SU(2) can be regarded as the unit three-sphere S^3 in
$\mathbb{R}^4$. The three-sphere is connected and simply connected; therefore SU(2)
is the universal covering Lie group. Also, comparison with the formula
(8.12) shows that in the language of quaternions these relations just state
that the Lie group SU(2) consists of the quaternions of length one.

Next we show that SO(3) is *not* simply connected; this implies that
SU(2) and SO(3) are indeed different, and that SU(2) is the universal
covering group of SO(3). To this end we characterize any rotation by a
vector $\vec{v} \in \mathbb{R}^3$ as follows: The axis of the rotation is the direction of $\vec{v}$, and
the angle of the rotation is its length $|\vec{v}|$. Since this angle is a number in
the interval $[-\pi, \pi]$, with π and $-\pi$ to be identified, the group manifold
of all rotations is isomorphic to a massive ball in three dimensions with
antipodal points of its surface identified. To show that this space is not
simply connected it is sufficient to find a non-contractible closed path in
the manifold. To do so, fix a vector $\vec{v}$ of length π and consider the path

$$[-1, 1] \ni t \mapsto t\vec{v}. \tag{9.14}$$

Due to the identification of antipodal points on the surface this path
is indeed closed. Also, any deformation of this path contains two antipodal points on the surface of the sphere, and in fact the path is non-
contractible.

Taking this path twice yields a contractible loop (it is a good exercise
to check this simple geometric fact). Accordingly, SU(2) is a *two*fold
covering of SO(3). This can also be understood as follows. To any vector

$\vec{v}$ in $\mathbb{R}^3$ we associate the linear combination

$$M(\vec{v}) := \sum_{i=1}^{3} v^i \sigma_i \qquad (9.15)$$

of Pauli matrices. The mapping $\vec{v} \mapsto M(\vec{v})$ is clearly one-to-one. The norm of $\vec{v}$ can be expressed through the determinant of $M(\vec{v})$ as $|\vec{v}|^2 = -\det M(\vec{v})$. For any element γ of SU(2) the matrix product $\gamma M(\vec{v}) \gamma^{-1}$ is again a real linear combination of Pauli matrices, and hence of the form $M(\vec{v}')$ for some vector $\vec{v}'$. Since γ has determinant 1, the determinants of $M(\vec{v})$ and $M(\vec{v}')$ are equal, and hence so are the norms of $\vec{v}$ and $\vec{v}'$. Also, the mapping $\phi_\gamma : \vec{v} \mapsto \vec{v}'$ is linear and hence an element of the group O(3). Since SU(2) is connected, any $\gamma \in$ SU(2) belongs to the connected component of the identity of SU(2); this implies that any ϕ_γ belongs to the identity component of O(3) and hence is an element of SO(3). However, this correspondence between SU(2) and SO(3) is not one-to-one but rather two-to-one, because precisely the two elements γ and $-\gamma$ of SU(2) result in the same element of SO(3). (For more details, see section 20.5.)

According to a general principle (compare equation (5.20)), the existence of a group homomorphism

$$\phi : \quad \mathrm{SU}(2) \to \mathrm{SO}(3) , \qquad (9.16)$$

implies that any representation R of SO(3) leads to a representation $R \circ \phi$ of SU(2). The converse, however, is not true. A necessary and sufficient condition for a representation of SU(2) to induce a representation of SO(3) is that the element $-\mathbb{1}_{2 \times 2}$ of SU(2) is represented as the identity matrix. Now according to section 5.7 any finite-dimensional irreducible representation of SU(2) can be labelled by an integer Λ, the highest weight. The Pauli matrix σ_3 is represented on the irreducible representation with highest weight Λ by the diagonal matrix $\mathrm{diag}(\Lambda, \Lambda-2, \ldots, -\Lambda+2, -\Lambda)$. Hence the matrix $\mathrm{e}^{i\pi\sigma_3} = -\mathbb{1}_{2 \times 2} \in$ SU(2) is represented by $+\mathbb{1}_{(\Lambda+1) \times (\Lambda+1)}$ if Λ is even and by $-\mathbb{1}_{(\Lambda+1) \times (\Lambda+1)}$ if Λ is odd. Thus the representations of SO(3) are precisely the representations of SU(2) with Λ even, i.e. the representations with integer spin $j = \Lambda/2$. By a slight abuse of terminology, one often refers to representations of SU(2) that are not representations of SO(3) as *spinor representations* of SO(3).

Spinor representations arise naturally in quantum mechanics (compare section 12.2). Since half-integer spin representations are not representations of SO(3), their appearance in physics cannot be interpreted in terms of the orbital angular momentum. Rather, half-integer spin particles possess some intrinsic angular momentum which constitutes an internal degree of freedom. Heuristically, such particles 'spin' with respect to an internal degree of freedom. This is the origin of the term spin.

9.7 The classical Lie groups

The connection between the real Lie algebra $\mathfrak{g}$ and the Lie group G is most easily made explicit if the elements of the Lie algebra $\mathfrak{g}$ are given as finite-dimensional matrices, e.g. for the matrix Lie algebras described in section 5.6. In this case the exponential mapping EXP is just the usual exponential power series of matrices, $\text{EXP}(x) = \exp(x) \equiv \sum_{n=0}^{\infty} x^n/n!$, with x^n denoting the n-fold matrix product; also, the unit element e of G is the unit matrix, and the inverse of $\gamma = \exp(x)$ is $\gamma^{-1} = \exp(-x)$.

With a given parametrization of G as a manifold, this yields an explicit relation between the group element EXP(x) and the Lie algebra element x. For instance, for $\mathfrak{g} = \mathfrak{sl}(2, \mathbb{R})$ we have

$$x = \xi E_+ + \eta H + \zeta E_- = \begin{pmatrix} \eta & \xi \\ \zeta & -\eta \end{pmatrix} \mapsto \exp(x) = \cosh(\tau)\mathbf{1} + \frac{1}{\tau}\sinh(\tau)\,x\,,$$

where $\mathbf{1}$ is the 2×2 unit matrix and $\tau = \sqrt{\eta^2 + \xi\zeta} = \sqrt{-\det x}$.

The fact that every (finite-dimensional) matrix in a neighborhood of the identity can be expressed as $\exp(x)$ with exp the exponential power series is an immediate consequence of the inverse function theorem. Now the inverse function theorem only holds for finite-dimensional spaces. It is therefore not surprising that in the infinite-dimensional case the relation between Lie algebras and Lie groups is considerably more involved, as we have already pointed out.

The exponential power series of matrices relates the various families of matrix Lie groups to the classical simple Lie algebras that we encountered in the classification of section 5.6; correspondingly these Lie groups are known as the *classical Lie groups*. We have already seen that for any vector space V the set of all invertible linear mappings of V forms a Lie group, called the *general linear group* GL(V), which is isomorphic to the group of all dim$V \times$dim$V$-matrices with non-zero determinant. The group product is ordinary matrix multiplication. The Lie algebra $\mathfrak{gl}(V)$ of this group is just the Lie algebra of all dim$V \times$dim$V$-matrices, where the Lie bracket is given by the commutator. However, the Lie algebra $\mathfrak{gl}(V)$ is not simple (see exercise 8.6). To identify subgroups of GL(V) associated with simple Lie algebras, one has to impose additional conditions. For example, one can restrict oneself to the subgroup U(V) of unitary matrices, which obey $UU^\dagger = \mathbf{1}$, the *unitary group* of V. The Lie algebra of U(V) consists of the anti-hermitian matrices. This subgroup is not yet simple; but requiring in addition that U is unimodular, i.e. that the determinant of U is equal to one, one obtains a simple Lie group, the *special unitary group* SU(V). Its Lie algebra consists of all anti-hermitian matrices with trace zero. If the dimension of V is n, one also writes SU(n) for SU(V). The complexification of the Lie algebra of SU(n) is the complex simple Lie algebra A_{n-1} (to show this, one can use the matrix representation of A_{n-1} given in section 5.6).

Instead of complex vector spaces one can work with real vector spaces, too. The analogue of the unitarity condition in the complex case is then to require the matrices to be orthogonal, i.e. that the inverse of the matrix should be equal to its transpose, $M^{-1} = M^t$. This yields the orthogonal groups $O(n)$, where n denotes the *real* dimension of V. Again imposing the additional requirement that M should be unimodular yields a simple Lie group, the *special orthogonal group* $SO(n)$. The structure of $SO(n)$ depends strongly on whether n is odd or even. If $n = 2r + 1$ is odd, the complexification of the Lie algebra of $SO(2r + 1)$ is isomorphic to B_r, while for $SO(2r)$ one obtains D_r.

A common feature of $SO(n)$ and $SU(n)$ is that both preserve a non-degenerate bilinear inner product which in this case is a symmetric scalar product. To obtain also the last of the four series of classical Lie groups which is still missing, one considers the subgroup of matrices that preserve a so-called symplectic form. A *symplectic form* η is by definition a non-degenerate antisymmetric bilinear form; thus $\eta(x, y) = -\eta(y, x)$, and for any $x \in V$ there is a $y \in V$ such that $\eta(x, y) \neq 0$. Symplectic forms only exist on even-dimensional vector spaces. By a suitable basis transformation they can be brought to the standard form

$$\eta = \begin{pmatrix} 0 & \mathbb{1} \\ -\mathbb{1} & 0 \end{pmatrix}, \tag{9.17}$$

where $\mathbb{1}$ is the $n \times n$ unit matrix (just like to any symmetric inner product one can associate an orthonormal basis in which the matrix for the inner product is the unit matrix $\mathbb{1}$). The Lie group that preserves a symplectic structure on a vector space of dimension $2n$ will be denoted by $SP(n)$ (another common convention is to denote it by $SP(2n)$); its Lie algebra is spanned by all $2n \times 2n$-matrices that fulfill the relation $M^t = \eta M \eta$ with η as in (9.17). This coincides with the relation (5.25), and hence the complexification of this Lie algebra is isomorphic to the complex simple Lie algebra C_n. Note that in the case of $SU(n)$ and $SO(n)$ we had to impose the additional condition that the determinant is one to obtain a simple Lie group; since any symplectic matrix has determinant one (see exercise 9.10), this is not necessary in the $SP(n)$ case.

In classical mechanics the phase space is an n-dimensional symplectic manifold (compare chapter 1), i.e. any tangent space is equipped with a symplectic form Ω. Time evolution according to Hamilton's equations can described by a continuous family of symplectic transformations on phase space which preserve Ω. Along with Ω also any multiple wedge product $\Omega^k := \underbrace{\Omega \wedge \Omega \wedge \Omega \ldots \wedge \Omega}_{k \text{ factors}}$ is preserved. It can be shown that for $k \leq n/2$ these forms do not vanish. In particular for $k = n/2$ one obtains an invariant volume form on the phase space, the Liouville volume. For more information we refer to section 44 of [Arnold 1978].

Information

Let us discuss a few common features of the four series of classical Lie groups. First we note that in terms of the matrix elements, the relations needed above to define the various subgroups of $\mathrm{GL}(n)$ are all non-linear. Via the inverse of the exponential mapping, all these relations get linearized, e.g. in place of the non-linear condition $\det U = 1$ needed to obtain $\mathrm{SU}(n)$ one gets the linear condition $\operatorname{tr} u = 0$. However, all these relations are algebraic, i.e. can be written in terms of polynomial equations for the matrix elements. Thus in particular all classical Lie groups are algebraic groups.

We also remark that the groups $\mathrm{SU}(n)$ and $\mathrm{SO}(n)$ are compact, whereas $\mathrm{SP}(n)$ is not. For example, $\mathrm{SP}(2, \mathbb{R})$ is isomorphic as a Lie group to $\mathrm{SL}(2, \mathbb{R})$; a convenient parametrization of its group elements is

$$\gamma = \gamma(\phi, \xi, \zeta) := \left(\begin{smallmatrix} \cos\phi & \sin\phi \\ -\sin\phi & \cos\phi \end{smallmatrix} \right) \cdot \exp \left(\begin{smallmatrix} \xi & \zeta \\ \zeta & -\xi \end{smallmatrix} \right). \tag{9.18}$$

Hence as a manifold, the group manifold is isomorphic to $S^1 \times \mathbb{R}^2$ and therefore not compact. The universal covering group is isomorphic as a manifold to $\mathbb{R}^3$, and the fundamental group is $\mathbb{Z}$. In fact, the symplectic groups allow for spinor representations, i.e. representations of the universal covering group which are not representations of $\mathrm{SP}(n, \mathbb{R})$.

We also warn the reader that the classical Lie groups listed above are *not* always the universal covering groups. For example, the argument given for $\mathrm{SO}(3)$ can be easily generalized to show that the Lie groups $\mathrm{SO}(n)$ are doubly rather than simply connected i.e. $\pi_1(\mathrm{SO}(n)) = \mathbb{Z}_2$; the universal covering Lie group, denoted by $\mathrm{Spin}(n)$, does not correspond to any classical matrix Lie group, except for $n = 3, 4, 5, 6$ (for the latter cases, see section 20.10). The relation between $\mathrm{SO}(n)$ and $\mathrm{Spin}(n)$ is $\mathrm{SO}(n) = \mathrm{Spin}(n)/\mathbb{Z}_2$, with $\mathbb{Z}_2$ a subgroup of the center of $\mathrm{Spin}(n)$ (compare also section 20.4). In contrast, the group $\mathrm{SU}(n)$ is simply connected; its center $\mathcal{Z}$ consists of the multiples $\zeta^m \mathbb{1}$ of the unit matrix, where ζ is an nth root of unity, $\zeta = \exp(2\pi\mathrm{i}/n)$. Hence the center is isomorphic to $\mathbb{Z}_n$, and the group is simply connected (compare the list of the groups $\mathcal{Z}(\mathrm{G})$ that will be presented after equation (11.33)).

9.8 BCH formulæ

The multiplication of matrices is a non-commutative operation. This implies in particular that the matrices $\exp(X)\exp(Y)$, $\exp(X+Y) = \exp(Y+X)$ and $\exp(Y)\exp(X)$, with X and Y $n{\times}n$-matrices, are generically three distinct objects. An analogous statement then of course applies to the exponentials $\mathrm{Exp}(x)$ of elements of a Lie algebra. The relations between such exponentials are given by the so-called *Baker–Campbell–Hausdorff* formulæ. Different orderings of exponentials correspond to different elements of the Lie group.

Baker–Campbell–Hausdorff (BCH) type relations arise naturally when one describes Lie group elements by exponentials of Lie algebra elements. Namely, while the representation matrices for the Lie group elements can in principle be obtained via the exponential mapping from those of the Lie algebra elements, such direct calculations tend to be rather complicated. Combining the BCH relations with the triangular decomposition of the Lie algebra simplifies such computations considerably. Note in particular that the entries of the exponential of an upper or lower triangular matrix (and hence for the representation matrices of $\mathfrak{g}_+$ and $\mathfrak{g}_-$ in a standard matrix realization) are just polynomials in the entries of the original matrix, rather than transcendental functions.

The basic Baker–Campbell–Hausdorff identity says that the product of two exponentials can be written as $\text{Exp}(x)\,\text{Exp}(y) = \text{Exp}(x * y)$ with

$$x * y := x + y + \tfrac{1}{2}\,[x, y] + \tfrac{1}{12}\,[x, [x, y]] + \tfrac{1}{12}\,[[x, y], y] + \dots \,. \qquad (9.19)$$

Here the ellipsis stands for terms which are of higher order in x and y. In particular, if the element $w := [x, y]$ has vanishing bracket with both x and y, then $\text{Exp}(x)\,\text{Exp}(y) = \text{Exp}(x + y + w/2)$, and hence e.g. $\text{Exp}(x)\text{Exp}(2y)\text{Exp}(x) = \text{Exp}(2x + 2y)$. (Note that in particular the commutator $[x, y]$ has vanishing bracket with x and y whenever it is a central element, or, in physicists' language, a $\mathbb{C}$-number. This observation is helpful in many applications; e.g. canonical commutation relations are typically given by commutators of this type.)

Note that in (9.19) only the operation of taking the Lie bracket is used, i.e. there is no need to refer e.g. to an underlying associative product. Nevertheless one may expand (9.19) in a formal power series of, say, y; one then obtains

$$x * y := x + \frac{\text{ad}_x}{1 - \exp(\text{ad}_{-x})}\,(y) + \mathcal{O}(y^2)\,. \qquad (9.20)$$

The identity (9.19) can be derived as follows. One first shows that differentiation of the exponential of a square matrix $Z(\xi)$ that can be written as a power series in some parameter ξ obeys

$$\frac{\partial}{\partial \xi}\,e^{\eta Z(\xi)} = \int_0^\eta dt\; e^{(\eta - t)Z}\,\frac{\partial Z}{\partial \xi}\,e^{tZ}\,. \qquad (9.21)$$

This follows by regarding each side of (9.21) as a function $f = f(\eta)$; then each side satisfies the differential equation $\partial f / \partial \eta = (\partial Z/\partial \xi)\,e^{\eta Z} + Z\,f(\eta)$ together with the initial condition $f(0) = 1$, and hence both sides indeed coincide. To proceed, we apply the result (9.21) with $\eta = 1$, and with t replaced by $1 - t$, to the function $\exp(Z(\xi) := \exp(\xi X)\exp(\xi Y)$ of ξ; this yields

$$X + e^{\xi X}\,Y\,e^{-\xi X} = \int_0^1 dt\; e^{tZ}\,\frac{\partial Z}{\partial \xi}\,e^{-tZ}\,. \qquad (9.22)$$

We now wish to expand both sides of (9.22) in powers of ξ. To this end, we first evaluate the expression $e^X Y e^{-X}$ by applying the following (frequently used) trick: One regards some quantity of interest as the value of a function $f(t)$ at $t = 1$ and subtracts the value of that function at $t = 0$, which allows us to employ the identity $f(1) - f(0) = \int_0^1 dt\,(df/dt)$. In the present case we have $f(t) = e^{tX} Y e^{-tX}$ and obtain

$$
e^X Y e^{-X} - Y = \int_0^1 dt\,(X e^{tX} Y e^{-tX} - e^{tX} Y e^{-tX} X)
$$

$$
= \sum_{n,m=0}^{\infty} \frac{1}{n!m!}(-1)^m (X^{n+1}Y X^m - X^n Y X^{m+1}) \int_0^1 dt\, t^{n+m} \tag{9.23}
$$

$$
= \sum_{n,m=0}^{\infty} \frac{1}{n!m!(n+m+1)}(-1)^m (X^{n+1}Y X^m - X^n Y X^{m+1}).
$$

Each term in the last expression is just a multiple commutator, so that we finally arrive at

$$
e^X Y e^{-X} = Y + \sum_{n=1}^{\infty} \frac{1}{n!}(\mathrm{ad}_X)^n(Y). \tag{9.24}
$$

Applying the result (9.24) both to the left hand side of equation (9.22) and to the integrand on the right hand side and afterwards performing the t-integration termwise, one obtains

$$
X + Y + \sum_{j=1}^{\infty} \frac{\xi^j}{j!}(\mathrm{ad}_X)^j(Y) = \frac{\partial Z}{\partial \xi} + \sum_{j=1}^{\infty} \frac{1}{(j+1)!}(\mathrm{ad}_Z)^j\left(\frac{\partial Z}{\partial \xi}\right). \tag{9.25}
$$

Finally we replace $Z(\xi)$ on the right hand side by its power series expansion $Z(\xi) =: \sum_{m=0}^{\infty} Z_m \xi^m$, and $\partial Z/\partial \xi$ by $\sum_{m=1}^{\infty} m Z_m \xi^{m-1}$; the result (9.19) then follows by comparing the coefficients of powers of ξ on the two sides of (9.25).

Let us also mention that for the case of $\mathfrak{su}(2)$ the relation (9.24) specializes to

$$
e^{-i\vec{\eta}\cdot\vec{L}}(\vec{\zeta}\cdot\vec{L})e^{i\vec{\eta}\cdot\vec{L}} = (1-\cos|\vec{\eta}|)\frac{\vec{\eta}\cdot\vec{\zeta}}{|\vec{\eta}|}\frac{\vec{\eta}\cdot\vec{L}}{|\vec{\eta}|} + (\cos|\vec{\eta}|)\vec{\zeta}\cdot\vec{L}
$$
$$
+ (\sin|\vec{\eta}|)\frac{\vec{\eta}\times\vec{\zeta}}{|\vec{\eta}|}\cdot\vec{L}. \tag{9.26}
$$

Summary:

The finite-dimensional vector space of left-invariant vector fields on a Lie group over $\mathbb{R}$ or equivalently, the tangent space at the unit element, carries the structure of a Lie algebra. The Lie algebra is related to the Lie group by an exponential mapping. Correspondingly, real Lie algebras encode most of the information about Lie groups. The classical Lie groups are obtained as groups of matrices which satisfy certain algebraic equations. Their Lie algebras are reals forms of the classical Lie algebras described in section 5.6.

Keywords:

Lie group, left- and right-translations, exponential map, tangent space, universal covering group, Maurer–Cartan theory; spin; unitary, orthogonal and symplectic groups; Baker–Campbell–Hausdorff formulæ.

Exercises:

a) Show that the fact that the multiplication of a Lie group is differentiable implies that the inversion is differentiable as well.

Exercise 9.1

Hint: Define the map $f\colon G \times G \to G \times G$, $f(\gamma, \gamma') := (\gamma, \gamma\gamma')$. Compute its inverse, show that (e, e) is a regular value of f, and conclude that locally a differentiable inverse exists. Use the fact that left-translation is differentiable to show the claim globally.

b) Show that the action of a (Lie) group on itself by left or right multiplication is transitive and free. Is the same true for the adjoint action of G on itself, i.e. for the action defined by $R_\gamma(\beta) := \gamma\beta\gamma^{-1}$?

Check that the definition (9.4) of the Lie derivative is covariant under coordinate transformations $y = y(x)$ and turns the space of all vector fields on a manifold $\mathcal{M}$ into a Lie algebra.

Exercise 9.2

Explain how the Lie derivative (9.4) can be interpreted as the commutator of the differential operators $A = \sum_a A^a \, \partial/\partial\xi^a$ acting on functions on $\mathcal{M}$.

Show that on a Lie group the bracket of two left-invariant vector fields is left-invariant as well.

Show that the Lorentz group $SO(n, 1)$ is not compact.

Exercise 9.3

Hint: In a compact manifold any sequence of points has an accumulation point. Find a suitable sequence of pseudo-rotations (boosts).

Express the exponentials $\exp(\xi M_+)$ and $\exp(\xi M_-)$ of the matrices

Exercise 9.4

$$M_\pm = \begin{pmatrix} 0 & 1 \\ \pm 1 & 0 \end{pmatrix} \qquad (9.27)$$

in terms of trigonometric or hyperbolic functions. Which range of values of the variable ξ is needed to describe these exponentials, and how is this related to the sign of $\mathrm{tr}((M_\pm)^2)$?

a) Compute the exponential of the matrix M whose entries are

Exercise 9.5

$$M_{ij} = \xi\delta_{i,j} + \delta_{j,i+1}, \qquad (9.28)$$

with ξ some complex number.

b) Show that $\exp(NMN^{-1}) = N \exp(M) N^{-1}$ for finite-dimensional matrices M, N.

c) Any finite-dimensional matrix $\tilde{M}$ can be written as $\tilde{M} = NMN^{-1}$ for some matrices M and N, such that M is block-diagonal and each of its diagonal blocks $M_{(k)}$ is of the form (9.28). (The matrix M is known as the Jordan normal form of $\tilde{M}$, and its blocks $M_{(k)}$ as Jordan blocks.) Combine this information with the results of a) and b) to describe the exponentiation of arbitrary finite-dimensional matrices.

A maximal torus G_o of a Lie group G is a Lie subgroup whose Lie algebra is a Cartan subalgebra of the Lie algebra of G.
Describe a maximal torus for each of the classical Lie groups.

Exercise 9.6

Derive the Maurer–Cartan structure equation (9.10), and show that the relation $d^2\zeta_a = 0$ is equivalent to the Jacobi identity.

Exercise 9.7

Show that the Lie algebra of the Lie group SO(3) has the commutation relations (9.12) and is spanned by the Pauli matrices. Compute the Killing form to conclude that this real form of $\mathfrak{sl}(2)$ is compact.

Exercise 9.8

Consider the motion of a rigid body in the absence of external forces. Explain why the motion can be described by two angular coordinates. Hint: Count the number of conservation laws to determine the dimension of invariant subspaces of the phase space. Note that the only two-dimensional compact orientable manifold that admits a global vector field like the one induced by the Hamiltonian is the torus.

Exercise 9.9

Consider a symplectic vector space V of dimension n, with standard symplectic form Ω as in (9.17). Let M be an $n \times n$-matrix and denote by $x^{(i)}$, $i = 1, 2, \dots, n$, the vector of V whose components in the standard symplectic basis are given by the entries of the ith column of M. Show that the determinant of M is

Exercise 9.10

$$\det(M) = \frac{1}{2^{n/2}(n/2)!} \sum_{\sigma \in S_n} \text{sign}(\sigma)\, \Omega(x^{(\sigma(1))}, x^{(\sigma(2))})$$
$$\Omega(x^{(\sigma(3))}, x^{(\sigma(4))}) \cdots \Omega(x^{(\sigma(n-1))}, x^{(\sigma(n))}) \,.$$

(9.29)

Show that under a linear transformation S of V, the determinants of M and the matrix M' constructed from $x^{(i)'} = Sx^{(i)}$ are related as $\det(M') = \det(S) \det(M)$. Then use the formula (9.29) to prove that any symplectic transformation is unimodular.

a) Show that the exponential map for $\mathfrak{sl}(2, \mathbb{R})$ is not surjective.

Exercise 9.11

b) The group $SL(2, \mathbb{R})$ of real 2×2-matrices, which has fundamental group $\mathbb{Z}$, has the complexification $SL(2, \mathbb{C})$, which is simply connected. Show that the universal covering group G of $SL(2, \mathbb{R})$ does not possess a complexification.
Hint: Any complexification would be a universal covering group of $SL(2, \mathbb{C})$ as well.

10
Symmetries of the root system. The Weyl group

10.1 The Weyl group

Already from the diagrams (3.15) and (3.42) for the root systems of simple Lie algebras of rank 2 it is apparent that the root system of a simple Lie algebra has a high degree of symmetry, and also that there exist many choices for the basis of simple roots, which are all equivalent. In this chapter we will have a closer look at these structures, which are of great importance for various applications. In accordance with our general considerations in chapter 1, symmetries of the root system of a Lie algebra – which are in fact symmetries of a structure which can itself play the rôle of a symmetry of some other system – form a group, which is called the *automorphism group* $Aut(\Phi)$ of the root system Φ. For finite-dimensional simple Lie algebras, $Aut(\Phi)$ is a finite group. In many applications, a subgroup W of the automorphism group, known as the *Weyl group*, is particularly relevant.

A distinguished element of this symmetry group can be inferred from the fact that for any root α also $-\alpha$ is a root: The mapping

$$\alpha \mapsto -\alpha \tag{10.1}$$

for all roots α is clearly a symmetry of the root system. In the case of the Lie algebra A_1 where there are only two roots $\pm\alpha$, each of which can take the rôle of the simple root, this is the only available non-trivial map. Now for any weight λ of a finite-dimensional representation of A_1 also $-\lambda$ is a weight of that representation; therefore this reflection also leaves the weight system of any such representation invariant. Together with the identity map, the mapping (10.1) forms the group $\mathbb{Z}_2 \cong \{\pm 1\}$, and hence the automorphism group of the A_1 root system is isomorphic to $\mathbb{Z}_2$.

In the general case, there are however many more symmetries. Before starting to describe them, we should make clear which requirements a map on the set of roots must satisfy in order to qualify as a symmetry of the root system. There are two conditions: First, since the roots are linear functions on the Cartan subalgebra, the map must be linear and invertible; second, the map must permute the roots. From the second property it follows that the automorphism group $Aut(\Phi)$ must be contained in the symmetric group $\mathcal{S}_{d-r}$ of permutations of the $d - r = \dim\mathfrak{g} - \operatorname{rank}\mathfrak{g}$ roots. Further, together the two properties imply that the image of the chosen system Φ_s of simple roots is another allowed set Φ'_s of simple roots, corresponding to a different convention for dividing Φ into positive and negative roots. But an arbitrary permutation of the roots cannot be described by a linear mapping, nor will the image of the basis of simple roots typically any longer satisfy the requirements on simple roots. Consequently in the general case the automorphism group is much smaller than $\mathcal{S}_{d-r}$. For example, a permutation which exchanges any two prescribed roots α and $\beta \neq \pm\alpha$ and leaves all other roots fixed is never an automorphism of Φ, and for non-simply laced algebras any permutation which exchanges roots of different length is not an automorphism either.

Despite these restrictions, it is not difficult at all to describe more general automorphisms than (10.1) explicitly. Namely, consider for any fixed root α the reflection w_α with respect to the hyperplane (through zero) in root space that is perpendicular to α, i.e. subtract from each root $\beta \in \Phi$ twice the component in α-direction:

$$w_\alpha: \quad \beta \mapsto w_\alpha(\beta) := \beta - 2\frac{(\beta,\alpha)}{(\alpha,\alpha)}\alpha. \tag{10.2}$$

This is indeed a permutation of Φ: For any root β, the inner product $m := (\beta, \alpha^\vee) \equiv 2(\beta, \alpha)/(\alpha, \alpha)$ is an integer. This shows that $w_\alpha(\beta)$ is an element of the root lattice. It is even a root of $\mathfrak{g}$, which can be seen as follows. We decompose the Lie algebra $\mathfrak{g}$ into modules of the $\mathfrak{sl}(2)_\alpha$-subalgebra that is associated to the root α; such a module consists of the weight spaces for all weights that belong to a root string in α-direction. Now w_α acts as the reflection of weight space with respect to the hyperplane perpendicular to α; hence on any root string it acts like the unique Weyl reflection of $\mathfrak{sl}(2)_\alpha$. We have already seen that this reflection maps the weights of an $\mathfrak{sl}(2)_\alpha$-module to weights of the same module; hence w_α leaves the root string invariant. In other words, along with β, $w_\alpha(\beta)$ is also a root of $\mathfrak{g}$.

We still have to convince ourselves that the image of the simple roots is again a system of simple roots. This can be seen directly from the defining properties of simple roots; namely, suppose that $w_\alpha(\alpha^{(i)})$ can be written as a positive linear combination of images $w_\alpha(\alpha)$ of positive roots,

$$w_\alpha(\alpha^{(i)}) = \sum_{\beta>0} n_\beta\, w_\alpha(\beta), \qquad n_\beta \geq 0. \tag{10.3}$$

Then by applying w_α to this equation and using $w_\alpha^2 = id$, one would find that the root $\alpha^{(i)}$ can be written as a positive linear combination of positive roots as well, which clearly contradicts the assumption that $\alpha^{(i)}$ is a simple root. Thus we can conclude that the transformation (10.2) is a symmetry in the required sense, i.e. an element of the automorphism group $Aut(\Phi)$.

As is already implicit in the argument just given, the reflections w_α can be multiplied; the product is just given by the composition of maps, i.e.

$$w_\alpha\, w_{\alpha'} \equiv w_\alpha \circ w_{\alpha'}\,. \tag{10.4}$$

Further, since w_α can be interpreted geometrically as a reflection, in particular any such transformation possesses an inverse, namely itself, and of course there is also a unit element for the composition of reflections, namely the identity map. As a consequence, the maps w_α, $\alpha \in \Phi$, generate a discrete group which is a subgroup of the automorphism group. This group is known as the *Weyl group* of the root system $\Phi \equiv \Phi(\mathfrak{g})$, or as the Weyl group of the Lie algebra $\mathfrak{g}$; it is denoted by $W(\mathfrak{g})$, or shortly by W. Note that the composition of reflections leads again to reflections as well as to rotations. Thus a generic element of W is *not* of the form (10.2).

By linearity, the action of the Weyl group can be extended from the root *system* to the whole root (or weight) *space* of $\mathfrak{g}$. Then for any $\mathfrak{g}$-weight λ the Weyl reflection w_α acts as

$$\lambda \mapsto w_\alpha(\lambda) := \lambda - (\lambda, \alpha^\vee)\,\alpha\,. \tag{10.5}$$

Since an arbitrary element w of the Weyl group is the product of reflections, it must leave the inner product on weight space invariant. To check this, we compute

$$\begin{aligned}
(w_\alpha(\lambda), w_\alpha(\mu)) &= (\lambda, \mu) - (\lambda, \alpha^\vee)\,(\mu, \alpha) - (\mu, \alpha^\vee)\,(\lambda, \alpha) \\
&\quad + (\lambda, \alpha^\vee)\,(\mu, \alpha^\vee)\,(\alpha, \alpha) \\
&= (\lambda, \mu) + (-1 - 1 + 2)\,(\lambda, \alpha^\vee)\,(\mu, \alpha) = (\lambda, \mu)\,.
\end{aligned} \tag{10.6}$$

Thus any Weyl group element is an isometry. When we describe the action of a Weyl group element w on the weights λ by a matrix M_w acting on the Dynkin components λ^i as $(w(\lambda))^i = \sum_j (M_w)^i{}_j \lambda^j$, this property means that the matrix M_w is an element of the group $O(r)$ of orthogonal $r \times r$ matrices, and hence it satisfies

$$M_w M_w^t = \mathbb{1}\,, \tag{10.7}$$

where $\mathbb{1}$ is the $r \times r$ unit matrix.

As it turns out, the Weyl group W already exhausts the automorphisms of the root system to a large extent. More precisely, W acts *transitively* as well as *freely* on the set $\{\Phi_s\}$ of bases of simple roots. Transitivity means that any basis of simple roots can be obtained from a given one by applying a suitable element w_α of W; that the action is free means that this Weyl transformation w_α is unique (in particular, $w_\alpha(\Phi_s) = \Phi_s$ if and only if $w_\alpha = id$). As a consequence we have:

> Any automorphism of the root system is a composition
> of a Weyl group element and a permutation which acts
> completely within the set Φ_s of simple roots.

Further, from the relation (6.34) between the simple roots and the Cartan matrix of $\mathfrak{g}$, we can conclude that among the latter only those are allowed which correspond to a symmetry of the Cartan matrix, and hence to a symmetry of the Dynkin diagram of $\mathfrak{g}$. In short, the non-Weyl group part of the automorphism group $Aut(\Phi)$ is given by a subgroup $\Gamma \subseteq Aut(\Phi)$ which describes the ambiguity in ordering the simple roots within a given basis and hence is isomorphic to the group of symmetries of the Dynkin diagram of $\mathfrak{g}$.

If $\pi \in Aut(\Phi)$ is a permutation that is induced by a symmetry of the Dynkin diagram, then for any Weyl transformation $w \in W$ the automorphism w' obtained by *conjugation* of w by π, i.e. $w' = \pi w \pi^{-1}$, is again a Weyl transformation. For any subgroup H of G one defines the (right) *coset space* G/H as the set $\{[\gamma] \mid \gamma \in G\}$ of all equivalence classes $[\gamma] := \{\gamma\gamma' \mid \gamma' \in H\}$ of elements $\gamma \in G$ modulo H. This set is again a group (called the *factor group* of G modulo H) if and only if H is a *normal* subgroup of G. (Recall that a subgroup H of a group G is called a normal subgroup if $\gamma\gamma'\gamma^{-1} \in H$ for all $\gamma' \in H$ and all $\gamma \in G$, or equivalently, if left and right cosets coincide.) Thus the Weyl group W is a normal subgroup of Aut, and $\Gamma = Aut/W$ is again a group. Another way to describe this situation is by saying that Aut has the structure of a *semidirect product* of W and Γ, which is denoted as

$$Aut(\Phi) = \Gamma \ltimes W .\tag{10.8}$$

10.2 Fundamental reflections. Length and sign

Since the size of the root system of simple Lie algebras grows rapidly with the rank, the description of the Weyl group as being generated by the reflections with respect to all roots is not too convenient. But in fact already the whole Weyl group is generated by a small number of specific transformations of the type (10.2), namely by the reflections corresponding to the simple roots (see exercise 10.3 for an example). These are called the *fundamental* or *simple reflections*. That the whole Weyl group is generated by the fundamental reflections is most fortunate, because (just as for any other group that is given in terms of generators and relations) it allows for a convenient two-step procedure for checking statements about the group, namely first to check that an assertion is valid for the generators, and then to verify that it extends to arbitrary products.

The simple reflections are denoted by

$$w_{(i)} \equiv w_{\alpha^{(i)}} \tag{10.9}$$

for $i = 1, 2, \ldots, r$. Thus any $w \in W$ can be written as a 'word'

$$w = w_{(i_1)} w_{(i_2)} \cdots w_{(i_n)}, \tag{10.10}$$

also called the *Weyl word* of w, in the 'letters' $w_{(i)}$. However, any given Weyl transformation may be described by many different words. Particularly important are those decompositions of w which consist of a minimal number of letters; such a decomposition is known as a *reduced* Weyl word, and the minimal possible number of letters in a word for w is called the *length* $\ell(w)$ of the Weyl group element. Generically, also the reduced Weyl word is not unique.

If $w_{(i_1)} w_{(i_2)} \cdots w_{(i_n)}$ is a word for w, then $w_{(i_n)} w_{(i_{n-1})} \cdots w_{(i_1)}$ is a word for w^{-1}; this implies that the length ℓ obeys $\ell(w) = \ell(w^{-1})$. Further, we have the inequalities (see exercise 10.4)

$$\ell(w) + \ell(w') \geq \ell(w' \circ w) \geq |\ell(w) - \ell(w')| \tag{10.11}$$

for any two Weyl group elements w, w'. Also, for finite-dimensional simple Lie algebras the length is bounded from above. The 'longest' element is unique, i.e. there exists a $w_{\max} \in W$ such that

$$\ell(w_{\max}) > \ell(w) \quad \text{for all } w \in W \setminus \{w_{\max}\}. \tag{10.12}$$

This element maps Φ_s to $-\Phi_s$; it is sometimes called the *opposition* element of W. If $-id$ is an element of the Weyl group, then $w_{\max} = -id$. This holds for all finite-dimensional simple Lie algebras except for A_r with $r \geq 2$, D_r with r even and E_6, where $w_{\max}$ acts as $\lambda \mapsto -\lambda^+$; here λ^+, called the weight *conjugate* to λ, is a weight of the conjugate representation R^+ – defined as in section 5.3 – if λ is a weight of R (see also the end of section 13.3). In short, in all cases $w_{\max}$ sends a weight to minus its conjugate weight. Finally there is a rather explicit formula for the length: It is the number of all positive roots whose pre-image under the transformation is a negative root, i.e.

$$\ell(w) = |\{\alpha \in \Phi \mid \alpha > 0, \, w^{-1}(\alpha) < 0\}|. \tag{10.13}$$

In particular, each non-trivial Weyl transformation sends at least one positive root to a negative root. A related property of the length function, which often proves to be useful, is the following: one has $\ell(w_{(i)} w) = \ell(w) + 1$ if and only if $w^{-1}(\alpha^{(i)})$ is a positive root.

For many applications, it only matters whether the length of a Weyl transformation is even or odd. Therefore one introduces the *sign* of a

Weyl group element,

$$\text{sign}(w) := (-1)^{\ell(w)}. \tag{10.14}$$

That $\text{sign}(w)$ depends on the length $\ell(w)$ only modulo 2 is quite fortunate, since the number of letters modulo 2 in a Weyl word for $w \in W$ does not depend on the choice of the word. This implies that if $w_n w_{n-1} \cdots w_2 w_1$ is *any* word for w, not necessarily a reduced one, we can already conclude that

$$\text{sign}(w) = (-1)^n. \tag{10.15}$$

In particular it is now easy to show that

$$\text{sign}(w \circ w') = \text{sign}(w) \cdot \text{sign}(w') \tag{10.16}$$

for any two Weyl transformations w, w'. Elements of the Weyl group with positive signs are pure rotations, while elements with negative sign are the product of a rotation and a reflection. The pure rotations form a normal subgroup of the Weyl group. In terms of the orthogonal matrices $M_w \in O(\text{r})$ (10.7), the sign of the Weyl transformation w is precisely the determinant,

$$\text{sign}(w) = \det(M_w) \in \{\pm 1\}. \tag{10.17}$$

As examples, let us now describe the Weyl groups of the algebras A_1 and A_2 in terms of fundamental reflections. For A_1, the only roots are $\pm \alpha$, and correspondingly there is a single generator $w \equiv w_{(1)}$ acting on weights $\lambda \in \mathbb{Z}$ as multiplication by -1,

$$w(\lambda) = -\lambda. \tag{10.18}$$

The Weyl group is therefore isomorphic to $\mathcal{S}_2 = \mathbb{Z}_2 = \{\pm 1\}$, as already noted above.

For A_2, the Weyl group looks a bit more complicated. It is now generated by two fundamental reflections. They act on weights $\lambda \equiv (\lambda^1, \lambda^2)$ as (see exercise 10.1)

$$\begin{aligned} w_{(1)}: \quad & (\lambda^1, \lambda^2) \mapsto (-\lambda^1, \lambda^1 + \lambda^2), \\ w_{(2)}: \quad & (\lambda^1, \lambda^2) \mapsto (\lambda^1 + \lambda^2, -\lambda^2). \end{aligned} \tag{10.19}$$

From this formula one can directly read off the matrix representations for the action of the Weyl group on weight space. It also follows (compare exercise 10.1, and also the picture (10.21) below) that besides these two elements of length 1, there are two elements of length 2, and one element of length 3, such that the Weyl group of A_2 has six elements,

$$W(A_2) = \{id, w_{(1)}, w_{(2)}, w_{(1)} \circ w_{(2)}, w_{(2)} \circ w_{(1)}, w_{(1)} \circ w_{(2)} \circ w_{(1)}\}, \tag{10.20}$$

and that the group multiplication rules are those of the permutation group $\mathcal{S}_3$.

10.3 Weyl chambers

When removing from the root space all hyperplanes (through the origin) that are perpendicular to some root, the root space gets divided into a fan of open cones. These cones are called *Weyl chambers*. The Weyl chambers are all congruent, but having made a choice for the simple roots, nevertheless one of them is distinguished. This is the unique chamber whose points have only positive Dynkin labels; it is called the *fundamental* or *dominant* Weyl chamber, or also the fundamental Weyl domain.

The effect of Weyl transformations on the chambers is analogous as for the root system, that is, W acts transitively and freely on the Weyl chambers. Thus in particular by acting with the Weyl group on the fundamental Weyl chamber including its boundary, one obtains the whole root space. Conversely, to any weight λ which does not lie on the boundary of some Weyl chamber, there is associated a unique Weyl transformation w_λ such that the image $w_\lambda(\lambda)$ belongs to the fundamental chamber, while to any λ on the boundary of some chamber there exists a (non-unique) Weyl transformation which maps it to the boundary of the fundamental chamber. Because of the transitive and free action of Weyl group on the Weyl chambers, one can label the chambers by the elements of W; the fundamental chamber then corresponds to the identity element of W. This labelling is displayed for $\mathfrak{g} = A_2$ in the following picture.

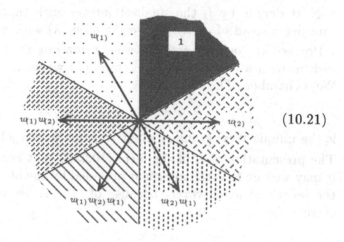

$$(10.21)$$

The points of the weight space that stay fixed under some Weyl transformation are precisely the points in the hyperplanes perpendicular to the roots, or in other words, in the union of the boundaries of all Weyl chambers (this is the reason why one defines the chambers as *open* cones). Thus the fixed points include in particular the weights on the boundary of the fundamental Weyl chamber. The latter are precisely those weights

λ for which one or more Dynkin labels λ^i vanish and the rest of the Dynkin labels are positive; the transformations which leave such a weight invariant are the corresponding fundamental reflections $w_{(i)}$:

$$w_{(i)}(\lambda) = \lambda \qquad \text{if } \lambda^i = 0. \tag{10.22}$$

The size of the Weyl group depends rather strongly on the rank of the Lie algebra; we will see in the next section that e.g. for A_n the Weyl group has $n!$ elements. As a consequence, already for Lie algebras with rather small rank the implementation of the entire Weyl group on a computer is a (computer) time consuming task. Fortunately, for many applications it is not necessary to know all Weyl group elements individually, but rather it is sufficient to be able to construct for any given weight λ which does not lie on the boundary the unique element $w_\lambda \in W$ as a product of fundamental Weyl reflections $w_{(l)}$. This construction is achieved by the following simple and efficient algorithm, which can be easily implemented on a computer (see exercise 10.5 for an example):

• To start, let us denote by $j_1 \in \{1, 2, \dots, \text{rank } \mathfrak{g}\}$ the smallest integer such that the j_1th Dynkin component of λ is the 'first' negative one, i.e. $\lambda^{j_1} < 0$, but $\lambda^i \geq 0$ for $i < j_1$ (if there is no negative Dynkin label, then λ already lies in the fundamental Weyl chamber, so that $w_\lambda = id$). Then consider instead of λ the Weyl-transformed weight $w_1(\lambda)$, where $w_1 := w_{(j_1)}$.

• Next denote by j_2 the smallest integer such that $(w_1(\lambda))^{j_2} < 0$, and consider instead of $w_1(\lambda)$ the weight $w_2 w_1(\lambda)$ with $w_2 := w_{(j_2)}$.

• Proceed analogously with the weight now obtained, and so on, until you end up with a weight $w_n w_{n-1} \cdots w_2 w_1(\lambda)$ which lies in the fundamental Weyl chamber. Then

$$w = w_n w_{n-1} \cdots w_2 w_1 \tag{10.23}$$

is the unique Weyl group element which does the job.

The presentation of w in the form (10.23) is not necessarily reduced, i.e. it may well be that the integer n provided by the algorithm is larger than the length of w; however, because of (10.15) we still can read off the correct sign of w.

10.4 The Weyl group in the orthogonal basis

Since the elements of the Weyl group are orthogonal transformations, the structure of the Weyl group becomes particularly evident in the orthonormal basis of the weight space that was described in table VII. Let us consider the case of A_r in some detail. In this case inspection of the list of the roots in the orthogonal basis shows (see exercise 10.3) that the fundamental reflection $w_{(i)}$ acts as

$$w_{(i)}: \quad \begin{cases} \ell^i \leftrightarrow \ell^{i+1}, \\ \ell^j \mapsto \ell^j \quad \text{for } j \neq i, i+1 \end{cases} \tag{10.24}$$

on the coordinates ℓ^j of a weight λ in this basis. As a consequence any Weyl group element acts as a permutation of the $r+1$ coordinates ℓ^j, and hence the Weyl group is isomorphic to the symmetric group on $r+1$ objects,

$$W(A_r) \cong S_{r+1}. \tag{10.25}$$

For the other classical algebras, the Weyl group induces again permutations of the coordinates in the orthonormal basis, but in addition also sign changes $\ell^j \mapsto -\ell^j$ appear. Let us explain this for the case of D_r.

First recall from table VII that in the orthonormal basis the roots are the vectors with components $(\pm 1, \pm 1, 0, ..., 0)$ or $(\pm 1, \mp 1, 0, ..., 0)$, as well as the vectors obtained from these by arbitrary permutations of the coordinates. Also, a basis of r simple roots is given by

$$\alpha^{(1)} = (1, -1, 0, ..., 0), \quad \alpha^{(2)} = (1, 0, -1, 0, ..., 0), \quad \cdots$$
$$\cdots, \quad \alpha^{(r-1)} = (0, ..., 0, 1, -1) \tag{10.26}$$

together with

$$\alpha^{(r)} = (0, ..., 0, 1, 1). \tag{10.27}$$

The automorphism group $Aut(\Phi)$ acts by permuting the components in the orthogonal basis and multiplying them by signs. Among these mappings, only those with an even number of sign flips are elements of the Weyl group. This can be seen as follows. Assume that there existed a Weyl group element with an odd number of sign changes. By multiplying it with suitable permutations we could then obtain an element of W for which the permutation part is trivial. Further, multiplying with elements having an even number of signs, we would then obtain an element which maps the last coordinate to minus itself and leaves any other coordinate fixed. This element however maps the set (10.26) and (10.27) of simple roots on itself and merely exchanges the simple roots $\alpha^{(r-1)}$ and $\alpha^{(r)}$. But the Weyl group acts freely on the set $\{\Phi_s\}$ of bases of simple roots, and hence this automorphism of the root system cannot be an element of W (rather, it corresponds to the symmetry of the D_r Dynkin diagram which exchanges the $(r-1)$th and rth node). Thus the assumption of having a Weyl transformation with an odd number of sign flip leads to a contradiction.

From these considerations it follows that the Weyl group of D_r has the structure of a semidirect product

$$\mathcal{S}_r \ltimes (\mathbb{Z}_2)^{r-1} , \tag{10.28}$$

where the first factor describes the permutation part and the second factor the normal subgroup given by the possible assignments of an even number of sign flips to the components.

In the following table we display the number $|W|$ of elements of the Weyl groups W of the simple Lie algebras, and also the group structure of W, except for the E-series.

$\mathfrak{g}$	$|W|$	W
A_r	$(r+1)!$	$\mathcal{S}_{r+1}$
B_r, C_r	$2^r \cdot r!$	$\mathcal{S}_r \ltimes (\mathbb{Z}_2)^r$
D_r	$2^{r-1} \cdot r!$	$\mathcal{S}_r \ltimes (\mathbb{Z}_2)^{r-1}$
E_6	$2^7 \cdot 3^4 \cdot 5$	
E_7	$2^{10} \cdot 3^4 \cdot 5 \cdot 7$	
E_8	$2^{14} \cdot 3^5 \cdot 5^2 \cdot 7$	
F_4	$2^7 \cdot 3^2$	$\mathcal{S}_3 \ltimes \mathcal{S}_4 \ltimes (\mathbb{Z}_2)^3$
G_2	$2^2 \cdot 3$	$\mathcal{D}_6$

$$\tag{10.29}$$

In (10.29), $\mathcal{S}_n$ denotes the symmetric group of n objects, and $\mathcal{D}_6$ the dihedral group (see exercise 10.2 for the explicit structure of the group $\mathcal{D}_6$).

This description of the Weyl group provides another algorithm for determining the weight $w_\lambda(\lambda)$ in the fundamental Weyl chamber that lies on the same Weyl group orbit as some given weight λ. The basic observation is that according to the formulæ (7.26) the weights in the fundamental Weyl chamber can be characterized in terms of their components ℓ^i in the orthonormal basis of weight space as follows. For $\mathfrak{g} = A_r$ they satisfy $\ell^1 \geq \ell^2 \geq \ldots \geq \ell^r$, in the case of $\mathfrak{g} = B_r$ or C_r the requirement is $\ell^1 \geq \ell^2 \geq \ldots \geq \ell^r \geq 0$, and for $\mathfrak{g} = D_r$ one needs $\ell^1 \geq \ell^2 \geq \ldots \geq \ell^{r-1} \geq 0$. Accordingly one just starts by computing the components of λ in the orthonormal basis. Afterwards, in the case of A_r all that remains to be done is to permute these components until they form a monotonically decreasing sequence. Similarly, for B_r or C_r one first flips the signs of the components until all of them are positive and then permutes them until they form a monotonically decreasing sequence. In the case of D_r one has to apply the same procedure, but only to the first $r-1$ components.

Note that the Weyl groups of B_r and C_r are isomorphic. This is an immediate consequence of the fact that the respective root systems differ from each other only by exchanging long and short roots, since the root length cancels out in the transformation (10.2).

The structure of the Weyl groups of the E-series is quite complicated. In the orthonormal basis of the weight space the reflections with respect to those roots α which have integral (rather than half-integral) components act again by signed permutations of the components, but generic Weyl group elements act in a more complicated manner. One can also show that any Weyl group element can be written as the product of a reflection with respect to a root with integral components and at most two reflections with respect to roots which also have half-integral components.

A different possibility for studying these Weyl groups consists in analyzing their coset decomposition with respect to the Weyl group of a suitable classical Lie subalgebra, compare e.g. [King and Al-Qubanchi 1981]. Yet another description of the Weyl groups of the E-series is in terms of Lie-type groups over the finite field of characteristic 2.

10.5 The Weyl vector

In this section a few more properties of the Weyl group are listed which have some relevance in applications. In particular we derive the identity $\rho = \sum_{i=1}^{r} \Lambda_{(i)}$ (see equation (7.22)) for the Weyl vector ρ of $\mathfrak{g}$.

First, for any root α, the image $W(\alpha)$ spans the whole root space, and for any basis Φ_s of simple roots, the image under the Weyl group consists of the whole root system,

$$W(\Phi_s) = \Phi. \tag{10.30}$$

Second, the fundamental reflection with respect to the simple root $\alpha^{(i)}$ acts on the Dynkin components of weights λ as (see exercise 10.3)

$$\left(w_{(i)}(\lambda)\right)^j = \lambda^j - A^{ij}\,\lambda^i, \tag{10.31}$$

where A^{ij} are the entries of the Cartan matrix of $\mathfrak{g}$. It follows in particular that the reflection $w_{(i)}$ sends the simple root $\alpha^{(i)}$ to $-\alpha^{(i)}$. Moreover, since the inner product of any two simple roots is non-positive, $w_{(i)}$ permutes the rest of the positive roots:

$$w_{(i)}(\Phi_+ \backslash \{\alpha^{(i)}\}) = \Phi_+ \backslash \{\alpha^{(i)}\}. \tag{10.32}$$

An important consequence of (10.32) is that the reflection $w_{(i)}$ subtracts $\alpha^{(i)}$ from the Weyl vector $\rho = \frac{1}{2} \sum_{\alpha>0} \alpha$ of $\mathfrak{g}$:

$$w_{(i)}(\rho) = \rho - \alpha^{(i)}. \tag{10.33}$$

Together with the invariance (10.6) of scalar products, this result is the key to the derivation of the formula $\rho = \sum_{i=1}^{r} \Lambda_{(i)}$. Namely, when considering (10.6) for the special case $w = w_{(i)}$ and $\mu = \rho$, $\lambda = \alpha^{(i)\vee}$ and inserting (10.33), one obtains

$$\begin{aligned} 0 &= (\alpha^{(i)\vee}, \rho) - (w_{(i)}(\alpha^{(i)\vee}), w_{(i)}(\rho)) = (\alpha^{(i)\vee}, \rho + w_{(i)}(\rho)) \\ &= (\alpha^{(i)\vee}, 2\rho - \alpha^{(i)}) = 2\,(\rho^i - 1). \end{aligned} \tag{10.34}$$

This holds for arbitrary values $i \in \{1, 2, ..., \text{rank } \mathfrak{g}\}$, and hence indeed the formula (7.22) follows.

Another property of the Weyl group is that it can be endowed with a partial ordering, called the *Bruhat ordering* and denoted by $w \preceq w'$. (A partial ordering '$\preceq$' of a set S is a relation with the property that $a \preceq a$, and that the relations $a \preceq b$ and $b \preceq c$ for $a, b, c \in S$ imply $a \preceq c$; but for arbitrary $a, b \in S$ it is allowed that neither the relation $a \preceq b$ nor $b \preceq a$ holds.) To this end one introduces the notation '$w_1 \leftarrow w_2$' for the situation that the respective lengths are related by $\ell(w_1) = \ell(w_2) + 1$ and that $w_1 = w_\alpha \circ w_2$ for some positive root α. Then the ordering relation $w \preceq w'$ is said to hold if and only if there exist Weyl group elements $w_1, w_2, ..., w_n$ such that $w \leftarrow w_1 \leftarrow w_2 \leftarrow \cdots \leftarrow w_n \leftarrow w'$.

Finally we mention that a distinguished element of W is the *Coxeter element* w_c which is defined as the product $w_c := w_{(1)} w_{(2)} \cdots w_{(r)}$ of all fundamental reflections. While the Coxeter element itself depends on the choice of a basis of simple roots, the order of w_c, i.e. the lowest integer n such that $(w_c)^n = id$, is the same for any basis of simple roots; it is equal to the Coxeter number of $\mathfrak{g}$.

10.6 The normalizer of the Cartan subalgebra

Above we have introduced the Weyl group as the group generated by the reflections that leave the root system invariant. However, the Weyl group arises also in a different context, which plays an important rôle in many physical applications as well.

Given a simple Lie group G with Lie algebra $\mathfrak{g}$, for any group element $\gamma \in G$ and any Lie algebra element $x \in \mathfrak{g}$ the formula

$$R_\gamma(x) := \gamma \, x \, \gamma^{-1} \qquad\qquad (10.35)$$

defines an action of G on the vector space $\mathfrak{g}$. The operation (10.35) is called *conjugation* of the Lie algebra element x by the group element γ. This terminology, which originates from the description of products of the type (10.35) in groups, makes sense because both γ and x are here considered as matrices, so that the product in (10.35) is just ordinary matrix multiplication. But even if G is not a matrix Lie group, (10.35) can be given a sense. Namely, given a local one-parameter group $\{\gamma'(t)\}$ in G (for the definition see section 1.2), for any group element $\gamma \in G$ we can define a map from an interval of the real axis around zero to the group by

$$t \mapsto R_\gamma(\gamma'(t)) := \gamma \, \gamma'(t) \, \gamma^{-1} \,. \qquad\qquad (10.36)$$

The reader should check that $\{R_\gamma(\gamma'(t))\}$ is again a local one-parameter group. Taking the derivative of (10.36) on both sides at $t = 0$ supplies us with a linear map defined on the Lie algebra. In the case of matrix Lie algebras, this is just the map given in (10.35).

Often one is not so much interested in individual elements of the Lie algebra $\mathfrak{g}$, but rather in orbits with respect to conjugation, i.e. one wants to consider two elements of $\mathfrak{g}$ that are connected by conjugation as equivalent. It is then necessary to have a good characterization of the orbit space at hand. A first reduction in the analysis of this space is given by the following theorem: Fix any Cartan subalgebra $\mathfrak{g}_\circ$ of $\mathfrak{g}$; then for any $x \in \mathfrak{g}$ there exists some group element $\gamma \in G$ such that $y := \gamma x \gamma^{-1}$ is an element of $\mathfrak{g}_\circ$. To describe the conjugation orbits, we can therefore restrict our attention to the Cartan subalgebra.

However, different elements of the Cartan subalgebra can still belong to the same equivalence class, and hence for given $x \in \mathfrak{g}$ there are different elements $y_i = \gamma_i x \gamma_i^{-1} \in \mathfrak{g}_\circ$ in the Cartan subalgebra that can be obtained via conjugation with different group elements γ_i. To get an idea of how these different elements are related, we consider the subset $\mathcal{N}_G(\mathfrak{g}_\circ)$ of those group elements which by conjugation map the Cartan subalgebra $\mathfrak{g}_\circ$ to itself. It is easy to see that $\mathcal{N}_G(\mathfrak{g}_\circ)$ is in fact a subgroup of G; it is called the *normalizer* of the Cartan subalgebra $\mathfrak{g}_\circ$.

But not all elements of the normalizer induce different linear mappings of the Cartan subalgebra by conjugation. The elements of $\mathcal{N}_G(\mathfrak{g}_\circ)$ that obey $\gamma y \gamma^{-1} = y$ for all $y \in \mathfrak{g}_\circ$, i.e. for which the associated conjugation is just the identity map, form a normal subgroup of $\mathcal{N}_G(\mathfrak{g}_\circ)$, called the *centralizer* $\mathcal{C}_G(\mathfrak{g}_\circ)$ of the Cartan subalgebra. The different non-trivial mappings of the Cartan subalgebra that relate elements which are equivalent under conjugation therefore form the quotient group $\mathcal{N}_G(\mathfrak{g}_\circ)/\mathcal{C}_G(\mathfrak{g}_\circ)$. With considerable effort one can show that this group is a discrete group and is in fact isomorphic to the Weyl group of $\mathfrak{g}$:

$$\mathcal{N}_G(\mathfrak{g}_\circ)/\mathcal{C}_G(\mathfrak{g}_\circ) \cong W(\mathfrak{g}) . \tag{10.37}$$

Indeed, not only are these two sets isomorphic as groups, but also their action on the Cartan subalgebra $\mathfrak{g}_\circ$, respectively on the weight space $\mathfrak{g}_\circ^\star$, are dual. Namely, via the group isomorphism (10.37) we associate to any element $w^\star \in \mathcal{N}_G(\mathfrak{g}_\circ)/\mathcal{C}_G(\mathfrak{g}_\circ)$ a Weyl group element w such that for all $\lambda \in \mathfrak{g}_\circ^\star$ and all $h \in \mathfrak{g}_\circ$ we have

$$w(\lambda)\,(w^\star(h)) = \lambda(h) . \tag{10.38}$$

(In exercise 10.7 the reader can work out this correspondence in more detail for the case $\mathfrak{g} = A_1$.) As a consequence, also the Cartan subalgebra $\mathfrak{g}_\circ$ can be decomposed into Weyl chambers and we can now give an easy characterization of the regular elements of the Cartan subalgebra as described in section 6.1: They are precisely the elements in the interior of a Weyl chamber. From this characterization it is obvious that the regular elements form a dense subset of $\mathfrak{g}_\circ$. (Thus any regular element h can be used to define a decomposition of the root system into positive and negative roots: A root α is positive if $\alpha(h) > 0$ and negative if $\alpha(h) < 0$.)

A quite non-trivial theorem states that any two elements in $\mathfrak{g}_\circ$ that are conjugate are related by the action of some Weyl group element. In other words, the intersection of the conjugation orbits with the Cartan subalgebra coincides with the orbits of the Weyl group action on $\mathfrak{g}_\circ$.

The so-called moduli space of vacua of a gauge theory with structure group G is parametrized by 'vacuum expectation values' $\phi_\circ \equiv \langle\phi\rangle$ of Lorentz-scalar fields ϕ (Higgs fields) which carry non-trivial representations of the Lie algebra $\mathfrak{g}$ of G. However, only the gauge invariant information contained in these vacuum expectation values is relevant.

In particular, for a Higgs field carrying the adjoint representation of $\mathfrak{g}$ one can use the conjugation with elements of G such that $\phi_\circ$ becomes a diagonal matrix, i.e. belongs to the Cartan subalgebra of $\mathfrak{g}$. The remaining redundancy is then given by the Weyl group, and it follows that in this case the moduli space has the structure $\mathbb{C}^r/W$.

Information

10.7 The Weyl group of affine Lie algebras

In complete analogy to simple Lie algebras, one defines the *Weyl reflection* w_α for an affine Lie algebra $\mathfrak{g}$ by

$$w_\alpha: \quad \lambda \mapsto w_\alpha(\lambda) := \lambda - (\lambda, \alpha^\vee)\, \alpha. \tag{10.39}$$

However, because of the appearance of the coroot $\alpha^\vee \equiv 2\alpha/(\alpha, \alpha)$ in the formula (10.39), one must require that $(\alpha, \alpha) \neq 0$. While for simple Lie algebras this is automatic, in the affine case it does not hold for all roots. As we will see in section 12.7, there is a specific root, denoted by δ, such that the roots with $(\alpha, \alpha) = 0$ are all of the form $\alpha = n\delta$ with $n \in \mathbb{Z}$; these roots are called *imaginary* or lightlike roots. All other roots satisfy $(\alpha, \alpha) > 0$; they are called *real* roots. The reflections (10.39) for real roots generate a group, which is again called the *Weyl group* of the affine algebra $\mathfrak{g}$. Just like the Weyl group of a simple Lie algebra, it acts on the weight space by linear mappings.

Many properties of this Weyl group are analogous to those of the Weyl group of a simple Lie algebra. In particular, W is generated by the reflections $w_{(i)} \equiv w_{\alpha^{(i)}}$ with respect to simple roots (which are all real), where now $i = 0, 1, ..., r$ with $r = \text{rank}\,\mathfrak{g} \equiv \text{rank}\,\bar{\mathfrak{g}}$ (recall from chapter 7 that $\bar{\mathfrak{g}}$ is the simple subalgebra of $\mathfrak{g}$ that is obtained by deleting an appropriate node of the Dynkin diagram of $\mathfrak{g}$). Also, each of these elementary reflections $w_{(i)}$ permutes the set of positive roots. There are, however, also several new features which are related to the existence of imaginary roots. The imaginary root δ satisfies $(\alpha, \delta) = 0$ for any real root α, and hence it follows that

$$w_\alpha(\delta) = \delta - (\delta, \alpha^\vee)\, \alpha = \delta. \tag{10.40}$$

Thus any Weyl transformation acts on the set $\Phi_i = \{n\delta \mid n \neq 0\}$ of imaginary roots as the identity map. As any Weyl transformation is an automorphism of the root system, this means that the Weyl group also maps the set Φ_r of real roots onto itself.

It will be seen in chapter 12 that there is a close relation (see (12.35)) between the simple roots of $\mathfrak{g}$ and those of its horizontal subalgebra $\bar{\mathfrak{g}}$. As a consequence, it is not difficult to describe the action of the fundamental reflections $w_{(i)}$ on the $\bar{\mathfrak{g}}$-part $\bar{\lambda}$ of $\mathfrak{g}$-weights λ. In fact, for $i \in \{1, 2, \ldots, \operatorname{rank} \bar{\mathfrak{g}}\}$ they act precisely as the corresponding fundamental reflections of $\bar{\mathfrak{g}}$ and hence are in particular linear mappings on the weight space of $\bar{\mathfrak{g}}$. In contrast, the reflection $w_{(0)}$ with respect to the zeroth simple root does not correspond to a Weyl reflection in the weight space of $\bar{\mathfrak{g}}$, but rather to a reflection – namely with respect to the hyperplane (through zero) perpendicular to the highest $\bar{\mathfrak{g}}$-root $\bar{\theta}$ – supplemented by a certain translation. This shows in particular that $w_{(0)}$ does not act on the weight space of the *horizontal* algebra any more as a linear mapping, but rather as an affine mapping. In short, the Weyl group of the affine Lie algebra $\mathfrak{g}$ acts on the weight space of the horizontal subalgebra $\bar{\mathfrak{g}}$ by means of affine mappings. This is in fact the origin of the qualification 'affine' in the term 'affine Lie algebra'.

The vector by which the weights are translated is the $\bar{\mathfrak{g}}$-weight $k^{\vee} \cdot \bar{\theta}$, where $k^{\vee}$ is the so-called level (see equation (13.41)) of the affine weight λ, which together with the $\bar{\mathfrak{g}}$-part $\bar{\lambda}$ characterizes the weight λ. Alternatively, the mapping provided by the zeroth fundamental reflection can also be described as reflection with respect to the appropriately *shifted* hyperplane. These shifted planes are displayed in the following figure for $\bar{\mathfrak{g}} = A_2$ and $k^{\vee} = 1, 2, 3, 4$:

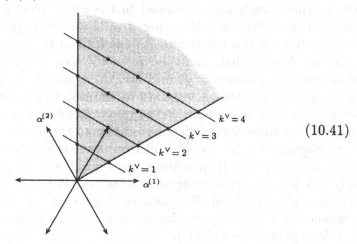

$$(10.41)$$

The above results show that a generic element $w \in W$ is of the form

$$w = \bar{w} \circ t, \qquad (10.42)$$

where $\bar{w}$ is an element of the $\bar{\mathfrak{g}}$-Weyl group $\overline{W}$ and $t \equiv t_{\bar{\beta}}$ denotes the translation by the $k^\vee$-fold of some weight $\bar{\beta}$, which must be an element of the coroot lattice $L^\vee$ of $\bar{\mathfrak{g}}$. Moreover, for all $w \in W$ the relation

$$w \circ t_{\bar{\beta}} \circ w^{-1} = t_{w(\bar{\beta})} \tag{10.43}$$

holds (see exercise 10.8); this means that the abelian group T of translations by $k^\vee$-multiples of elements of the coroot lattice $L^\vee$ is a normal subgroup of W. Furthermore, because the translations satisfy $t_{\bar{\alpha}} \circ t_{\bar{\beta}} = t_{\bar{\alpha}+\bar{\beta}}$, one may actually identify the translation group T with the ($k^\vee$-fold) coroot lattice, $T \cong k^\vee L^\vee$. Since none of these translations belongs to the Weyl group of the subalgebra $\bar{\mathfrak{g}}$, it then follows (together with the formula (10.43)) that W is isomorphic to the semidirect product of $\overline{W}$ and the translation group T:

$$W \cong \overline{W} \ltimes T \cong \overline{W} \ltimes k^\vee L^\vee . \tag{10.44}$$

Note that as an abstract group, $k^\vee L^\vee$ is of course isomorphic to the coroot lattice $L^\vee$. Nevertheless it makes sense to stick to the notation $k^\vee L^\vee$ in (10.44); this emphasizes the important fact that, while the affine Weyl group is independent of the level, its action on those weights which have a definite value $k^\vee$ of the level depends in a rather non-trivial manner on $k^\vee$. (In applications one often indeed only considers weights which all possess the same level, compare e.g. section 12.12 and section 14.11). The coroot lattice of $\bar{\mathfrak{g}}$ is in fact generated by the highest coroot $\bar{\theta}^\vee$ together with its images $\overline{w}(\bar{\theta}^\vee)$ with respect to Weyl transformations $\overline{w} \in \overline{W}$ (thus in the conventional normalization $(\bar{\theta}, \bar{\theta}) = 2$, $L^\vee$ can also be characterized as the lattice generated by the long roots of $\bar{\mathfrak{g}}$). This shows again that all possible translations contained in the affine Weyl group W are the result of the reflection with respect to the zeroth simple root $\alpha^{(0)}$ and of combinations of this reflection with elements of $\overline{W}$.

It also follows that, analogously to the case of $\overline{W}$, the affine Weyl group permutes transitively and freely the *affine Weyl chambers*. Just as for simple Lie algebras, these are by definition those open subsets of the weight space which are obtained by removing all hyperplanes that are left invariant by some Weyl reflection. Owing to the presence of the translation subgroup $T \subset W$, the Weyl chambers are now, however, polytopes of finite volume rather than infinite cones, and hence in particular contain only a finite number of integral weights. Instead of Weyl chambers, one therefore also speaks of Weyl *alcoves*. Note that because of the level dependence of $T \cong k^\vee L^\vee$, the size of the chambers depends on the level as well (compare figure (10.41)).

Among the Weyl chambers, there is again a distinguished one, the *fundamental* or *dominant* affine Weyl chamber, consisting of weights for which all Dynkin components, including the zeroth one, are strictly positive. In the following picture we display schematically the affine Weyl chambers of $A_2^{(1)}$ (or more precisely, their horizontal projection, corresponding to the $\bar{\mathfrak{g}}$-part $\bar{\lambda}$ of $\mathfrak{g}$-weights λ); the shaded region is the fundamental Weyl chamber.

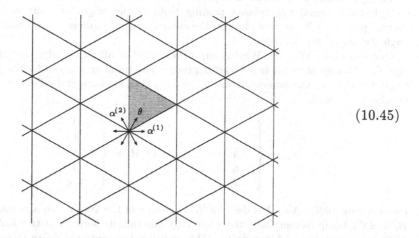

$$(10.45)$$

Moreover, to any weight λ which does not lie on the boundary of some chamber there is associated a unique Weyl transformation w_λ such that $w_\lambda(\lambda)$ belongs to the fundamental affine Weyl chamber. The map w_λ can be determined by an algorithm that is both simple and efficient (and is easily implemented on a computer). This algorithm works precisely as in the case of simple $\mathfrak{g}$, for which it has been described in the paragraph preceding the result (10.23); the only new aspect is that now one must consider also the zeroth Dynkin component, and correspondingly allow also for reflections with respect to the zeroth simple root. Note that due to the translation subgroup, the Weyl group of an affine algebra is an *infinite* group; nevertheless the algorithm still works, and in fact terminates after a finite number of steps. (This result is a key ingredient which allows us to extend the so-called Racah–Speiser algorithm for decomposing tensor products of finite-dimensional simple Lie algebras (to be presented in section 15.7) to compute the so-called fusion products for the associated affine Lie algebras, see section 22.13.)

*10.8 Coxeter groups

Weyl groups are rather special discrete groups; they belong to the class of so-called Coxeter groups. A *Coxeter group* is by definition a group which possesses the following presentation through generators modulo relations: All group elements can be written as products of a finite set $\{\sigma_{(i)} \,|\, i = 1, 2, \dots, n\}$ of generators, which are subject to the relations

$$(\sigma_{(i)})^2 = e, \qquad (\sigma_{(i)}\sigma_{(j)})^{m_{ij}} = e \qquad (10.46)$$

for $i, j = 1, 2, \dots, n$, but are otherwise independent. Here e denotes the unit element of the group, and m_{ij} are positive integers or 'infinity' ∞, where the convention is that $x^\infty \equiv e$ for any x (so that in the case $m_{ij} = \infty$ the second relation in (10.46) is trivial). In particular, $\sigma_{(i)}\sigma_{(j)} = \sigma_{(j)}\sigma_{(i)}$ for $m_{ij} = 2$.

The generators $\sigma_{(i)}$ can be interpreted geometrically as reflections with respect to $(n - 1)$-dimensional hyperplanes passing through the origin of some n-dimensional vector space over $\mathbb{R}$. In this picture, the element $\sigma_{(i)}\sigma_{(j)}$ amounts to a rotation by an angle $2\pi/m_{ij}$ if m_{ij} is finite.

One can check that the Weyl group of a simple or affine Lie algebra (and in fact, any Kac-Moody algebra) is a Coxeter group. The number n given by the size of the Cartan matrix A, the generators are $\sigma_{(i)} = w_{(i)}$ (the fundamental reflections), and the numbers m_{ij} read

$$
m_{ij} = \begin{cases} 2 \\ 3 \\ 4 \\ 6 \\ \infty \end{cases} \quad \text{for} \quad A^{ij}A^{ji} = \begin{cases} 0 \\ 1 \\ 2 \\ 3 \\ \geq 4 \end{cases} \tag{10.47}
$$

(see exercise 10.9). Also, besides the Weyl groups of the simple Lie algebras (those of B_r and C_r being isomorphic), there are only one infinite series and two isolated cases of Coxeter groups which are finite. The additional exceptional finite Coxeter groups are the groups H_3 and H_4 of symmetries of the regular icosahedron and of a certain regular 4-dimensional polytope, while the additional infinite series $I_2(N)$ describes the reflection groups of the regular N-gons in the plane.

Summary:

The root system of a simple Lie algebra is invariant with respect to certain reflections and rotations. The Weyl group is a subgroup of these symmetries of the root system; it can also be interpreted as the quotient of the normalizer of the Cartan subalgebra by its centralizer. The full symmetry group is the semidirect product of the Weyl group and the symmetry group of the Dynkin diagram. The Weyl group is generated by the fundamental reflections.

The Weyl group of an affine Lie algebra $\mathfrak{g}$ is a semidirect product of the Weyl group of its horizontal subalgebra $\bar{\mathfrak{g}}$ and translations by the coroot lattice of $\bar{\mathfrak{g}}$. The explicit form of the translations is level-dependent. The affine Weyl chambers are polytopes of finite volume.

Keywords:

Automorphisms of the root system, fundamental reflection, Weyl group, length and sign of a Weyl transformation, Bruhat ordering, Coxeter element;
Weyl chamber, fundamental chamber, transitive and free action of a group, reflection groups and Coxeter groups;
zeroth affine reflection, coroot lattice, affine Weyl chamber.

Exercises:

Use the formula (10.2) for the action of a Weyl reflection to derive equations (10.19) for the algebra A_2. Write the result in matrix notation, and use matrix multiplication to determine the structure of the Weyl group. By providing an explicit isomorphism, check that this group is isomorphic to the symmetric group S_3 (and hence in particular non-abelian).
To identify the matrix that describes the reflection with respect to the highest root $\theta = \alpha^{(1)} + \alpha^{(2)}$, show that this reflection acts as

$$w_\theta : \quad (\lambda^1, \lambda^2) \mapsto (-\lambda^2, -\lambda^1) . \qquad (10.48)$$

Also show that it is the element of maximal length.
Check that the elements of $W(A_2)$ satisfy the formula (10.13) for the length.

Exercise 10.1

Starting from the picture of the G_2 root system in (3.42), determine the Weyl group structure of G_2 geometrically. Use the formula (10.31) to obtain a matrix description of this group. (This yields an explicit realization of the group $\mathcal{D}_6$.)

Exercise 10.2

Prove that for any simple Lie algebra $\mathfrak{g}$ the fundamental reflections act as claimed in (10.31).
For $\mathfrak{g} = A_r$ show that the action of the fundamental reflections in the orthogonal basis is given by (10.24), and that they generate the entire Weyl group.

Exercise 10.3

Prove that the length of Weyl words satisfies the inequalities (10.11). Hint: first show that $\ell(w) + \ell(w') \geq \ell(w' \circ w)$. The second inequality $\ell(w' \circ w) \geq |\ell(w) - \ell(w')|$ then follows from the first one with the help of the identity $w = (w')^{-1} \circ (w'w)$.

Exercise 10.4

Implement the algorithm described after equation (10.22) for $\mathfrak{g} = A_2$ on a computer.

Exercise 10.5

Consider the number

Exercise 10.6

$$S(\lambda, \mu) := \sum_{w \in W} \text{sign}(w) \, \exp[-\tfrac{2\pi i}{h} \, (w(\lambda + \rho), \mu + \rho)], \qquad (10.49)$$

defined for any two weights λ and μ of a simple Lie algebra $\mathfrak{g}$. Here ρ denotes the Weyl vector of $\mathfrak{g}$, h is some positive integer, and the summation is over the Weyl group W of $\mathfrak{g}$.
Show that $S(\lambda, \mu) = S(\mu, \lambda)$, and that $S(\lambda, \mu)$ vanishes if $\lambda + \rho$ or $\mu + \rho$ is a weight on a boundary of an affine Weyl chamber at level h.
(The formula (10.49) is, up to a normalization, the so-called Kac–Peterson formula for the modular S-matrix, see section 14.11.)

The action of the Weyl group of A_1 can be extended from an action on root space to an action on A_1 itself by representing the non-trivial element of $W \cong \mathbb{Z}_2$ by

Exercise 10.7

$$E^{\pm} \mapsto -E^{\mp}, \qquad H \mapsto -H. \qquad (10.50)$$

Check that this is an automorphism of A_1.
Notice that the matrix

$$\gamma := \begin{pmatrix} 0 & -1 \\ 1 & 0 \end{pmatrix} \qquad (10.51)$$

is an element of the Lie group SU(2), i.e. that it is unitary and has determinant 1. Show that the automorphism (10.50) is the conjugation with this group element γ. Find out which other elements of the Lie group SU(2) implement the same automorphism via conjugation, and thereby determine the centralizer.
(Automorphisms of Lie algebras that can be described by the conjugation with a group element are called inner automorphisms, see chapter 11.)

Verify the formula (10.43) for the conjugation of an affine Weyl translation by an arbitrary Weyl group element.

Exercise 10.8

a) Prove that the Weyl group of a simple or affine Lie algebra is a Coxeter group.

Exercise 10.9

Hint: For any i, j show that the subspace $\mathbb{R}\alpha^{(i)} \oplus \mathbb{R}\alpha^{(j)}$ of the weight space is invariant with respect to both $w_{(i)}$ and $w_{(j)}$. Compute the matrices describing these reflections in the basis $\{\alpha^{(i)}, \alpha^{(j)}\}$ and then the eigenvalues of the product of these matrices.
b) Deduce from the relations (10.46) that the number of simple reflections in two different realizations of a Weyl group element as a Weyl word differs by an even integer.

Find the normalizer $\mathcal{N}_G(G_\circ)$ of a maximal torus $G_\circ$ of the Lie group $\mathrm{SL}(n, \mathbb{C})$ and determine the quotient $\mathcal{N}_G(G_\circ)/G_\circ$.

Exercise 10.10

Compare with the description (10.37) of the Weyl group of $\mathfrak{sl}(n)$.

11
Automorphisms of Lie algebras

11.1 The group of automorphisms

As was already mentioned in chapter 4, an automorphism ω of a Lie algebra $\mathfrak{g}$ is by definition a map from $\mathfrak{g}$ to itself which satisfies the following requirements. First, it respects the structure of the Lie algebra; thus it is linear and it is compatible with the Lie bracket, i.e. obeys

$$\omega([x, y]) = [\omega(x), \omega(y)]. \tag{11.1}$$

And second, it is both injective (one-to-one) and surjective (onto). The set of all automorphisms of $\mathfrak{g}$ will be denoted by $Aut(\mathfrak{g})$.

The composition $\omega \circ \omega'$ of two automorphisms $\omega, \omega' \in Aut(\mathfrak{g})$ is again an automorphism of $\mathfrak{g}$. In other words, the composition of maps endows the set $Aut(\mathfrak{g})$ with a multiplication. With respect to this product, there exists a unit element, namely the trivial automorphism that is provided by the identity map id. Furthermore, each automorphism ω has an inverse ω^{-1} satisfying

$$\omega \circ \omega^{-1} = id = \omega^{-1} \circ \omega. \tag{11.2}$$

Finally, the composition of maps is associative. Together these properties show that $Aut(\mathfrak{g})$ is a group. (On the other hand, the composition of maps is usually not commutative, so that in general $Aut(\mathfrak{g})$ is non-abelian.)

11.2 Automorphisms of finite order

For multiple compositions of an automorphism ω with itself one uses the notation

$$\omega^n \equiv \underbrace{\omega \circ \omega \circ \cdots \circ \omega}_{n \text{ factors}}. \tag{11.3}$$

181

If there exists a natural number N such that

$$\omega^N = id \,, \tag{11.4}$$

then N is called the *order* of ω, while if such a number does not exist, ω is said to be of infinite order. If ω is of finite order N, then as a vector space the *complex* Lie algebra $\mathfrak{g}$ splits into the direct sum

$$\mathfrak{g} = \bigoplus_{i=0}^{N-1} \mathfrak{g}_{(i)} \tag{11.5}$$

of eigenspaces of ω, with

$$\mathfrak{g}_{(j)} = \{x \in \mathfrak{g} \mid \omega(x) = \mathrm{e}^{2\pi \mathrm{i} \cdot j/N} x\} \,. \tag{11.6}$$

This is so because, first, no element of $\mathfrak{g}_{(j)}$ can be expressed as a linear combination of elements of the spaces $\mathfrak{g}_{(k)}$ with $k \neq j$, and second, any $x \in \mathfrak{g}$ can be decomposed as a linear combination

$$x = \sum_{j=0}^{N-1} x_{(j)} \tag{11.7}$$

with $x_{(j)}$ an element of $\mathfrak{g}_{(j)}$, namely

$$x_{(j)} = \frac{1}{N} \sum_{k=0}^{N-1} \mathrm{e}^{-2\pi \mathrm{i} \cdot jk/N} \omega^k(x) \,. \tag{11.8}$$

The automorphism property of ω implies (see exercise 11.2) that upon taking Lie brackets the subspaces $\mathfrak{g}_{(j)}$ behave as

$$[\mathfrak{g}_{(j)}, \mathfrak{g}_{(k)}] \subseteq \mathfrak{g}_{(j+k \bmod N)} \,. \tag{11.9}$$

The formulæ (11.6) and (11.9) are actually special versions of equations that we have already written down in chapter 4 (namely (4.34) and (4.35)) when we introduced the concept of a gradation. In terms of that concept, these formulæ state that any order N automorphism ω supplies a $\mathbb{Z}_N$-gradation of $\mathfrak{g}$. Conversely, given any $\mathbb{Z}_N$-gradation of a Lie algebra $\mathfrak{g}$, the linear map on $\mathfrak{g}$ defined by multiplying all elements of $\mathfrak{g}_{(j)}$ by $\exp(2\pi \mathrm{i} j/N)$ is an automorphism of order N. Among the subspaces $\mathfrak{g}_{(j)}$ of $\mathfrak{g}$, only the eigenspace $\mathfrak{g}_{(0)}$ with eigenvalue 1 is a sub*algebra*; $\mathfrak{g}_{(0)}$ is called the *fixed point algebra* of the automorphism.

This gradation also implies a specific structure of the Killing form κ of the Lie algebra $\mathfrak{g}$. The Killing form is invariant under any automorphism (see exercise 11.3). Since it is also bilinear, it follows that

$$\kappa(x_{(i)}, x_{(j)}) = \kappa(\omega(x_{(i)}), \omega(x_{(j)})) = \mathrm{e}^{2\pi \mathrm{i} \cdot (i+j)/N} \kappa(x_{(i)}, x_{(j)}) \tag{11.10}$$

for $x_{(i)} \in \mathfrak{g}_{(i)}$ and $x_{(j)} \in \mathfrak{g}_{(j)}$, so that in particular

$$\kappa(\mathfrak{g}_{(i)}, \mathfrak{g}_{(j)}) = 0 \qquad \text{for } i + j \neq 0 \bmod N \,, \tag{11.11}$$

i.e. the subspaces $\mathfrak{g}_{(i)}$ and $\mathfrak{g}_{(N-i)}$ are paired with respect to the Killing form. If $\mathfrak{g}$ is a finite-dimensional semisimple Lie algebra, so that κ is non-degenerate, then (11.11) implies that to each $x \in \mathfrak{g}_{(i)}$ there exists a $y \in \mathfrak{g}_{(N-i)}$ such that $\kappa(x, y) \neq 0$; in particular, the restriction of κ to $\mathfrak{g}_{(0)} \times \mathfrak{g}_{(0)}$ is again non-degenerate, so that the fixed point algebra $\mathfrak{g}_{(0)}$ is a reductive subalgebra of $\mathfrak{g}$. (These arguments work analogously for *any* gradation of $\mathfrak{g}$.)

11.3 Inner and outer automorphisms

Examples of automorphisms can be constructed from suitable endomorphisms of $\mathfrak{g}$. By the exponential $\mathrm{EXP}(\varphi)$ of an endomorphism φ of $\mathfrak{g}$ one means the exponential power series

$$\mathrm{EXP}(\varphi) := \sum_{n=0}^{\infty} \tfrac{1}{n!}\, \varphi^n \,. \tag{11.12}$$

This is merely a formal object, and generically does *not* correspond to a well-defined endomorphism. There are, however, situations where the expression (11.12) does make sense, and in many of these cases it provides a convenient tool for manipulating endomorphisms. Consider, for instance, any element $x \in \mathfrak{g}$ for which there exists a positive integer m such that the adjoint map ad_x (4.16) satisfies $(\mathrm{ad}_x)^m = 0$. Such elements x of $\mathfrak{g}$ are called *ad-nilpotent*, or shortly *nilpotent*; among the nilpotent elements of a semisimple Lie algebra $\mathfrak{g}$, there are in particular all step operators E^α. If x is nilpotent, then the power series in the expression

$$\mathrm{Ad}_x := \mathrm{EXP}(\mathrm{ad}_x)\,, \tag{11.13}$$

terminates so that $\mathrm{EXP}(\mathrm{ad}_x)$ is a polynomial in ad_x, and as a consequence (11.13) is a well-defined map from $\mathfrak{g}$ to $\mathfrak{g}$. In fact (see exercise 11.5), it also satisfies

$$\mathrm{Ad}_x([y, z]) = [\mathrm{Ad}_x(y), \mathrm{Ad}_x(z)] \tag{11.14}$$

for all $y, z \in \mathfrak{g}$, and hence Ad_x is indeed an endomorphism of the Lie algebra $\mathfrak{g}$. Ad_x is even an *auto*morphism of $\mathfrak{g}$, as can e.g. be seen by noticing that $\mathrm{EXP}(-\mathrm{ad}_x) = \mathrm{EXP}(\mathrm{ad}_{-x})$ is inverse to Ad_x. At the other extreme, Ad_x is again a well-defined automorphism if ad_x is diagonalizable, as happens in particular for any x in a Cartan subalgebra of a semisimple Lie algebra.

Any automorphism of $\mathfrak{g}$ which can be written as the product of automorphisms of the type (11.13) is called an *inner automorphism* of $\mathfrak{g}$, while any other automorphism is called *outer*. The set of all inner automorphisms is denoted by *Int*($\mathfrak{g}$). In fact, by construction, *Int*($\mathfrak{g}$) is again a group, and hence a subgroup of *Aut*($\mathfrak{g}$); it is called the *adjoint group* of $\mathfrak{g}$.

Further, for any automorphism $\omega \in Aut(\mathfrak{g})$ one has

$$\omega \circ \mathrm{ad}_x \circ \omega^{-1} = \mathrm{ad}_{\omega(x)} \, ; \tag{11.15}$$

for nilpotent x this implies that also

$$\omega \circ \mathrm{Ad}_x \circ \omega^{-1} = \mathrm{Ad}_{\omega(x)} \, . \tag{11.16}$$

Thus $Int(\mathfrak{g})$ is even a normal subgroup of $Aut(\mathfrak{g})$, so that the coset space

$$Out(\mathfrak{g}) := Aut(\mathfrak{g})/Int(\mathfrak{g}) \tag{11.17}$$

of outer modulo inner automorphisms is again a group. (The factor group $Out(\mathfrak{g})$ must not be confused with the set of all outer automorphisms which is *not* a group.)

Recalling that the adjoint maps ad_x are specific examples of derivations of $\mathfrak{g}$, it is quite natural to perform an analogous analysis also for other derivations. If a derivation δ is nilpotent, i.e. satisfies $\delta^m = 0$ for some positive integer m, then one can define a map $\mathrm{Exp}(\delta)$ from $\mathfrak{g}$ to $\mathfrak{g}$ by the exponential power series. One has (see exercise 11.5)

$$[\mathrm{Exp}(\delta(x)), \mathrm{Exp}(\delta(y))] = \mathrm{Exp}(\delta([x,y])) \, ; \tag{11.18}$$

accordingly $\mathrm{Exp}(\delta)$ is an endomorphism of $\mathfrak{g}$; again it is in fact an automorphism. This automorphism is outer unless $\delta = \mathrm{ad}_x$ for some $x \in \mathfrak{g}$. (Consequently, the adjoint maps are called inner derivations, and all other derivations outer.)

The space of all derivations of a Lie algebra $\mathfrak{g}$ is denoted by $Der(\mathrm{g})$. Since the commutator of two derivations is again a derivation (see exercise 4.8), the derivations form a subalgebra of the Lie algebra $\mathfrak{gl}(\mathfrak{g})$ of linear maps from $\mathfrak{g}$ to itself. Recalling that the map $x \mapsto \mathrm{ad}_x$ defines the adjoint representation R_{ad} of $\mathfrak{g}$, it follows that the image $R_{\mathrm{ad}}(\mathfrak{g})$ of $\mathfrak{g}$ under the adjoint representation is a subalgebra of $Der(\mathrm{g})$.

If the Lie algebra $\mathfrak{g}$ is finite-dimensional, then the automorphism group $Aut(\mathfrak{g})$ is a Lie group. Its Lie algebra is the algebra $Der(\mathrm{g})$ of derivations of $\mathfrak{g}$, and $Int(\mathfrak{g})$ is a Lie subgroup of $Aut(\mathfrak{g})$ whose Lie algebra is the image of $\mathfrak{g}$ under R_{ad}.

In the rest of this section we will derive two important facts:

• Any two Cartan subalgebras of a simple Lie algebra are conjugate by some *inner* automorphism.

• The group $Int(\mathfrak{g})$ of inner automorphisms is generated by those automorphisms Ad_x for which the elements x are the step operators in an arbitrarily chosen triangular decomposition.

The first statement implies the claim made in section 6.1 that the choice of a Cartan subalgebra does not introduce any arbitrariness in the description of a simple Lie algebra.

To arrive at these results, we consider a definite Cartan subalgebra $\mathfrak{g}_\circ$ and the associated root space decomposition of $\mathfrak{g}$. Then we fix an element $h \in \mathfrak{g}_\circ$ of the Cartan subalgebra and examine the polynomial mapping

$$x_1 \otimes \cdots \otimes x_{d-r} \otimes h \; \mapsto \; \left(\mathrm{Ad}_{x_1} \circ \cdots \circ \mathrm{Ad}_{x_{d-r}} \right)(h) \tag{11.19}$$

from $\bigoplus_\alpha \mathfrak{g}_\alpha$ to $\mathfrak{g}$, where $x_i \in \mathfrak{g}_{\alpha_i}$ with α_i the distinct roots of $\mathfrak{g}$ (with respect to $\mathfrak{g}_\circ$). For the following argument we also assume that the Cartan subalgebra element h is such that $\{x \in \mathfrak{g} \mid [x, h] = 0\} = \mathfrak{g}_\circ$ (such an element is called a *regular* element of $\mathfrak{g}_\circ$; as remarked in section 6.1, the regular elements form an open and dense subset of $\mathfrak{g}_\circ$). One can then show that the derivative of the map (11.19) with respect to h is surjective in the neighborhood of $0 \otimes 0 \otimes \cdots \otimes 0 \otimes h$. Next consider the maps φ and $\tilde{\varphi}$ defined by (11.19) for two different Cartan subalgebras $\mathfrak{g}_\circ$ and $\tilde{\mathfrak{g}}_\circ$, where we have chosen regular elements $h \in \mathfrak{g}_\circ$ and $\tilde{h} \in \tilde{\mathfrak{g}}_\circ$. Using tools from algebraic geometry, one can show that the images of these two maps have non-vanishing intersection. Denoting by E the subgroup of $Int(\mathfrak{g})$ that is generated by automorphisms of the type Ad_x, with x a step operator for a root with respect to $\mathfrak{g}_\circ$, and the analogous subgroup obtained with $\tilde{\mathfrak{g}}_\circ$ by $\tilde{E}$, this means that there exist automorphisms $\omega \in E$ and $\tilde{\omega} \in \tilde{E}$ and (regular) elements $h \in \mathfrak{g}_\circ$ and $\tilde{h} \in \tilde{\mathfrak{g}}_\circ$ such that $\omega(h) = \tilde{\omega}(\tilde{h})$. But this implies that

$$\begin{aligned} \omega(\mathfrak{g}_\circ) &= \{x \in \mathfrak{g} \mid [\omega^{-1}(x), h] = 0\} = \{x \in \mathfrak{g} \mid [x, \omega(h)] = 0\} \\ &= \{x \in \mathfrak{g} \mid [x, \tilde{\omega}(\tilde{h})] = 0\} = \tilde{\omega}(\tilde{\mathfrak{g}}_\circ). \end{aligned} \tag{11.20}$$

In short, for any two Cartan subalgebras $\mathfrak{g}_\circ$ and $\tilde{\mathfrak{g}}_\circ$ the inner automorphism $\sigma := \tilde{\omega}^{-1}\omega$ of $\mathfrak{g}$ maps $\mathfrak{g}_\circ$ to $\tilde{\mathfrak{g}}_\circ$, $\tilde{\mathfrak{g}}_\circ = \sigma(\mathfrak{g}_\circ)$: one says that $\mathfrak{g}_\circ$ and $\tilde{\mathfrak{g}}_\circ$ are *conjugate* with respect to the inner automorphism σ.

Similarly one has $E \equiv E_{\mathfrak{g}_\circ} = \omega^{-1}E_{\mathfrak{g}_\circ}\omega = E_{\omega(\mathfrak{g}_\circ)} = E_{\tilde{\omega}(\tilde{\mathfrak{g}}_\circ)} = \tilde{E}$, and this implies that both of these subgroups of $Aut(\mathfrak{g})$ are actually identical to the adjoint group $Int(\mathfrak{g})$. Thus $Int(\mathfrak{g})$ is already generated by those automorphisms Ad_x for which x are the step operators in an arbitrarily chosen triangular decomposition.

11.4 An $\mathfrak{sl}(2)$ example

To illustrate some of the concepts mentioned above, we consider the following automorphism of the Lie algebra $\mathfrak{sl}(2)$:

$$\omega := \mathrm{Ad}_{E_+} \circ \mathrm{Ad}_{-E_-} \circ \mathrm{Ad}_{E_+}. \tag{11.21}$$

By definition, this is an inner automorphism (in fact, $\mathfrak{sl}(2)$ does not possess any outer automorphism). It acts on the generators H and $E_\pm$ of $\mathfrak{sl}(2)$ as (see exercise 11.7)

$$\omega(H) = -H, \qquad \omega(E_\pm) = -E_\mp. \tag{11.22}$$

In particular, ω is of order two, and its eigenspaces are

$$\mathfrak{g}_{(0)} = \mathrm{span}_{\mathbb{C}}\{E_+ - E_-\}, \qquad \mathfrak{g}_{(1)} = \mathrm{span}_{\mathbb{C}}\{H, E_+ + E_-\}. \tag{11.23}$$

In the realization of $\mathfrak{sl}(2)$ as the matrix Lie algebra of traceless 2×2-matrices, the transformations (11.23) can be obtained as

$$\omega(x) = M_\omega \cdot x \cdot M_\omega^{-1} \quad \text{with} \quad M_\omega := \begin{pmatrix} 0 & 1 \\ -1 & 0 \end{pmatrix}. \tag{11.24}$$

The matrix M_ω appearing here satisfies

$$M_\omega = \mathrm{EXP}(E_+) \cdot \mathrm{EXP}(-E_-) \cdot \mathrm{EXP}(E_+), \tag{11.25}$$

with $E_\pm$ considered as 2×2-matrices.

Finally we observe that $\mathfrak{g}_{(0)}$ is in fact a Cartan subalgebra of $\mathfrak{sl}(2)$. The corresponding Cartan–Weyl basis $\tilde{\mathcal{B}}$ is given by

$$\tilde{H} = \mathrm{i}(E_+ - E_-), \qquad \tilde{E}_\pm = \tfrac{1}{2}\left[H \mp \mathrm{i}(E_+ + E_-)\right]. \tag{11.26}$$

On this specific basis, the automorphism acts as

$$\omega(\tilde{H}) = \tilde{H}, \qquad \omega(\tilde{E}_\pm) = -\tilde{E}_\pm, \tag{11.27}$$

i.e. all generators in $\tilde{\mathcal{B}}$ are eigenvectors of ω.

The result (11.24) found in the example above generalizes as follows to arbitrary inner automorphisms Ad_x of finite-dimensional matrix Lie algebras. We can write ad_x as $\mathrm{m}_x^{(l)} + \mathrm{m}_{-x}^{(r)}$, where $\mathrm{m}_x^{(l)}$ and $\mathrm{m}_x^{(r)}$ stand for the left and right multiplication with x, respectively. It then follows that

$$\mathrm{Exp}(\mathrm{ad}_x) = \mathrm{Exp}(\mathrm{m}_x^{(l)} + \mathrm{m}_{-x}^{(r)}) = \mathrm{Exp}(\mathrm{m}_x^{(l)}) \circ \mathrm{Exp}(\mathrm{m}_{-x}^{(r)}) = \mathrm{m}_{\mathrm{Exp}(x)}^{(l)} \circ \mathrm{m}_{\mathrm{Exp}(-x)}^{(r)}$$

or more explicitly,

$$\mathrm{Ad}_x(y) = \mathrm{Exp}(x) \cdot y \cdot \mathrm{Exp}(-x) = \mathrm{Exp}(x) \cdot y \cdot (\mathrm{Exp}(x))^{-1}. \tag{11.28}$$

In short, Ad_x is the conjugation by the Lie *group* element $\gamma = \mathrm{Exp}(x)$, and Ad describes the adjoint action of the Lie group on its Lie algebra (compare equation (10.35)). We have already seen in section 10.6 that the intersection of the conjugation orbits with a fixed Cartan subalgebra $\mathfrak{g}_\circ$ are just the Weyl group orbits on $\mathfrak{g}_\circ$. We conclude that also the intersection of the orbits of the action of $Int(\mathfrak{g})$ with $\mathfrak{g}_\circ$ are precisely the orbits of the Weyl group.

11.5 Dynkin diagram symmetries

Let now $\mathfrak{g}$ be a finite-dimensional simple or an affine Lie algebra. We have already remarked that the Dynkin diagrams are multi-purpose graphs which can visualize various features of these Lie algebras. It is therefore reasonable to expect that the symmetries of the Dynkin diagram of $\mathfrak{g}$ can teach us something about the automorphisms of $\mathfrak{g}$. Let us therefore consider any permutation

$$\dot{\omega} : \; i \; \mapsto \; \dot{\omega}(i) \tag{11.29}$$

of the labels $i \in \{1, 2, ..., \mathrm{r}\}$ corresponding to the nodes of the Dynkin diagram of $\mathfrak{g}$. If this permutation of the nodes is a symmetry of the Dynkin diagram, it is called an automorphism of the diagram.

As an example, consider the Dynkin diagram of E_6; this has an order two automorphism which acts as $1 \leftrightarrow 5$, $2 \leftrightarrow 4$, $3 \mapsto 3$, $6 \mapsto 6$, as depicted in figure (11.30).

$$\tag{11.30}$$

Because of the correspondence between the Dynkin diagram and the Cartan matrix of $\mathfrak{g}$, a permutation $\dot\omega$ is an automorphism precisely if the Cartan matrix satisfies

$$A^{\dot\omega(i)\,\dot\omega(j)} = A^{ij} . \tag{11.31}$$

Taking into account the Chevalley–Serre construction of $\mathfrak{g}$ which only requires the knowledge of the Cartan matrix A, it follows that any Dynkin diagram automorphism $\dot\omega$ gives rise to an automorphism ω of $\mathfrak{g}$ via

$$\omega : \quad H^i \mapsto H^{\dot\omega(i)} , \quad E^i_\pm \mapsto E^{\dot\omega(i)}_\pm . \tag{11.32}$$

(More precisely, to the Dynkin diagram symmetry one can associate a whole class of automorphisms of $\mathfrak{g}$, which differ by inner automorphisms of the form Ad_h with $h \in \mathfrak{g}_\circ$; (11.32) is a particularly convenient representative of this class.) Thus all generators for simple roots are mapped to generators for simple roots, and as a consequence also all generators for positive roots are mapped to generators for positive roots. As we will see below, this implies that the automorphism (11.32) is an outer automorphism.

While the set of all outer automorphisms is not a group, the special outer automorphisms that are of the type (11.32) do form a (finite) group, which we denote by $\Gamma(\mathfrak{g})$. In many cases $\Gamma(\mathfrak{g})$ is trivial, i.e. contains only the identity map. Among the simple Lie algebras, this happens for A_1, B_r, C_r, and for all exceptional simple Lie algebras other than E_6, among the untwisted affine Lie algebras for $E_8^{(1)}$, $F_4^{(1)}$ and $G_2^{(1)}$, and among the twisted affine Lie algebras for $A_1^{(2)}$, $\tilde B_\mathrm{r}^{(2)}$, $F_4^{(2)}$ and $G_2^{(3)}$. In the remaining cases $\Gamma(\mathfrak{g})$ is as listed in table X. In table X, $\mathbb{Z}_n = \mathbb{Z}/n\mathbb{Z}$ denotes the cyclic group of order n, $\mathcal{S}_n$ the symmetric group consisting of all permutations of n objects, and $\mathcal{D}_n$ stands for the *dihedral group* of n objects. The latter is by definition the symmetry group of a regular polygon with n edges (thus e.g. $\mathcal{D}_4$ is the symmetry group of the square); it is generated by two elements, a reflection and a rotation by an angle $2\pi/n$, so that it is a semidirect product of $\mathbb{Z}_n$ and $\mathbb{Z}_2$.

Along with the action of ω on $\mathfrak{g}_\circ$ we also have a dual action on the space dual to $\mathfrak{g}_\circ$, i.e. the weight space $\mathfrak{g}_\circ^\star$. Hence the Dynkin diagram automorphism $\dot\omega$ induces an automorphism of the weight space and also of the weight lattice of $\mathfrak{g}$, which acts as $\Lambda_{(i)} \mapsto \Lambda_{(\dot\omega(i))}$ for $i = 1, 2, ..., \mathrm{r}$. (This can also immediately be seen from the visualization of dominant weights through Dynkin diagrams, see figure (13.9) in chapter 13.) Whenever the conjugation $\Lambda \mapsto \Lambda^+$ (compare (13.16) below) of the dominant weights of a finite-dimensional simple Lie algebra is non-trivial (which happens for A_r with $\mathrm{r} \geq 2$, D_r with r even and E_6), this conjugation can be described as such an automorphism of $\mathfrak{g}_\circ^\star$ which is induced by a permutation $\dot\omega$ of order two. This permutation is the unique non-trivial automorphism of

Table X.   *Simple and affine Lie algebras with non-trivial Dynkin diagram automorphisms*

$\mathfrak{g}$	$\Gamma(\mathfrak{g})$
A_r (r≥2)	$\mathbb{Z}_2$
D_4	$\mathcal{S}_3$
D_r (r>4)	$\mathbb{Z}_2$
E_6	$\mathbb{Z}_2$

$\mathfrak{g}$	$\Gamma(\mathfrak{g})$
$A_r^{(1)}$	$\mathcal{D}_{r+1}$
$B_r^{(1)}$, $C_r^{(1)}$	$\mathbb{Z}_2$
$D_4^{(1)}$	$\mathcal{S}_4$
$D_r^{(1)}$ (r>4)	$\mathcal{D}_4$
$E_6^{(1)}$	$\mathcal{S}_3$
$E_7^{(1)}$	$\mathbb{Z}_2$
$B_r^{(2)}$, $C_r^{(2)}$	$\mathbb{Z}_2$

Γ, except for D_4, where it acts by permuting the third and the fourth node while leaving all other nodes fixed.

According to table X, for all untwisted affine Lie algebras $\mathfrak{g}^{(1)}$ the automorphism group $\Gamma(\mathfrak{g}^{(1)})$ contains the group $\Gamma(\mathfrak{g})$ of automorphisms of the Dynkin diagram of the simple Lie algebra $\mathfrak{g}$, and is in fact larger than $\Gamma(\mathfrak{g})$ except for $E_8^{(1)}$, $F_4^{(1)}$ and $G_2^{(1)}$. It turns out that the gain in symmetry is described by the center $\mathcal{Z}$ of the universal covering group G which has $\mathfrak{g}$ as its Lie algebra. Namely, $\mathcal{Z}$ is a normal subgroup of group $\Gamma(\mathfrak{g}^{(1)})$ such that

$$\Gamma(\mathfrak{g}) = \Gamma(\mathfrak{g}^{(1)})/\mathcal{Z} . \tag{11.33}$$

$\mathcal{Z}$ is always a cyclic group, i.e. $\mathcal{Z} \cong \mathbb{Z}_n$ for some integer n, except for $D_r^{(1)}$ with r odd, in which case $\mathcal{Z} \cong \mathbb{Z}_2 \times \mathbb{Z}_2$. In more detail, the situation looks as follows. The universal covering group associated to A_r is isomorphic to $SU(r+1)$ (the group of complex $(r+1)\times(r+1)$-matrices of determinant 1) which has $\mathbb{Z}_{r+1}$ as its center (namely the multiples $c\mathbb{1}$ of the unit matrix with c an $(r+1)$th root of unity); thus the relevant quotient is $\mathcal{D}_{r+1}/\mathbb{Z}_{r+1} \cong \mathbb{Z}_2$ for $r \geq 2$, and $\mathcal{D}_2/\mathbb{Z}_2 \cong \mathbb{Z}_2/\mathbb{Z}_2 = \{1\}$ for $r = 1$. Similarly, the covering group of E_6 has center $\mathbb{Z}_3$, and $\mathcal{S}_3/\mathbb{Z}_3 \cong \mathbb{Z}_2$ is indeed the symmetry of the E_6 Dynkin diagram. Next consider $\mathfrak{g} = D_r$; its covering group Spin(2r) has center $\mathbb{Z}_4$ for r odd, and $\mathbb{Z}_2 \times \mathbb{Z}_2$ for r even, respectively; thus $\mathcal{S}_4/\mathbb{Z}_4 \cong \mathcal{S}_3$ is the symmetry of the Dynkin diagram of D_4, whereas for $r \geq 4$ the Dynkin diagram symmetry is $\mathcal{D}_4/\mathcal{Z} \cong \mathbb{Z}_2$. Finally, for $\mathfrak{g} = B_r$, C_r or E_7 one has $\mathcal{Z} \cong \mathbb{Z}_2 \cong \Gamma(\mathfrak{g}^{(1)})$ so that there is no symmetry left for the Dynkin diagram of $\mathfrak{g}$.

11.6 Finite order inner automorphisms of simple Lie algebras

Consider now any inner automorphism of a *simple* Lie algebra (over $\mathbb{C}$, or of its compact real form). Denote by $\mathfrak{h}_{(0)}$ a maximal abelian subalgebra of the fixed point algebra $\mathfrak{g}_{(0)}$. It can be shown that the centralizer $\mathcal{C}_{\mathfrak{g}}(\mathfrak{h}_{(0)})$ of $\mathfrak{h}_{(0)}$ in $\mathfrak{g}$ is a Cartan subalgebra of $\mathfrak{g}$ (in particular the fixed point algebra is not zero), and also that $\mathfrak{h}_{(0)}$ contains a regular element x which has the property that its centralizer $\mathcal{C}_{\mathfrak{g}}(\{x\})$ is an ω-invariant Cartan subalgebra of $\mathfrak{g}$. We will work with this particular Cartan subalgebra. Further, by applying ω to the bracket relations $[H^i, E^\alpha] = \alpha^i E^\alpha$ for $i = 1, 2, ..., \mathrm{r}$ it follows from $\omega(H^i) = H^i$ that $\omega(E^\alpha)$ is an eigenvector of the maps ad_{H^i} with the same eigenvalues as E^α. Since the root spaces of a simple Lie algebra are non-degenerate, this implies that $\omega(E^\alpha)$ is proportional to E^α. Hence each step operator E^α relative to this ω-invariant Cartan subalgebra is an eigenvector of the automorphism ω.

Summarizing these results, we learn that for any inner automorphism ω of finite order of a simple Lie algebra $\mathfrak{g}$, there exists a Cartan–Weyl basis consisting of eigenvectors of ω such that the fixed point algebra contains the Cartan subalgebra. This implies (compare exercise 11.10) that in this specific basis ω has the form

$$\omega = \mathrm{Ad}_h \qquad (11.34)$$

for some Cartan subalgebra element h. Put differently, since all Cartan subalgebras of $\mathfrak{g}$ are conjugate with respect to $Int(\mathfrak{g})$, any finite order inner automorphism ω of $\mathfrak{g}$ is conjugate to an automorphism of the form (11.34). The automorphisms of the type (11.34) can be parametrized by writing h as $h = 2\pi i \sum_{j=1}^{\mathrm{r}} (\zeta_\omega)_j H^j$, so that

$$\omega = \mathrm{Exp}\Big(\frac{2\pi i}{N} \sum_{j=1}^{\mathrm{r}} (\zeta_\omega)_j \, \mathrm{ad}_{H^j}\Big). \qquad (11.35)$$

The weight ζ_ω defined this way is called the *shift vector* associated to the automorphism. Because of $\mathrm{ad}_h(E^\alpha) = (2\pi i(\zeta_\omega, \alpha)/N) E^\alpha$ one has $\omega(E^\alpha) = \exp(2\pi i(\zeta_\omega, \alpha)/N) E^\alpha$. Thus the fact that ω has order N means that the shift vector obeys

$$(\zeta_\omega, \alpha) \in \mathbb{Z} \quad \text{for all} \quad \alpha \in \Phi(\mathfrak{g}). \qquad (11.36)$$

These results imply that, up to conjugation, the finite order inner automorphisms of a simple Lie algebra can be classified by shift vectors obeying (11.36). Now for (11.36) to hold for all roots, it is in fact sufficient that it holds for the simple roots. For these the automorphism acts as

$$\omega(E_+^i) = \exp\Big(\frac{2\pi i}{N} (\tilde{\zeta}_\omega)^i\Big) E_+^i, \qquad (11.37)$$

where

$$\tilde{\zeta}_\omega^i := \zeta_\omega^i \cdot \tfrac{1}{2}\,(\alpha^{(i)}, \alpha^{(i)})\,. \tag{11.38}$$

Thus the requirement is that the weight $\tilde{\zeta}_\omega$ with Dynkin components (11.38) is an element of the weight lattice, i.e.

$$\tilde{\zeta}_\omega^i \in \mathbb{Z} \quad \text{for } i = 1, 2, \dots, r\,. \tag{11.39}$$

The integers $\tilde{\zeta}_\omega^i$ are defined only modulo the order N, and hence can be chosen to lie in the range from 0 to $N-1$. Then the largest of the numbers (ζ_ω, α) is given by

$$(\zeta_\omega, \theta) = \sum_{j=1}^{r} a_j \tilde{\zeta}_\omega^j\,, \tag{11.40}$$

with a_j the components of the highest root θ in the basis of simple roots, i.e. the Coxeter labels of $\mathfrak{g}$. Moreover, one can show that the order of the automorphism is never smaller than the number (11.40), and correspondingly we set

$$N = (\zeta_\omega, \theta) + \tilde{\zeta}_\omega^0 = \sum_{j=0}^{r} a_j \tilde{\zeta}_\omega^j \tag{11.41}$$

with some non-negative integer $\tilde{\zeta}_\omega^0$. It can then be shown that, up to conjugation, the finite order inner automorphisms are classified precisely by those $(r+1)$-tuples of non-negative integers $\tilde{\zeta}_\omega^0, \tilde{\zeta}_\omega^1, \dots, \tilde{\zeta}_\omega^r$ which do not possess any common prime factor.

As an illustration, consider the automorphism (11.21) of $\mathfrak{sl}(2)$. On the Cartan–Weyl basis defined in (11.26), the action of this automorphism is given by equation (11.27). This corresponds to the shift vector with entries $\tilde{\zeta}_\omega^0 = \tilde{\zeta}_\omega^1 = 1$.

The fixed point algebra of an inner automorphism described by the shift vector ζ_ω is the direct sum of an abelian Lie algebra $\mathfrak{u}$ and a semisimple Lie algebra $\tilde{\mathfrak{g}}_{(0)}$, which can be specified with the help of the *extended Dynkin diagram* of $\mathfrak{g}$; the latter is by definition identical to the Dynkin diagram of that untwisted affine Lie algebra $\mathfrak{g}^{(1)} = X_r^{(1)}$ which corresponds to the simple Lie algebra $\mathfrak{g} = X_r$ for $X \in \{A, B, \dots, G\}$ (compare the tables IV and VIII in chapter 7), with the additional node being the 'zeroth node' which is labelled by '0'. Namely, $\tilde{\mathfrak{g}}_{(0)}$ is the Lie algebra whose Dynkin diagram is obtained from the extended Dynkin diagram of $\mathfrak{g}$ by deleting all nodes i for which ζ_ω^i is non-zero, while $\mathfrak{u} \cong (\mathfrak{u}(1))^{p-1}$ with p the number of the deleted nodes.

11.7 Automorphisms of the root system

Instead of using shift vectors as in the previous section, it is often more convenient to describe automorphisms in a basis of $\mathfrak{g}$ in which the Cartan subalgebra is not left pointwise fixed. Given a Cartan–Weyl basis of a simple Lie algebra $\mathfrak{g}$, let us consider the specific inner automorphisms $\omega_{(i)}$ of $\mathfrak{g}$ defined by

$$\omega_{(i)} := \mathrm{Ad}_{E_+^i} \circ \mathrm{Ad}_{-E_-^i} \circ \mathrm{Ad}_{E_+^i}, \tag{11.42}$$

which generalizes the $\mathfrak{sl}(2)$-automorphism (11.21). Such an automorphism maps the Cartan subalgebra to itself, though it does not leave it pointwise fixed:

$$\omega_{(i)}(H^j) = H^j - A^{ij} H^i \tag{11.43}$$

(note that on the right hand side there is no summation on i). Since the rows of the Cartan matrix are the Dynkin components of the simple roots, this action is isomorphic to the action $(w_{(i)}(\lambda))^j = \lambda^j - (\alpha^{(i)})^j \lambda^i$ of a simple Weyl reflection on the weight space. Thus we have:

> The subgroup of $Aut(\mathfrak{g})$ that is generated by the inner auto-morphisms (11.42) is isomorphic to the Weyl group W of $\mathfrak{g}$.

More generally, if ω is an arbitrary automorphism which leaves the Cartan subalgebra $\mathfrak{g}_\circ$ invariant, then via the identification of the Cartan subalgebra with the weight space the restriction of ω to $\mathfrak{g}_\circ$ defines an automorphism of the root system of $\mathfrak{g}$. Conversely, each automorphism of the root system Φ can be extended to an automorphism of $\mathfrak{g}$ which leaves $\mathfrak{g}_\circ$ invariant. This shows that $Aut(\Phi)$ can be embedded into $Aut(\mathfrak{g})$. It can also be shown that this embedding is a group homomorphism. We then obtain another group homomorphism by projecting on the factor group:

$$Aut(\Phi) \rightarrow Aut(\mathfrak{g})/Int(\mathfrak{g}) \equiv Out(\mathfrak{g}). \tag{11.44}$$

The kernel of this automorphism consists of those automorphisms of the root system which are induced by inner automorphisms that leave the Cartan subalgebra fixed; we have seen in section 10.6 that this is precisely the Weyl group. Dividing out this kernel, we obtain an injective group homomorphism

$$Aut(\Phi)/W \hookrightarrow Aut(\mathfrak{g})/Int(\mathfrak{g}). \tag{11.45}$$

This embedding is also surjective, and hence in fact an isomorphism:

$$Out(\mathfrak{g}) \equiv \frac{Aut(\mathfrak{g})}{Int(\mathfrak{g})} \cong \frac{Aut(\Phi)}{W}. \tag{11.46}$$

Indeed, as seen above any two Cartan subalgebras are conjugate under some *inner* automorphism; therefore for any automorphism we can find another automorphism in the same class which leaves the Cartan subalgebra fixed. The latter automorphisms just induce automorphisms of the root system.

As we have learned in chapter 10, any non-trivial Weyl transformation freely permutes the Weyl chambers, so that the factor group $Aut(\Phi)/W$ consists of those automorphisms of the root system which leave some chosen Weyl chamber, say the fundamental one, invariant. Hence such automorphisms permute the simple roots, so that they precisely correspond to the group $\Gamma(\mathfrak{g})$ of automorphisms of the Dynkin diagram of $\mathfrak{g}$. Thus the symmetry group of the Dynkin diagram, which was listed in table X, satisfies

$$\Gamma(\mathfrak{g}) \cong Aut(\mathfrak{g})/Int(\mathfrak{g}) \equiv Out(\mathfrak{g}) . \tag{11.47}$$

In other words, it is possible to write any automorphism $\omega \in Aut(\mathfrak{g})$ uniquely as

$$\omega = \omega_{\text{int}} \circ \sigma \tag{11.48}$$

with $\omega_{\text{int}} \in Int(\mathfrak{g})$ and σ an automorphism induced by a symmetry of the Dynkin diagram as in equation (11.32).

11.8 Finite order outer automorphisms of simple Lie algebras

Making use of the relation (11.47), the outer automorphisms of a simple Lie algebra $\mathfrak{g}$ which have finite order can be classified similarly to the inner ones. We describe the result only briefly. Given any non-trivial outer automorphism σ of $\mathfrak{g}$, which may be taken to be of the form (11.32), denote by $\mathfrak{s}$ its fixed point algebra and by n the rank of $\mathfrak{s}$. The Lie algebra $\mathfrak{s}$ is always simple. Up to conjugation, the outer automorphisms of $\mathfrak{g}$ of finite order are then classified by shift vectors referring to the fixed point algebra $\mathfrak{s}$, i.e. by $(n+1)$-tuples $\tilde{\zeta}_\sigma^0, \tilde{\zeta}_\sigma^1, ..., \tilde{\zeta}_\sigma^n$ of non-negative integers which do not possess any common factor. The order of such an automorphism is

$$N = \ell \cdot \sum_{j=0}^{n} a_j \tilde{\zeta}_\sigma^j \tag{11.49}$$

with ℓ the order of σ and a_j the Coxeter labels of $\mathfrak{s}$.

There is again a close relation to affine Dynkin diagrams. First, if $\mathfrak{g} = X_r$ is a simple Lie algebra, let $\mathfrak{g}_{\text{aff}}$ be that twisted affine Lie algebra which is called $X_r^{(\ell)}$ in the notation of the *second* column of table VIII B. Then the simple Lie algebra $\mathfrak{s}$ is given by $\mathfrak{s} = Y_n$ such that $Y_n^{(\ell)}$ is the notation used for $\mathfrak{g}_{\text{aff}}$ in the *first* column of table VIII B. (The fact that the algebra $\mathfrak{g}_{\mathrm{aff}}$ is related in this manner to the two distinct Lie algebras $\mathfrak{g}$ and $\mathfrak{s}$ is the origin of the two different notations used for $\mathfrak{g}_{\mathrm{aff}}$; see also chapter 12). For example, the only simple Lie algebra which possesses an outer automorphism σ of order three is $\mathfrak{g} = D_4$; in this case $\mathfrak{g}_{\mathrm{aff}} = D_4^{(3)} \equiv G_2^{(3)}$, and hence $\mathfrak{s} = G_2$.

Second, the fixed point algebra $\mathfrak{g}_{(0)}$ of σ is again the direct sum of an abelian Lie algebra $\mathfrak{u}$ and a semisimple Lie algebra $\tilde{\mathfrak{g}}_{(0)}$; $\tilde{\mathfrak{g}}_{(0)}$ is the Lie algebra whose Dynkin diagram is obtained from the Dynkin diagram of $\mathfrak{g}_{\mathrm{aff}}$ by deleting all nodes i for which ζ_σ^i is non-zero, and $\mathfrak{u} \cong (\mathfrak{u}(1))^{p-1}$ with p the number of deleted nodes.

11.9 Involutive automorphisms, conjugations, real forms

Of special interest are automorphisms which are of order two; these are called *involutive* automorphisms. The eigenvalues of an involutive automorphism are ± 1. The decomposition (11.5) into eigenspaces reads $\mathfrak{g} = \mathfrak{g}_{(0)} \oplus \mathfrak{g}_{(1)}$, and their Lie brackets obey

$$[\mathfrak{g}_{(0)}, \mathfrak{g}_{(0)}] \subseteq \mathfrak{g}_{(0)}, \qquad [\mathfrak{g}_{(0)}, \mathfrak{g}_{(1)}] \subseteq \mathfrak{g}_{(1)}, \qquad [\mathfrak{g}_{(1)}, \mathfrak{g}_{(1)}] \subseteq \mathfrak{g}_{(0)}. \qquad (11.50)$$

According to equation (11.11), the subspaces $\mathfrak{g}_{(0)}$ and $\mathfrak{g}_{(1)}$ are orthogonal with respect to the Killing form. Thus, in short, $\mathfrak{g}_{(0)}$ is a subalgebra of $\mathfrak{g}$ and $\mathfrak{g}_{(1)}$ is its orthogonal complement.

If $\mathfrak{g}$ is a semisimple Lie algebra over the real numbers $\mathbb{R}$, then as a consequence of the relations (11.50) the vector space direct sum

$$\mathfrak{g}^* := \mathfrak{g}_{(0)} \oplus i\, \mathfrak{g}_{(1)} \qquad (11.51)$$

is a real subalgebra of the complexification $\mathfrak{g}_{\mathbb{C}} = \mathfrak{g} \oplus i\,\mathfrak{g}$ of $\mathfrak{g}$. Starting from a *compact* real form $\mathfrak{g}$, by considering all involutive automorphisms of its complexification $\mathfrak{g}_{\mathbb{C}}$ one obtains this way all real forms of $\mathfrak{g}_{\mathbb{C}}$. In other words:

> The finite-dimensional semisimple real Lie algebras are in one-to-one correspondence with the pairs $(\mathfrak{h}, \omega)$, where $\mathfrak{h}$ is a finite-dimensional semisimple complex Lie algebra and ω an involutive automorphism of $\mathfrak{h}$.

If $\breve{\mathfrak{g}}$ is a real form of the complex Lie algebra $\mathfrak{g}$, then the map σ defined by

$$\sigma: \quad x + iy \;\mapsto\; x - iy \qquad \text{for } x, y \in \breve{\mathfrak{g}} \qquad (11.52)$$

is called the *conjugation* of $\mathfrak{g}$ with respect to $\breve{\mathfrak{g}}$. (This notion of conjugation must of course not be mixed up with the conjugation $\Lambda \mapsto \Lambda^+$ of $\mathfrak{g}$-weights.) The conjugation satisfies $\sigma^2 = id$, i.e. is involutive, as well as $\sigma([x, y]) = [\sigma(x), \sigma(y)]$. However, although it is $\mathbb{R}$-linear and hence obeys $\sigma(x + y) =$

$\sigma(x) + \sigma(y)$, it is *not* $\mathbb{C}$-linear, since

$$\sigma(\xi x) = \bar{\xi}\,\sigma(x) \tag{11.53}$$

for $\xi \in \mathbb{C}$ ($\bar{\xi}$ denotes the complex conjugate of ξ). Therefore σ is not an automorphism. A map from $\mathfrak{g}$ to itself which satisfies (11.53), but otherwise has all properties of an automorphism, is called an *anti-automorphism*.

Thus any conjugation is an involutive anti-automorphism. Conversely, it can be shown that for any involutive anti-automorphism σ of a complex Lie algebra $\mathfrak{g}$ the fixed point set $\breve{\mathfrak{g}}$ is a real form of $\mathfrak{g}$, such that σ is the conjugation of $\mathfrak{g}$ with respect to $\breve{\mathfrak{g}}$. We also note that for any conjugation the Killing form κ of $\mathfrak{g}$ obeys

$$\kappa(\sigma(x), \sigma(y)) = \overline{\kappa(x, y)}\,. \tag{11.54}$$

If $\breve{\mathfrak{g}}$ is the compact real form of a semisimple Lie algebra $\mathfrak{g}$, then according to chapter 8 (see e.g. (8.20)) a basis of $\breve{\mathfrak{g}}$ is given by

$$\mathrm{i}\,H^i,\quad \mathrm{i}\,(E^i_+ + E^i_-),\quad E^i_+ - E^i_- \tag{11.55}$$

for $i = 1, 2, \ldots, \mathrm{r}$, with $\{H^i, E^i_\pm\}$ a Chevalley–Serre basis of $\mathfrak{g}$ (this is the generalization of the basis (2.5) of $\mathfrak{su}(2)$). The conjugation of $\mathfrak{g}$ with respect to $\breve{\mathfrak{g}}$ thus acts as

$$H^i \mapsto -H^i,\qquad E^i_\pm \mapsto -E^i_\mp \tag{11.56}$$

for $i = 1, 2, \ldots, \mathrm{r}$. Of course, this map can also be extended linearly to all of $\mathfrak{g}$, instead of anti-linearly as in (11.52); then one obtains an involutive automorphism ω of $\mathfrak{g}$. This automorphism is known as the *Weyl automorphism* or *Chevalley involution* of $\mathfrak{g}$. In the case of $\mathfrak{g} = A_1$, this automorphism is the one that was introduced in (11.21) above.

* 11.10 Involutive automorphisms and symmetric spaces

An important geometric application of automorphisms of simple Lie algebras is the description of symmetric spaces. A *geodesic symmetry* τ of a neighborhood of a point $P \in \mathcal{M}$ in a Riemannian manifold $\mathcal{M}$ is defined as follows. If $\gamma(t)$ is a geodesic through P parametrized such that $P = \gamma(0)$, then let

$$\tau:\quad \gamma(t) \;\mapsto\; \gamma(-t)\,. \tag{11.57}$$

Clearly, $\tau^2 = id$ and $\tau(P) = P$. The map τ is called a geodesic symmetry if it leaves the Riemannian metric of $\mathcal{M}$ invariant. If a geodesic symmetry exists, then $\mathcal{M}$ is called a locally symmetric space, and if the locally defined map (11.57) can be extended to a global map which still leaves the metric invariant, then $\mathcal{M}$ is called a *globally symmetric space*. Roughly, in a globally symmetric space 'each point looks like every other point'.

The maximal subspaces of a real Lie algebra $\mathfrak{g}$ on which the Killing form is positive definite, or negative definite, respectively, yield via the exponential map globally symmetric Riemannian manifolds which are compact, or non-compact, respectively. Conversely, it can be shown that all globally symmetric Riemannian spaces are coset spaces, i.e. spaces $\mathcal{M} = \mathrm{G/H}$ whose elements are the orbits on a Lie group manifold G with respect to a closed Lie subgroup H of G. (Such spaces are also called *homogeneous spaces*.) For example, $\mathrm{SU}(2)/\mathrm{U}(1) \cong \mathrm{SO}(3)/\mathrm{SO}(2)$ is the sphere S^2, which is compact, and $\mathrm{SL}(2,\mathbb{R})/\mathrm{U}(1)$ is a non-compact space, namely a sheet of a two-sheeted hyperboloid. The Lie algebra $\mathfrak{h}$ of H corresponds to the infinitesimal diffeomorphisms of $\mathcal{M}$ which leave the point $P \in \mathcal{M}$ fixed; correspondingly H is known as the isotropy group of P. The infinitesimal diffeomorphisms of $\mathcal{M}$ which leave the metric invariant and induce a parallel translation along a geodesic define another Lie algebra $\mathfrak{m}$. These Lie algebras satisfy $\mathfrak{h} \oplus \mathfrak{m} = \mathfrak{g}$ (the Lie algebra of G) and

$$[\mathfrak{h},\mathfrak{h}] \subseteq \mathfrak{h}, \qquad [\mathfrak{h},\mathfrak{m}] \subseteq \mathfrak{m}, \qquad [\mathfrak{m},\mathfrak{m}] \subseteq \mathfrak{h}. \tag{11.58}$$

The splitting $\mathfrak{g} = \mathfrak{h} \oplus \mathfrak{m}$ is called the *Cartan decomposition* of the real Lie algebra $\mathfrak{g}$ (not to be confused with the triangular decomposition (6.32) for which this term is also sometimes used). Comparison with (11.50) shows that $\mathfrak{h}$ and $\mathfrak{m}$ are the eigenspaces of an involutive automorphism ω of $\mathfrak{g}$: $\omega|_{\mathfrak{h}} = id$, $\omega|_{\mathfrak{m}} = -id$. Also, $\mathfrak{m}$ can be considered as the tangent space of the manifold $\mathcal{M}$ in the point P, and hence according to (11.58) this tangent space is endowed with an action of the Lie algebra $\mathfrak{h}$.

Further analysis shows that globally symmetric spaces are obtained precisely in the following cases. Either $\mathfrak{g}$ is a compact simple Lie algebra over $\mathbb{R}$, and ω is an arbitrary involutive automorphism of $\mathfrak{g}$; $\mathcal{M}$ is then called *type I*. Or else, $\mathfrak{g}$ is the direct sum of two isomorphic compact simple Lie algebras over $\mathbb{R}$, and ω is the automorphism which interchanges the two direct summands; then $\mathcal{M}$ is said to be of *type II*.

A globally symmetric space $\mathcal{M}$ is called a *hermitian symmetric space* if there exists a complex structure on $\mathcal{M}$ which is invariant under all geodesic symmetries. The hermitian symmetric spaces are all of type I, and the relevant subalgebras $\mathfrak{h} \subset \mathfrak{g}$ are of the form $\mathfrak{h} = \mathfrak{h}' \oplus \mathfrak{u}(1)$, where $\mathfrak{h}'$ is a semisimple Lie algebra whose Dynkin diagram is obtained from the extended Dynkin diagram of $\mathfrak{g}$ by deleting precisely two nodes both of which correspond to a Coxeter label which is equal to one. Also, for a hermitian symmetric space the Lie algebra $\mathfrak{m}$ can be decomposed into a vector space direct sum of two subalgebras $\mathfrak{m}_\pm$ which are both abelian,

$$\mathfrak{m} = \mathfrak{m}_+ \oplus \mathfrak{m}_-, \qquad [\mathfrak{m}_+,\mathfrak{m}_+] = 0 = [\mathfrak{m}_-,\mathfrak{m}_-]. \tag{11.59}$$

Summary:

The automorphisms of a simple Lie algebra form a (typically non-abelian) group $Aut(\mathfrak{g})$. The inner automorphisms, i.e. those automorphisms which are given by the exponential of the adjoint action of some Lie algebra element, form a normal subgroup $Int(\mathfrak{g})$ of $Aut(\mathfrak{g})$. The factor group $Out(\mathfrak{g}) := Aut(\mathfrak{g})/Int(\mathfrak{g})$ is isomorphic to the subgroup of isomorphisms that is induced by the symmetries of the Dynkin diagram, and also to the automorphism group of the root system modulo the Weyl group.

Inner automorphisms that map the Cartan subalgebra to itself can be described by shift vectors. Involutive automorphisms of a complex Lie algebra classify its real forms.

Keywords:

Automorphism, adjoint group, order, fixed point subalgebra, inner and outer automorphisms, conjugate Cartan subalgebras, shift vector, Dynkin diagram symmetry, extended Dynkin diagram, Chevalley involution, involutive automorphism, conjugation, symmetric space.

Exercises:

Check that the linear combination $x_{(j)}$ defined by (11.8) is an eigenvector of the order N automorphism ω with eigenvalue $e^{2\pi i \cdot j/N}$.

Exercise 11.1

Verify the Lie bracket relation (11.9), and conclude that $\mathfrak{g}_{(0)}$ is a subalgebra of $\mathfrak{g}$.

Are there other proper subalgebras of $\mathfrak{g}$ which are of the form $\bigoplus_i \mathfrak{g}_{(i)}$ (with $\mathfrak{g}_{(i)}$ the eigenspaces as in (11.6))?

Exercise 11.2

Show that the Killing form of a Lie algebra $\mathfrak{g}$ is invariant under any automorphism of $\mathfrak{g}$.

Exercise 11.3

Show that a Lie algebra $\mathfrak{g}$ is nilpotent if and only if all $x \in \mathfrak{g}$ are ad-nilpotent. (This is known as *Engel's theorem*.)

Exercise 11.4

If for any $y \in \mathfrak{g}$ there exists a positive integer m_y (that may depend on y) such that $(\mathrm{ad}_x)^{m_y}(y) = 0$, then $x \in \mathfrak{g}$ is called *locally nilpotent*. Of course, nilpotency of x implies local nilpotency.

Show that for a finite-dimensional Lie algebra the two concepts are equivalent. Convince yourself that local nilpotency of x is sufficient for Ad_x (11.13) to be a well-defined endomorphism of $\mathfrak{g}$, even when $\mathfrak{g}$ is infinite-dimensional.

Prove the property (11.14) of Ad_x, and the analogous property (11.18) of $\mathrm{Exp}(\delta)$ for arbitrary derivations δ.

Hint: Start by showing, by induction, that

$$\delta^n([x,y]) = \sum_{j=0}^{n} \frac{n!}{j!(n-j)!} \left[\delta^j(x), \delta^{n-j}(y)\right]. \qquad (11.60)$$

Check the other properties which are required for Ad_x to be an automorphism.

Exercise 11.5

Prove the formula (11.15) for the conjugation of an inner automorphism by some other automorphism.

Exercise 11.6

Derive the action (11.22) of the $\mathfrak{sl}(2)$-automorphism (11.21). Check that in terms of 2×2-matrices, this action is realized as described in (11.24).

Exercise 11.7

Work out the multiplication table of the cyclic group $\mathbb{Z}_n$, and, for small values of n, of the symmetric group $\mathcal{S}_n$ and the dihedral group $\mathcal{D}_n$.

Exercise 11.8

Explain why the fixed point algebra of a finite order automorphism of a semisimple Lie algebra can be reductive (i.e. contain abelian ideals), in spite of the fact that the Killing form of the fixed point algebra is non-degenerate.
Hint: Analyze the simple example of the automorphism $H \mapsto H,\ E_\pm \mapsto -E_\pm$ of $\mathfrak{sl}(2)$. What is the adjoint representation of $\mathfrak{u}(1)$?

Exercise 11.9

Show that each automorphism ω of a finite-dimensional simple Lie algebra $\mathfrak{g}$ which leaves the Cartan subalgebra $\mathfrak{g}_\circ$ invariant acts on the step operators as $\omega(E^\alpha) = \eta_\alpha E^\alpha$ with $|\eta_\alpha| = 1$ and $\eta_{-\alpha} = \overline{\eta}_\alpha$.
Conclude that ω leaves $\mathfrak{g}_\circ$ pointwise fixed if and only if $\omega = \mathrm{Ad}_h$ for some $h \in \mathfrak{g}_\circ$.

Exercise 11.10

Show that the factor group $Aut(\mathfrak{g})/Int(\mathfrak{g})$ is isomorphic to the group of symmetries of the Dynkin diagram.
Hint: Proceed as follows:
1. Use the fact that all Cartan subalgebras are conjugate to show that for each $\omega \in Aut(\mathfrak{g})$ and each Cartan subalgebra $\mathfrak{g}_\circ$ there is an inner automorphism ϖ such that $\omega_1 := \varpi \circ \omega$ maps $\mathfrak{g}_\circ$ to itself.
2. Construct an inner automorphism σ such that $\omega_2 := \sigma \omega_1$ preserves the whole triangular decomposition of $\mathfrak{g}$.
3. Find an automorphism τ of the type (11.32) such that $\omega_3 := \omega_2 \circ \tau$ is an inner automorphism of the form (11.34).
4. Conclude that ω can be written as the product of an inner automorphism and the automorphism τ^{-1} that is induced by a Dynkin diagram symmetry.

Exercise 11.11

12
Loop algebras and central extensions

12.1 Central extensions

In the application of Lie algebras to physical problems, one frequently encounters the following situation. On one hand, the symmetries of a physical system are described at the level of *classical* mechanics or field theory by some Lie algebra $\mathfrak{g}$. On the other hand, in the *quantum* theoretic description of the same system, the Lie brackets ('commutation relations') of $\mathfrak{g}$ are not recovered completely. Rather, in addition to the terms that were already present in the classical formulation certain further terms appear for the Lie brackets, which however are not arbitrary, but are 'constant' terms in the sense that in any irreducible representation of $\mathfrak{g}$, they act by numerical constants (which may depend however on the specific representation).

In order to describe this new situation entirely in Lie algebraic terms, one must interpret these numbers as the eigenvalues of some new operators K_i which have constant eigenvalue on any irreducible module of $\mathfrak{g}$. These additional generators K_i extend the Lie algebra $\mathfrak{g}$ to some closely related algebra $\hat{\mathfrak{g}}$. A situation in which such a re-interpretation is required has in fact already been encountered in chapter 2: When one wants to regard the commutation relations (2.29) of the Heisenberg algebra as proper Lie brackets, then one must interpret the right hand side as a Lie algebra element of the type just described.)

By Schur's Lemma, the property to have constant eigenvalues on ir-reducible modules is certainly fulfilled for those elements K_i which have zero Lie bracket ('commute') with all elements of $\mathfrak{g}$, i.e. for all elements of the center $\mathcal{Z}(\mathfrak{g})$ of $\mathfrak{g}$. Accordingly for any arbitrary Lie algebra $\mathfrak{g}$, we can construct an algebra $\hat{\mathfrak{g}}$ whose dimension exceeds the dimension of $\mathfrak{g}$ by ℓ, by simply adjoining ℓ additional generators K^j, $j = 1, 2, ..., \ell$, to a basis $\{T^a\}$ of $\mathfrak{g}$ and imposing the relations

$$[K^i, K^j] = 0, \qquad [T^a, K^j] = 0 \tag{12.1}$$

for $i, j = 1, 2, \ldots, \ell$ and $a = 1, 2, \ldots, d$, while keeping the original values $f^{ab}{}_c$ of those structure constants which involve only the generators T^a. Then the most general form of the brackets among these $\hat{\mathfrak{g}}$-generators reads

$$[T^a, T^b] = \sum_{c=1}^{d} f^{ab}{}_c T^c + \sum_{i=1}^{\ell} f^{ab}{}_i K^i, \tag{12.2}$$

with $f^{ab}{}_c$ the structure constants of $\mathfrak{g}$ in the basis $\{T^a\}$. A Lie algebra $\hat{\mathfrak{g}}$ with brackets of the form (12.1) and (12.2) is called an (ℓ-dimensional) *central extension* of $\mathfrak{g}$. (In this context, recall from section 4.8 that the center of a Lie algebra $\mathfrak{g}$ is an abelian ideal of $\mathfrak{g}$. Simple Lie algebras do not possess any non-trivial ideals and hence their center is trivial. The same is true also for semisimple Lie algebras which only have simple ideals.)

The new additional structure constants $f^{ab}{}_i$ cannot be chosen arbitrarily, but are restricted by the Jacobi identity. An obvious solution to the constraints implied by the Jacobi identity is $f^{ab}{}_i \equiv 0$; $\hat{\mathfrak{g}}$ is then nothing but the direct sum of $\mathfrak{g}$ and an abelian algebra,

$$\hat{\mathfrak{g}} \cong \mathfrak{g} \oplus (\mathfrak{u}(1))^\ell. \tag{12.3}$$

Now $f^{ab}{}_i \equiv 0$ is typically *not* the only solution for which $\hat{\mathfrak{g}}$ is isomorphic to a direct sum as in (12.3). Rather, $\hat{\mathfrak{g}}$ is of this form whenever there exists a choice of basis elements

$$\tilde{T}^a = T^a + \sum_{i=1}^{\ell} u_i^a K^i \tag{12.4}$$

with suitable coefficients u_i^a in which the transformed structure constants

$$\tilde{f}^{ab}{}_i = f^{ab}{}_i - \sum_{c=1}^{d} f^{ab}{}_c u_i^c \tag{12.5}$$

are zero.

In particular, if $\mathfrak{g}$ is a reductive Lie algebra (i.e. the direct sum of simple and abelian ideals), then $\hat{\mathfrak{g}}$ is again reductive, so that the extension does not yield a new type of Lie algebras. We can therefore consider extensions of the type (12.3) as trivial, and we will count only those extensions as genuine central extensions which are not of that form. The number of independent genuine central extensions of $\mathfrak{g}$ is then the dimension of the space of solutions to the constraints which the Jacobi identity implies for the $f^{ab}{}_i$ modulo the space of those solutions which satisfy (12.3).

It is in general a difficult question whether a given Lie algebra allows for non-trivial central extensions or not. As the reader can show in exercise 12.1, finite-dimensional simple (and, more generally, semisimple) Lie algebras do not possess non-trivial central extensions. In section 12.3 we will investigate a class of Lie algebras which do admit non-trivial central extensions.

*12.2 Ray representations and cohomology

In a basis independent formulation, the extension of a complex Lie algebra (or of the compact real form of a semisimple Lie algebra) by central elements is described as follows. Denoting the Lie bracket of $\mathfrak{g}$ by $[\cdot,\cdot]_{\text{old}}$, the bracket of $\hat{\mathfrak{g}}$ is given by $[x,y]_{\text{new}} = [x,y]_{\text{old}} + \Omega(x,y)\,K$ and $[K,x] = 0$, with Ω a bilinear complex function which is defined on $\mathfrak{g}\times\mathfrak{g}$. Since the new Lie bracket must again be antisymmetric and bilinear, Ω has these properties, too, while the Jacobi identity implies that

$$0 = \Omega(x,[y,z]_{\text{old}}) + \Omega(y,[z,x]_{\text{old}}) + \Omega(z,[x,y]_{\text{old}}). \qquad (12.6)$$

A function Ω with these properties is called a two-*cocycle* of $\mathfrak{g}$. The central extension is trivial precisely if this cocycle is a linear function of $[x,y]_{\text{old}}$, because then the equality (12.6) directly follows from the Jacobi identity of $\mathfrak{g}$; in this case $\Omega(x,y)$ is called a coboundary. The inequivalent non-trivial central extensions are then described by the vector space of two-cocycles modulo coboundaries, which is called the second (Lie algebra) cohomology $H^2(\mathfrak{g},\mathbb{R})$. (This terminology stems from cohomology theory, which is beyond the scope of this book; for references about Lie algebra cohomology, see the Epilogue of the book.)

The relation between $\hat{\mathfrak{g}}$ and $\mathfrak{g}$ can also be described by requiring that

$$0 \;\to\; \mathfrak{u}(1) \;\to\; \hat{\mathfrak{g}} \;\to\; \mathfrak{g} \;\to\; 0\,, \qquad (12.7)$$

where '0' stands for the trivial zero-dimensional Lie algebra, is a so-called *exact sequence* of maps. Such a sequence is to be read as follows. Any arrow denotes a Lie algebra homomorphism. Moreover, it is understood that for each pair of consecutive mappings the image of the first homomorphism coincides with the kernel of the second homomorphism. Applying this prescription to the $\mathfrak{u}(1)$-entry in (12.7), we see that the image of 0, which is just the zero vector of $\mathfrak{u}(1)$, must be the kernel of the second map; in other words, this map has to be injective. Similarly, the map from $\hat{\mathfrak{g}}$ to $\mathfrak{g}$ must be surjective, and the kernel of this map must be isomorphic to $\mathfrak{u}(1)$. Exact sequences are particularly useful when one deals with many homomorphisms between various different objects, and when they are combined in commuting diagrams.

Central extensions also arise naturally when one describes symmetries in quantum theory. In quantum mechanics, physical states are not described by individual Hilbert space vectors, but rather by rays of such vectors. (A *ray* in a vector space V is by definition a one-dimensional subspace of V.) Now groups G of symmetries act on physical states, and hence when describing symmetries in terms of individual Hilbert space vectors one generically does not deal with a representation of G; rather, the putative representation matrices $R(\gamma)$ only need to satisfy $R(\gamma)R(\gamma') = \omega(\gamma,\gamma')\,R(\gamma\gamma')$ for some complex number $\omega(\gamma,\gamma')$. This number $\omega(\gamma,\gamma')$ is then called a multiplier, and R is called a ray representation or projective representation of G. (The origin of the latter name is as follows. Associated to any vector space V there is the space $\mathbb{P}V$ of all one-dimensional subspaces, i.e. of all rays of V. The space $\mathbb{P}V$ is a smooth manifold; it is called the *projective* space over V. Any ray representation on V induces a genuine representation on $\mathbb{P}V$ which, however, is *not* a *linear* representation any more. Points of the projective space $\mathbb{P}V$ can be described by coordinates ξ^i of the vector space V, then referred to as *homogeneous* 'coordinates', provided one adopts the convention that the values (ξ^i) and $\zeta(\xi^i)$ describe one and the same point when ζ is any non-zero complex number, independent of i. Thus for each j one obtains a coordinate patch, defined where $\xi^j \neq 0$, by taking ξ^i/ξ^j $(i \neq j)$ as coordinates; these patches cover $\mathbb{P}V$.)

The associativity of the group product implies that the multipliers $\omega(\gamma, \gamma')$ satisfy

$$\omega(\gamma, \gamma')\,\omega(\gamma\gamma', \gamma'') = \omega(\gamma', \gamma'')\,\omega(\gamma, \gamma'\gamma'') \tag{12.8}$$

for all $\gamma, \gamma', \gamma'' \in G$. Since one is only interested in rays, one can use any function φ on $\mathfrak{g}$ to redefine the representation on vectors by $R(\gamma) \mapsto \varphi(\gamma)R(\gamma)$. This implies that two ray representations R and $\tilde{R}$ which differ only in a specific manner by the values of the multipliers, namely such that

$$\tilde{\omega}(\gamma, \gamma') = \omega(\gamma, \gamma') \cdot \varphi(\gamma\gamma') / \varphi(\gamma)\varphi(\gamma'), \tag{12.9}$$

should be considered as equivalent. In particular, whenever the multipliers can be written as $\omega(\gamma, \gamma') = \varphi(\gamma\gamma')/\varphi(\gamma)\varphi(\gamma')$, the ray representation is equivalent to an ordinary representation. Just like in the case of equation (12.6), this again gives rise to cohomology problem. (Technically speaking, when G is a compact Lie group, then any ray representation on $\mathbb{P}V$ is induced by a linear representation on the vector space V if and only if the (Lie group) cohomology $H^2(G, U(1))$ is trivial.)

The relation with the central extension of the Lie *algebra* $\mathfrak{g}$ of G with cocycle Ω is, not surprisingly, given by an exponential mapping:

$$\Omega(x, y) = \frac{\mathrm{d}^2}{\mathrm{d}t^2}\Big|_{t=0} \omega(e^{tx}, e^{ty}). \tag{12.10}$$

In agreement with this exponential relationship, one can show that for a finite-dimensional Lie group G which is connected and simply connected, the Lie group cohomology coincides with the cohomology of the Lie algebra $\mathfrak{g}$ of G in the sense that $H^2(G, U(1)) = H^2(\mathfrak{g}, \mathbb{R})$. On the other hand, for non-simply connected G the Lie group cohomology $H^2(G, U(1))$ can be non-trivial even when $H^2(\mathfrak{g}, \mathbb{R}) = 0$.

When the multipliers are non-trivial one does not have an ordinary representation. However, one can extend any Lie group G to a larger group $\widehat{G}$ and 'lift' the ray representation to $\widehat{G}$ in such a way that it becomes an ordinary representation of $\widehat{G}$. This is achieved by considering pairs $(\gamma; \varpi)$ of group elements $\gamma \in G$ and elements ϖ of the abelian group K that is generated by all the multipliers $\omega(\gamma, \gamma')$. The group $\widehat{G}$ is then the set of all such pairs, with multiplication defined by

$$(\gamma; \varpi)\,(\gamma'; \varpi') := (\gamma\gamma'; \varpi\varpi'\omega(\gamma, \gamma')). \tag{12.11}$$

This implies that $\widehat{G}$ is a semidirect product of G and the abelian group K, that is, a central extension of G by K. Also, one can show that if G is a *Lie* group, then so is $\widehat{G}$.

When G is a finite-dimensional compact semisimple Lie group, any ray representation is in particular also a ray representation of its universal covering group $\tilde{G}$. It can be shown that for all simply connected compact simple Lie groups $\tilde{G}$ the homology group $H^2(\tilde{G}, U(1))$ is trivial, so that any ray representation of $\tilde{G}$ is equivalent to a honest representation of $\tilde{G}$. It follows that any projective representation of a compact simple Lie group G can be lifted to a honest representation of its universal covering group. This means that the centrally extended group $\widehat{G}$ is in fact isomorphic to the universal covering group $\tilde{G}$ of G. (As a consequence, universal covering groups play an important rôle in quantum mechanics.) As a converse of this construction, those representations of $\tilde{G}$ which are not honest representations of G, like e.g. the spinor representations of $\mathfrak{so}(n)$

for G the non-simply connected group $G = SO(n)$, are nonetheless ray representations of G. (In the case of $SO(n)$ the extended group is the covering group $\text{Spin}(n)$, compare also section 20.4.)

Note, however, that these arguments do *not* apply to all Lie groups. For instance, the central extension of the symmetry group of classical mechanics, the ten-dimensional Galilei group (generated by three rotations in space, four translations in space and time and three motions with constant velocity) is an eleven-dimensional group.

12.3 Loop algebras

An interesting class of infinite-dimensional Lie algebras is provided by the so-called loop algebras. These can be used to construct the affine Lie algebras introduced in chapter 7 in a rather explicit manner.

By a *loop* in some topological space $\mathcal{M}$ one means the smooth embedding of a circle into $\mathcal{M}$ (together with a chosen parametrization). The loop algebra associated to a given Lie algebra $\bar{\mathfrak{g}}$ consists of the space of analytic mappings from the circle S^1 to $\bar{\mathfrak{g}}$. (For future convenience, we use the notation $\bar{\mathfrak{g}}$ rather than $\mathfrak{g}$ for the Lie algebra we are starting with.) If $\{T^a \,|\, a = 1, 2, \ldots, d = \dim \bar{\mathfrak{g}}\}$ is a basis of $\bar{\mathfrak{g}}$ and S^1 is considered as the unit circle in the complex plane with coordinate $z = \mathrm{e}^{2\pi i t}$, then Fourier analysis shows that a topological basis of the vector space of these maps is given by

$$\mathcal{B} = \{\tilde{T}_n^a \,|\, a = 1, 2, \ldots, d; \; n \in \mathbb{Z}\}, \qquad (12.12)$$

where

$$\tilde{T}_n^a := T^a \otimes z^n \equiv T^a \otimes \mathrm{e}^{2\pi i n t}. \qquad (12.13)$$

(A *topological* or *analytic* basis, sometimes also called Hilbert space basis, of a vector space V is characterized by the property that the closure in the topology of V of the set of all *finite* linear combinations of basis vectors is the whole space V. Since one takes the closure, such a basis is in general *not* a vector space basis of V.)

Moreover, this space inherits a natural bracket operation from $\bar{\mathfrak{g}}$, namely $[\tilde{T}_m^a, \tilde{T}_n^b] = [T^a \otimes z^m, T^b \otimes z^n] := [T^a, T^b] \otimes (z^m z^n)$, i.e.

$$[\tilde{T}_m^a, \tilde{T}_n^b] = \sum_{c=1}^{d} f^{ab}{}_c \, T^c \otimes z^{m+n} = \sum_{c=1}^{d} f^{ab}{}_c \, \tilde{T}_{m+n}^c, \qquad (12.14)$$

where $f^{ab}{}_c$ are the structure constants of $\bar{\mathfrak{g}}$. Thus the structure constants in the basis $\mathcal{B}$ read $f^{ab}{}_c \delta_{m+n,l}$. With this bracket operation the space of analytic maps from S^1 to $\bar{\mathfrak{g}}$ becomes a Lie algebra itself (see exercise 12.2). This infinite-dimensional Lie algebra is denoted by $\bar{\mathfrak{g}}_{\text{loop}}$ and called the *loop algebra* over $\bar{\mathfrak{g}}$. Moreover, under the Lie bracket (12.14) the index n of $\tilde{T}_n^a$ is additive and hence according to section 4.11 provides a $\mathbb{Z}$-gradation of $\bar{\mathfrak{g}}_{\text{loop}}$. The subset of $\mathfrak{g}_{\text{loop}}$ generated by the generators $\tilde{T}_0^a$

is therefore a Lie subalgebra; it is called the *zero mode subalgebra* of $\bar{\mathfrak{g}}_{\text{loop}}$. The zero mode subalgebra is isomorphic to the original Lie algebra $\bar{\mathfrak{g}}$.

Loop algebras – or rather their subalgebras which are spanned by the generators with non-negative grading – have been applied e.g. to the description of symmetries in extended supergravity models (compare [Ellis *et al.* 1983]) and in the self-dual Yang–Mills equations (see [Dolan 1985] and, for a more detailed analysis in the context of loop groups, [Crane 1987]). It must be noted that in the cited papers the loop algebras are inaccurately referred to as Kac–Moody algebras.

Although loop algebras look quite promising for the description of the symmetries of physical systems such as the ones just mentioned, they have a serious drawback. Namely, for applications in physics a Lie algebra usually must possess modules into which the Hilbert space of the physical system can be decomposed. Now a closer look reveals that the only unitary representation of a loop algebra $\bar{\mathfrak{g}}_{\text{loop}}$ is the trivial one-dimensional representation. Ray representations do exist, however. In accordance with the general strategy outlined in the previous section, we will therefore investigate in the next section central extensions of loop algebras. This will lead us to the construction of Lie algebras which do possess unitary representations, namely of affine Lie algebras.

12.4 Towards untwisted affine Lie algebras

The construction just described works for any Lie algebra. In the special case where the original Lie algebra $\bar{\mathfrak{g}}$ is simple, the loop algebra appears at an intermediate step in the construction of affine Lie algebras. (From here on we also assume that $\bar{\mathfrak{g}}$ is a Lie algebra over $\mathbb{C}$.) However, the loop algebra itself is not yet an affine Lie algebra, for the simple reason that the center of the loop algebra based on a simple Lie algebra is trivial, whereas affine Lie algebras possess a non-zero center. To see this, consider the element

$$K := \sum_{i=0}^{r} a_i^{\vee} H^i \qquad (12.15)$$

of an affine Lie algebra $\mathfrak{g}$, where H^i are the canonical Chevalley generators of $\mathfrak{g}$ which appear e.g. in the relation (7.1) and $a_i^{\vee}$ are the dual Coxeter labels of $\mathfrak{g}$. Since the H^i are in the Cartan subalgebra, K satisfies $[K, H^i] = 0$ for $i = 0, 1, ..., r$. Further, owing to the fact that the dual Coxeter labels form an eigenvector of the Cartan matrix with eigenvalue zero (see equation (7.41)), the brackets of K with the step operators associated to simple roots vanish as well:

$$[K, E_{\pm}^i] = \pm \sum_{j=0}^{r} a_j^{\vee} A^{ij} E_{\pm}^i = 0. \qquad (12.16)$$

Thus K is a central element, $K \in \mathcal{Z}(\mathfrak{g})$. Because of the condition $\det A_{\{i\}} > 0$ (7.30), affine Cartan matrices possess only a single zero eigenvalue; it follows that $\mathcal{Z}(\mathfrak{g})$ is in fact one-dimensional, i.e. all central elements are scalar multiples of K. The generator K defined by (12.15) is therefore called the *canonical central element* of $\mathfrak{g}$.

What is required to construct affine Lie algebras is to combine the two new concepts introduced above, i.e. consider central extensions of the loop algebras of simple Lie algebras. (We reserve the symbol $\mathfrak{g}$ for the affine algebra in question, which is the reason why we chose to denote the original simple Lie algebra, which is isomorphic to the zero mode subalgebra of the loop algebra, by $\bar{\mathfrak{g}}$.) In the case of loop algebras, the general ansatz (12.2) for the new Lie brackets can be written as

$$[T_m^a, T_n^b] = \sum_{c=1}^{\mathrm{d}} f^{ab}{}_c \, T_{m+n}^c + \sum_{i=1}^{\ell} (f^{ab}{}_i)_{mn} \, K^i \,. \qquad (12.17)$$

Here we use the notation T_m^a etc. for the generators of the centrally extended algebra, to be distinguished from the generators $\tilde{T}_m^a$ of the unextended loop algebra they stem from. Since $\bar{\mathfrak{g}}$ does not admit any non-trivial central extension, by a suitable redefinition of generators we can put $(f^{ab}{}_i)_{00}$ to zero; in fact, extending the arguments of exercise 12.1, this is even possible for $(f^{ab}{}_i)_{m0}$ for all $m \in \mathbb{Z}$. Then for any fixed n one has

$$[T_0^a, T_n^b] = \sum_{c=1}^{\mathrm{d}} f^{ab}{}_c \, T_n^c \,, \qquad (12.18)$$

so that the vector space spanned by $\{T_n^a \,|\, a = 1, 2, \dots, \mathrm{d}\}$ (with fixed $n \in \mathbb{Z}$) transforms like the adjoint module of the zero mode algebra

$$\mathfrak{g}_{[0]} := \mathrm{span}\{T_0^a \,|\, a = 1, 2, \dots, \mathrm{d}\} \qquad (12.19)$$

which is again isomorphic to $\bar{\mathfrak{g}}$. Accordingly with respect to the the upper indices a, b the structure constants $(f^{ab}{}_i)_{mn}$ form a so-called invariant tensor of the adjoint representation of $\bar{\mathfrak{g}}$. As we will see in chapter 17, there is a unique (up to normalization) such tensor, namely the Killing form $\bar{\kappa}$ of $\bar{\mathfrak{g}}$. Hence there are only one-dimensional non-trivial central extensions of $\bar{\mathfrak{g}}_{\mathrm{loop}}$; thus the index i is superfluous, and $(f^{ab}{}_i)_{mn}$ reduces to $(f^{ab})_{mn} = \bar{\kappa}^{ab} f_{mn}$. Because of the symmetry of the Killing form, f_{mn} must be antisymmetric in m, n, and in fact there is a unique solution for f_{mn} compatible with the Jacobi identity, namely $f_{mn} = m \, \delta_{m+n,0}$. One can show that this extension is indeed non-trivial.

In summary, for simple $\bar{\mathfrak{g}}$ there is a unique non-trivial central extension $\hat{\mathfrak{g}}$ of $\bar{\mathfrak{g}}_{\mathrm{loop}}$; its brackets read

$$[K, T_n^a] = 0 \,, \qquad [T_m^a, T_n^b] = \sum_{c=1}^{\mathrm{d}} f^{ab}{}_c \, T_{m+n}^c + m \, \delta_{m+n,0} \bar{\kappa}^{ab} \, K \,. \qquad (12.20)$$

This construction of $\hat{\mathfrak{g}}$ can be summarized as

$$\hat{\mathfrak{g}} = \bar{\mathfrak{g}}_{\text{loop}} \oplus \mathbb{C}K = \mathbb{C}\left[z, z^{-1}\right] \otimes_{\mathbb{C}} \bar{\mathfrak{g}} \oplus \mathbb{C}K . \tag{12.21}$$

Analogously, it can be shown that loop *groups*, i.e. groups of maps from the circle S^1 to a finite-dimensional compact Lie group G, possess a unique central extension by the Lie group U(1). If the original Lie group G is simple, then the Lie algebra of this centrally extended loop group is an untwisted affine Lie algebra.

Another notation that is sometimes used for stating the bracket relations (12.20) is to employ *Laurent polynomials*, i.e. polynomials in the coordinate z on the circle and its inverse z^{-1}. The elements of $\hat{\mathfrak{g}}$ are then written as sums of elements $x \otimes \mathcal{P}(z) + \zeta K$ which consist of the tensor product of $x \in \bar{\mathfrak{g}}$ with a Laurent polynomial $\mathcal{P}$ and some multiple of the central element K. In this notation the Lie brackets of $\hat{\mathfrak{g}}$ are determined by

$$[x \otimes \mathcal{P}(z) + \zeta K, y \otimes \mathcal{Q}(z) + \eta K]$$
$$= [x, y] \otimes (\mathcal{P}(z)\mathcal{Q}(z)) + \bar{\kappa}(x, y) K \otimes \text{Res}\,(\mathcal{P}(z)\partial \mathcal{Q}(z)), \tag{12.22}$$

where $\partial \equiv \frac{d}{dz}$, and 'Res' denotes the residue, i.e. the coefficient of z^{-1}, of a Laurent polynomial.

Affine Lie algebras made their first appearance in physics as symmetries of certain two-dimensional conformal field theories, which are nowadays called WZW theories (for a few more details, see section 12.12 below). In string theory, conformal field theories based on affine Lie algebras give e.g. rise to the gauge bosons in a heterotic string compactification [Green *et al.* 1987]. Also, affine Lie algebras are the starting point of several other constructions of conformal field theories, like the coset construction (compare e.g. [Goddard *et al.* 1985], [Schellekens and Yankielowicz 1990] and [Bouwknegt and Schoutens 1993]), irrational conformal field theories [Halpern *et al.* 1996], and Hamiltonian reduction [Fehér *et al.* 1992]. More recently, affine Lie algebras have made their appearance also in several other contexts. For example, the generating function of the Euler numbers of the instanton moduli spaces on ALE surfaces is related to a character of an affine Lie algebra [Vafa and Witten 1994].

Information

12.5 The derivation

The algebra $\hat{\mathfrak{g}}$ is not yet an affine Lie algebra, but it already comes rather close. Namely, for $\bar{\mathfrak{g}} = X_r$ the untwisted affine algebra $\mathfrak{g} = X_r^{(1)}$ is obtained from $\hat{\mathfrak{g}} \equiv \hat{X}_r$ by adding just one further generator D:

$$\mathfrak{g} = \hat{\mathfrak{g}} \oplus \mathbb{C}D = \mathbb{C}\left[z, z^{-1}\right] \otimes_{\mathbb{C}} \bar{\mathfrak{g}} \oplus \mathbb{C}K \oplus \mathbb{C}D . \tag{12.23}$$

The prescription (12.23) defines $\mathfrak{g}$ as a vector space. To obtain also its Lie algebra structure, we need to specify the Lie brackets; those involving the new generator D read

$$[D, T_m^a] = m\, T_m^a , \qquad [D, K] = 0 , \tag{12.24}$$

while the brackets (12.20) among the other generators remain unchanged. Thus the element D measures the mode number m of the generators T_m^a, or in other words, the degrees with respect to the gradation are given by the eigenvalues of D.

The center of the algebra obtained this way is still one-dimensional, and hence the generator K introduced in (12.20) as a central extension of $\bar{\mathfrak{g}}_{\text{loop}}$ can be identified with K as defined in (12.15). In particular, already the subalgebra $\hat{\mathfrak{g}}$ contains the Cartan subalgebra generator H^0 that is associated to the zeroth simple root, namely $H^0 = K - \sum_{i=1}^{r} a_i^\vee H^i$. We will identify the other generators of a Chevalley–Serre basis later in equation (12.37).

The realization (12.23) of the untwisted affine algebras shows in particular that, just like the loop algebras, they are infinite-dimensional. In contrast, from the abstract definition in terms of generalized Cartan matrices given in chapter 7, this is not at all obvious, but rather can only be seen by explicitly constructing the root system.

The zero mode subalgebra $\mathfrak{g}_{[0]}$ of $\mathfrak{g}$ can be characterized as being spanned by those generators T_m^a which have a vanishing bracket with D. Because of $\mathfrak{g}_{[0]} \cong \bar{\mathfrak{g}}$ one may view the simple Lie algebra $\bar{\mathfrak{g}}$ that was the starting point of the loop construction as a subalgebra of the affine Lie algebra $\mathfrak{g}$. We will do so from now on, and correspondingly refer to $\bar{\mathfrak{g}}$ as the zero mode subalgebra or *horizontal* subalgebra of $\mathfrak{g}$.

The generator D does not appear on the right hand side of any of the brackets (12.20) and (12.24) of $\mathfrak{g}$, and it is the only generator with this property. Thus we have

$$[\mathfrak{g}, \mathfrak{g}] = \hat{\mathfrak{g}}, \tag{12.25}$$

i.e. the centrally extended loop algebra $\hat{\mathfrak{g}}$ is the derived algebra of $\mathfrak{g}$. The generator D is therefore referred to as the *derivation* of the affine algebra $\mathfrak{g}$. In terms of the notation used in (12.22), D acts as $[D, x \otimes \mathcal{P}(z)] = x \otimes z\, \partial \mathcal{P}(z)$ on the subalgebra $\bar{\mathfrak{g}}_{\text{loop}}$, i.e. as the differential operator $z\frac{\mathrm{d}}{\mathrm{d}z}$ (D is therefore also called the *scaling element* of $\mathfrak{g}$).

An important rôle played by the derivation D is to make the Killing form κ on $\mathfrak{g}$ non-degenerate. Note that the definition (6.10), i.e. $\text{tr}(\text{ad}_x \circ \text{ad}_y)$, of the Killing form used for finite-dimensional Lie algebras cannot be taken over to infinite-dimensional algebras like the affine ones because the trace over $\text{ad}_x \circ \text{ad}_y$ will typically be ill-defined. To circumvent this problem, we try to fix κ (up to normalization) by imposing the properties of symmetry, bilinearity and invariance, which in the semisimple case characterize the Killing form completely. By analyzing the invariance constraint $\kappa([w, x], y) = \kappa(w, [x, y])$ for specific combinations of generators and simplifying the resulting formulæ with the help of the symmetry and bilinearity of κ, one finds (see exercise 12.4) $\kappa(T_m^a, T_n^b) = \bar{\kappa}(T^a, T^b)\delta_{m+n,0}$,

$\kappa(T^a_m, K) = 0 = \kappa(K, K)$, as well as

$$\kappa(T^a_m, D) = 0, \qquad \kappa(K, D) = 1. \qquad (12.26)$$

Since D does not appear on the right hand side of the bracket relations, it is not possible to obtain also $\kappa(D, D)$ in this way. However, suppose that $\kappa(D, D) = \eta$ is non-vanishing; then instead of D one may as well take the combination $D' := D - \frac{1}{2}\eta K$ as a generator of $\mathfrak{g}$. D' has the same Lie brackets as D, and using the bilinearity of κ it follows that $\kappa(D', D') = 0$. Therefore without loss of generality we can assume that $\kappa(D, D) = 0$. Summarizing, we get

$$\kappa = \begin{pmatrix} \bar{\kappa}\,\delta_{m+n,0} & 0 & 0 \\ 0 & 0 & 1 \\ 0 & 1 & 0 \end{pmatrix}, \qquad (12.27)$$

where the rows and columns correspond to T^a_m, K and D, respectively. In particular, from the fact that the Killing form $\bar{\kappa}$ of the simple Lie algebra $\bar{\mathfrak{g}}$ is non-degenerate, it follows that κ is non-degenerate, too. In contrast, the analogous form on the centrally extended loop algebra $\bar{\mathfrak{g}}_{\text{loop}}$ has one row (the one corresponding to K) identically zero and hence is degenerate. It is precisely the introduction of the derivation which allows for a non-degenerate inner product on the algebra. (Also, by using $D + K$ and $D - K$ instead of K and D as basis elements, one could diagonalize the 2×2-matrix in the lower right hand corner. However, this would obscure the structure of the bracket relations of $\mathfrak{g}$, because then all generators would appear on their right hand sides.)

12.6 The Cartan–Weyl basis

The affine Lie algebras have been introduced in section 7.9 via a Chevalley–Serre type description, i.e. in terms of the generators H^i and $E^i_\pm$ ($i = 0, 1, ..., r$) and the relations satisfied by them. As described above, in addition the derivation D must be included in a proper manner. But even ignoring the latter, it is a difficult task to construct the analogue of the Cartan-Weyl basis, i.e. all step operators rather than only those for the simple roots, for untwisted affine Lie algebras directly from the Chevalley–Serre relations. It is much easier to obtain the affine Cartan-Weyl basis by employing the realization of $\mathfrak{g}$ as a centrally extended loop algebra with derivation. Namely, we only have to recall that the generators T^a_m of $\mathfrak{g}$ were obtained in the loop algebra construction as in (12.13), where T^a are the generators of the horizontal subalgebra $\bar{\mathfrak{g}}$.

Choosing a Cartan–Weyl basis $\{H^i \mid i = 1, 2, ..., r\} \cup \{E^\alpha\}$ of $\bar{\mathfrak{g}}$ (and also recalling the specific form of $\bar{\kappa}$ in a Cartan–Weyl basis, compare the equations (6.61) and (6.62)), the Cartan–Weyl basis version of the brackets

(12.20) thus reads as follows:

$$[H_m^i, H_n^j] = m\,\bar{G}^{ij}\delta_{m+n,0}\,K\,,$$

$$[H_m^i, E_n^{\bar{\alpha}}] = \bar{\alpha}^i E_{m+n}^{\bar{\alpha}}\,,$$

$$[E_m^{\bar{\alpha}}, E_n^{\bar{\beta}}] = e_{\bar{\alpha},\bar{\beta}} E_{m+n}^{\bar{\alpha}+\bar{\beta}} \qquad \text{for } \bar{\alpha}+\bar{\beta} \text{ a } \bar{\mathfrak{g}}\text{-root,}$$

$$[E_n^{\bar{\alpha}}, E_{-n}^{-\bar{\alpha}}] = \sum_{i=1}^{r} \bar{\alpha}_i H_0^i + nK\,. \tag{12.28}$$

Here $i,j \in \{1,2,...,r\}$, $\bar{\alpha}$ are arbitrary elements of the root system $\overline{\Phi} \equiv \Phi(\bar{\mathfrak{g}})$, and $\bar{G}$ is the symmetrized Cartan matrix of $\bar{\mathfrak{g}}$.

It must be stressed that even though the notations H_m^i and $E_m^{\bar{\alpha}}$ introduced here are quite suggestive, at this point we do not yet know whether they define a triangular decomposition of $\mathfrak{g}$, and if so, which of these generators span the Cartan subalgebra and which correspond to roots. In order to determine a Cartan subalgebra, we start from a Cartan subalgebra of $\mathfrak{g}_{[0]}$, generated by H_0^i with $i = 1,2,...,r$. In addition any Cartan subalgebra certainly contains the central generator K, since $[H_0^i, H_n^j] = 0 = [K, H_n^j]$. Moreover, the derivation D is ad-diagonalizable by its definition, and it satisfies $[H_0^i, D] = 0 = [K, D]$; therefore the Cartan subalgebra contains at least still the further generator D. The bracket relation $[D, H_n^j] \neq 0$ for $n \neq 0$ then tells us that there is no other commuting generator left. Thus a Cartan subalgebra of $\mathfrak{g}$ is $\mathfrak{g}_\circ = \text{span}\{K, D, H_0^i \,|\, i = 1,2,...,r\}$, or what is the same,

$$\mathfrak{g}_\circ = \text{span}\{D, H_0^i \,|\, i = 0,1,...,r\}\,. \tag{12.29}$$

To save space, we will use the short-hand notation $\bar{H}$ for $\{H^i \,|\, i = 1,2,...,r\}$. Inspection of the bracket relations (12.28) shows (see exercise 12.5) that the roots with respect to $(\bar{H}, K, D)$ are given by

$$\alpha = (\bar{\alpha}, 0, n) \qquad \text{for } \bar{\alpha} \in \overline{\Phi},\ n \in \mathbb{Z} \tag{12.30}$$

and

$$\alpha = (0, 0, n) \qquad \text{for } n \in \mathbb{Z}\backslash\{0\}\,. \tag{12.31}$$

They correspond to the generators $E_n^{\bar{\alpha}}$ with $n \in \mathbb{Z}$ and to H_n^j with $n \neq 0$, respectively. While the roots (12.30) appear with multiplicity one, i.e. are non-degenerate, the roots (12.31) do not depend on the label j of H_n^j and hence have multiplicity r. Note that the degenerate roots are all of the form $\alpha = n\delta$ with

$$\delta := (0, 0, 1) \tag{12.32}$$

and $n \in \mathbb{Z}\backslash\{0\}$.

The roots of the derived algebra $[\mathfrak{g}, \mathfrak{g}] = \hat{\mathfrak{g}} = \bar{\mathfrak{g}}_{\text{loop}}$ are all infinitely degenerate, because in $\hat{\mathfrak{g}}$ there is no Cartan subalgebra generator distinguishing between different labels n. Thus another rôle of the derivation D is to avoid such infinite multiplicities. A simple, but important consequence of this observation is that in an affine Lie algebra traces over any single root space are well-defined, which is not true for (centrally extended) loop algebras. However, while the infinite degeneracy of the roots of the derived algebra is removed by including the derivation D, there is no way to get rid of the remaining finite degeneracy of the roots (12.31).

12.7 The root system

We denote the set of roots of $\mathfrak{g}$ by Φ. According to equation (12.30), we can embed the root system $\overline{\Phi}$ of $\bar{\mathfrak{g}}$ into Φ via the identification $(\bar{\alpha}, 0, 0) \equiv \bar{\alpha}$. The next step in the construction of a triangular decomposition of $\mathfrak{g}$ is then to define positive roots. Having made a choice for distinguishing between positive and negative roots of $\bar{\mathfrak{g}}$, the positive roots ($\alpha > 0$) of $\mathfrak{g}$ can be defined by

$$\Phi_+ := \{\alpha = (\bar{\alpha}, 0, n) \in \Phi \mid n > 0 \text{ or } (n = 0, \bar{\alpha} \in \overline{\Phi}_+)\} \qquad (12.33)$$

and the negative ones ($\alpha < 0$) by $\Phi_- = \Phi \backslash \Phi_+$. This way we arrive at the triangular decomposition

$$\mathfrak{g} = \mathfrak{g}_+ \oplus \mathfrak{g}_\circ \oplus \mathfrak{g}_- , \qquad (12.34)$$

where $\mathfrak{g}_+ = \text{span}\{E^\alpha \mid \alpha > 0\}$ and $\mathfrak{g}_- = \text{span}\{E^{-\alpha} \mid \alpha > 0\}$. The subspaces $\mathfrak{g}_\pm$ of $\mathfrak{g}$ are nilpotent subalgebras, while $\mathfrak{b}_\pm := \mathfrak{g}_\circ \oplus \mathfrak{g}_\pm$ are Borel subalgebras of $\mathfrak{g}$.

The simple roots are those $\mathfrak{g}$-roots $\alpha^{(i)}$ for which the decomposition $\alpha = \sum_{i=0}^r b_i \alpha^{(i)}$ implies that, for all $i = 0, 1, ..., \mathrm{r}$, $b_i \geq 0$ if $\alpha > 0$ and $b_i \leq 0$ if $\alpha < 0$. With the choice (12.33) of positive roots, the simple roots are

$$\alpha^{(i)} = (\bar{\alpha}^{(i)}, 0, 0) \equiv \bar{\alpha}^{(i)} \qquad \text{for } i = 1, 2, ..., \mathrm{r} \qquad (12.35)$$

and

$$\alpha^{(0)} = (-\bar{\theta}, 0, 1) = \delta - \bar{\theta} , \qquad (12.36)$$

with $\bar{\theta}$ the highest root of $\bar{\mathfrak{g}}$. The generators E^α corresponding to these simple roots are precisely the generators that in the Chevalley–Serre formulation are denoted by E^i_+, $i = 0, 1, ..., \mathrm{r}$. The relation to the E^α_n-notation is thus

$$E^i_+ = E^{\bar{\alpha}^{(i)}}_0 \quad \text{for } i = 1, 2, ..., \mathrm{r}, \qquad E^0_+ = E^{-\bar{\theta}}_1 . \qquad (12.37)$$

As an example, consider the case of $\mathfrak{g} = A_1^{(1)}$. Then there are two simple roots $\alpha^{(1)} = \bar{\theta} = \bar{\alpha}$ and $\alpha^{(0)} = \delta - \bar{\theta} = \delta - \bar{\alpha}$, with $\bar{\alpha}$ the single positive root of the horizontal subalgebra A_1. It follows that the root system Φ is of the form that was already obtained in equation (7.38) by studying the Chevalley–Serre basis of $A_1^{(1)}$. The origin of the appearance of $\bar{\theta}$ in $\alpha^{(0)}$ is the fact that any root of $\bar{\mathfrak{g}}$ can be obtained by adding simple roots to $-\bar{\theta}$. Thus by adding simple roots $\alpha^{(i)}$, $i = 1, 2, \ldots, r$, to $\alpha^{(0)}$ one can obtain all roots of the form $(\bar{\alpha}, 0, 1)$, so that the identification of $\alpha^{(0)}$ as a simple root is precisely what is needed to have $(\bar{\alpha}, 0, 1) \in \Phi_+$ for each $\bar{\alpha} \in \bar{\Phi}$, which is required by (12.33). By iteration, one obtains analogously all other positive roots.

The restriction of the Killing form κ of $\mathfrak{g}$ to the Cartan subalgebra induces a metric G for the root space and for its dual, the weight space. The specific form (12.27) of κ implies that G looks like

$$G = \begin{pmatrix} \bar{G} & 0 & 0 \\ 0 & 0 & 1 \\ 0 & 1 & 0 \end{pmatrix}. \tag{12.38}$$

Unlike the quadratic form matrix $\bar{G}$ of $\bar{\mathfrak{g}}$, this metric is of lorentzian nature; in addition to r eigenvectors with positive eigenvalues that it inherits from $\bar{G}$, it has two eigenvectors $(0, 0, \ldots, 0, 1, \pm 1)$ with eigenvalues ± 1.

According to (12.38) the scalar product of two weights $\lambda = (\bar{\lambda}, k, n)$ and $\lambda' = (\bar{\lambda}', k', n')$ is

$$(\lambda \,|\, \lambda') = (\bar{\lambda}, \bar{\lambda}') + kn' + k'n. \tag{12.39}$$

In particular for roots the scalar product equals the one of their $\bar{\mathfrak{g}}$-components:

$$(\alpha \,|\, \alpha') \equiv ((\bar{\alpha}, 0, n) \,|\, (\bar{\alpha}', 0, n')) = (\bar{\alpha}, \bar{\alpha}'). \tag{12.40}$$

Thus the non-degenerate roots, for which $\bar{\alpha} \neq 0$, have positive length squared, $(\alpha \,|\, \alpha) > 0$, while the degenerate roots have length zero: $(\alpha \,|\, \alpha) = 0$ for $\alpha = n\delta$. The non-degenerate roots are therefore called *real roots*, whereas the degenerate ones are called *imaginary roots*. (This latter term is a bit misleading; because of the Lorentzian structure of the metric (12.38) the term *lightlike* roots which is also sometimes used is certainly more appropriate.)

The simple roots are non-degenerate and hence real. Therefore it makes sense to define simple *coroots* by $\alpha^{(i)^\vee} := (2/(\alpha^{(i)}, \alpha^{(i)}))\, \alpha^{(i)}$. The inner products

$$A^{ij} := (\alpha^{(i)}, \alpha^{(j)^\vee}) \qquad \text{for } i, j = 0, 1, \ldots, r \tag{12.41}$$

are then precisely the entries of the affine Cartan matrix as it was introduced in chapter 7. This observation in fact proves that the centrally extended loop algebras with derivation are indeed isomorphic to the untwisted affine algebras defined via their generalized Cartan matrix. The only check that still has to be made is to verify that the matrix (12.41) gives rise to the correct Dynkin diagram. Let us do this for two examples, namely $\bar{\mathfrak{g}} = A_r$ and $\bar{\mathfrak{g}} = E_6$. In the former case, one has $\bar{\theta} = \bar{\Lambda}_{(1)} + \bar{\Lambda}_{(r)}$ and hence $(\alpha^{(0)}, \alpha^{(i)}) = -(\bar{\theta}, \bar{\alpha}^{(i)}) = -\delta_{i,1} - \delta_{i,r}$; therefore the additional node in the Dynkin diagram must be connected by a single line with the first and last nodes of the A_r Dynkin diagram; we then get indeed the Dynkin diagram of $A_r^{(1)}$. In the case of E_6, one has $\bar{\theta} = \bar{\Lambda}_{(6)}$ and hence $(\alpha^{(0)}, \alpha^{(i)}) = -\delta_{i,6}$, again leading to the correct Dynkin diagram.

12.8 Twisted affine Lie algebras

So far we have constructed only the untwisted affine algebras. But along similar lines one can also obtain a realization of the twisted ones. The only modification is to give up the single-valuedness of the maps from S^1 to $\bar{\mathfrak{g}}$ that were used in the loop construction of section 12.3, i.e. one must no longer require that $\mathcal{P}(e^{2\pi i} z) = \mathcal{P}(z)$. Rather, one has to impose the *twisted* boundary conditions

$$x \otimes \mathcal{P}(e^{2\pi i} z) = \omega(x) \otimes \mathcal{P}(z) \tag{12.42}$$

for all $x \in \bar{\mathfrak{g}}$, where ω is an automorphism of the horizontal subalgebra $\bar{\mathfrak{g}}$ of finite order N. In other words, $\mathcal{P}$ is now no longer a function on the circle S^1, but rather on an N-fold covering of the circle. Now according to chapter 11, $\bar{\mathfrak{g}}$ is the vector space direct sum of the eigenspaces $\bar{\mathfrak{g}}_{(j)}$ of ω, and the index j corresponds to a $\mathbb{Z}_N$-gradation of $\mathfrak{g}$. In order for (12.42) to hold, the maps $\mathcal{P}$ must therefore behave as

$$\mathcal{P}(e^{2\pi i} z) = e^{2\pi i \cdot j/N} \mathcal{P}(z) \tag{12.43}$$

if they are to be multiplied with $x \in \bar{\mathfrak{g}}_{(j)} \subseteq \bar{\mathfrak{g}}$. Hence a basis for the vector space of the maps satisfying (12.42) is given by

$$\{T^a_{m+j/N}\} \equiv \{T^a \otimes z^{m+j/N} \,|\, T^a \in \bar{\mathfrak{g}}_{(j)}, j = 0, 1, \dots, N-1; \; m \in \mathbb{Z}\}. \tag{12.44}$$

As in the untwisted case, the vector space generated by this basis is naturally endowed with a Lie algebra structure, namely via

$$[T^a_{m+j/N}, T^b_{n+j'/N}] = \sum_{c=1}^{d} f^{ab}{}_c \, T^c_{m+n+(j+j')/N}, \tag{12.45}$$

and again there exists a unique non-trivial central extension, denoted by $\hat{\mathfrak{g}}^\omega$. The twisted loop algebra $\hat{\mathfrak{g}}^\omega$ is generated by $\{T^a_{m+j/N}\}$ together with a central generator K; its Lie brackets read

$$
\begin{aligned}
[T^a_{m+j/N}, T^b_{n+j'/N}] = {} & \textstyle\sum_{c=1}^{d} f^{ab}{}_c \, T^c_{m+n+(j+j')/N} \\
& + \left(m + \tfrac{j}{N}\right) \bar{\kappa}^{ab} \delta_{m+n+(j+j')/N,0} \, K
\end{aligned}
\tag{12.46}
$$

and $[T^a_{m+j/N}, K] = 0$. The twisted affine algebra $\mathfrak{g}^\omega$ is obtained from $\hat{\mathfrak{g}}^\omega$ by again including a derivation D. Note that unlike for untwisted affine algebras, now the subalgebra $\bar{\mathfrak{g}}_{(0)}$ of $\bar{\mathfrak{g}}$, rather than $\bar{\mathfrak{g}}$ itself, can be identified with the zero mode (or horizontal) subalgebra $\mathfrak{g}_{[0]} \equiv \{T^a_m \,|\, m = 0\}$ of $\mathfrak{g}$. We thus have the proper inclusion $\mathfrak{g}_{[0]} \subset \bar{\mathfrak{g}}$.

If ω is an inner automorphism, the so obtained Lie algebra $\mathfrak{g}^\omega$ is actually isomorphic to an untwisted algebra. Namely, an inner automorphism of $\bar{\mathfrak{g}}$ of order N can always be characterized by a shift vector ζ_ω as in (11.35). Such automorphisms leave a Cartan subalgebra $\bar{\mathfrak{g}}_\circ$ of $\bar{\mathfrak{g}}$ invariant. As a consequence, the generators appearing in (12.46) with $T^a \in \bar{\mathfrak{g}}_\circ$ are all 'integer moded', i.e. are of the form H_n^i with $n \in \mathbb{Z}$. One may then verify that the following generators satisfy the bracket relations (12.28) of an untwisted affine algebra:

$$\tilde{E}_n^{\bar{\alpha}} := E_{n+(\zeta_\omega, \bar{\alpha})}^{\bar{\alpha}}, \qquad \tilde{H}_n^i := H_n^i + \zeta_\omega^i \, \delta_{n,0} \, K \, ,$$

$$\tilde{K} := K \, , \qquad \tilde{D} := D - \sum_{i=1}^r (\zeta_\omega)_i H_0^i \, . \tag{12.47}$$

In contrast, if the automorphism ω is outer, then $\mathfrak{g}^\omega$ is a new type of algebra. Recalling from chapter 11 that any outer automorphism of a simple Lie algebra $\bar{\mathfrak{g}}$ can be decomposed into an inner automorphism and an automorphism of the Dynkin diagram of $\bar{\mathfrak{g}}$, it follows that in order to obtain all genuinely twisted algebras up to isomorphism, we only need to consider automorphisms induced by Dynkin diagram symmetries; these are of order $N = 2$ in the cases $\bar{\mathfrak{g}} = A_r \, (r \geq 2)$, $D_r \, (r \geq 5)$ and E_6, and of order $N = 3$ for $\bar{\mathfrak{g}} = D_4$. The twisted affine algebras appearing in the classification in section 7.9 are precisely the Lie algebras obtained in these cases from $\hat{\mathfrak{g}}^\omega$ by adding the derivation D.

In view of this particular realization of $\mathfrak{g}$ as a central extension of the ℓ-twisted loop algebra of $\bar{\mathfrak{g}} = X_r$ it is natural to denote the twisted affine algebra by $\mathfrak{g} = X_r^{(\ell)}$. This is the notation that was used in the *second* column of table VIII B. The notation for $\mathfrak{g}$ employed in the *first* column of table VIII B, on the other hand, is better suited for memorizing the Dynkin diagram of $\mathfrak{g}$ (compare also section 11.8).

By scaling the mode index $m + j/N$ of the generators $T_{m+j/N}^a$ by a factor of N, one obtains an algebra with integer mode numbers only which is still isomorphic to the algebra (12.45). On the other hand, this algebra is manifestly a subalgebra of the untwisted affine Lie algebra $X_r^{(1)}$, where X is taken from the *second* column of table VIII B. Thus the twisted affine Lie algebras can also be described as specific subalgebras of untwisted affine algebras.

*12.9 The root system of twisted affine Lie algebras

The root system of the twisted affine Lie algebras can be obtained similarly as in the untwisted case, but the details are a bit more involved. The relevant outer automorphism ω of $\bar{\mathfrak{g}}$ corresponds to an automorphism $\dot{\omega}$ of the Dynkin diagram of $\bar{\mathfrak{g}}$, and hence acts as $\omega(\bar{\alpha}^{(i)}) = \bar{\alpha}^{(\dot{\omega}(i))}$, and analogously on the Chevalley generators as $\omega(H^i) = H^{\dot{\omega}(i)}$, $\omega(E_\pm^i) = E_\pm^{\dot{\omega}(i)}$ for $i = 1, 2, \ldots, r$. From this one can deduce that the ladder operator corresponding to an arbitrary root obeys

$$\omega(E^{\bar{\alpha}}) = \varepsilon_{\bar{\alpha}} E^{\omega(\bar{\alpha})} \tag{12.48}$$

with $\varepsilon_{\bar{\alpha}} \in \{\pm 1\}$ and $\varepsilon_{-\bar{\alpha}} = \varepsilon_{\bar{\alpha}}$. Also one has $\omega(\overline{\Phi}_+) = \overline{\Phi}_+$, so that in particular $\omega(\bar{\alpha}) \neq -\bar{\alpha}$ for any root $\bar{\alpha}$. (The origin of the signs $\varepsilon_{\bar{\alpha}}$ in (12.48) can best be understood by studying the following simple situation. Let $\alpha^{(i)}$ and $\alpha^{(j)}$ be simple roots such that $\omega(\alpha^{(i)}) = \alpha^{(j)}$ and $\omega(\alpha^{(j)}) = \alpha^{(i)}$, and such that also $\alpha^{(i)} + \alpha^{(j)}$ is a $\mathfrak{g}$-root, so that $E^{\alpha^{(i)}+\alpha^{(j)}} = [E^i, E^j]$ is a non-zero element of $\mathfrak{g}_+$. The automorphism property of ω then implies that $\omega(E^{\alpha^{(i)}+\alpha^{(j)}}) = [\omega(E^i), \omega(E^j)] = [E^j, E^i] = -E^{\alpha^{(i)}+\alpha^{(j)}}$.)

Consider now first the algebras of type $\mathfrak{g} = X_r^{(2)}$. In this case ω is of order two, and one finds that

$$(\bar{\alpha}^{(i)}, H) \in \bar{\mathfrak{g}}_{(0)} \qquad\qquad \text{if } \dot{\omega}(i) = i \, ,$$

$$\left. \begin{array}{l} \tfrac{1}{2} \, (\bar{\alpha}^{(i)} + \bar{\alpha}^{(\dot{\omega}(i))}, H) \in \bar{\mathfrak{g}}_{(0)} \\[4pt] \tfrac{1}{2} \, (\bar{\alpha}^{(i)} - \bar{\alpha}^{(\dot{\omega}(i))}, H) \in \bar{\mathfrak{g}}_{(1)} \end{array} \right\} \quad \text{otherwise} \, , \tag{12.49}$$

where $\bar{\mathfrak{g}}_{(0)}$ and $\bar{\mathfrak{g}}_{(1)}$ are the eigenspaces of ω to the eigenvalues $+1$ and -1, respectively. Similarly, from $\omega(E^{\bar{\alpha}} \pm E^{\omega(\bar{\alpha})}) = \pm\varepsilon_{\bar{\alpha}}(E^{\bar{\alpha}} \pm E^{\omega(\bar{\alpha})})$ one concludes that

$$
\begin{aligned}
E^{\bar{\alpha}} &\in \bar{\mathfrak{g}}_{(0)} && \text{if } \omega(\bar{\alpha}) = \bar{\alpha} \text{ and } \varepsilon_{\bar{\alpha}} = 1, \\
E^{\bar{\alpha}} &\in \bar{\mathfrak{g}}_{(1)} && \text{if } \omega(\bar{\alpha}) = \bar{\alpha} \text{ and } \varepsilon_{\bar{\alpha}} = -1,
\end{aligned}
$$

$$
\left.
\begin{aligned}
(E^{\bar{\alpha}} + \varepsilon_{\bar{\alpha}} E^{\omega(\bar{\alpha})}) &\in \bar{\mathfrak{g}}_{(0)} \\
(E^{\bar{\alpha}} - \varepsilon_{\bar{\alpha}} E^{\omega(\bar{\alpha})}) &\in \bar{\mathfrak{g}}_{(1)}
\end{aligned}
\right\} \quad \text{if } \omega(\bar{\alpha}) \neq \bar{\alpha}.
\tag{12.50}
$$

Thus the set of real roots of $\bar{\mathfrak{g}}_{(0)}$ is the union of $\{\bar{\alpha} \in \bar{\Phi} \,|\, \omega(\bar{\alpha}) = \bar{\alpha}, \, \varepsilon_{\bar{\alpha}} = 1\}$ and $\{\frac{1}{2}(\bar{\alpha} + \omega(\bar{\alpha})) \,|\, \omega(\bar{\alpha}) \neq \bar{\alpha}, \, \varepsilon_{\bar{\alpha}} = 1\}$, while $\bar{\mathfrak{g}}_{(1)}$ carries a representation of $\bar{\mathfrak{g}}_{(0)}$ whose weight system is the union of $\{\bar{\alpha} \in \bar{\Phi} \,|\, \omega(\bar{\alpha}) = \bar{\alpha}, \, \varepsilon_{\bar{\alpha}} = -1\}$ and $\{\frac{1}{2}(\bar{\alpha} + \omega(\bar{\alpha})) \,|\, \omega(\bar{\alpha}) \neq \bar{\alpha}, \, \varepsilon_{\bar{\alpha}} = -1\}$.

A basis of simple roots of $\bar{\mathfrak{g}}_{(0)}$ is given by $\bar{\alpha}_{(0)}{}^{(i)} = \frac{1}{2}(\bar{\alpha}^{(i)} + \bar{\alpha}^{(\omega(i))})$. The simple roots of the affine algebra $\mathfrak{g}$ can then be chosen to be the simple roots $\bar{\alpha}_{(0)}{}^{(i)}$ of $\bar{\mathfrak{g}}_{(0)}$ together with $\frac{1}{2}\delta - \bar{\Lambda}_1$ where $\delta = (0, 0, 1)$ and $\bar{\Lambda}_1$ is the highest weight of the $\bar{\mathfrak{g}}_{(0)}$-module $V_{\bar{\Lambda}_1}$ furnished by $\bar{\mathfrak{g}}_{(1)}$. The real roots are then either of the form $\alpha = \bar{\alpha} + n\delta$ with $\bar{\alpha} \in \bar{\Phi}(\bar{\mathfrak{g}}_{(0)})$ and $n \in \mathbb{Z}$, or else of the form

$$
\alpha = \bar{\lambda}_1 + (n + \tfrac{1}{2})\delta
\tag{12.51}
$$

with $\bar{\lambda}_1$ a weight of $V_{\bar{\Lambda}_1}$ and $n \in \mathbb{Z}$; they are not degenerate. The imaginary roots are $\alpha = \frac{1}{2}n\delta$ with $n \in \mathbb{Z}\setminus\{0\}$ and have multiplicity

$$
\text{mult}\,(\tfrac{1}{2}n\delta) = \begin{cases} \text{rank}(\bar{\mathfrak{g}}_{(0)}) & \text{for} \quad n \in 2\mathbb{Z} \\ r - \text{rank}(\bar{\mathfrak{g}}_{(0)}) & \text{for} \quad n \in 2\mathbb{Z}+1. \end{cases}
\tag{12.52}
$$

In the case of the triply twisted algebra $G_2^{(3)}$, the algebra $\bar{\mathfrak{g}}$ on which the automorphism ω acts is D_4. The root system of $\bar{\mathfrak{g}}_{(0)}$ is now the union of $\{\bar{\alpha} \in \bar{\Phi} \,|\, \omega(\bar{\alpha}) = \bar{\alpha}\}$ with $\{\frac{1}{3}(\bar{\alpha} + \omega(\bar{\alpha}) + \omega^2(\bar{\alpha})) \,|\, \omega(\bar{\alpha}) \neq \bar{\alpha}\}$. Furthermore, from $(\bar{\alpha}, \omega(\bar{\alpha})) = 0$ for $\bar{\alpha} \neq \omega(\bar{\alpha})$ it follows that the relative length squared of the roots in these two subsets is 3, and $\bar{\mathfrak{g}}_{(0)}$ is the simple Lie algebra G_2. A basis of simple roots for G_2 is given by $\bar{\alpha}^{(4)}$ and $\frac{1}{3}(\bar{\alpha}^{(1)} + \bar{\alpha}^{(2)} + \bar{\alpha}^{(3)})$. The simple root that one must add to obtain a basis of simple roots for $G_2^{(3)}$ is $\frac{1}{3}\delta - \bar{\Lambda}$, with $\bar{\Lambda} = \frac{1}{3}(\bar{\alpha}^{(1)} + \bar{\alpha}^{(2)} + \bar{\alpha}^{(3)}) = \bar{\Lambda}_{(2)}$, which is the maximal weight of the form $\frac{1}{3}(\bar{\alpha} + \omega(\bar{\alpha}) + \omega^2(\bar{\alpha}))$ with $\omega(\bar{\alpha}) \neq \bar{\alpha}$. Thus the real roots of $G_2^{(3)}$ are of either of the three forms

$$
\alpha = \bar{\alpha} + n\delta \quad \text{or} \quad \alpha = \bar{\lambda} + (n \pm \tfrac{1}{3})\delta
\tag{12.53}
$$

with $n \in \mathbb{Z}$, where $\bar{\alpha}$ are the roots of G_2 and $\bar{\lambda}$ the weights of the 7-dimensional G_2-module $V_{\bar{\Lambda}_{(2)}}$. The real roots are non-degenerate, while the imaginary roots, which are of the form $\alpha = \frac{1}{3}n\delta$ with $n \in \mathbb{Z}\setminus\{0\}$, have multiplicity one if $n \in 3\mathbb{Z}$, and two otherwise.

Inspection also shows that in most cases the non-zero weights of the modules furnished by $\bar{\mathfrak{g}}_{(1)}$ (and, for $G_2^{(3)}$, by $\bar{\mathfrak{g}}_{(2)}$ as well), are precisely the short roots of $\bar{\mathfrak{g}}_{(0)}$. The only exceptions are $A_1^{(2)}$ and $\tilde{B}_r^{(2)}$ with $r \geq 2$, for which the non-zero weights of $\bar{\mathfrak{g}}_{(1)}$ are the roots of $\bar{\mathfrak{g}}_{(0)}$ together with the short roots multiplied by two.

12.10 Heisenberg algebras

In the construction of the affine Lie algebras described above, the finite-dimensional algebra $\bar{\mathfrak{g}}$ has to be simple because the affine Cartan matrix is required to be indecomposable. If indecomposability is not imposed, then the same construction still works, but with $\bar{\mathfrak{g}}$ now semisimple. In fact one can also extend the construction easily to the case of reductive $\bar{\mathfrak{g}}$. One then has to introduce an independent central extension for each simple subalgebra and also for each $\mathfrak{u}(1)$-subalgebra. However, the central extensions of the $\mathfrak{u}(1)$-algebras can all be identified with each other owing to the fact that the irreducible modules of $\mathfrak{u}(1)$ are one-dimensional. Writing $\bar{\mathfrak{g}} = \bar{\mathfrak{s}} \oplus \bar{\mathfrak{h}}$ with $\bar{\mathfrak{s}}$ semisimple and $\bar{\mathfrak{h}}$ abelian, the central extension of $\bar{\mathfrak{g}}_{\text{loop}}$ is therefore of the form of a direct sum $\hat{\mathfrak{g}} = \hat{\mathfrak{s}} \oplus \hat{\mathfrak{h}}$, with the Lie brackets of $\hat{\mathfrak{s}}$ given by the formulæ (12.20) (and in the twisted case by (12.46), respectively), and those of $\hat{\mathfrak{h}}$ by

$$[t_m^a, t_n^b] = d^{ab}\, m\, \delta_{m+n,0} K\,. \tag{12.54}$$

Here $m, n \in \mathbb{Z}$, $a, b = 1, 2, \ldots, \dim \bar{\mathfrak{h}}$, and d^{ab} is a constant symmetric matrix. By a suitable change of basis, the matrix d^{ab} can be brought to the form δ^{ab}. This means that the Lie algebra defined by (12.54) is isomorphic to the direct sum of *Heisenberg* algebras (compare chapter 2, equation (2.29)) with identified centers.

The Heisenberg algebra, i.e. the Lie algebra with brackets (12.54) without the index a, can be regarded as a central extension of the loop algebra of $\bar{\mathfrak{g}} = \mathfrak{u}(1)$; correspondingly it is often denoted by $\hat{\mathfrak{u}}(1)$. It is also sometimes referred to as the '$\mathfrak{u}(1)$ Kac–Moody algebra'; this is a misnomer, since the Heisenberg algebra can*not* be described in terms of a (generalized) Cartan matrix. (There is, however, a generalization of the notion of a Kac–Moody algebra, the so-called generalized Kac–Moody algebras or Borcherds algebras. This larger class of Lie algebras does contain the Heisenberg algebra, which is then described by a generalized Cartan matrix which is just the 1×1-matrix $[0]$. This is in accordance with the fact that the Heisenberg algebra possesses a triangular decomposition, and is the deeper reason why we could treat the two examples of chapter 2 on a very similar footing.)

Actually any affine Lie algebra contains a sum of Heisenberg algebras as a subalgebra, namely the one generated by $\{H_m^i\}$ and K, with brackets $[H_m^i, K] = 0$ and

$$[H_m^i, H_n^j] = \bar{\kappa}(H^i, H^j)\, m\, \delta_{m+n,0}\, K\,. \tag{12.55}$$

Just as for centrally extended loop algebras, one also introduces a derivation D, acting as $[D, H_m^i] = -m H_m^i$.

*12.11 Gradations

While it is essential to adjoin D to the centrally extended loop algebras $\hat{g}$, there remains in fact a large freedom in the choice of the brackets $[D, x]$ for $x \in \hat{g}$, and the particular prescription adopted in (12.24) is only one among many different possibilities. A consistent prescription for $[D, x]$ is said to define a *gradation* of g, because any such choice can be shown to correspond to the decomposition of an underlying simple Lie algebra $\bar{g}$ into eigenspaces with respect to some automorphism ω. The prescription (12.24) can be characterized by

$$[D, x] = 0 \qquad \text{for all } x \in g_{[0]} \subset g\,, \tag{12.56}$$

and is called a *homogeneous gradation* of g.

For untwisted algebras, (12.56) fixes the gradation uniquely. In contrast, for some of the twisted algebras, there is still some further freedom left: The algebras $C_r^{(2)}$, $F_4^{(2)}$, and $G_2^{(3)}$ admit two inequivalent homogeneous gradations, and the algebras $B_r^{(2)}$ admit $[\frac{r}{2}] + 1$ inequivalent ones. The homogeneous gradation used so far corresponds to identifying the zero mode subalgebra $\bar{g}_{[0]}$ of g as the subalgebra $\bar{g}$ that is obtained by deleting a node with $a_i = a_i^\vee = 1$ from the Dynkin diagram of g. For the other homogeneous gradations the zero mode subalgebra is obtained by deleting a node obeying $a_i = 1$ but $a_i^\vee \neq 1$ from the Dynkin diagram. Inspection of the various Dynkin diagrams (see table VIII) shows that for $C_r^{(2)}$, $F_4^{(2)}$, $G_2^{(3)}$ and $B_r^{(2)}$, the zero mode subalgebras in these other homogeneous gradations are, respectively, D_r instead of C_r, C_4 instead of F_4, A_2 instead of G_2, and $B_n \oplus B_{r-n}$ for $n = 1, 2, ..., [\frac{r}{2}]$ instead of B_r. This is summarized in table XI.

Table XI. *Homogeneous gradations of the twisted affine Lie algebras*

$\bar{g}$	g	$\bar{g}_{[0]}$	$\bar{g}_{[1]}$	$\bar{g}_{[2]}$	$\dim \bar{g}_{[1]}$
A_2	$A_1^{(2)}$	A_1	$4\bar{\Lambda}_{(1)}$		5
E_6	$F_4^{(2)}$	F_4	$\bar{\Lambda}_{(4)}$		26
		C_4	$\bar{\Lambda}_{(4)}$		42
D_{r+1}	$B_r^{(2)}$	B_r	$\bar{\Lambda}_{(1)}$		$2r + 1$
		$B_n \oplus B_{r-n}$	$(\bar{\Lambda}_{(1)}; \bar{\Lambda}_{(1)})$		$(2n+1)(2r-2n+1)$
A_{2r}	$\tilde{B}_r^{(2)}$	B_r	$2\bar{\Lambda}_{(1)}$		$r(2r + 3)$
A_{2r-1}	$C_r^{(2)}$	C_r	$\bar{\Lambda}_{(2)}$		$(2r + 1)(r - 1)$
		D_r	$2\bar{\Lambda}_{(1)}$		$(2r - 1)(r + 1)$
D_4	$G_2^{(3)}$	G_2	$\bar{\Lambda}_{(2)}$	$\bar{\Lambda}_{(2)}$	7
		A_2	$3\bar{\Lambda}_{(1)}$	$3\bar{\Lambda}_{(2)}$	10

In the different homogeneous gradations of course also the root system looks quite different. As an example, consider $G_2^{(3)}$, for which the root system obtained with the first gradation in table XI has been displayed in (12.53) and thereafter. In the gradation where A_2 instead of G_2 is the zero mode subalgebra, the real roots are

$$\{\bar{\alpha} + n\,\delta\} \cup \{\bar{\lambda} + (n + \tfrac{1}{3})\,\delta\} \cup \{\bar{\lambda}^+ + (n - \tfrac{1}{3})\,\delta\}, \qquad (12.57)$$

where $\bar{\alpha}$ are now the roots of A_2, $\bar{\lambda}$ the weights of the 10-dimensional A_2-module $V_{3\bar{\Lambda}_{(1)}}$, and $\bar{\lambda}^+$ the weights of its conjugate module $V_{3\bar{\Lambda}_{(2)}}$. The real roots are not degenerate. The imaginary roots are again of the form $\frac{1}{3}\,n\delta$ with n a non-zero integer, and their multiplicity is one if n is divisible by three, and two otherwise.

The independent gradations of $\mathfrak{g}$ can be labelled by the conjugacy classes of the Weyl group $\overline{W}$ of $\bar{\mathfrak{g}}$. The homogeneous gradation for which the zero mode subalgebra is obtained by deleting a node with $a_i = a_i^\vee = 1$ from the Dynkin diagram corresponds to the class of the identity element of $\overline{W}$. Another gradation that is often of interest is the one that corresponds to the class of the Coxeter element w_c of $\overline{W}$; this is called the *principal* gradation.

12.12 The Virasoro algebra

Above we have constructed untwisted affine Lie algebras starting from loop algebras and extending them by a central element K and a derivation D. Now in fact one can extend an untwisted affine Lie algebra even further, and this proves to be quite important for applications. To see this, we introduce another central element C and infinitely many generators L_n with $n \in \mathbb{Z}$, which among themselves have the bracket relations (compare exercise 8.2)

$$[L_n, L_m] = (n - m)L_{n+m} + \tfrac{1}{12}\,n(n^2 - 1)\delta_{n+m,0}\,C\,, \quad [L_n, C] = 0\,. \quad (12.58)$$

These bracket relations define a Lie algebra, which is called the *Virasoro algebra*.

The relations involving both the L_n and the generators of $\mathfrak{g}$ read

$$[L_m, T_n^a] = -n\,T_{m+n}^a\,, \qquad [L_n, K] = [C, T_n^a] = [C, K] = 0\,. \qquad (12.59)$$

By comparison with (12.24), it is natural to identify the derivation D of $\mathfrak{g}$ with the Virasoro generator $-L_0$. Similar remarks apply to the Heisenberg algebra that is spanned by the generators u_n with relations

$$[u_n, u_m] = n\delta_{n+m,0}\,K\,, \qquad [K, u_n] = 0\,. \qquad (12.60)$$

In this case (12.59) gets replaced by

$$[L_m, u_n] = -n\,u_{n+m}\,, \qquad [L_n, K] = [C, u_n] = [C, K] = 0\,. \qquad (12.61)$$

Thus one is dealing with Lie algebras which have the structure of a semidirect sum of the untwisted affine Lie algebra and the Heisenberg algebra, respectively, with the Virasoro algebra.

Note that for this semidirect sum the Virasoro zero mode L_0 does appear on the right hand side of the bracket relations and hence is contained in the derived algebra of the semidirect sum. It follows that, while the eigenvalue of the derivation D of the highest weight of a module of an affine Lie algebra can be chosen at will (as long as the level is non-zero, see the remarks after equation (13.40) below), this is no longer possible for the generator L_0 of the semidirect sum.

Let us now list a few facts about the Virasoro algebra $\mathcal{V}ir$. A crucial observation is that $\mathcal{V}ir$ possesses a triangular decomposition

$$\mathcal{V}ir_+ \oplus \mathcal{V}ir_\circ \oplus \mathcal{V}ir_- \qquad (12.62)$$

into the subalgebras generated by the positive, zero, and negative modes $\mathcal{V}ir_+ = \mathrm{span}\{L_n \,|\, n > 0\}$, $\mathcal{V}ir_\circ = \mathrm{span}\{L_0, C\}$, $\mathcal{V}ir_- = \mathrm{span}\{L_n \,|\, n < 0\}$, respectively. $\mathcal{V}ir_\circ$ is a maximal abelian subalgebra, and $\mathcal{V}ir_\pm \oplus \mathcal{V}ir_\circ$ are Borel subalgebras. Owing to the existence of a triangular decomposition, the representation theory of the Virasoro algebra parallels to a large extent the one of affine Lie algebras; in particular there are highest weight modules and Verma modules. We will not discuss the representation theory in detail, but only summarize a few basic facts.

Highest weight vectors are labelled as $v_{c,\Delta}$ by the eigenvalues c with respect to C and Δ with respect to L_0:

$$R(C)\, v_{c,\Delta} = c \cdot v_{c,\Delta} \,, \qquad R(L_0)\, v_{c,\Delta} = \Delta \cdot v_{c,\Delta} \,. \qquad (12.63)$$

Since C is a central generator, all vectors in the Verma module obtained from $v_{c,\Delta}$ possess the same eigenvalue c; it is called the Virasoro central charge or, for reasons that will become clear soon, conformal anomaly. The Virasoro algebra admits an involution $\omega(L_n) = -L_{-n}$, and we define unitarity with respect to this involution. It turns out that unitarizable highest weight modules occur for all values c, Δ with $c \geq 1$ and $\Delta > 0$; in contrast, for $c < 1$ it is necessary that $c = 1 - 6/(l+2)(l+3)$ for some positive integer l, and in these cases only a finite set of rational values of Δ is allowed.

The Virasoro algebra is intimately linked with conformal transformations in two dimensions. In terms of complex coordinates $z := x + iy$ and $\tilde{z} := x - iy$ on the complex plane $\mathbb{C}$ (or rather its one-point compactification, the Riemann sphere), the finite conformal transformations are generated by the mappings $z \mapsto f(z)$ and $\tilde{z} \mapsto \tilde{f}(\tilde{z})$ with independent meromorphic functions f and $\tilde{f}$. These generate an infinite-dimensional Lie group which is the direct product of two isomorphic factors. The Lie algebra of each factor is the so-called *Witt algebra*, i.e. the Lie algebra of smooth vector fields on the unit circle S^1, with generators

$$L_n^{(c)} = -z^{n+1} \frac{\mathrm{d}}{\mathrm{d}z} \,. \qquad (12.64)$$

The Lie brackets of these vector fields are defined as commutators, which yields

$$[L_m^{(c)}, L_n^{(c)}] = (m - n)\, L_{m+n}^{(c)} \,. \qquad (12.65)$$

In a *conformal field theory*, i.e. a quantum field theory which is covariant under conformal transformations, one needs unitarizable representations of the symmetry algebra. However, just as loop algebras, the Witt algebra does not possess any non-trivial unitary representations, but only projective representations. Therefore one must again introduce a central extension. The form of this extension is again determined by solving a cohomology problem; but it can also be obtained with the help of some basic principles of quantum field theory. Indeed, under very general conditions (namely validity of the Wightman axioms, dilatation invariance, and existence of an energy-momentum tensor which is symmetric and a conserved Noether current) the so-called *Lüscher–Mack theorem* holds, which asserts that the extension just consists of the addition of a $\mathbb{C}$-number term on the right hand side of the bracket (12.65). (In physicists' language, one says that the conformal symmetry develops an anomaly.) Dimensional analysis shows that this term must be of the form appearing in the Virasoro algebra relations (12.58). According to the general strategy outlined in section 4.12, one has to interpret the $\mathbb{C}$-number term as the eigenvalue of a central element C, and hence ends up with the Virasoro algebra.

Affine Lie algebras made their first appearance in physics as symmetries of two-dimensional conformal field theory models [Bardakçi and Halpern 1971]. In this context one regards $x \otimes \mathcal{P}(z)$, with $x \in \bar{\mathfrak{g}}$, z a complex coordinate, and $\mathcal{P}$ a Laurent polynomial, as a component of a quantum field. This field satisfies the conservation law $\frac{\partial}{\partial \bar{z}}(x \otimes \mathcal{P}(z)) = 0$ and hence is a conserved current. It can be obtained as the Noether current for a symmetry of these quantum field theories. The central extension shows up as a non-ultralocal term, i.e. a term with a derivative of the delta function $\delta(z - z')$, in the equal-time commutator (or classically, the Poisson bracket) of the currents, which is called a Schwinger term.

The models obtained this way form a particularly important subclass of two-dimensional conformal field theories, the so-called WZW theories. In any two-dimensional conformal field theory the generators L_n of the Virasoro algebra play the rôle of the components of the energy–momentum tensor which is the Noether current for the conformal symmetry. A distinctive feature of WZW theories is that L_n can be written as an infinite sum over bilinears in the affine currents:

$$L_n := \frac{1}{2(k^{\vee}+g^{\vee})} \sum_{m \in \mathbb{Z}} \kappa_{ab} : J^a_{m+n} J^b_{-m} : \qquad (12.66)$$

(the colons stand for a suitable a normal-ordering prescription; $k^{\vee}$ is the eigenvalue of the central element K (multiplied by $2/(\theta, \theta)$), $g^{\vee}$ the dual Coxeter number of the horizontal subalgebra $\bar{\mathfrak{g}}$ of $\mathfrak{g}$, and κ the Killing form of $\bar{\mathfrak{g}}$).

In particular, one deals with representations of $\mathfrak{g}$ which have a fixed eigenvalue k of K, and the eigenvalue of the Virasoro central charge C on an irreducible highest weight module can be expressed in terms of $k^{\vee} \equiv k \cdot 2/(\bar{\theta}, \bar{\theta})$ as $c(\mathfrak{g}, k^{\vee}) = k^{\vee} \dim \bar{\mathfrak{g}}/(k^{\vee} + g^{\vee})$. The relation (12.66) is known as the Sugawara construction. For more details, see e.g. chapter 3.2 of [Fuchs 1992].

Information

Summary:

Loop algebras, i.e. the algebras of maps from the circle S^1 to some simple Lie algebra $\mathfrak{g}$, can be used to construct untwisted affine Lie algebras explicitly. To this end, a non-trivial central extension and a derivation have to be included. Twisted affine Lie algebras are obtained by implementing in addition an automorphism of the Dynkin diagram of the relevant simple Lie algebra and twisted boundary conditions. This explicit realization of an affine Lie algebra allows one to read off its root system.

Keywords:

Central extension, loop algebra, zero mode subalgebra, canonical central element, horizontal subalgebra, derivation, real and imaginary roots, (homogeneous) gradation;
Heisenberg algebra, Virasoro algebra.

Exercises:

Show that simple Lie algebras do not possess any non-trivial central extensions.

Exercise 12.1

Hints: First use the Jacobi identity and the fact that the Killing form is non-degenerate to deduce that

$$f^{ab}{}_i = \sum_{c,d,e=1}^{d} f^{ab}{}_c f^{cd}{}_e f_d{}^e{}_i .$$
(12.67)

Choose $u^a_i = \sum_{b,c=1}^{d} f^{ab}{}_c f_c{}^b{}_i$ in (12.5) to show the claim.

Show that the loop algebra (12.14) is indeed a Lie algebra, and that its center is trivial.

Exercise 12.2

Determine the unique (up to normalization) invariant form on the centrally extended loop algebra $\hat{\mathfrak{g}}$. Show that this form is degenerate.

Exercise 12.3

Hint: Compare with the similar situation for the Virasoro algebra (exercise 8.2).

Derive the ingredients used in the text to obtain the formula (12.27) for the Killing form κ.

Exercise 12.4

Hint: Consider the invariance property $\kappa([w, x], y) = \kappa(w, [x, y])$ of κ for $w = T_m^a$, $x = D$, $y = T_n^b$ and for $w = T_m^a$, $x = T_0^b$, $y = T_{-m}^c$ to deduce that $\kappa(T_m^a, T_n^b)$ is proportional to $\delta_{m+n,0}$. Determine the constant of proportionality by considering the triple $w = T_m^a$, $x = T_1^b$, $y = T_{-m-1}^c$. Next consider the triples $w = T_m^a$, $x = T_n^b$, $y = K$ and $w = T_m^a$, $x = T_{-m}^b$, $y = K$, as well as $w = T_m^a$, $x = T_n^b$, $y = D$.

Verify that the roots of an untwisted affine Lie algebra with respect to the Cartan subalgebra (12.29) are given by (12.30).

Exercise 12.5

Find a non-trivial ideal of the loop algebra (12.14), i.e. show that loop algebras are not simple.

Exercise 12.6

Hint: The loop algebra is isomorphic to $\mathbb{C}[z, z^{-1}] \otimes \mathfrak{g}$, where $\mathbb{C}[z, z^{-1}]$ is the algebra of Laurent polynomials. For any complex number ζ, this algebra has the ideal $(z - \zeta)\mathbb{C}[z, z^{-1}]$.

As mentioned in exercise 3.5, the components of angular momentum $\vec{L}$, of the Runge–Lenz vector $\vec{R}$ and of the hydrogen Hamiltonian $H = \vec{p}^2 - 1/r$ do not span a Lie algebra. In the treatment of exercise 3.5 the structure of a Lie algebra was obtained by restricting to a definite energy level (compare also exercise 8.1). There is however also a different, and somewhat less artificial, way out. Namely, one defines [Daboul *et al.* 1993] the operators

Exercise 12.7

$$L_j^{(n)} := H^n L_j, \qquad R_j^{(n)} := H^n R_j, \qquad (12.68)$$

where arbitrary powers H^n, $n \in \mathbb{Z}_{\geq 0}$, of the Hamiltonian appear.

a) Compute the commutation relations of these operators and conclude that they span an infinite-dimensional Lie algebra.

b) Verify that this Lie algebra can be interpreted as the subalgebra of non-negative modes of a twisted loop algebra of $\mathfrak{sl}(2) \oplus \mathfrak{sl}(2)$, where the twisting automorphism acts as $L_j \mapsto L_j$, $R_j \mapsto R_j$. Show that starting from the real forms $\mathfrak{so}(4)$ and $\mathfrak{so}(3,1)$ one obtains real forms of this algebra which are isomorphic even as real Lie algebras.

13
Highest weight representations

13.1 Highest weight representations of $\mathfrak{sl}(2)$

As we have already emphasized in chapter 5, Lie algebras play their rôle in physics not as abstract algebras, but through their representations which act on suitable representation spaces. Among these, the so-called highest weight representations form a particularly interesting subclass. The main reason for this is that any finite-dimensional representation of a simple Lie algebra belongs to this subclass. In the case of infinite-dimensional algebras, such as the Heisenberg algebra described in chapter 2, there is also another reason: Frequently, the Hamiltonian of a physical system under consideration is an element of these algebras. In this case, the highest weight property reflects the physical principle that the energy should be bounded from below. (To stay closer to this physical picture, in the literature sometimes lowest weight representations are considered, because they typically are 'positive energy' representations, i.e. correspond to having the Hamiltonian itself as a basis element of the Cartan subalgebra. Here we follow the usual treatment in mathematics, which in these situations corresponds to taking *minus* the Hamiltonian as a basis element.)

In this chapter we first consider the highest weight representations of finite-dimensional semisimple Lie algebras $\mathfrak{g}$ over $\mathbb{C}$ (or of their compact real forms). But in fact many of our considerations apply with minor modifications also to more general Kac–Moody algebras. As a consequence, one can make much more general statements; these will be obtained in the next chapter.

The crucial idea for the investigation of these highest weight representations is to describe them in terms of the $\mathfrak{sl}(2)_\alpha$-subalgebras of $\mathfrak{g}$ that correspond to the simple roots. Accordingly, we start by recalling from chapter 2 and section 5.7 some facts about the representation theory of $\mathfrak{sl}(2) \cong A_1$.

The generators $E_\pm$ and H of A_1 satisfy $[H, E_\pm] = \pm 2E_\pm$ and $[E_+, E_-] = H$, so that H spans the Cartan subalgebra. Accordingly the representations R of A_1 which appear in applications have the non-trivial property that the associated representation space V possesses a basis on which the generator $R(H)$ acts diagonally. Thus each module V of A_1 decomposes into *weight spaces* $R_{(\lambda)}$ according to

$$V = \bigoplus_\lambda V_{(\lambda)}, \qquad V_{(\lambda)} = \{v \in V \mid R(H)\,v = \lambda \cdot v\}. \tag{13.1}$$

The eigenvalues $\lambda \in \mathbb{C}$ appearing in (13.1) are called the weights of the module V, and the dimensionality of $V_{(\lambda)}$ is called the *multiplicity* of the weight λ. Further, if $v \in V_{(\lambda)}$, then $R(E_\pm)\,v \in V_{(\lambda \pm 2)}$. For *finite-dimensional* representations there must then necessarily exist a weight Λ such that $V_{(\Lambda)} \neq \emptyset$ but $V_{(\Lambda+2)} = \emptyset$; such a weight Λ is called a *maximal weight*, and any element $v_\Lambda \in V_{(\Lambda)}$ a maximal weight vector. The vector v_Λ is unique up to scalar multiplication if the module V is irreducible; v_Λ is then called the *highest weight vector* of the module, Λ is called the *highest weight*, and V an irreducible *highest weight module* and is denoted by V_Λ.

All weights of a highest weight module are of the form $\lambda = \Lambda - n\alpha = \Lambda - 2n$ with n a non-negative integer. For V_Λ to be finite-dimensional, it is thus in addition necessary that $V_{(\Lambda-2n)} = \emptyset$ for sufficiently large n. This happens if and only if Λ is itself a non-negative integer, in which case the weights of V_Λ are $\lambda = -\Lambda, -\Lambda+2, \ldots, \Lambda-2, \Lambda$. Each of these occurs with multiplicity one, and one obtains all possible weights by subtracting the root α up to Λ times from the highest weight; while these are rather simple observations, they will be the basis of various non-trivial results to be obtained below. Any finite-dimensional highest weight module is irreducible. Conversely, for each $\Lambda \in \mathbb{Z}_{\geq 0}$, there exists precisely one (up to isomorphism) irreducible module of dimension $\Lambda + 1$, namely the highest weight module V_Λ.

13.2 Highest weight modules of simple Lie algebras

The key idea in the analysis of finite-dimensional representations of semisimple Lie algebras is to reduce the problem to the representation theory of $\mathfrak{sl}(2)$. This is possible because each generator H^i of the Cartan subalgebra $\mathfrak{g}_\circ$ of a semisimple algebra $\mathfrak{g}$ spans the Cartan subalgebra of an $\mathfrak{sl}(2)$-subalgebra of $\mathfrak{g}$, namely of the $\mathfrak{sl}(2)_\alpha$-subalgebra that is spanned by $E_\pm^i$ and H^i with fixed i, corresponding to the simple root $\alpha = \alpha^{(i)}$. Applying the representation theory of $\mathfrak{sl}(2)$ to these subalgebras for all $i = 1, 2, \ldots, \mathrm{r}$, it follows that any $\mathfrak{g}$-module V has a basis on which the whole Cartan subalgebra $\mathfrak{g}_\circ$ acts diagonally.

Thus there is a decomposition

$$V = \bigoplus_{\lambda} V_{(\lambda)} \tag{13.2}$$

into weight spaces $V_{(\lambda)}$ analogous to (13.1), such that

$$R(H^i) v_\lambda = \lambda^i \cdot v_\lambda \tag{13.3}$$

for all $v_\lambda \in V_{(\lambda)}$ and for all $i = 1, 2, \dots, r$. The eigenvalues λ^i can be grouped into an r-dimensional vector $\lambda \equiv (\lambda^i)_{i=1,\dots,r}$. Such vectors λ are called the *weights* of the module V, and the collection of all weights of V is called the weight *system* of V. (Thus the dimension of the space of weights – which must not be mixed up with the vector space V which underlies the $\mathfrak{g}$-module V – is given by the rank r of $\mathfrak{g}$. Also recall from section 6.8 that arbitrary elements of the weight space, i.e. the space dual to the root space, were referred to as weights. Now the action (13.3) of $H^i \in \mathfrak{g}_\circ$ on weight vectors v_λ tells us once more that the Cartan subalgebra $\mathfrak{g}_\circ$ can be identified with the weight space, and hence its dual $\mathfrak{g}_\circ^*$ with the complexified root space $\mathrm{span}_{\mathbb{C}}(\Phi)$ of $\mathfrak{g}$. Therefore the weights introduced above are indeed also weights in the sense of chapter 6, and the eigenvalues λ^i are the components of λ in the basis of fundamental weights (the Dynkin basis), i.e. the Dynkin labels of λ.)

If V is finite-dimensional, then it must be finite-dimensional as a module with respect to the $\mathfrak{sl}(2)$-subalgebra corresponding to each simple root $\alpha^{(i)}$ as well. Now we learned above that in this case $H \equiv H^i$ has integer eigenvalues only, and hence we have

$$\lambda^i \equiv (\alpha^{(i)\vee}, \lambda) \in \mathbb{Z} \qquad \text{for } i = 1, 2, \dots, r. \tag{13.4}$$

Comparison of this result with the definition $(\Lambda_{(i)}, \alpha^{(j)\vee}) = \delta_i^j$ (see equation (6.39)) of the fundamental weights $\Lambda_{(i)}$ of $\mathfrak{g}$ shows that λ is a weight of a finite-dimensional module only if it is an integral linear combination of the fundamental weights,

$$\lambda = \sum_{i=1}^{r} \lambda^i \Lambda_{(i)} \qquad \text{with } \lambda^i \in \mathbb{Z} \text{ for } i = 1, 2, \dots, r, \tag{13.5}$$

or in other words, only if it lies on the weight *lattice* L_w. Thus the weights of a finite-dimensional module have integral Dynkin labels; such weights are called *integral weights*.

Instead of considering the $\mathfrak{sl}(2)$-subalgebra corresponding to a simple root, one can also investigate the subalgebra $\mathfrak{sl}(2)_\alpha$ of $\mathfrak{g}$ that is generated by E^α, $E^{-\alpha}$ and $H^\alpha \equiv (\alpha^\vee, H)$ for *any* root α; with respect to $\mathfrak{sl}(2)_\alpha$, all weights must be weights of a finite-dimensional module, too. Thus we have

$$(\alpha^\vee, \lambda) \in \mathbb{Z} \qquad \text{for all } \alpha \in \Phi. \tag{13.6}$$

Further, by the same argument as in the case of $\mathfrak{sl}(2)$, it follows that for every *finite-dimensional* module of $\mathfrak{g}$, there exist maximal weights Λ such that

$$R(E^\alpha)\, v_\Lambda = 0 \qquad \text{for all } \alpha > 0 \tag{13.7}$$

and all $v_\Lambda \in V_{(\Lambda)}$. If there is exactly one weight with this property, it is called the *highest weight* of the module. The Dynkin labels of any highest weight of a finite-dimensional module are non-negative integers,

$$\Lambda^i \in \mathbb{Z}_{\geq 0} \qquad \text{for } i = 1, 2, \ldots, r. \tag{13.8}$$

Weights with this property are called *dominant integral* weights. If the module is also irreducible, the highest weight appears with multiplicity one. The irreducible finite-dimensional $\mathfrak{g}$-module with maximal weight Λ is referred to as the *irreducible highest weight module* with highest weight Λ and is denoted by V_Λ; the corresponding $\mathfrak{g}$-representation is called an *irreducible highest weight representation R_Λ*.

> An irreducible finite-dimensional module of a semisimple Lie algebra $\mathfrak{g}$ is thus characterized by the fact that it has a highest weight of multiplicity one and that this weight is dominant integral. Conversely, each dominant integral weight is the highest weight of a unique irreducible finite-dimensional $\mathfrak{g}$-module.

As we have seen in chapter 7, the nodes of the Dynkin diagram can be regarded as representing the fundamental weights of $\mathfrak{g}$. More generally, one may visualize highest weight modules of $\mathfrak{g}$ by attaching the appropriate Dynkin labels of the highest weight to the nodes of the Dynkin diagram. For instance, the modules 3, $\bar{3}$ and 8 of $\mathfrak{sl}(3)$ then look like

$$\underset{1 \qquad 0}{\bullet\!\!-\!\!\bullet} \qquad \underset{0 \qquad 1}{\bullet\!\!-\!\!\bullet} \qquad \underset{1 \qquad 1}{\bullet\!\!-\!\!\bullet} \tag{13.9}$$

This visualization can be helpful in the case of the exceptional simple Lie algebras, because sometimes a different numbering for the nodes of their Dynkin diagrams is used. It also manifestly displays which representations are related by symmetries of the Dynkin diagram like e.g. conjugation.

That Λ has multiplicity one means that the dimension of the subspace $V_{(\Lambda)}$ of an irreducible highest weight module V_Λ is one-dimensional, i.e. up to scalar multiplication there is a unique eigenvector v_Λ of the action of $\mathfrak{g}_\circ$ with weight Λ; such a vector is called a *highest weight vector* of the module V_Λ. All other elements of V_Λ can be obtained by applying step operators for negative roots to v_Λ, i.e. any $v \in V_\Lambda$ is a linear combination of vectors of the form

$$v_{\{\beta\}} = R(E^{-\beta_1})R(E^{-\beta_2})\cdots R(E^{-\beta_m})\,v_\Lambda, \qquad (13.10)$$

where $m \in \mathbb{Z}_{\geq 0}$ and β_i, $i = 1, 2, \ldots, m$, are positive $\mathfrak{g}$-roots. However, not all elements of the form (13.10) are linearly independent. One obvious source of linear dependencies is that the step operators $E^{-\beta}$ obey non-trivial relations among themselves, like e.g. the Serre relations (7.3). (A detailed investigation of this type of linear dependence of the vectors of the form (13.10) leads to the concept of Verma modules; this will be described in chapter 14.) It turns out that the resulting space is a $\mathfrak{g}$-module, and that this structure strongly depends on whether the highest weight is dominant integral or not. The $\mathfrak{g}$-module spanned by the vectors $v_{\{\beta\}}$ is infinite-dimensional. As long as the highest weight is not dominant integral, this module is also irreducible. In contrast, for dominant integral highest weight the span of the $v_{\{\beta\}}$ is reducible, but not fully reducible, and the quotient of the infinite-dimensional module by a its maximal submodule (which is unique) is a finite-dimensional irreducible module. A precise description of this situation will be given in chapter 14; it will lead us to the important concept of null vectors.

After identifying the root and weight space of a simple Lie algebra $\mathfrak{g}$ via the inner product, the $\mathfrak{g}$-roots are integral weights. In particular, the highest $\mathfrak{g}$-root θ is dominant integral and hence the highest weight of an irreducible $\mathfrak{g}$-module V_θ. The associated representation R_θ is the adjoint representation of $\mathfrak{g}$ (see exercise 13.2).

13.3 The weight system

In any (linear) representation R, the Lie brackets of a Lie algebra $\mathfrak{g}$ are represented as commutators. The bracket relation $[H^i, E^\alpha] = \alpha^i E^\alpha$ therefore implies that $R(H^i)(R(E^\alpha)v_\lambda) = (\alpha^i + \lambda^i)\,R(E^\alpha)v_\lambda$, or in other words, that

$$R(E^\alpha)\,v_\lambda \propto v_{\lambda+\alpha}. \qquad (13.11)$$

In view of (13.10) this means that any weight of the module V_Λ is of the form $\lambda = \Lambda - \beta$, with β a sum of positive roots. Moreover, since any positive root is a non-negative integral linear combination of simple roots, the same holds for β, i.e. $\beta = \sum_{i=1}^r n_i \alpha^{(i)}$ with $n_i \in \mathbb{Z}_{\geq 0}$. Whenever two weights λ, μ satisfy $\lambda - \mu = \beta$ with β of this special form, one writes $\mu \leq \lambda$; this provides a partial ordering on the weight system of V_Λ. The height of the root lattice element β, i.e. the number $\sum_{i=1}^r n_i$, is called the *depth* of the weight λ.

For any irreducible highest weight module V_Λ there is a unique weight of depth zero, the highest weight Λ. The fact that all weights of a finite-dimensional irreducible $\mathfrak{sl}(2)$-module are obtained from the highest weight Λ by subtracting the positive root of $\mathfrak{sl}(2)$ up to Λ times implies the following simple algorithm for calculating the weights of a highest weight module, not counting the multiplicities. From $\Lambda = \sum_{i=1}^r \Lambda^i \Lambda_{(i)}$, all weights of depth one are obtained by subtracting those simple roots $\alpha^{(i)}$ for which $\Lambda^i > 0$. More generally, given an arbitrary weight λ of V_Λ, then, provided that $\lambda^i > 0$, subtracting from λ the simple root $\alpha^{(i)}$ up to λ^i times yields weights which all belong again to the weight system of V_Λ. Proceeding recursively, one constructs in this manner all distinct weights of V_Λ, and finally arrives at some value of the depth for which all weights have non-positive Dynkin labels so that the procedure terminates.

As an application of this prescription, consider the weight systems of the two three-dimensional modules of the Lie algebra $A_2 \cong \mathfrak{sl}(3)$, which were already displayed in the figure (3.36). These are the irreducible highest weight modules with highest weight $\Lambda_{(1)}$ and $\Lambda_{(2)}$, respectively. In the former case, owing to $\Lambda^1 = 1$ one can subtract once the simple root $\alpha^{(1)}$ which yields the weight $\lambda = \Lambda_{(2)} - \Lambda_{(1)}$, from which in turn because of $\lambda^2 = 1$ one can subtract $\alpha^{(2)}$, thus obtaining the weight $\mu = -\Lambda_{(2)}$; then, because of μ^1, $\mu^2 \leq 0$, no more roots can be subtracted. For the second module the situation is completely analogous, just with the superscripts 1 and 2 interchanged. The weight systems are thus as displayed in the following figure.

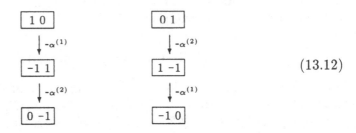

$$(13.12)$$

Here each weight is symbolized by a box containing the Dynkin labels of the weight.

As further examples, consider the adjoint modules of $\mathfrak{g} = A_2$ and $\mathfrak{g} = B_2$. The weight system of these modules is displayed in figure (13.13) below; multiple boxes indicate the multiplicity of the weight. The non-zero weights are thus precisely the roots of A_2 and B_2, respectively (compare the pictures (3.15) and (3.42) in chapter 3). More generally, the weights of the adjoint module of any semisimple Lie algebra $\mathfrak{g}$ are the $\mathfrak{g}$-roots, each occurring with multiplicity one, and in addition the weight $\lambda = 0$ with multiplicity $r = \text{rank}\,\mathfrak{g}$.

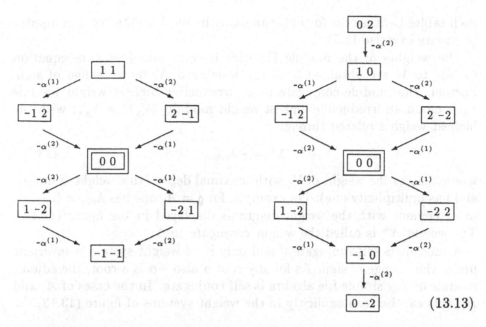

$$(13.13)$$

As the latter examples show, the multiplicity

$$\mathrm{mult}_\Lambda(\lambda) := \dim V_{(\lambda)} \qquad (13.14)$$

of a (non-highest) weight λ can be larger than one even for irreducible modules. In other words, in general the 'quantum numbers' λ^i do not characterize a weight vector v_λ completely (such weight vectors, or also their weights, are sometimes called degenerate). In fact, for a simple Lie algebra of rank r and dimension d, a complete specification of the weight vectors (up to scalar multiples) requires $\frac{1}{2}(d - r) = |\Phi_+|$ different labels, i.e. $\frac{1}{2}(d - 3r)$ quantum numbers in addition to the Dynkin labels λ^i.

By analyzing the decomposition of the highest weight module V_Λ into irreducible modules with respect to all $\mathfrak{sl}(2)_\alpha$-subalgebras, one can in principle compute the multiplicities $\mathrm{mult}_\Lambda(\lambda)$ of all weights λ of V_Λ. In practice, it is much more convenient to employ the *Freudenthal recursion formula*

$$\mathrm{mult}_\Lambda(\lambda) = 2\left[(\Lambda + \rho, \Lambda + \rho) - (\lambda + \rho, \lambda + \rho)\right]^{-1}$$

$$\cdot \sum_{\alpha > 0} \sum_{\substack{m > 0 \\ v_{\lambda + m\alpha} \in V_\Lambda}} (\lambda + m\alpha, \alpha) \cdot \mathrm{mult}_\Lambda(\lambda + m\alpha) \qquad (13.15)$$

for $\lambda \neq \Lambda$, by means of which the multiplicities can be computed recursively, starting from $\mathrm{mult}_\Lambda(\Lambda) = 1$. In equation (13.15) $\rho = \frac{1}{2}\sum_{\alpha > 0} \alpha = \sum_i \Lambda_{(i)}$ is the Weyl vector, i.e. half the sum of positive roots, of $\mathfrak{g}$. In the past, the Freudenthal formula has been employed to compile extensive tables of the weight systems for irreducible highest weight modules of simple Lie algebras. Nowadays it is no longer necessary to resort to such tables because the formula can easily be implemented on a computer (compare exercise 13.5).

The weights of the module V^+ that is conjugate (compare equation (5.16)) to V are obtained from the weights of V by a change of sign. Further, the module conjugate to an irreducible highest weight module V_Λ is again an irreducible highest weight module, $(V_\Lambda)^+ = V_{\Lambda^+}$, with the highest weights related through

$$\Lambda^+ = -\lambda_{\min} , \tag{13.16}$$

where $\lambda_{\min}$ is the weight of V_Λ with maximal depth (this weight is unique and has multiplicity one). For example, for $\mathfrak{g} = A_2$ one has $\Lambda_{(2)} = (\Lambda_{(1)})^+$, in agreement with the weight diagrams displayed in the figure (13.12). The weight Λ^+ is called the weight *conjugate* to Λ.

A module is self-conjugate if and only if its weight system is invariant under the change of sign. As for any root α also $-\alpha$ is a root, the adjoint module of any simple Lie algebra is self-conjugate. In the cases of A_2 and B_2, this can be seen explicitly in the weight systems of figure (13.13).

13.4 Unitarity

In quantum theory one often encounters the situation that the application of some kind of step operators to an energy eigenstate produces another eigenstate. Depending on the system under consideration, the energy eigenvalue may or may not change in this process; in the former case (an example of which is provided by the Heisenberg algebra of a scalar field treated in chapter 2) one is in fact typically dealing with Lie algebra representations for which the energy corresponds essentially to *minus* the depth of the weights, so that raising and lowering operators lower and raise the energy, respectively. The requirement that the energy is bounded from below is then the source of the particular relevance of highest weight representations. But even if application of a step operator does not change the energy, one is often still dealing with highest weight representations. This comes about because another fundamental principle of quantum mechanics is that the representation spaces carry a Hilbert space structure, with (finite) symmetry transformations acting by a unitary representation of a compact group G.

That a representation U of a *group* G on a complex vector space V is unitary means that, first, there is a (non-degenerate) hermitian product $V \times V \to \mathbb{C}$, $v \otimes w \mapsto (v \,|\, w)$ (so that when completed with respect to the associated norm, the module V is a *Hilbert* space); and second, that this hermitian product is G-invariant, i.e. that

$$(U(\gamma)\, v \,|\, U(\gamma)\, w) = (v \,|\, w) \tag{13.17}$$

for all $v, w \in V$ and all representation matrices $U(\gamma)$, $\gamma \in G$. When combined with the definition $(U(\gamma)\,v\,|\,w) = (v\,|\,(U(\gamma))^\dagger\,w)$ of the adjoint map $(U(\gamma))^\dagger$, (13.17) implies that the linear mappings $U(\gamma)$ are unitary:

$$(U(\gamma))^{-1} = (U(\gamma))^\dagger. \tag{13.18}$$

If V is a finite-dimensional complex vector space, the map $U(\gamma)$ is described by a unitary matrix and the 'dagger' symbol ' $\dagger$ ' stands for the matrix adjoint, i.e. transposition combined with complex conjugation.

Assume now that G is a compact simple Lie group, with compact real Lie algebra $\mathfrak{g}$ spanned by generators T^a. To see what unitarity implies for the $\mathfrak{g}$-representations which describe infinitesimal symmetries, use the exponential map (9.7) to write an element $\gamma \in G$ (in the connected component of the identity element) as $\gamma = \mathrm{Exp}(\sum_{a=1}^{d} \xi^a T^a)$ with real parameters ξ^a. Then $U(\gamma) = U_R(\gamma) := \exp(\sum_{a=1}^{d} \xi^a R(T^a))$ for some $\mathfrak{g}$-representation R. The condition (13.18) thus reads

$$\exp\left[-\sum_a \xi^a R(T^a)\right] = \left(\exp\left[\sum_a \xi^a R(T^a)\right]\right)^\dagger = \exp\left[\sum_a \xi^a \,(R(T^a))^\dagger\right].$$

This implies that

$$R(T^a) = -(R(T^a))^\dagger \tag{13.19}$$

for all generators T^a, $a = 1, 2, \ldots, \dim \mathfrak{g}$, of the compact real form. Thus the representation matrices for the generators T^a must be anti-selfadjoint matrices; as $\mathfrak{g}$ is a Lie algebra over $\mathbb{R}$, it follows that this must hold for *all* representation matrices $R(x)$, $x \in \mathfrak{g}$. Representations R of $\mathfrak{g}$ on a Hilbert space which satisfy the relation (13.19) are called *unitary* (or *unitarizable*) representations of $\mathfrak{g}$. The Hilbert space V on which a unitary $\mathfrak{g}$-representation R acts is called a unitary module of $\mathfrak{g}$. In physics, it is common to write the exponential relation between G and $\mathfrak{g}$ with an additional factor of the imaginary unit, i.e. as $\gamma = \mathrm{Exp}(\mathrm{i}\sum_{a=1}^{d} \xi^a \tilde{T}^a)$; the condition for unitarity is then that the representation matrices $\tilde{T}^a$ are selfadjoint (whereas the T^a are anti-selfadjoint) and the coefficients ξ are real.

Equation (13.19) is written down for the generators of the compact real form and extends linearly over the *real* numbers to all of the real form $\mathfrak{g}$. We would like to extend (13.19) to the complexification $\mathfrak{g}_{\mathbb{C}}$ of $\mathfrak{g}$ as well. Now we realize that while hermitian conjugation is $\mathbb{R}$-linear, it is *anti*-linear over the complex numbers, and hence (13.19) cannot be naively continued to the complex Lie algebra. Moreover, hermitian conjugation reverses the order of the arguments of a commutator:

$$([R(T^a), R(T^b)])^\dagger = [R(T^b)^\dagger, R(T^a)^\dagger]. \tag{13.20}$$

It is therefore natural to look for an anti-linear map of $\mathfrak{g}_{\mathbb{C}}$ that obeys $\omega([x, y]) = [\omega(y), \omega(x)]$ and that acts as minus the identity on the compact real form. To find such a map, we recall from section 11.9 that there is a unique anti-automorphism of the complex Lie algebra, the Chevalley involution ω_0, that leaves precisely the compact real form fixed. According to the equations (11.56) and (11.52) the Chevalley involution is defined on a Chevalley–Serre basis of the complexification $\mathfrak{g}_{\mathbb{C}}$ of $\mathfrak{g}$ by $H^i \mapsto -H^i$, $E^\alpha \leftrightarrow -E^{-\alpha}$ and is anti-linearly extended as an anti-automorphism to all of $\mathfrak{g}_{\mathbb{C}}$. The map ω we are after is then the one which is defined by $H^i \mapsto H^i$ and $E^\alpha \leftrightarrow E^{-\alpha}$ on the generators of $\mathfrak{g}$ (and hence of $\mathfrak{g}_{\mathbb{C}}$) and is anti-linearly extended to all of $\mathfrak{g}_{\mathbb{C}}$. Note that this anti-involution acts like minus the identity on the generators $\mathrm{i}H^i$, $\mathrm{i}(E^\alpha + E^{-\alpha})$ and $E^\alpha - E^{-\alpha}$ of the real form $\mathfrak{g}$ (compare equation (11.55), and also (8.20)), and hence on all of $\mathfrak{g}$. Hence we have

$$(v \mid R(x)\, w) \equiv ((R(x))^\dagger\, v \mid w) = (R(\omega(x))\, v \mid w) \tag{13.21}$$

for all $x \in \mathfrak{g}_{\mathbb{C}}$ and all $v, w \in V$. We stress that in the notion of unitarity the choice of a real form is implicit; if one starts with a different, non-compact real form, one arrives at another notion of unitarity. In the sequel we will always refer to the notion of unitarity that is induced by the compact real form.

One can use the relation (13.21) to construct a candidate hermitian product on any highest weight module V by declaring the length squared $(v_\Lambda \mid v_\Lambda)$ of the highest weight vector v_Λ to be equal to one and obtaining all other products by writing any $v \in V$ in the form (13.10). For semisimple Lie algebras, the hermitian form obtained this way is indeed non-degenerate provided that one is dealing with a *finite-dimensional* highest weight module (see exercise 13.4). Thus all finite-dimensional highest weight modules of semisimple complex or compact real algebras are unitary. In contrast, if one uses this procedure to compute $(v \mid v)$ also for the infinite-dimensional highest weight modules which have non-dominant integral highest weight, or in the finite-dimensional case for vectors v that are formally obtained from the lowest weight vector $v_{-\Lambda^+}$ of the irreducible module by acting with further lowering operators, one finds that the hermitian form becomes degenerate.

The inner product on a unitary module can be used to restrict attention to weight vectors v of a specific length, say $(v \mid v) = 1$. In the case of an irreducible highest weight module V_Λ these normalized weight vectors satisfy e.g. (with a convenient choice of phase factors)

$$\begin{aligned} R_\Lambda(E^{-\alpha})\, v_\Lambda &= \sqrt{(\Lambda, \alpha)} \cdot v_{\Lambda-\alpha}\,, \\ R_\Lambda(E^\alpha)\, v_{\Lambda-\alpha} &= \sqrt{(\Lambda, \alpha)} \cdot v_\Lambda \end{aligned} \tag{13.22}$$

for all positive roots α.

13.5 Characters and dimensions

We would like to encode in a convenient way the information contained in the weight system of a module V, including the multiplicities $\text{mult}_V(\lambda)$ of all the weights of the module. A convenient trick in such a situation where one has to keep track of many numbers is the introduction of a *generating function*. To explain the idea, let us assume that we are given a set of numbers a_n, indexed by $n \in \mathbb{Z}$. We pick some indeterminate q; then the formal Laurent series $f(q) := \sum_n a_n q^n$, regarded as a function of q, contains the same information as the infinite set of the a_n. If we set $q = e^{2\pi i \tau}$, we can introduce a function of the variable τ whose Fourier coefficients do the same job: $f = \sum_n a_n e^{2\pi i n \tau}$.

In the case of our interest, where we want to describe the multiplicities of the weight spaces of some module, we cannot use directly the abelian group $\mathbb{Z}$; rather we have to generalize the formalism slightly. To this end we introduce for all weights λ formal exponentials e^λ which are required to satisfy the usual properties of exponentials, i.e. $e^{-\lambda} e^\lambda = 1 = e^\lambda e^{-\lambda}$ and $e^\lambda e^\mu = e^{\lambda+\mu}$. In more mathematical terms, this can also be described as follows. The weight lattice L_w forms an abelian group under vector addition of weights. As we will see in section 19.5, to any group G one can associate an associative algebra, its group algebra. A basis $\{v_\gamma\}$ of this algebra is labelled by the elements γ of G, and the algebra product is induced by the group product, $v_\gamma \circ v_{\gamma'} := v_{\gamma\gamma'}$. Thus the formal exponentials just constitute such a basis for the group algebra of the weight lattice L_w.

Since we are dealing with elements of an algebra, we can also add up these exponentials, and hence we can consider the generating function

$$\chi_V := \sum_\lambda \text{mult}_V(\lambda) \, e^\lambda, \tag{13.23}$$

which keeps track of the multiplicities of the weight spaces, as an element of this group algebra (or, in the case of infinite-dimensional modules, of some completion of the group algebra). Now we recall that the weights have been introduced as linear functions on the Cartan subalgebra $\mathfrak{g}_\circ$; hence also e^λ can be regarded as a function on the Cartan subalgebra. The function χ_V, defined on the Cartan subalgebra, is called the *character* of the $\mathfrak{g}$-module V. (The characters of the Lie algebra $\mathfrak{g}$ should not be confused with the characters of the corresponding Lie *group* G; the latter will be defined as functions on the group manifold in section 21.10.)

Since the Cartan subalgebra $\mathfrak{g}_\circ$ can be identified with the weight space, we can interpret the e^λ also as functions from the weight space to the complex numbers, by setting

$$e^\lambda : \quad \mu \mapsto e^\lambda(\mu) := \exp[(\lambda, \mu)] \tag{13.24}$$

for any weight μ. Correspondingly, the character can also be seen as a function on weight space. In the sequel, we will use both interpretations interchangeably.

Up to isomorphism, the weight system, and hence also the character χ_Λ, characterizes an irreducible highest weight module uniquely. It follows in particular that when V is fully reducible, its character χ_V can be written in a unique way as a linear combination of characters $\chi_\Lambda \equiv \chi_{V_\Lambda}$ of irreducible highest weight modules V_Λ, which reflects the decomposition of V into a direct sum of irreducible modules. The explicit form of the functions χ_Λ looks particularly simple for $\mathfrak{g} = \mathfrak{sl}(2)$, where one obtains (compare exercise 13.7)

$$\chi_\Lambda(\mu) = \frac{\sinh(\mu(\Lambda + 1)/2)}{\sinh(\mu/2)} \, . \tag{13.25}$$

When the module V is unitary, we can write

$$\chi_V(\mu) \equiv \sum_\lambda \mathrm{mult}_V(\lambda) \, \exp((\lambda, \mu)) = \sum_{v_\lambda} (v_\lambda | v_\lambda) \, \exp((\lambda, \mu)) \tag{13.26}$$

with $\{v_\lambda\}$ a basis of V consisting of vectors of unit length. This can be rewritten as

$$\chi_V(\mu) = \sum_{v_\lambda} (v_\lambda \mid \exp[R(h_\mu)] \cdot v_\lambda) = \mathrm{tr}_V\big(\exp[R(h_\mu)]\big) \, , \tag{13.27}$$

where R is the $\mathfrak{g}$-representation associated to V, h_μ is the Cartan subalgebra element $h_\mu := (\mu, H) \equiv \sum_i \mu_i H^i$, and the symbol 'tr' denotes the trace of matrices. This interpretation is unproblematic if V is finite-dimensional; for infinite-dimensional V it is still possible if all weight spaces $V_{(\lambda)} \subset V$ are finite-dimensional.

Now for any finite-dimensional highest weight module V_Λ of a semisimple Lie algebra $\mathfrak{g}$ consider again the $\mathfrak{sl}(2)_\alpha$-subalgebras corresponding to the simple $\mathfrak{g}$-roots $\alpha = \alpha^{(i)}$. The weight system of V_Λ decomposes into a finite sum of finite-dimensional highest weight modules of $\mathfrak{sl}(2)_\alpha$. The non-trivial Weyl group element w of $\mathfrak{sl}(2)_\alpha$ acts on $\mathfrak{sl}(2)_\alpha$-weights λ as $\lambda \mapsto -\lambda$, and hence in particular maps the weight system of any finite-dimensional highest weight module to itself. Combining this result (for all simple roots $\alpha^{(i)}$, $i = 1, 2, ..., \mathrm{r}$) with the fact that the fundamental reflections $w_{(i)}$ generate the whole Weyl group W of $\mathfrak{g}$, it follows that the weight system of V_Λ, including the multiplicities, is invariant under all of W, i.e.

$$\mathrm{mult}_\Lambda(w(\lambda)) = \mathrm{mult}_\Lambda(\lambda) \tag{13.28}$$

for all $w \in W$. The same arguments show that any weight λ of a finite-dimensional highest weight module of $\mathfrak{g}$ lies on the W-orbit of a unique dominant integral weight which is again a weight of V_Λ. It follows that the weight system of a finite-dimensional highest weight module is the union of orbits with respect to the Weyl group, each of which contains precisely one dominant integral weight. The orbits of dominant integral weights other than the highest weight typically appear with multiplicity larger than one.

When combined with the definition (13.23) of χ_V and the natural action $w(e^\lambda) := e^{w(\lambda)}$ of the Weyl group on the formal exponentials e^λ, the equality (13.28) implies that the character χ_Λ of an irreducible highest weight module is symmetric with respect to Weyl transformations, i.e.

$$\chi_\Lambda(w(\mu)) = \chi_\Lambda(\mu) \qquad (13.29)$$

for all $w \in W$. When combined with several other tools (for the details see chapter 14), this property leads to the so-called *Weyl character formula* for the character of V_Λ, which reads

$$\chi_\Lambda(\mu) = \frac{\sum_{w \in W} \text{sign}(w) \exp[(w(\Lambda + \rho), \mu)]}{\sum_{w \in W} \text{sign}(w) \exp[(w(\rho), \mu)]} \,. \qquad (13.30)$$

Here $\text{sign}(w) \in \{\pm 1\}$ is the sign (10.14) of a Weyl group element w, and ρ is the Weyl vector of $\mathfrak{g}$. Note that the character formula expresses the quantity χ_Λ which is symmetric with respect to W as the quotient of two quantities Σ_i which are both antisymmetric with respect to W, i.e. satisfy $\Sigma_i(w(\mu)) = \text{sign}(w) \cdot \Sigma_i(\mu)$. Another important relation (for the proof see again chapter 14), called the *denominator identity*, expresses the denominator of (13.30) as a product:

$$\sum_{w \in W} \text{sign}(w) \exp[(w(\rho), \mu)]$$
$$= \prod_{\alpha > 0} \left(\exp[\tfrac{1}{2}(\alpha, \mu)] - \exp[-\tfrac{1}{2}(\alpha, \mu)] \right). \qquad (13.31)$$

Evaluation of the character (13.26) at the weight zero yields the sum of the dimensions of all weight spaces, i.e. the dimension of the module: $\dim(V) = \chi_V(0)$. Now directly setting μ to zero in the character formula (13.30) yields an ill-defined expression, but with the help of the denominator identity the limit can be performed properly (see exercise 13.7), leading to the *Weyl dimension formula*

$$d_\Lambda \equiv \dim(V_\Lambda) = \prod_{\alpha > 0} \frac{(\Lambda + \rho, \alpha)}{(\rho, \alpha)} \,. \qquad (13.32)$$

For $\mathfrak{g} = A_1$ this reduces to the formula $d_\Lambda = \Lambda + 1$ that we already obtained in chapter 2. As a less trivial example, take $\mathfrak{g} = A_2$; then (13.32) yields

$$d_\Lambda(A_2) = (\Lambda^1 + 1)(\Lambda^2 + 1)(\tfrac{1}{2}(\Lambda^1 + \Lambda^2) + 1). \qquad (13.33)$$

For the direct sum $V = \bigoplus_i V_i$ of $\mathfrak{g}$-modules V_i, the definition (13.23) of the character immediately gives

$$\chi_V = \sum_i \chi_{V_i}. \tag{13.34}$$

Similarly, the character of the tensor product $V \otimes V'$ of $\mathfrak{g}$-modules obeys

$$\chi_{V \otimes V'} = \chi_V \cdot \chi_{V'}, \tag{13.35}$$

which can be seen as follows. If $\{v_\lambda\}$ and $\{v_{\lambda'}\}$ are bases of V and V', respectively, then a basis of the tensor product module is given by $\{v_\lambda \otimes v_{\lambda'}\}$. Thus the weights of $V \otimes V'$ are of the form $\lambda + \lambda'$ with λ a weight of V and λ' a weight of V', and appear with multiplicity

$$\mathrm{mult}_{V \otimes V'}(\lambda + \lambda') = \sum_{\substack{\mu, \mu' \\ \mu + \mu' = \lambda + \lambda'}} \mathrm{mult}_V(\mu) \cdot \mathrm{mult}_{V'}(\mu'). \tag{13.36}$$

These are precisely the multiplicities which one reads off the right hand side of (13.35) if one multiplies the formal functions χ_V and $\chi_{V'}$.

13.6 Defining modules and basic modules

The highest weight module of a simple Lie algebra $\mathfrak{g}$ with highest weight $\Lambda = 0$ is one-dimensional; it is called the *trivial module* or *singlet* of $\mathfrak{g}$. On the singlet, all elements $x \in \mathfrak{g}$ act as $R_0(x) = 0$; thus it is definitely the 'simplest' highest weight module of $\mathfrak{g}$. In contrast, there is no natural identification of a 'simplest' non-trivial highest weight module. For instance, on one hand the lowest-dimensional modules may be regarded as particularly simple. On the other hand the adjoint module is of course most directly related to the algebra. Finally one might consider those modules as the simplest ones from which all others can be obtained by building tensor products. We will now list all these special modules for all finite-dimensional simple Lie algebras.

• In table XII we display the highest weights and dimensions of the lowest-dimensional (non-trivial) modules V_Λ, as well as some properties of these modules which refer to concepts to be introduced later on, namely the (normalized) second order Dynkin index I_Λ (see chapter 14, equation (14.33)) and the information whether V_Λ is an orthogonal (o), symplectic (s) or non-selfconjugate (c) module (see chapter 17). (Here we use the numbering of fundamental weights as indicated in the last column of table IV. Recall that some authors use a different numbering.) The modules of table XII are often called the *defining* modules of $\mathfrak{g}$ (or also, in particular for B_r and D_r, the *vector* modules), because in the case of the classical algebras the associated representation coincides with the matrix realization of the algebra which preserves the relevant symmetric or symplectic form.

Table XII.   *Lowest-dimensional modules of the finite-dimensional simple Lie algebras*

$\mathfrak{g}$	Λ	d_Λ	$\tilde{I}_\Lambda$	o, s, c
A_1	$\Lambda_{(1)}$	2	1/2	s
$A_r,\ {}_{r\geq 2}$	$\Lambda_{(1)}, \Lambda_{(r)}$	$r+1$	1/2	c
$B_r,\ {}_{r\geq 3}$	$\Lambda_{(1)}$	$2r+1$	1	o
C_r	$\Lambda_{(1)}$	$2r$	1/2	s
D_4	$\Lambda_{(1)}, \Lambda_{(r-1)}, \Lambda_{(r)}$	8	1	o
$D_r,\ {}_{r\geq 5}$	$\Lambda_{(1)}$	$2r$	1	o
E_6	$\Lambda_{(1)}, \Lambda_{(5)}$	27	3	c
E_7	$\Lambda_{(6)}$	56	6	s
E_8	$\Lambda_{(1)}$	248	30	o
F_4	$\Lambda_{(4)}$	26	3	o
G_2	$\Lambda_{(2)}$	7	1	o

• The highest weight of the adjoint module is the highest root θ of $\mathfrak{g}$. It can be read off the extended Dynkin diagram of $\mathfrak{g}$ as follows. According to chapter 12, the zeroth root of the untwisted affine Lie algebra $X_r^{(1)}$ which has $\mathfrak{g} = X_r$ as its horizontal subalgebra corresponds to minus θ. Recalling the definition of the fundamental weights $\Lambda_{(i)}$, it follows that θ is the sum of those $\Lambda_{(i)}$ for which the entry A^{i0} of the affine Cartan matrix is non-zero. This yields the weights displayed in table XIII, which also provides the expression in the orthonormal basis that was e.g. also used in table VII.

• The following modules, from which all finite-dimensional modules can be obtained by building tensor products, are sometimes referred to as the *basic* modules of $\mathfrak{g}$: In the case of the classical algebras A_r and C_r and of all exceptional algebras, this is just the defining module, while for B_r it is the module of highest weight $\Lambda_{(r)}$ (the so-called *spinor* module, which has dimension 2^r); for D_r two distinct irreducible modules, with highest weights $\Lambda_{(r)}$ and $\Lambda_{(r-1)}$, respectively, are needed (the *spinor* and *conjugate spinor* modules, each of dimension 2^{r-1}).

Table XIII. *The highest root of finite-dimensional simple Lie algebras*

$\mathfrak{g}$	θ	
A_r	$\Lambda_{(1)} + \Lambda_{(r)}$	$e_1 - e_{r+1}$
B_2	$2\Lambda_{(2)}$	$e_1 + e_2$
$B_r\ {}_{(r \geq 3)},\ D_r$	$\Lambda_{(2)}$	$e_1 + e_2$
C_r	$2\Lambda_{(1)}$	$2e_1$
E_6	$\Lambda_{(6)}$	$e_8 - e_6 - e_7 + \sum_{j=1}^{5} e_j$
E_7	$\Lambda_{(1)}$	$e_8 - e_7$
E_8	$\Lambda_{(1)}$	$e_7 + e_8$
F_4	$\Lambda_{(1)}$	$e_1 + e_2$
G_2	$\Lambda_{(1)}$	$-e_1 - e_2 + 2e_3$

*13.7 The dual Weyl vector

The maximal depth of a weight in V_Λ can be written as the inner product $(\Lambda, 2\rho^\vee)$ of the highest weight Λ with twice the so-called *dual Weyl vector* or *level vector* $\rho^\vee$, which is by definition half the sum of positive coroots, $\rho^\vee = \frac{1}{2} \sum_{\alpha > 0} \alpha^\vee$. (Note that for non-simply laced algebras this differs from $2\rho/(\rho,\rho)$.) The components of $\rho^\vee$ in the Dynkin basis and in the basis of simple coroots are displayed in table (13.37).

$\mathfrak{g}$	$\frac{1}{2}(\theta,\theta)\cdot(\rho^\vee)^i$		$(\rho^\vee)_i$	
A_r	1		$\frac{1}{2}i\,(r - i + 1)$	
B_r	$\begin{cases} 1 & \text{for } i = 1,\dots,r-1 \\ 2 & \text{for } i = r \end{cases}$		$\begin{cases} \frac{1}{2}i(2r-i+1) & \text{for } i = 1,\dots,r-1 \\ \frac{1}{4}r(r+1) & \text{for } i = r \end{cases}$	
C_r	$\begin{cases} 2 & \text{for } i = 1,\dots,r-1 \\ 1 & \text{for } i = r \end{cases}$		$\frac{1}{2}i\,(2r - i)$	
D_r	1		$\begin{cases} \frac{1}{2}i(2r-i-1) & \text{for } i = 1,\dots,r-2 \\ \frac{1}{4}r(r-1) & \text{for } i = r-1, r \end{cases}$	
E_6	1		$(8, 15, 21, 15, 8, 11)$	
E_7	1		$(17, 33, 48, \frac{75}{2}, 26, \frac{27}{2}, \frac{49}{2})$	
E_8	1		$(29, 57, 84, 110, 135, 91, 46, 68)$	
F_4	$(1, 1, 2, 2)$		$(11, 21, 15, 8)$	
G_2	$(1, 3)$		$(5, 3)$	

(13.37)

More generally, the function $D(\lambda) := (\lambda, 2\rho^\vee)$ defines a partial order on the weight lattice of $\mathfrak{g}$. According to (13.37) the components $(\rho^\vee)_i$ of $\rho^\vee$ in the basis of simple coroots are half-integral, so that on the weight lattice $D(\lambda)$ is integer-valued. (For elements β of the root lattice, already $(\beta, \rho^\vee)$ is an integer, as can e.g. be seen by interpreting $\rho^\vee(\mathfrak{g})$ as the Weyl vector of the dual Lie algebra $\mathfrak{g}^\vee$ (compare section 7.4) and β as a sum of coroots of $\mathfrak{g}^\vee$.)

13.8 Highest weight modules of affine Lie algebras

In contrast to the situation with simple Lie algebras, all non-trivial modules of an affine Lie algebra $\mathfrak{g}$ are infinite-dimensional. But still, the most interesting modules are again highest weight modules, because they allow for an easy investigation of unitarity and because they display many other similarities with the finite-dimensional case. (On the other hand, there is at least one interesting module which is not a highest weight module, namely the adjoint module. The weights of the adjoint module are the roots, and an affine Lie algebra does not possess a highest root.) Analogously to the case of simple Lie algebras, a highest weight module V_Λ with highest weight Λ is characterized by two requirements. First, there is a highest weight vector $v_\Lambda \in V_\Lambda$ which is annihilated by all step operators corresponding to positive roots, which in the case of untwisted affine algebras means that $R_\Lambda(T_n^a) v_\Lambda = 0$ for all $n > 0$, as well as $R_\Lambda(E_0^{\bar\alpha}) v_\Lambda = 0$ for all positive roots $\bar\alpha$ of the horizontal subalgebra $\bar{\mathfrak{g}}$ of $\mathfrak{g}$. And second, each vector v in the module is a linear combination of the vectors

$$v = R_\Lambda(T_{-n_1}^{a_1}) R_\Lambda(T_{-n_2}^{a_2}) \cdots R_\Lambda(T_{-n_m}^{a_m}) v_\Lambda \tag{13.38}$$

that are obtained by applying an arbitrary number of generators corresponding to negative roots to v_Λ. In (13.38), each $T_{-n_p}^{a_p}$ stands either for $E_{-n_p}^{-\bar\alpha_p}$ with $\bar\alpha_p > 0$ and $n_p \geq 0$, or for $H_{-n_p}^{i_p}$ or $E_{-n_p}^{\bar\alpha_p}$ with $n_p > 0$.

The highest weight vector is an eigenvector of the Cartan subalgebra of $\mathfrak{g}$. Its eigenvalues constitute the highest weight Λ, which can be written as a triple

$$\Lambda = (\bar\Lambda, k, n_0), \tag{13.39}$$

with

$$R_\Lambda(H_0^i) v_\Lambda = \bar\Lambda^i \cdot v_\Lambda \quad \text{for } i = 1, 2, ..., \mathrm{r}, \quad R_\Lambda(K) v_\Lambda = k \cdot v_\Lambda \tag{13.40}$$

and $R_\Lambda(D) v_\Lambda = n_0 v_\Lambda$. By the substitution $R_\Lambda(D) \mapsto R_\Lambda(D) - n_0 R_\Lambda(K)/k$, for $k \neq 0$ the D-eigenvalue can be redefined to zero, and hence from now on for simplicity we set $n_0 = 0$. Then a highest weight module of $\mathfrak{g}$ is characterized by a highest weight $\bar\Lambda$ of the horizontal subalgebra $\bar{\mathfrak{g}}$ together with an eigenvalue k of the central generator K.

The normalized eigenvalue

$$k^\vee := k \cdot \frac{2}{(\bar\theta, \bar\theta)}, \tag{13.41}$$

with $\bar\theta$ the highest root of $\bar{\mathfrak{g}}$, is called the *level* of the highest weight module V_Λ. Because of $[K, T_n^a] = 0$, k is in fact also the K-eigenvalue of all other vectors in V_Λ. Similarly, the bracket relation $[D, T_n^a] = n T_n^a$ implies that

$$R_\Lambda(D)\, v_\lambda = -(\textstyle\sum_p n_p) \cdot v_\lambda, \tag{13.42}$$

where n_p are the integers appearing in the expression (13.38) that relates v_λ to the highest weight vector v_Λ; their sum $\sum_p n_p$ is called the *grade* of the weight λ.

It is worth stressing that the description of a $\mathfrak{g}$-module strongly depends on the specific choice of the gradation. Here we always work in the particular homogeneous gradation for which the zero mode subalgebra $\mathfrak{g}_{[0]}$ of $\mathfrak{g}$ is isomorphic to the simple Lie algebra $\bar{\mathfrak{g}}$ which is the basis of the loop algebra construction. If a different gradation is chosen, then generically for the same $\mathfrak{g}$-module V_Λ a different highest weight with respect to the zero mode subalgebra is obtained. In contrast, the level of a module is independent of the choice of gradation.

13.9 Integrable highest weight modules

While all non-trivial modules of $\mathfrak{g}$ are infinite-dimensional, in special cases the subspaces into which they decompose with respect to the $\mathfrak{sl}(2)$-subalgebras of all *real* roots are finite-dimensional. Loosely speaking, such $\mathfrak{g}$-modules V are 'less infinite-dimensional' than a generic $\mathfrak{g}$-module. From the results of chapter 12 we know that the $\mathfrak{sl}(2)$-subalgebra of $\mathfrak{g}$ that corresponds to the real root $\alpha = (\bar\alpha, 0, n)$ is generated by $E^\alpha \equiv E_n^{\bar\alpha}$, $E^{-\alpha} \equiv E_{-n}^{-\bar\alpha}$ and

$$H^\alpha = \frac{2}{(\alpha,\alpha)}\, [E_n^{\bar\alpha}, E_{-n}^{-\bar\alpha}] = \frac{2}{(\bar\alpha,\bar\alpha)} \Big(\sum_{i=1}^{r} \bar\alpha_i H_0^i + nK \Big). \tag{13.43}$$

It follows that

$$R_\Lambda(H^\alpha)\, v_\lambda = \frac{2}{(\bar\alpha,\bar\alpha)}\, ((\bar\alpha, \bar\lambda) + nk) \cdot v_\lambda = (\lambda, \alpha^\vee) \cdot v_\lambda \tag{13.44}$$

for any weight $\lambda = (\bar\lambda, k, m)$. For the modules of $\mathfrak{sl}(2)_\alpha$ to be finite-dimensional, we need $(\lambda, \alpha^\vee) \in \mathbb{Z}$, and all of them must possess a highest (and a lowest) weight. This implies that also the $\mathfrak{g}$-module V must possess a weight Λ which is a highest weight with respect to all the $\mathfrak{sl}(2)_\alpha$-subalgebras, so that $(\Lambda, \alpha^\vee) \in \mathbb{Z}_{\geq 0}$ for any real root α. The infinitely many real roots yield an infinite number of conditions, but only those for the $r + 1$ simple roots are independent; these just state that V must be a highest weight module V_Λ with highest weight Λ which satisfies

$$\Lambda^i \equiv (\Lambda, \alpha^{(i)\vee}) \in \mathbb{Z}_{\geq 0} \quad \text{for all } i = 0, 1, \dots, r. \tag{13.45}$$

As for infinite-dimensional Lie algebras in general, affine Lie algebras cannot be naively exponentiated to an associated Lie *group* (compare also section 9.4). Rather, the level has to be kept fixed, and for each level one obtains a different infinite-dimensional Lie group G_k (G_k is then known as a Kac–Moody group; the study of these groups is a field in its own right and will not be pursued here). The $\mathfrak{g}$-representation R_Λ associated to an irreducible module V_Λ can be 'integrated' to a representation of a corresponding Lie group G by means of an exponential mapping $R(x) \mapsto \exp[R(x)]$ if and only if V_Λ obeys (13.45); the latter modules are therefore called *integrable* highest weight modules. If a module V_Λ is integrable, then for each real root α, the representations of the $\mathfrak{sl}(2)$ subalgebra $\mathfrak{sl}(2)_\alpha$ of $\mathfrak{g}$, generated by E^α, $E^{-\alpha}$ and H^α, into which the representation R_Λ decomposes, can be integrated up to representations of some Lie *group* G_α isomorphic to SU(2). (If $\mathfrak{g}$ is simple, then the group G generated by all these groups G_α is isomorphic to the universal covering Lie group of $\mathfrak{g}$, while for affine Lie algebras, $G = G_k$ depends on the level $k^\vee$ of the representation R_Λ.) In contrast, for a generic infinite-dimensional representation integration to a group is not possible, i.e. the formal power series $\exp[R(x)]$ cannot be given a well-defined meaning for all $x \in \mathfrak{g}$.

Taking $\alpha^{(i)} = (\bar\alpha^{(i)}, 0, 0)$ in (13.45), we learn that $\bar\Lambda^i \in \mathbb{Z}_{\geq 0}$ for any integrable highest weight module V_Λ, i.e. $\bar\Lambda$ is a dominant integral $\bar{\mathfrak{g}}$-weight, and the choice $\alpha^{(i)} = \alpha^{(0)} \equiv (-\bar\theta, 0, 1)$ leads to the requirement that the level (13.41) must be a non-negative integer, $k^\vee \in \mathbb{Z}_{\geq 0}$, and furthermore there is the bound

$$0 \leq \sum_{i=1}^{r} a_i^\vee \bar\Lambda^i = (\bar\Lambda, \bar\theta^\vee) \leq k^\vee. \tag{13.46}$$

The inequality (13.46) provides a simple recipe for enumerating all integrable highest weight modules which have a given value of the level. In particular, for any fixed value of the level only a *finite* number of highest weight modules is integrable. Namely, in the case of $\bar{\mathfrak{g}} = A_r$ and C_r where all dual Coxeter labels are equal to one, (13.46) tells us that the allowed dominant integral $\bar{\mathfrak{g}}$-weights $\bar\Lambda$ at arbitrary level are those for which the sum of the Dynkin labels $\bar\Lambda^i$ of $\bar\Lambda$ does not exceed $k^\vee$, $\sum_{i=1}^{r} \bar\Lambda^i \leq k^\vee$, while for the other untwisted affine Lie algebras (13.46) leads to a slightly more complicated prescription. For instance, the value $k = 0$ is possible only for $\bar\Lambda = 0$, which implies $\Lambda = (0, 0, 0)$ so that V_Λ is the trivial one-dimensional module; thus the level is strictly positive for any non-trivial integrable highest weight module. At $k^\vee = 1$ the allowed highest weights of $\bar{\mathfrak{g}}$ are, besides $\bar\Lambda = 0$, precisely those fundamental $\bar{\mathfrak{g}}$-weights $\bar\Lambda_{(i)}$ for which the dual Coxeter labels $a_i^\vee$ are equal to one; these are called *minimal* fundamental weights. According to table VIII, the minimal fundamental weights are all fundamental weights for $\bar{\mathfrak{g}} = A_r$ and C_r, the fundamental weights of the spinor and vector modules of B_r and D_r, and the fundamental weights of the defining modules of E_6, E_7, F_4 and G_2, while none of the fundamental weights of E_8 is allowed.

As in the case of semisimple algebras, the *fundamental weights* $\Lambda_{(i)}$, $i = 0, 1, \ldots, r$ of $\mathfrak{g}$ are defined by the property that they are dual to the simple coroots, i.e. $(\Lambda_{(i)} \mid \alpha^{(j)^\vee}) = \delta_{ij}$ for all $j = 0, 1, \dots, r$. The explicit form of $\Lambda_{(i)}$ is

$$\Lambda_{(i)} = (\bar{\Lambda}_{(i)}, \tfrac{1}{2}(\bar{\theta}, \bar{\theta})a_i^\vee, 0) \tag{13.47}$$

for $i = 1, 2, \dots, r$, while $\Lambda_{(0)} = (0, \tfrac{1}{2}(\bar{\theta}, \bar{\theta}), 0)$. (The factors of $(\bar{\theta}, \bar{\theta})$ appearing here are necessary to make the results independent of the normalization of the inner product on the weight space of $\bar{g}$.) Because of $a_0^\vee = 1$ the formula (13.47) thus also holds for $i = 0$ provided we adopt the convention that $\bar{\Lambda}_{(0)} \equiv 0$, which is also useful in other situations. Any integrable highest weight is of the form $\Lambda = \sum_{i=0}^{r} \Lambda^i \Lambda_{(i)}$ with $\Lambda^i \in \mathbb{Z}_{\geq 0}$ and hence satisfies

$$\Lambda \equiv (\bar{\Lambda}, k, 0) = (\bar{\Lambda}, \tfrac{1}{2}(\bar{\theta}, \bar{\theta})\sum_{i=0}^{r} a_i^\vee \Lambda^i, 0). \tag{13.48}$$

This means in particular that

$$k^\vee = \sum_{i=0}^{r} a_i^\vee \Lambda^i, \tag{13.49}$$

or in other words, that $k^\vee = \Lambda^0 + (\bar{\theta}^\vee, \bar{\Lambda})$. In terms of the Dynkin label Λ^0 the integrability requirement (13.46) thus reads $\Lambda^0 \geq 0$, in agreement with (13.45).

*13.10 Twisted affine Lie algebras

When $g = g^\omega$ is a twisted affine Lie algebra, the vectors in a module that have grade zero again furnish a module of the zero mode subalgebra $g_{[0]} \subset g$, but $g_{[0]}$ no longer coincides with the simple algebra $\bar{g}$ which is the starting point of the loop construction. Highest weight modules V_Λ are thus labelled by a highest weight $\bar{\Lambda}$ of $g_{[0]}$ together with a value of the level $k^\vee$. For integrability, $(\Lambda, \alpha^\vee)$ must again be a non-negative integer for all real roots α. Fundamental weights are again defined by the property that $(\Lambda_{(i)}, \alpha^{(j)^\vee}) = \delta_{ij}$, where now the simple roots are given by $\alpha^{(i)} = (\alpha_{[0]}^{(i)}, 0, 0)$ with $i = 1, 2, \dots, \mathrm{rank}\, g_{[0]}$ (and with $\alpha_{[0]}^{(i)}$ the simple roots of $g_{[0]}$), supplemented by $\alpha^{(0)} = (-\bar{\Lambda}_1, 0, 1/\ell)$. Here ℓ is the order of the relevant automorphism ω of $\bar{g}$ that is used in the loop construction, and $\bar{\Lambda}_1$ the highest weight of that $g_{[0]}$-module that is furnished by the subspace $\bar{g}_{[1]}$, respectively, for $\ell = 3$, by $\bar{g}_{[1]}$ and $\bar{g}_{[2]}$ (compare table XI). The explicit form of the fundamental weights is

$$\Lambda_{(0)} = \left(0, \tfrac{\ell}{2}(\bar{\Lambda}_1, \bar{\Lambda}_1), 0\right),$$
$$\Lambda_{(i)} = \left(\bar{\Lambda}_{(i)}, \tfrac{\ell}{2}(\bar{\Lambda}_1, \bar{\Lambda}_1)\tfrac{a_i^\vee}{a_0^\vee}, 0\right) \quad \text{for } i = 1, 2, \dots, r. \tag{13.50}$$

The dual Coxeter labels $a_i^\vee$ of g that appear in these expressions coincide up to an overall factor $2a_0^\vee/(\bar{\theta}, \bar{\theta})$ (with $\bar{\theta}$ the highest root of $\bar{g}$) with the expansion coefficients of the weight $\bar{\Lambda}_1$ in the basis of simple coroots of $g_{[0]}$. Note that $a_0^\vee = 2$ for $A_1^{(2)}$ and $\tilde{B}_r^{(2)}$, while $a_0^\vee = 1$ otherwise.

The general solution to the integrability condition is again that $\Lambda = (\bar{\Lambda}, k, 0)$ must be dominant integral, i.e. a linear combination of the fundamental weights with non-negative integral coefficients Λ^i. The level of V_Λ is given by the formula

$$k^\vee = k_0 \sum_{i=0}^{r} a_i^\vee \Lambda^i \qquad (13.51)$$

with $k_0 := \ell(\bar{\Lambda}_1, \bar{\Lambda}_1)/a_0^\vee(\bar{\theta}, \bar{\theta})$. Explicitly, $k_0 = 4$ for $A_1^{(2)}$, $k_0 = 2$ for $\tilde{B}_r^{(2)}$, and $k_0 = 1$ otherwise, so that the level of an integrable module must be even for $\tilde{B}_r^{(2)}$ and a multiple of four for $A_1^{(2)}$, while it can be any non-negative integer for all other twisted affine algebras.

13.11 Triangular decomposition

The only property of simple and affine Lie algebras that is needed for the concept of highest weight modules is in fact the existence of a triangular decomposition

$$\mathfrak{g} = \mathfrak{g}_- \oplus \mathfrak{g}_\circ \oplus \mathfrak{g}_+ \qquad (13.52)$$

of the algebra into subalgebras which satisfy

$$[\mathfrak{g}_\circ, \mathfrak{g}_\circ] = 0\,,$$

$$[\mathfrak{g}_+, \mathfrak{g}_-] \subseteq \mathfrak{g}_\circ\,, \qquad (13.53)$$

$$[\mathfrak{g}_\pm, \mathfrak{g}_\circ \oplus \mathfrak{g}_\pm] \subseteq \mathfrak{g}_\pm\,.$$

Given such a decomposition, one may define a highest weight vector v_Λ by the condition that it is an eigenvector of the action of the abelian subalgebra $\mathfrak{g}_\circ$ with eigenvalue Λ, and then construct a highest weight module by acting on v_Λ with all possible combinations of lowering operators, i.e. elements of (say) $\mathfrak{g}_-$, as in equations (13.10) and (13.38).

Highest weight modules therefore exist for any Lie algebra $\mathfrak{g}$ which possesses a triangular decomposition (13.52), not just for simple and affine algebras. Among the Lie algebras which belong to this class are all Kac–Moody algebras, that is, the Lie algebras which are defined via a generalized Cartan matrix A. This class of Lie algebras also includes:

• The Heisenberg algebra which we described in section 2.5 and section 12.10;

• the Virasoro algebra (section 12.12);

• the so-called $\mathcal{W}$-algebras (see e.g. [Bouwknegt and Schoutens 1993]) of conformal field theory;

• and also certain generalized Kac–Moody algebras which are known as *Borcherds* algebras as well as all so-called *vertex operator* algebras; these are e.g. described in [Frenkel *et al.* 1988] and [Gebert 1993].

Summary:

Highest weight representations constitute a particularly important subclass of representations. Any finite-dimensional representation of a simple Lie algebra is a highest weight representation.

A highest weight representation contains a unique (up to a phase) vector, the highest weight vector, which is annihilated by the Borel subalgebra $\mathfrak{g}_+$. Unitary highest weight modules have positive integral highest weights.

The generating function for the multiplicities of the weight spaces, the character, can be computed by use of the Weyl character formula.

Keywords:

Highest weight representation, irreducible highest weight module, weight system, weight space, multiplicity, highest weight, dominant integral weight, unitary representation, character, Weyl character and dimension formulæ, defining module, basic module, grade, level.

Exercises:

Check the action (13.11) of the step operators $R(E^\alpha)$ in the representation R.

Exercise 13.1

Show that the roots of a simple Lie algebra are integral weights and that the highest root θ is dominant integral. Show that the associated irreducible highest weight representation is the adjoint representation.

Exercise 13.2

Verify that the prescription for constructing the weight system that is described after equation (13.11) yields the weight systems displayed in (13.12) and (13.13). Hint: Recall that in the Dynkin basis the simple roots coincide with the rows of the Cartan matrix.

Exercise 13.3

Construct the weight system of the adjoint module of the simple Lie algebra G_2 and check that it reproduces the figure (3.42). How many distinct orbits of the Weyl group are contained in this weight system? Construct the weight system of the highest weight module with highest weight $\Lambda = \Lambda_{(1)}$ of the Lie algebra E_6. Compute the multiplicities of the so obtained weights by combining the weight spaces to suitable $\mathfrak{sl}(2)_\alpha$-modules.

Determine the orbits of the weights of all these modules with respect to the Weyl group.

Show that the prescription given after formula (13.21) for constructing a bilinear form on a highest weight module yields a non-degenerate hermitian product if and only if the module is finite-dimensional. Check that for any positive root α the normalized weight vectors v_Λ and $v_{\Lambda-\alpha}$ satisfy the relations (13.22). What is the normalization of the vectors v_λ of the $\mathfrak{sl}(2)$-module V_Λ that were chosen in (5.38)?

Exercise 13.4

Make a list of all ingredients that are needed to put the Freudenthal recursion formula (13.15) on a computer. Estimate the number of multiplications needed to compute the multiplicity of a weight of depth n. Set up a computer program. (It is convenient to work with a programming language which allows for dynamical memory allocation and for treating integers of arbitrary size.)

Exercise 13.5

How many distinct Weyl orbits does the weight system of the adjoint module of E_8 contain? How large is the orbit of the highest weight? Hint: The highest weight of the adjoint module is the highest root; deduce from the explicit form of the simple roots that besides the highest root the weight system contains only a single other dominant integral weight.

Exercise 13.6

a) Derive the result (13.25) for the character χ_Λ of finite-dimensional highest weight modules of $\mathfrak{sl}(2)$ directly from the definition (13.23) and (13.24) of the characters, and show that it is in agreement with the Weyl character formula (13.30).
b) Evaluate the character χ_Λ for an arbitrary simple Lie algebra on the specific weight $\mu = t\rho$, with ρ the Weyl vector and $t \in \mathbb{R}$.
c) Derive the Weyl dimension formula (13.32) from the Weyl character formula and the denominator identity.
Hint: Realize that the result of exercise 13.7b) allows to write the character formula in such a way that the limit $t \to 0$ can be performed with the help of l'Hospital's rule.
d) Show that the numerator and denominator of the Weyl character formula are W-antisymmetric.

Exercise 13.7

Check the dimension formula (13.33) for $\mathfrak{g} = A_2$, and work out the Weyl dimension formula explicitly for $\mathfrak{g} = G_2$.
Show that the dimension of the irreducible highest weight module of highest weight $\Lambda = \rho$ of a simple Lie algebra of rank r and dimension d is $\dim(V_\rho) = 2^{|\Phi_+|} = 2^{(d-r)/2}$.

Exercise 13.8

Check for all $i = 0, 1, \ldots, r$ that the weights (13.47) are indeed dual to the simple coroots of an untwisted affine Lie algebra.

Exercise 13.9

Consider the three-dimensional Lie algebra spanned by elements $E_\pm$, K with relations $[K, E_\pm] = 0$ and $[E_+, E_-] = K$. Classify all modules which are integrable in the sense that E_+ and E_- act as locally nilpotent operators.

Exercise 13.10

14

Verma modules, Casimirs, and the character formula

14.1 Universal enveloping algebras

As we have already observed in section 4.5, from any associative algebra $\mathfrak{A}$ one can construct a Lie algebra by simply considering the same vector space and defining the Lie bracket as the commutator with respect to the associative product in $\mathfrak{A}$. As a consequence, many aspects of the theory of associative algebras can be reduced to Lie algebra theory. In this chapter we are interested in the opposite direction, i.e. starting from a Lie algebra $\mathfrak{g}$ we construct an associative algebra $\mathfrak{A}$ which contains $\mathfrak{g}$ as a subspace and for which the commutator on this subspace reproduces the Lie bracket in $\mathfrak{g}$. Such an algebra is called an *enveloping algebra* of $\mathfrak{g}$. This situation can be analyzed under rather general assumptions; in particular, the statements in this section hold for *any* Lie algebra that has a basis which is countable, not just for simple or affine algebras.

It is in fact rather straightforward to construct such an enveloping algebra. One starts from the vector space V in terms of which $\mathfrak{g}$ is defined, and takes the direct sum TV of all vector spaces V^n, where $V^n = \bigotimes_{j=1}^{n} V$ is the nth tensor power of V:

$$ TV := \bigoplus_{n=0}^{\infty} V^n = \bigoplus_{n=0}^{\infty} \left(\bigotimes_{j=1}^{n} V \right). \tag{14.1} $$

The first summand V^0 is the one-dimensional vector space, i.e. just the base field F (we will assume that $F = \mathbb{C}$ or $\mathbb{R}$, but the construction works in fact for any field of characteristic zero). The space TV possesses a natural multiplication, the tensor product, and hence is called the tensor algebra of V. With respect to this product, we have

$$ V^n \otimes V^m \subseteq V^{n+m}, \tag{14.2} $$

244

i.e. the tensor algebra TV is graded over $\mathbb{Z}_{\geq 0}$. This product is in itself not too interesting, as it does not know anything about the Lie algebra structure of $\mathfrak{g}$. To implement that structure, one would like to consider the element $x_1 \otimes x_2 - x_2 \otimes x_1$ in V^2, for arbitrary elements x_1, x_2 of $\mathfrak{g} = V$, as the same element as $[x_1, x_2]$ in V^1. In fact one needs even more. Namely, an arbitrary element y of TV is of the form $\sum_k x_1^k \otimes x_2^k \otimes \cdots \otimes x_{n(k)}^k$ with some elements $x_i^k \in \mathfrak{g}$. Now whenever an element of V^n of the form

$$x_1 \otimes x_2 \otimes \cdots \otimes x_i \otimes x_{i+1} \otimes x_{i+2} \otimes \cdots \otimes x_n - x_1 \otimes x_2 \otimes \cdots \otimes x_{i+1} \otimes x_i \otimes x_{i+2} \otimes \cdots \otimes x_n$$

appears in y, one should identify it with

$$x_1 \otimes x_2 \otimes \cdots \otimes [x_i, x_{i+1}] \otimes x_{i+2} \otimes \cdots \otimes x_n \in V^{n-1}.$$

After having implemented these identifications, one is still left with an associative algebra (which also still has a unit element); this algebra is called the *universal enveloping algebra* of $\mathfrak{g}$ and is denoted by $\mathsf{U}(\mathfrak{g})$. Furthermore, by considering any $x \in \mathfrak{g}$ as an element $x \in V^1$, one embeds $\mathfrak{g}$ as a subspace into $\mathsf{U}(\mathfrak{g})$; this canonical injection from $\mathfrak{g}$ into $\mathsf{U}(\mathfrak{g})$ will be denoted by $\iota\colon \mathfrak{g} \hookrightarrow \mathsf{U}(\mathfrak{g})$.

Another important subspace $\mathsf{U}^+(\mathfrak{g})$ of $\mathsf{U}(\mathfrak{g})$ is the one spanned by all elements $x \in \mathsf{U}(\mathfrak{g})$ that can be written as the product of some element $y \in \mathfrak{g}$ of the Lie algebra $\mathfrak{g}$ with an element $z \in \mathsf{U}(\mathfrak{g})$: $\mathsf{U}^+(\mathfrak{g}) = \mathfrak{g}\,\mathsf{U}(\mathfrak{g})$. It is straightforward to check that $\mathsf{U}^+(\mathfrak{g})$ is an ideal of the associative algebra $\mathsf{U}(\mathfrak{g})$; this ideal is called the *augmentation ideal* of $\mathsf{U}(\mathfrak{g})$.

From (14.2) it follows that the tensor algebra is graded over $\mathbb{Z}$. The identifications which one has to implement to obtain the universal enveloping algebra $\mathsf{U}(\mathfrak{g})$ do not respect this grading; therefore $\mathsf{U}(\mathfrak{g})$ is not graded over $\mathbb{Z}$ any more. However, the tensor algebra has also the structure of a *filtration*. Namely, the vector spaces $\tilde{V}^n$ defined as the union $\tilde{V}^n := \bigcup_{p=0}^n V^p$ satisfy

$$\tilde{V}^n \otimes \tilde{V}^m \subseteq \tilde{V}^{n+m}. \tag{14.3}$$

This property survives also in the universal enveloping algebra, and hence $\mathsf{U}(\mathfrak{g})$ is in fact an algebra with a filtration.

14.2 Quotients of algebras

In more mathematical terms, the construction of $\mathsf{U}(\mathfrak{g})$ from TV can be described as follows. By definition, a two-sided ideal of an associative algebra $\mathfrak{A}$ is a subset $\mathcal{I}$ with the property that $x \circ \mathcal{I} \subseteq \mathcal{I}$ and $\mathcal{I} \circ x \subseteq \mathcal{I}$ for any $x \in \mathfrak{A}$. Now consider the (infinite) subset I of TV which consists of all elements of the form $x_1 \otimes x_2 - x_2 \otimes x_1 - [x_1, x_2]$, with arbitrary $x_1, x_2 \in \mathfrak{g}$. Then define $\mathcal{I}$ as the smallest two-sided ideal of the algebra TV that contains the set I. Such an ideal exists because the intersection of any number of ideals is again an ideal. The identifications one has to implement correspond precisely to 'setting to zero' the elements of $\mathcal{I}$. Mathematically, this 'setting to zero' is described by forming the *quotient* $\mathfrak{A}/\mathcal{I}$ of the algebra modulo the ideal; thus we have

$$\mathsf{U}(\mathfrak{g}) \equiv \mathcal{TV}/\mathcal{I}. \tag{14.4}$$

The elements of this quotient are by definition the *equivalence classes* $[a]$ of elements of $\mathcal{TV}$ relative to $\mathcal{I}$; each equivalence class $[a]$ consists of all those elements of $\mathcal{TV}$ which only differ by elements of $\mathcal{I}$, i.e.

$$[a] = \{b \in \mathcal{TV} \,|\, b = a + c \text{ for some } c \in \mathcal{I}\}, \tag{14.5}$$

or in short, $[a] = a + \mathcal{I}$. It follows that $\mathsf{U}(\mathfrak{g})$ is generated algebraically by a unit element $\mathbf{1}$ and (the image of) a basis of $\mathfrak{g}$. Also, if $\mathfrak{g}$ is abelian, then $\mathsf{U}(\mathfrak{g})$ is just the symmetric algebra of $\mathfrak{g}$, i.e. the subalgebra of $\mathcal{TV}$ that consists of symmetrized tensor products.

The set $\{[a]\}$ of equivalence classes, or more generally, the quotient $\mathfrak{A}/\mathfrak{B}$ of any vector space $\mathfrak{A}$ modulo an arbitrary sub-vector space $\mathfrak{B}$, carries the structure of a vector space: Scalar multiplication and vector addition are defined via representatives, i.e. $\lambda[a] = [\lambda a]$ and $[a] + [b] = [a + b]$, and these classes do not depend on the choice of representatives. Also, if $\mathcal{B}$ is a basis of $\mathfrak{A}$ and $\mathcal{B}' \subset \mathcal{B}$ a basis of $\mathfrak{B}$, then a basis of $\mathfrak{A}/\mathfrak{B}$ is given by the equivalence classes of the elements of $\mathcal{B} \setminus \mathcal{B}'$.

By defining a product through $[a] \circ [b] := [a \circ b]$, the vector space $\mathsf{U}(\mathfrak{g})$ as described by (14.4) acquires the structure of an (associative) algebra. The reader can show in exercise 14.1 that this product is well-defined, i.e. that it does not depend on the choice of representatives a, b of the equivalence classes. This is indeed a special example of the general fact (see again exercise 14.1) that the quotient of an algebra modulo an ideal is again an algebra. In contrast, the quotient $\mathfrak{A}/\mathfrak{B}$ of an algebra modulo an arbitrary subalgebra $\mathfrak{B}$ is generically no longer an algebra.

These statements apply not only to associative algebras, but analogously also to arbitrary algebras, and in particular to Lie algebras. Thus for any (proper) ideal $\mathfrak{k}$ of a Lie algebra $\mathfrak{g}$, there is a corresponding quotient $\mathfrak{g}/\mathfrak{k}$, which is again a Lie algebra. For instance, the existence of a Levi decomposition (4.31) of $\mathfrak{g}$ into its radical $\mathfrak{g}_{\mathrm{rad}}$ and a semisimple part tells us – since $\mathfrak{g}_{\mathrm{rad}}$ is an ideal of $\mathfrak{g}$ – that the semisimple part can be obtained as the quotient $\mathfrak{g}/\mathfrak{g}_{\mathrm{rad}}$. We also mention that if $\mathfrak{h}$ is a solvable ideal of a Lie algebra $\mathfrak{g}$ and the quotient $\mathfrak{g}/\mathfrak{h}$ is solvable, then also $\mathfrak{g}$ is solvable; similarly, if the quotient $\mathfrak{g}/\mathcal{Z}(\mathfrak{g})$ of a Lie algebra $\mathfrak{g}$ by its center is nilpotent, then so is $\mathfrak{g}$.

The counterpart of the concept of a quotient of algebras in the theory of groups is the notion of a factor group that we encountered in chapter 10 when studying the structure of the Weyl group. In particular, taking the coset space G/H of a compact Lie group G by some compact Lie subgroup H, the tangent space of G/H at the class of the unit element can be described by the quotient $\mathfrak{g}/\mathfrak{h}$ of the corresponding Lie algebras.

The algebra $\mathsf{U}(\mathfrak{g})$ is called the *universal* enveloping algebra because it possesses (see exercise 14.1) a so-called *universal property*: Given *any* unital associative algebra $\mathfrak{A}'$ for which there exists a linear map f from $\mathfrak{g}$ to $\mathfrak{A}'$ obeying $f(x)\,f(x') - f(x')\,f(x) = f([x,x'])$ for all elements x, x' of $\mathfrak{g}$, there is a *unique* homomorphism $\tilde{f}$ of associative algebras from $\mathsf{U}(\mathfrak{g})$ to $\mathfrak{A}'$ such that $f = \tilde{f} \circ \iota$, or in short, such that any linear map from $\mathfrak{g}$ to $\mathfrak{A}'$ can be uniquely continued to a linear map from $\mathsf{U}(\mathfrak{g})$ to $\mathfrak{A}'$.

Universal properties provide an elegant tool for characterizing various mathematical objects. For example, the tensor product $V \otimes W$ of two vector spaces V and W can be characterized this way. Namely, there is a map '$\otimes$' from the cartesian product $V \times W$ to the tensor product $V \otimes W$,

$$\otimes : \quad (v; w) \;\mapsto\; v \otimes w, \tag{14.6}$$

such that *any* bilinear map f from $V \times W$ to *any* vector space U can be *uniquely* written as $f = \tilde{f} \circ \otimes$, where $\tilde{f}$ is a linear map from $V \otimes W$ to U. In this sense the tensor product 'linearizes' bilinear (and, more generally, multilinear) maps.

In the case of the universal enveloping algebra, the uniqueness of the map $\tilde{f}$ means that the universal property characterizes the universal enveloping algebra completely (up to isomorphism). This can be seen by the following arguments, which in a similar way can be applied also in many other situations where universal properties play a rôle. Assume that there is a second associative algebra U' and a map ι' from $\mathfrak{g}$ to U' which have the same properties as $\mathsf{U}(\mathfrak{g})$ and ι. Due to the universal property of $\mathsf{U}(\mathfrak{g})$, applied to U' and $f := \iota'$, we are given a unique homomorphism $\tilde{f} : \; \mathsf{U}(\mathfrak{g}) \to \mathsf{U}'$, and analogously one finds a unique map $\tilde{f}' : \mathsf{U}' \to \mathsf{U}(\mathfrak{g})$. The composition $\tilde{f}'' := \tilde{f} \circ \tilde{f}'$ then obeys $\tilde{f}'' \circ \iota = \iota$, which by the universal property of $\mathsf{U}(\mathfrak{g})$ implies that it is the identity map $id_{\mathsf{U}(\mathfrak{g})}$. Interchanging the rôles of $\mathsf{U}(\mathfrak{g})$ and U' one also learns that $\tilde{f}' \circ \tilde{f} = id_{\mathsf{U}'}$. Together it follows that $\tilde{f}$ and $\tilde{f}'$ are isomorphisms.

As a nice application of the universal property of $\mathsf{U}(\mathfrak{g})$ we show that there is a one-to-one correspondence between representations of $\mathsf{U}(\mathfrak{g})$ and representations of $\mathfrak{g}$. First, since $\mathsf{U}(\mathfrak{g})$ contains $\mathfrak{g}$, the restriction of any representation of $\mathsf{U}(\mathfrak{g})$ to $\mathfrak{g}$ yields a representation of $\mathfrak{g}$. Conversely, any representation R of the Lie algebra $\mathfrak{g}$ can be regarded as a homomorphism from $\mathfrak{g}$ to $\mathfrak{gl}(V)$, the algebra of all endomorphisms of the vector space V on ·hich the representation acts. Now $\mathfrak{gl}(V)$ is clearly an associative algebra, with the associative product just given by the composition of maps. The fact that R is a representation therefore means that $R(x) \circ R(x') - R(x') \circ R(x) = R([x,x'])$ for all $x, x' \in \mathfrak{g}$. Thus the universal property of $\mathsf{U}(\mathfrak{g})$ can be invoked; it ensures that R can be extended uniquely to a homomorphism $\tilde{R}$ from $\mathsf{U}(\mathfrak{g})$ to $\mathfrak{gl}(V)$, which is nothing else than a representation of $\mathsf{U}(\mathfrak{g})$ acting on V. Since this extension is unique, the representations of $\mathfrak{g}$ and $\mathsf{U}(\mathfrak{g})$ are indeed in one-to-one correspondence. It follows that if V is any $\mathfrak{g}$-module, then the general linear algebra $\mathfrak{gl}(V)$ of V is a homomorphic image of $\mathsf{U}(\mathfrak{g})$. In this sense the universal enveloping algebra is the most comprehensive associative algebra containing $\mathfrak{g}$.

14.3 The Poincaré–Birkhoff–Witt theorem

The universal enveloping algebra is in particular a vector space. For practical purposes, it is extremely useful to be able to describe a basis of this space explicitly. Fortunately, there is a powerful result, the *Poincaré–Birkhoff–Witt theorem*, which provides an explicit prescription for constructing a basis of $U(\mathfrak{g})$.

Consider any basis $\{T^a \mid a \in I\}$ of the Lie algebra $\mathfrak{g}$. Provided that the basis of $\mathfrak{g}$ is countable, we can assume that the basis elements are indexed by some subset $I \subseteq \mathbb{Z}_{>0}$ of the positive integers. The Poincaré–Birkhoff–Witt theorem then states the following:

▮ The set of all elements of $U(\mathfrak{g})$ which are of the form

$$T^{a_1} T^{a_2} \cdots T^{a_k}, \tag{14.7}$$

▮ with a_k a monotonically increasing sequence of k elements of I, i.e.

$$a_1 \leq a_2 \leq a_3 \ldots \leq a_{k-1} \leq a_k, \tag{14.8}$$

▮ and with k any non-negative integer, is a basis of $U(\mathfrak{g})$.

Note that in (14.7) we only allow for finite products (which can however have arbitrarily many factors), since only such elements are contained in the universal enveloping algebra $U(\mathfrak{g})$.

This theorem, which holds for any Lie algebra with a countable basis, is indeed most powerful. For instance, it follows that $U(\mathfrak{g})$ has a subspace that is isomorphic to $\mathfrak{g}$ and on which the commutator of $U(\mathfrak{g})$ reproduces the Lie bracket of $\mathfrak{g}$; this important property can hardly be proven without the Poincaré–Birkhoff–Witt theorem.

Also, one should realize that if $\mathfrak{g}$ is simple or affine, one must e.g. account for the Serre relations (7.3) to obtain $U(\mathfrak{g})$, and generically it is extremely difficult to decide whether the Serre relations imply that a given set of vectors in $U(\mathfrak{g})$ is linearly independent or not. In fact, it is truly non-trivial to show that the vectors (14.7), subject to the restriction (14.8), are linearly independent (for the proof of this property, we refer the interested reader to chapter V.2 of [Jacobson 1962] and to [Dixmier 1996]).

In contrast, it is not too difficult to verify that these vectors span all of $U(\mathfrak{g})$. Namely, one uses the fact that the vector space TV (14.1) is graded according to (14.2), and proceeds by induction in the grading. Trivially, the statement is true for grade zero, where there is only the unit element. Further, assume that the statement has been proven for grade n. The space of elements of $U(\mathfrak{g})$ that come from elements of TV at grade $n + 1$ is clearly spanned by the set of *all* (that is, not necessarily monotonic) sequences of the form (14.7). Using the commutation relations, any non-monotonic sequence in TV at grade $n+1$ can be written as the sum of a monotonic sequence (with the same set of indices) and commutator terms. However, as elements of $U(\mathfrak{g})$, these commutator terms all come from elements of TV at grade n and hence can be expressed by assumption as a linear combination of monotonic sequences that correspond to grade n or less. This shows that the monotonic sequences span all of $U(\mathfrak{g})$.

14.4 Verma modules and null vectors

The universal enveloping algebra $U(\mathfrak{g})$ can be used as a tool to construct representations of $\mathfrak{g}$. For concreteness, the reader can think about $\mathfrak{g}$ as being a finite-dimensional simple (or an affine) Lie algebra, but most of the contents of the present and the next two sections holds in fact for a very large class of Lie algebras, namely for all Kac–Moody algebras, which (compare section 7.8) are associated to arbitrary generalized Cartan matrices.

According to chapter 13 a highest weight vector v_Λ of a $\mathfrak{g}$-module with highest weight Λ is characterized by the properties that it is annihilated by the subalgebra $\mathfrak{g}_+$ of step operators corresponding to positive roots, $\mathfrak{g}_+ v_\Lambda = 0$, and that the Cartan subalgebra $\mathfrak{g}_\circ$ acts by scalar multiplication, $h v_\Lambda \propto v_\Lambda$ for any $h \in \mathfrak{g}_\circ$. (Here and below, we will simplify the notation by writing just the Lie algebra element x in place of its image $R(x)$ in the representation R.) Also, for the moment, we do not require that Λ is a dominant integral weight. All other vectors in a highest weight module can be obtained by applying to the highest weight vector v_Λ suitable elements of the universal enveloping algebra $U(\mathfrak{g}_-)$ of the subalgebra $\mathfrak{g}_-$ of step operators corresponding to negative roots. However, without imposing any further requirements, like e.g. finite-dimensionality in the case of simple Lie algebras, the highest weight does *not* characterize a module uniquely. But one can construct a module $\mathcal{V}_\Lambda$ which automatically contains any highest weight module of $\mathfrak{g}$ with highest weight Λ. One simply applies *all* elements of $U(\mathfrak{g}_-)$ to v_Λ; in other words, as a vector space $\mathcal{V}_\Lambda$ is isomorphic to $U(\mathfrak{g}_-)$.

The subalgebra $\mathfrak{g}_-$ acts on $\mathcal{V}_\Lambda$ as it acts on $U(\mathfrak{g}_-)$: $y \in \mathfrak{g}_-$ maps the vector $x v_\Lambda$ with $x \in U(\mathfrak{g}_-)$ to the vector $y x v_\Lambda$. The action of the Cartan subalgebra $\mathfrak{g}_\circ$ of $\mathfrak{g}$ is such that as compared to the action of $\mathfrak{g}_\circ$ on $U(\mathfrak{g}_-)$ the eigenvalues are shifted by the highest weight Λ: If $h \in \mathfrak{g}_\circ$ and $x \in U(\mathfrak{g}_-)$ obey $[h, x] = -\lambda_x \cdot x$, we set $h(x v_\Lambda) = (\Lambda - \lambda_x) \cdot v_\Lambda$. The action of $\mathfrak{g}_+$ is uniquely determined by these properties and the requirement that $\mathfrak{g}_+$ annihilates the highest weight vector, i.e. $y v_\Lambda = 0$. Namely, when acting with $y \in \mathfrak{g}_+$ on the vector $x v_\Lambda$ with $x \in U(\mathfrak{g}_-)$, we can use the commutation relations in $U(\mathfrak{g})$ to permute y and x, and afterwards we can use the information that y annihilates v_Λ.

The module $\mathcal{V}_\Lambda$ is called the *Verma module* with highest weight Λ; it plays an important rôle in the analysis of the highest weight modules of $\mathfrak{g}$. To make the definition of $\mathcal{V}_\Lambda$ more concrete, let us describe it in the case of $\mathfrak{g} = \mathfrak{sl}(2)$. Then $\mathfrak{g}_-$ is one-dimensional and generated by a single element E_-, so that according to the Poincaré–Birkhoff–Witt theorem a basis of $U(\mathfrak{g}_-)$ is given by $(E_-)^n$, where n takes as values all non-negative integers. The Verma module $\mathcal{V}_\Lambda$ with highest weight Λ is therefore spanned by all vectors of the form

$$v_{\Lambda-2n} := (E_-)^n v_\Lambda , \quad n = 0, 1, 2, \ldots , \tag{14.9}$$

on which $\mathfrak{g}$ acts as (compare equation (2.17))

$$E_- v_{\Lambda-2n} = v_{\Lambda-2n-2} , \qquad H v_{\Lambda-2n} = (\Lambda - 2n)\, v_{\Lambda-2n} ,$$
$$E_+ v_{\Lambda-2n} = n(\Lambda - n + 1)\, v_{\Lambda-2n+2} . \tag{14.10}$$

The concept of a Verma module may appear unnecessarily complicated to some readers. Also, as long as one is interested only in finite-dimensional $\mathfrak{g}$-modules, which cover many interesting applications, one can easily dispense of this concept. However, already the proof of the Weyl–Kac character formula which will be addressed below shows that Verma modules constitute a rather deep and fundamental concept in the representation theory of Lie algebras, and the efforts needed to understand this concept are really worthwhile. Furthermore, the notion of a null vector, which is closely related to that of Verma modules, has turned out to be central for many applications in modern mathematical physics.

Verma modules are generically not irreducible, nor even fully reducible, i.e. cannot be written as a direct sum of irreducible modules. To see this, let us have a look at the Verma modules of $\mathfrak{g} = \mathfrak{sl}(2)$ which are described by equations (14.9) and (14.10). The last relation in (14.10) allows us to determine all highest weight modules contained in $\mathcal{V}_\Lambda$: We have $E_+ v_{\Lambda-2n} = 0$ if either $n = 0$ – this is the highest weight vector we started with – or else $n = \Lambda + 1$, which occurs only if the highest weight Λ is a non-negative integer. In the latter case, $v_{-\Lambda-2}$ is a highest weight vector, too, and the corresponding highest weight Verma module $\mathcal{V}_{-\Lambda-2}$ is nested in $\mathcal{V}_\Lambda$. Thus $\mathcal{V}_\Lambda$ has a submodule, and hence cannot be irreducible. In fact, a reducible Verma module is never fully reducible, since any submodule $\mathcal{V}'$ of $\mathcal{V}_\Lambda$ containing the highest weight vector v_Λ is already the whole Verma module (the latter argument does not use any special properties of $\mathfrak{sl}(2)$ and therefore applies also in the general case).

We have already described in chapter 4 how the irreducible module V_Λ looks like for $\mathfrak{sl}(2)$: It is spanned by the vectors $v_\Lambda, v_{\Lambda-2}, \ldots, v_{-\Lambda}$. Therefore it can be written as the *quotient*

$$V_\Lambda = \mathcal{V}_\Lambda / \mathcal{V}_{-\Lambda-2} \tag{14.11}$$

of $\mathcal{V}_\Lambda$ modulo its maximal submodule $\mathcal{V}_{-\Lambda-2}$. Such a quotient is constructed as the quotient of vector spaces, i.e. the vectors $v_\Lambda, v_{\Lambda-2}, \ldots, v_{-\Lambda}$ of the irreducible module are regarded as representatives of the equivalence classes of $\mathcal{V}_\Lambda$ modulo $\mathcal{V}_{-\Lambda-2}$; similarly as in the case of algebras, this quotient of vector spaces again carries the structure of a $\mathfrak{g}$-module. In a sense, in the transition from the Verma module to its irreducible quotient all vectors in $\mathcal{V}_{-\Lambda-2}$ are set to zero; they are therefore called *null vectors* of the module $\mathcal{V}_\Lambda$.

In this example we can make a few observations which may seem trivial, but which generalize to arbitrary $\mathfrak{g}$. First, since all v_l with $l \le -\Lambda - 2$ are null vectors, acting with an element of $\mathsf{U}(\mathfrak{g}_-)$ on a null vector yields another null vector. But there is a specific null vector, namely $v_{-\Lambda-2}$, on which the action of $\mathsf{U}(\mathfrak{g}_+)$ yields zero already in the Verma module, and hence in particular does not produce another null vector. Vectors with this property are called *primitive* null vectors. Finally, adding the Weyl vector (which for $\mathfrak{sl}(2)$ is simply $\rho = 1$) to both highest weights of $\mathcal{V}_\Lambda$ yields $\Lambda + \rho = \Lambda + 1$ and $-\Lambda - 2 + \rho = -\Lambda - 1$; these two weights are related just by a sign flip, or in other words, by the action of the Weyl group of $\mathfrak{sl}(2)$. This property turns out to be a crucial ingredient in the proof of the Weyl–Kac character formula. It is not at all surprising that these features generalize to arbitrary Kac–Moody algebras $\mathfrak{g}$. One simply has to apply the arguments (as we already did in various other situations) to the $\mathfrak{sl}(2)$-subalgebras of $\mathfrak{g}$ which for any $\mathfrak{g}$-root α are spanned by the step operators $E^{\pm\alpha}$ and the Cartan subalgebra element $H^\alpha = (\alpha^\vee, H)$.

Verma modules over a Lie algebra $\mathfrak{g}$ are *free* modules over $\mathfrak{g}_-$ (the algebra spanned by the step operators for all negative roots), i.e. the only relations one imposes are those of $\mathfrak{g}_-$ itself. As a consequence, Verma modules have a number of nice properties. For example, the vector space of homomorphisms from one Verma module $\mathcal{V}_\lambda$ to some other Verma module $\mathcal{V}_\mu$ is either zero- or one-dimensional, and any non-trivial homomorphism from $\mathcal{V}_\lambda$ to $\mathcal{V}_\mu$ is injective. As we have mentioned, Verma modules are not necessarily fully reducible. It is a non-trivial consequence of the triangular decomposition of Kac–Moody Lie algebras and of the fact that the weight space for the highest weight is one-dimensional that every Verma module of a Kac–Moody algebra has a unique irreducible quotient which is an irreducible highest weight module. Namely, any Verma module $\mathcal{V}_\Lambda$ of a Kac–Moody algebra has a unique maximal submodule $\mathcal{M}_\Lambda$, which is strictly contained in $\mathcal{V}_\Lambda$ (but might be the trivial submodule $\{0\}$); $\mathcal{M}_\Lambda$ is generically *not* a Verma module. Then the quotient $V_\Lambda := \mathcal{V}_\Lambda/\mathcal{M}_\Lambda$ cannot contain any non-trivial submodule and hence is irreducible. In fact, also the converse is true: Each irreducible highest weight module of a Kac–Moody algebra can be constructed this way.

As already pointed out, any vector $x\, v_\mu$ with v_μ a null vector and $x \in \mathsf{U}(\mathfrak{g}_-)$ is again a null vector. This holds in fact whenever $\mathfrak{g}$ has a triangular decomposition. Therefore the *primitive* null vectors play a crucial rôle also in the description of arbitrary Verma modules. These vectors are characterized by the fact that the action of $\mathsf{U}(\mathfrak{g}_+)$ does not yield another null vector. Acting on a primitive null vector with any element of $\mathfrak{g}_+$ must in fact yield zero; therefore any primitive null vector v_μ fulfills the defining requirements of a highest weight vector, and hence $\mathcal{V}_\Lambda$ contains the Verma module $\mathcal{V}_\mu$ as a submodule. Conversely, any highest weight vector in the Verma module $\mathcal{V}_\Lambda$ other than v_Λ is a primitive null vector either of $\mathcal{V}_\Lambda$ or of one of its Verma submodules.

14.5 The character of a Verma module

Just like an irreducible highest weight module, any Verma module of a

Kac–Moody algebra is characterized up to isomorphism by its character. This character can be computed with the help of the Poincaré–Birkhoff–Witt theorem. To this end we first fix a basis of $U(\mathfrak{g}_-)$; we consider all positive $\mathfrak{g}$-roots β (there may be infinitely many of them, as is e.g. the case when $\mathfrak{g}$ is affine) in some arbitrary but fixed order, i.e. write them as $\beta_1, \beta_2, \ldots$. A basis of the root space $\mathfrak{g}_{-\beta_\ell}$ is spanned by step operators E^{β_ℓ, p_ℓ}, where $1 \le p_\ell \le m_\ell := \operatorname{mult} \beta \equiv \dim(\mathfrak{g}_{-\beta_\ell})$. (For simple Lie algebras the roots are non-degenerate, so in this case we can drop the multiplicity index p_ℓ. In general, however, the roots can have multiplicities $\dim(\mathfrak{g}_{-\beta_\ell})$ larger than one. For example, in section 12.6 we have seen that the imaginary roots of affine Lie algebras have multiplicity $\mathrm{r} = \operatorname{rank} \bar{\mathfrak{g}}$, because all lowering operators H_n^i, $i = 1, 2, \ldots, \mathrm{r}$, for fixed n correspond to the same root $n\delta$. In the present framework, these step operators H_n^i are denoted by E^{β_ℓ, p_ℓ} with $\beta_\ell = n\delta$ and $p_\ell = i \in \{1, 2, \ldots, \mathrm{r}\}$.)

The Poincaré–Birkhoff–Witt theorem tells us that the vectors

$$(E^{-\beta_1, 1})^{n_{1,1}} \cdots (E^{-\beta_1, m_1})^{n_{1,m_1}} (E^{-\beta_2, 1})^{n_{2,1}} \cdots (E^{-\beta_2, m_2})^{n_{2,m_2}} \cdots v_\Lambda \quad (14.12)$$

that are obtained by applying any finite monotonic sequence of lowering operators to the highest weight vector v_Λ, form a basis of the Verma module $\mathcal{V}_\Lambda$. The vector (14.12) has weight

$$\lambda = \Lambda - (n_{1,1} + \ldots + n_{1,m_1})\,\beta_1 - (n_{2,1} + \ldots + n_{2,m_2})\,\beta_2 + \ldots \quad (14.13)$$

This implies the formula $\mathcal{X}_\Lambda = e^\Lambda \prod_{\alpha > 0} (1 + e^{-\alpha} + e^{-2\alpha} + \ldots)^{\operatorname{mult} \alpha}$ for the character of $\mathcal{V}_\Lambda$ (see exercise 14.6); here the exponents take care of the degeneracies of the roots. In this formula the geometric series in the formal exponential $e^{-\alpha}$ can be summed up for each factor, leading to

$$\mathcal{X}_\Lambda = e^\Lambda \prod_{\alpha > 0} (1 - e^{-\alpha})^{-\operatorname{mult} \alpha} . \quad (14.14)$$

In the case of $\mathfrak{sl}(2)$, this formula reduces to $\mathcal{X}_\Lambda(\mu) = e^{\Lambda \mu/2}/(1 - e^{-\mu})$, where $\mu \in \mathbb{C}$ is a weight of $\mathfrak{sl}(2)$. (Recall from section 6.12 that the Cartan matrix of $\mathfrak{sl}(2)$ is just the number 2, so that its inverse, the metric on the weight space, is equal to $\frac{1}{2}$. The scalar product (Λ, μ) can therefore be expressed as a product of numbers as $(\Lambda, \mu) = \frac{1}{2}\Lambda\mu$.) Furthermore, the character of a quotient V/W of modules is obtained by subtracting the character of the submodule W from the character of V. It follows that the character of the irreducible $\mathfrak{sl}(2)$-module V_Λ is given by

$$\chi_\Lambda(\mu) = \mathcal{X}_\Lambda(\mu) - \mathcal{X}_{-\Lambda-2}(\mu) = \frac{e^{(\Lambda+1)\mu/2} - e^{-(\Lambda+1)\mu/2}}{e^{\mu/2} - e^{-\mu/2}}, \quad (14.15)$$

in agreement with the formula (13.25).

Note that $\mathcal{X}_\Lambda$ depends on the highest weight Λ only by a multiplicative factor e^Λ. For the irreducible module constructed from a Verma module by quotienting out its maximal submodule, the dependence of the character on the highest weight is, however, much less trivial. As a first step towards finding that character, let us determine the primitive null vectors of V_Λ, which is also interesting in its own right.

Consider first the case of a simple Lie algebra $\mathfrak{g}$ and a dominant integral $\mathfrak{g}$-weight Λ. Then the relations to be quotiented out are just those which require that V_Λ splits into finite-dimensional modules with respect to the $\mathfrak{sl}(2)$-subalgebras of $\mathfrak{g}$ corresponding to any root. Of these, only the relations involving the simple $\mathfrak{g}$-roots are independent; it follows that the primitive null vectors are

$$v_{\mu,i} := \left(E^{-\alpha^{(i)}}\right)^{\Lambda^i+1} \cdot v_\Lambda \tag{14.16}$$

for $i = 1, 2, \ldots, r \equiv \mathrm{rank}\,\mathfrak{g}$. Thus the entire information on the null vector structure of V_Λ is encoded directly in its Dynkin labels Λ^i (this illustrates once more how convenient it is to describe weights in the Dynkin basis).

For affine Lie algebras $\mathfrak{g}$, there are no longer any non-trivial finite-dimensional modules. (In any finite-dimensional module each element of $\mathfrak{g}$ is represented by the zero map.) As described in section 13.9, the irreducible quotient module V_Λ of $\mathcal{V}_\Lambda$ is *integrable* if and only if it decomposes into finite-dimensional modules with respect to the $\mathfrak{sl}(2)$-subalgebras corresponding to the simple $\mathfrak{g}$-roots. As a consequence, the primitive null vectors of $\mathcal{V}_\Lambda$ are again given by the formula (14.16), where, however, i now takes values in $\{0, 1, \ldots, r\}$. Because of the relation between the affine Lie algebra $\mathfrak{g}$ and its zero mode subalgebra $\mathfrak{g}_{[0]}$, the quotienting of the null vectors $v_{\mu,i}$ with $i = 1, 2, \ldots, r$ can be implemented by just considering only irreducible rather than Verma modules with respect to $\mathfrak{g}_{[0]}$. Furthermore, the primitive null vector corresponding to the zeroth simple root can be written more concretely by expressing $\alpha^{(0)}$ and Λ^0 through the formulæ (12.36) and (13.49):

$$v_{\mu,0} = (E^{-\alpha^{(0)}})^{\Lambda^0+1} \cdot v_\Lambda = (E^{\bar\theta}_{-1})^{k^\vee - (\bar\theta^\vee, \bar\Lambda)+1} \cdot v_\Lambda . \tag{14.17}$$

14.6 Null vectors and unitarity

The existence of null vectors in the modules of simple or affine Lie algebras is closely tied to the unitarity property of the modules. Namely, for any hermitian product $(\cdot \mid \cdot)$ on $\mathcal{V}_\Lambda$ and any elements $v_\lambda, v_\nu \in \mathcal{V}_\Lambda$ one has

$$(v_\lambda \mid v_\mu) = (v_\Lambda \mid x^\dagger y\, v_\Lambda) \tag{14.18}$$

with some elements x, y in the enveloping algebra $\mathsf{U}(\mathfrak{g}_-)$, because v_λ and v_ν can be written as $v_\lambda = x\, v_\Lambda$, $v_\nu = y\, v_\Lambda$. Recall that in this chapter we use the short hand notation x in place of $R(x)$; thus $x^\dagger$ stands for $R^\dagger(x)$, the matrix adjoint of the matrix $R(x)$. If the module is unitary, then $R^\dagger(x) = R(\omega(x))$, with ω the Chevalley anti-involution of $\mathfrak{g}$ (compare (13.21)) which can be uniquely extended to $\mathsf{U}(\mathfrak{g})$; conversely, inserting this equality as an ansatz, one obtains restrictions on the highest weight Λ which are necessary conditions for the unitarity of the module. Now ω acts as $\omega(E_\pm^i) = E_\mp^i$ and $\omega(H^i) = H^i$, so that

$$(E_\pm^i)^\dagger = E_\mp^i, \qquad (H^i)^\dagger = H^i. \tag{14.19}$$

Since a primitive null vector is annihilated by any operator in the enveloping algebra of $\mathfrak{g}_+$, it follows that the product of a primitive null vector $v_\mu = x v_\Lambda$ with $x \in \mathsf{U}(\mathfrak{g}_-)$ (except for the highest weight vector v_Λ for which x would be 1) with any vector v_λ vanishes:

$$(v_\lambda \,|\, v_\mu) = (x\,v_\Lambda \,|\, v_\mu) = (v_\Lambda \,|\, x^\dagger\,v_\mu) = 0. \tag{14.20}$$

In particular, the norm $(v_\mu \,|\, v_\mu)$ of a primitive null vector v_μ vanishes; this is another reason for the term *null* vector. It also follows that on the Verma module $\mathcal{V}_\Lambda$ containing v_μ the hermitian product $(\,\cdot\,|\,\cdot\,)$ is degenerate. A necessary condition for the existence of a non-degenerate hermitian product on a highest weight module is therefore that all null vectors are projected out, i.e. that the module is the irreducible quotient module V_Λ.

On the other hand, starting from $(v_\Lambda \,|\, v_\Lambda) \neq 0$ one can calculate all products with the help of (14.18) (commuting $x^\dagger$ through y so that it acts on v_Λ which it annihilates). For example (see exercise 14.4), the length squared of $E_-^i\, v_\Lambda$ is $(E_-^i v_\Lambda \,|\, E_-^i v_\Lambda) = \Lambda^i \cdot (v_\Lambda \,|\, v_\Lambda)$. This way one can prove that for Λ dominant integral, irreducibility is also sufficient for the existence of a positive definite hermitian product. The same sort of calculation also shows that no positive definite hermitian product can be defined for highest weights which are not dominant integral. Thus the unitarizable highest weight modules of a simple or affine Lie algebra are precisely the irreducible modules with dominant integral highest weight.

14.7 The quadratic Casimir operator

Let us restrict now to the case of finite-dimensional simple $\mathfrak{g}$. Then the universal enveloping algebra contains a set of elements which are quite interesting because they can be used to label the representations of $\mathfrak{g}$. Recall from the theory of angular momentum in quantum mechanics (cf. chapter 2) that the square

$$\vec{L}^2 \equiv L_1^2 + L_2^2 + L_3^2 = \tfrac{1}{4}L_0^2 + \tfrac{1}{2}\left(L_+ L_- + L_- L_+\right) \tag{14.21}$$

of the angular momentum vector commutes with all generators L_k, and hence also with any element in the corresponding universal enveloping algebra. In other words, $\vec{L}^2$ is an element of the center of $\mathsf{U}(\mathfrak{g})$. Therefore $\vec{L}^2$ acts as the multiple $\tfrac{1}{4}\Lambda(\Lambda + 2)$ of the identity on all vectors of any highest weight module V_Λ; this is in particular true for Verma modules and their irreducible quotients. As a consequence, this element is very useful when one wants to decompose modules into irreducible modules: As soon as two eigenvectors of $\vec{L}^2$ in a module have different eigenvalues, one knows that this module is reducible. One can even distinguish *all* inequivalent irreducible representations of $\mathfrak{sl}(2)$ by the eigenvalue of $\vec{L}^2$. In short: The finite-dimensional irreducible representations of $\mathfrak{sl}(2)$ are completely characterized by their *spin* eigenvalue $j(j+1) = \frac{1}{4}\Lambda(\Lambda+2)$.

It is natural to try to generalize the operator $\vec{L}^2$ to arbitrary finite-dimensional simple Lie algebras. By comparing equation (14.21) to the explicit form of the Killing form of $\mathfrak{sl}(2)$ (see exercise 14.3), one is led to consider the element

$$\mathcal{C}_2 := \sum_{a,b=1}^{d} \kappa_{ab}\, T^a T^b \tag{14.22}$$

of $\mathsf{U}(\mathfrak{g})$, which is known as the *quadratic* (or second order) *Casimir operator* or Casimir element.

Let us verify that $\mathcal{C}_2$ commutes with any element of $\mathsf{U}(\mathfrak{g})$, or in other words, that it is indeed an element of the center of the associative algebra $\mathsf{U}(\mathfrak{g})$. We will perform this calculation in a Chevalley–Serre basis. Using the results of section 6.10 on how the Killing form looks like in this basis, and taking the Killing form normalized such that the constants n_α defined in (8.4) are

$$n_\alpha = n_{-\alpha} = \tfrac{1}{2}\,(\alpha,\alpha)\,, \tag{14.23}$$

we can rewrite the Casimir element (14.22) as

$$\mathcal{C}_2 = \mathcal{C}_2^{(0)} + \mathcal{C}_2^{(\Phi)}\,, \qquad \mathcal{C}_2^{(0)} := \sum_{i,j=1}^{r} G_{ij} H^i H^j\,, \qquad \mathcal{C}_2^{(\Phi)} := \sum_{\alpha\in\Phi} n_\alpha E^\alpha E^{-\alpha}\,. \tag{14.24}$$

We will compute the commutators of $\mathcal{C}_2^{(0)}$ and $\mathcal{C}_2^{(\Phi)}$ with the basis elements of $\mathfrak{g}$ separately. First we have $[H^i, \mathcal{C}_2^{(0)}] = 0$ because $\mathcal{C}_2^{(0)}$ contains only generators in the Cartan subalgebra. Moreover, according to the identity

$$[H^i, E^\alpha E^{-\alpha}] = [H^i, E^\alpha]E^{-\alpha} + E^\alpha[H^i, E^{-\alpha}] = \alpha^i(E^\alpha E^{-\alpha} - E^\alpha E^{-\alpha}) = 0\,,$$

the combination $\mathcal{C}_2^{(\Phi)}$ commutes with H^i, too, and hence so does $\mathcal{C}_2$. The commutators with the step operators E^α require a bit more work. One finds

$$[\mathcal{C}_2^{(0)}, E^\alpha] = [\textstyle\sum_{i,j} G_{ij} H^i H^j, E^\alpha] = (\alpha, H)\, E^\alpha + E^\alpha\, (\alpha, H) = \frac{(\alpha,\alpha)}{2}\,(H^\alpha E^\alpha + E^\alpha H^\alpha)\,,$$

$$[\mathcal{C}_2^{(\Phi)}, E^\alpha] = \textstyle\sum_{\beta\in\Phi} n_\beta\,[E^\beta E^{-\beta}, E^\alpha] = n_\alpha E^\alpha H^{-\alpha} + n_{-\alpha} H^{-\alpha} E^\alpha$$

$$+ \textstyle\sum_{\beta\neq\pm\alpha} n_\beta e_{-\beta,\alpha} E^\beta E^{\alpha-\beta} + \sum_{\beta\neq\pm\alpha} n_\beta e_{\beta,\alpha} E^{\alpha+\beta} E^{-\beta}\,.$$

Because of the formula (14.23) the first two terms in the second commutator cancel against the first commutator. Further, by shifting the summation variable from β to $\gamma := \alpha + \beta$ and then using the relations (6.67) for the structure constants $e_{\alpha,\beta}$, the last term can be rewritten as

$$\sum_\gamma n_{\gamma-\alpha} e_{\gamma-\alpha,\alpha} E^\gamma E^{\alpha-\gamma} = \sum_\gamma n_{\gamma-\alpha} \frac{(\gamma,\gamma)}{(\alpha-\gamma,\alpha-\gamma)} e_{\alpha,-\gamma} E^\gamma E^{\alpha-\gamma}$$

$$= -\sum_\beta n_{\beta-\alpha} \frac{(\beta,\beta)}{(\alpha-\beta,\alpha-\beta)} e_{-\beta,\alpha} E^\beta E^{\alpha-\beta}\,. \tag{14.25}$$

Using again (14.23), one also has

$$n_{\beta-\alpha} \frac{(\beta,\beta)}{(\alpha-\beta,\alpha-\beta)} = \frac{(\alpha-\beta,\alpha-\beta)}{2} \frac{(\beta,\beta)}{(\alpha-\beta,\alpha-\beta)} = \frac{(\beta,\beta)}{2} = n_\beta \,. \tag{14.26}$$

As a consequence, the remaining terms cancel each other as well. This shows that $C_2^{(\Phi)}$, and hence also C_2, commutes with all step operators. Thus the second order Casimir operator indeed commutes with any element of $\mathfrak{g}$, and hence of $U(\mathfrak{g})$.

14.8 Quadratic Casimir eigenvalues

The quadratic Casimir operator C_2 commutes with all elements of $\mathfrak{g}$; Schur's Lemma (see section 5.4) therefore implies that C_2 acts as a constant on each irreducible module. In particular,

$$C_2 \, v = C_\Lambda \cdot v \quad \text{for all } v \in V_\Lambda \tag{14.27}$$

for irreducible highest weight modules. To compute the eigenvalue C_Λ, we use the relation $[E^\alpha, E^{-\alpha}] = H^\alpha \equiv (\alpha^\vee, H)$ and the definition $\rho = \frac{1}{2}\sum_{\alpha>0} \alpha$ of the Weyl vector to write C_2 as

$$C_2 = \sum_{i,j} G_{ij} H^i H^j + \sum_\alpha n_\alpha E^\alpha E^{-\alpha} = (H,H) + \sum_{\alpha>0} n_\alpha E^{-\alpha} E^\alpha + 2\,(\rho,H)\,.$$

As C_Λ is constant on the whole module V_Λ, we can determine it by applying C_2 to any vector v in V_Λ. When taking the highest weight vector $v = v_\Lambda$, the terms with step operators do not contribute because v_Λ is annihilated by $\mathfrak{g}_+$. We then find that

$$C_\Lambda = (\Lambda, \Lambda + 2\rho)\,. \tag{14.28}$$

For the adjoint module $V_{\mathrm{ad}} \equiv V_\theta$, the quadratic Casimir eigenvalue can also be obtained by inserting the expression (5.14) for the generators $R_\theta(T^a)$ in terms of the structure constants into the definition (14.22) of the operator C_2. This leads to

$$C_\theta\,\delta_c^d = \sum_{a,b=1}^{\mathrm{d}} f^{ab}{}_c\, f_{ba}{}^d\,. \tag{14.29}$$

The Casimir eigenvalue C_θ in the adjoint module is related via

$$C_\theta = (\theta,\theta)\cdot g^\vee \tag{14.30}$$

to the dual Coxeter number $g^\vee$ of $\mathfrak{g}$ (see exercise 14.5). (When comparing the values for C_θ that are obtained when inserting the explicit values for $g^\vee$ from table V with the literature, one must be aware of the fact that the normalization (θ,θ) of the long roots used in physics is often $(\theta,\theta) = 1$, while in mathematics the normalization $(\theta,\theta) = 2$ is common.)

So far the Lie algebra $\mathfrak{g}$ was assumed to be finite-dimensional and simple. One might wonder whether an object like the second order Casimir operator exists also for more general Kac–Moody algebras, like e.g. affine Lie algebras. One ingredient that is definitely needed is an analogue of the quadratic form matrix G_{ij}, and hence we have to require that the Kac–Moody algebra is symmetrizable (see the discussion in section 7.8). Our considerations will therefore apply in particular to simple, affine and hyperbolic Lie algebras. In the general case the second term in equation (14.24) is an infinite sum, so that an analogue of the Casimir operator cannot be any more an element of the enveloping algebra $\mathsf{U}(\mathfrak{g})$, which by definition contains only *finite* linear combinations. Nevertheless, also in the general case the formal operator

$$\tilde{C}_2 := 2\,(\rho\,|\,H) + \sum_i (u^i\,|\,u_i) + 2\sum_{\alpha>0}\sum_{\ell} E_-^{\alpha,\ell} E_+^{\alpha,\ell} \qquad (14.31)$$

proves to be a sensible generalization. In (14.31) the u_i stand for the elements of an arbitrary basis of the Cartan subalgebra $\mathfrak{g}_\circ$ (and the u^i for its dual basis), and the sum over ℓ in the last term takes care of possible degeneracies of the roots. While already for affine $\mathfrak{g}$ the summation over the roots is infinite, only finitely many terms in the sum give a non-zero contribution when $\tilde{C}_2$ acts on any given vector in a highest weight module, so that this action is still well-defined. (On the other hand, on arbitrary modules which are not highest weight modules the action of $\tilde{C}_2$ need not be well-defined.) The generalized Casimir operator (14.31) plays in fact a crucial rôle in the representation theory of symmetrizable Kac–Moody algebras. Also, many properties that C_2 has in the simple case carry over; in particular the calculations in section 14.7 go through almost literally, and hence $\tilde{C}_2$ has the constant eigenvalue $(\Lambda+2\rho\,|\,\Lambda)$ on the Verma module $\mathcal{V}_\Lambda$ and on its irreducible quotient V_Λ.

*14.9 The second order Dynkin index

Comparing (14.29) with the expression (6.14) of the Killing form in terms of the structure constants, it follows that the normalization constant I_{ad} that was introduced in (6.12) coincides with the quadratic Casimir eigenvalue in the adjoint representation:

$$I_{\mathrm{ad}} \equiv I_\theta = C_\theta\,. \qquad (14.32)$$

As explained in section 8.1, instead of via the adjoint representation (as was done in chapter 6) one may describe the Killing form also through the representation matrices in any other highest weight representation, according to

$$\mathrm{tr}\,(R_\Lambda(T^a)R_\Lambda(T^b)) = \frac{I_\Lambda}{I_{\mathrm{ad}}}\,\kappa(T^a,T^b) \equiv I_\Lambda\,\kappa^{ab}\,. \qquad (14.33)$$

The constant of proportionality arising here, denoted by $I_\Lambda \equiv I_{R_\Lambda}$, is called the (second order) *Dynkin index* of the representation R_Λ. As can be checked by contracting (14.33) with κ_{ab}, the relation (14.32) between Dynkin index and Casimir eigenvalue generalizes to

$$I_\Lambda = \frac{d_\Lambda}{d}\,C_\Lambda\,, \qquad (14.34)$$

where d_Λ is the dimension of R_Λ. One can also show that

$$I_\Lambda = \frac{1}{r} \sum_\lambda \text{mult}_\Lambda(\lambda) \cdot (\lambda, \lambda) \,. \tag{14.35}$$

The quadratic Casimir eigenvalue in the adjoint representation can therefore be written as

$$C_\theta = \frac{1}{r} \sum_{\alpha \in \Phi} (\alpha, \alpha) = \frac{1}{r} (\theta, \theta) \cdot \left(n_L + \left(\tfrac{S}{L} \right)^2 n_S \right) , \tag{14.36}$$

where n_L and n_S are the numbers of long and short roots of the (simple) Lie algebra $\mathfrak{g}$, respectively, and S/L is the ratio of their lengths. For simply laced $\mathfrak{g}$ there are just $d - r$ long roots, so that this formula simplifies to $C_\theta = (\frac{d}{r} - 1)(\theta, \theta)$.

For applications in physics the second order Dynkin index is often a more natural quantity than the quadratic Casimir. For example, in four-dimensional gauge theories, I_Λ gives (up to a representation independent constant) the contribution of quantum fields carrying the representation R_Λ to the one-loop renormalization group β-function. I_Λ is also equal to the number of zero modes of the Dirac equation for (spin-$\frac{1}{2}$) fermions carrying the representation R_Λ in the background of an instanton.

<u>Information</u>

14.10 The character formula

The concepts of a Verma module and the Weyl group form the basic ingredients needed to prove a central result in the representation theory of Lie algebras, the Weyl–Kac character formula for irreducible highest weight modules with dominant integral highest weight. This formula is valid for a huge class of Lie algebras, both finite- and infinite-dimensional, namely for all Kac–Moody algebras whose generalized Cartan matrix is symmetrizable, so that the generalized second order Casimir operator exists. We will describe the proof for finite-dimensional simple Lie algebras, where it reduces to the Weyl character formula (13.30), but in a manner that can be translated almost literally into a proof for the general case (e.g. we include multiplicities $\text{mult}\,\alpha$ of the roots, even though for finite-dimensional simple Lie algebras all of them are equal to 1).

To start, we observe that the character $\mathcal{X}_\Lambda$ of the Verma module $\mathcal{V}_\Lambda$ can be written as the sum

$$\mathcal{X}_\Lambda = \sum_p c_p \, \mathcal{X}_{\lambda_p} \tag{14.37}$$

over various characters $\mathcal{X}_{\lambda_p}$ of irreducible modules, where the coefficients c_p are integers, and where λ_p are weights of $\mathcal{V}_\Lambda$ which are not necessarily dominant integral. This decomposition can be proven by induction on the depth of the weights (in fact, the integers c_p are non-negative, but to show this requires considerably more input). In order to contribute to the sum (14.37), λ_p must be a weight of $\mathcal{V}_\Lambda$, so that the difference $\Lambda - \lambda_p$ is a linear combination of simple roots with positive integral coefficients, i.e., using the notation '$\leq$' to denote this situation (compare section 13.3), $\lambda_p \leq \Lambda$. Furthermore, as the second order Casimir operator has constant value (14.28) on the whole Verma module $\mathcal{V}_\Lambda$, only those weights λ_p can contribute which obey $|\lambda_p + \rho|^2 = |\Lambda + \rho|^2$, where $|\mu|^2 \equiv (\mu, \mu)$. Therefore we introduce the set

$$B(\Lambda) := \{\lambda \mid \lambda \leq \Lambda, \ |\lambda + \rho|^2 = |\Lambda + \rho|^2\} \tag{14.38}$$

of weights. This set contains countably many weights; without loss of generality we can also assume that the weights in $B(\Lambda)$ are labelled by integers such that $\lambda_q \leq \lambda_p$ implies $p \leq q$.

An important observation is now that a decomposition like in (14.37) can be written down for the character of the Verma module also when the highest weight is any weight in $B(\Lambda)$, i.e. $\mathcal{X}_{\lambda_p} = \sum_q c_{pq} \mathcal{X}_{\lambda_q}$ with certain integers c_{pq}. These integers possess two crucial properties: First, since the highest weight of any Verma module has multiplicity 1, we have $c_{pp} = 1$ for all p; second, $c_{pq} \neq 0$ is only possible if λ_q is a weight in the Verma module with highest weight λ_p, i.e. if $\lambda_q \leq \lambda_p$, which with the chosen ordering of $B(\Lambda)$ implies that $p \leq q$. Thus the (infinite) matrix (c_{pq}) with integer entries is upper triangular, and all its diagonal entries are equal to 1. Such matrices are invertible, and the inverse contains again only integer entries and all its diagonal entries are equal to one. Applying this inversion to the particular weight $\Lambda \in B(\Lambda)$, one obtains

$$\mathcal{X}_\Lambda = \sum_{\lambda \in B(\Lambda)} \tilde{c}_\lambda \mathcal{X}_\lambda, \tag{14.39}$$

where the coefficients $\tilde{c}_\lambda$ are integers and $\tilde{c}_\Lambda = 1$. In short: The character of the irreducible highest weight module with highest weight Λ can be written as a linear combination of the characters of the Verma modules with highest weights in the set (14.38).

To compute the coefficients $\tilde{c}_\lambda$, we make use of the Weyl group W of $\mathfrak{g}$; more precisely, we apply elements $w \in W$ to both sides of equation (14.39). From the formula (13.29) we know already that $\mathcal{X}_\Lambda$ is symmetric with respect to the Weyl group, and hence the same must hold for the right hand side. Now we write the right hand side in the form

$$\sum_{\lambda \in B(\Lambda)} \tilde{c}_\lambda \mathcal{X}_\lambda = \Big(\sum_{\lambda \in B(\Lambda)} \tilde{c}_\lambda e^{\lambda + \rho} \Big) \cdot \big(e^{-\rho - \lambda} \mathcal{X}_\lambda \big), \tag{14.40}$$

where ρ is the Weyl vector (this bracketing is allowed since the second factor does not depend on λ any more, see equation (14.14)). Afterwards we focus our attention to the inverse of the second factor, which by the formula (14.14) for the Verma module character $\mathcal{X}_\lambda$ reads

$$e^\rho \prod_{\alpha > 0} (1 - e^{-\alpha})^{\text{mult}\,\alpha} =: \mathcal{P}. \tag{14.41}$$

Now recall from chapter 10 that the Weyl group is generated by the fundamental reflections $w_{(i)}$ (the reflections about hyperplanes perpendicular to the simple roots $\alpha^{(i)}$). To check how the Weyl group acts on the characters of the Verma module, we therefore only need to know how the simple reflections act. To this end, we note that these generators possess the following properties. First, the Weyl group preserves multiplicities, $\mathrm{mult}(w_{(i)}(\alpha)) = \mathrm{mult}(\alpha)$; second, if α is a positive root other than $\alpha^{(i)}$, then also the root $w_{(i)}(\alpha)$ is positive; third, $w_{(i)}$ acts on the Weyl vector as $w_{(i)}(\rho) = \rho - \alpha^{(i)}$. Combining this information, one finds

$$
\begin{aligned}
w_{(i)}(\mathcal{P}) &= w_{(i)}(\mathrm{e}^\rho)\, w_{(i)}(1 - \mathrm{e}^{-\alpha^{(i)}})\, w_{(i)}\Big(\prod_{\alpha>0\,;\,\alpha\neq\alpha^{(i)}} (1 - \mathrm{e}^{-\alpha})^{\mathrm{mult}\,\alpha} \Big) \\
&= \mathrm{e}^{\rho-\alpha^{(i)}} \cdot (1 - \mathrm{e}^{+\alpha^{(i)}}) \cdot \prod_{\alpha>0\,;\,\alpha\neq\alpha^{(i)}} (1 - \mathrm{e}^{-\alpha})^{\mathrm{mult}\,\alpha} \qquad (14.42) \\
&= -\mathrm{e}^\rho \prod_{\alpha>0}(1 - \mathrm{e}^{-\alpha})^{\mathrm{mult}\,\alpha} = -\mathcal{P} .
\end{aligned}
$$

For a general Weyl group element w one then obtains $w(\mathcal{P}) = \mathrm{sign}(w)\,\mathcal{P}$, where $\mathrm{sign}(w)$ is the sign of w (compare (10.14)). The same then holds for the inverse of $\mathcal{P}$, i.e.

$$
w(\mathrm{e}^{-\rho-\Lambda}\mathcal{X}_\Lambda) = \mathrm{sign}(w)\,\mathrm{e}^{-\rho-\Lambda}\mathcal{X}_\Lambda . \qquad (14.43)
$$

In short, the second factor in the product (14.40) is 'odd' under the action of the Weyl group W.

On the other hand, the product (14.40) is even under W, and hence the first factor on the right hand side, i.e. the sum $\sum_{\lambda\in B(\Lambda)} \tilde{c}_\lambda \mathrm{e}^{\lambda+\rho}$, must be odd under W. This implies that the coefficients $\tilde{c}_\lambda$ obey $\tilde{c}_\lambda = \mathrm{sign}(w)\tilde{c}_\mu$ whenever λ and μ are related by an element $w\in W$ like $w(\lambda+\rho) = \mu+\rho$. As a consequence, among the coefficients $\tilde{c}_\lambda$ we only need to know those for a single element out of each Weyl group orbit of $\lambda+\rho$. Assuming from now on that the highest weight Λ is dominant integral, we also know from chapter 13 that each Weyl group element permutes the weight system of V_Λ. Hence as the specific element of the Weyl orbit we can in particular choose the unique dominant integral weight on the orbit. But the set $B(\Lambda)$ (14.38) contains only a single dominant integral weight, namely the highest weight Λ of V_Λ itself (see exercise 14.6).

Thus $\tilde{c}_\lambda$ is non-zero only if $\lambda+\rho$ lies on the Weyl group orbit of $\Lambda+\rho$. Together with the information that $\tilde{c}_\Lambda = 1$, we therefore know all the coefficients $\tilde{c}_\lambda$, and find, using (14.14), that

$$
\mathcal{X}_\Lambda = \frac{\sum_{w\in W} \mathrm{sign}(w)\, \mathrm{e}^{w(\Lambda+\rho)-\rho}}{\prod_{\alpha>0}(1 - \mathrm{e}^{-\alpha})^{\mathrm{mult}(\alpha)}} . \qquad (14.44)
$$

This result is known as the *Weyl–Kac character formula*; our presentation of the proof closely followed a proof given by Kac.

An immediate consequence of the character formula is the so-called denominator identity

$$\prod_{\alpha>0}(1-e^{-\alpha})^{\text{mult }\alpha} = \sum_{w\in W} \text{sign}(w)\, e^{w(\rho)-\rho}, \tag{14.45}$$

which for simple Lie algebras was already written down in (13.31). This identity is obtained by applying the formula (14.44) to the trivial one-dimensional module with highest weight $\Lambda = 0$, which yields

$$1 = \chi_{\Lambda=0} = \frac{\sum_{w\in W}\text{sign}(w)\, e^{w(\rho)-\rho}}{\prod_{\alpha>0}(1-e^{-\alpha})^{\text{mult }\alpha}}. \tag{14.46}$$

Substituting (14.45) back into (14.44), one obtains the alternative form

$$\chi_\Lambda = \frac{\sum_{w\in W}\text{sign}(w)\, e^{w(\Lambda+\rho)}}{\sum_{w\in W}\text{sign}(w)\, e^{w(\rho)}} \tag{14.47}$$

of the Weyl–Kac character formula.

Another derivation of the Weyl–Kac character formula is based on the following description of irreducible modules. To any integrable highest weight Λ and any non-negative integer p one associates the direct sum

$$\mathcal{V}^{(p)} = \sum_{w\in W,\, \ell(w)=p} \mathcal{V}_{w(\Lambda+\rho)-\rho} \tag{14.48}$$

of Verma modules, where the sum is over all Weyl group elements of length p. One can show that there is an exact sequence

$$\ldots \to \mathcal{V}^{(p+1)} \to \mathcal{V}^{(p)} \to \ldots \to \mathcal{V}^{(0)} \equiv \mathcal{V}_\Lambda \to V_\Lambda \to 0 \tag{14.49}$$

of $\mathfrak{g}$-modules. This sequence is semi-infinite when the Weyl group W is infinite; it is called a *Bernstein–Gelfand–Gelfand resolution* or *BGG-resolution* of the irreducible module V_Λ. Such resolutions play an important rôle in Lie algebra cohomology.

One of the major problems in the theory of Verma modules is the computation of the multiplicities c_p (compare equation (14.37)) with which the irreducible characters χ_{λ_p} appear in a Verma module character $\mathcal{X}_\Lambda$. One finds that one can express these numbers with the help of the so-called Kazhdan–Lusztig polynomials $P_{w,w'}(q)$ which depend on pairs of elements w, w' of the Weyl group W and on a continuous variable q. These polynomials can be generalized to involve arbitrary Coxeter groups instead of W; these generalized polynomials also appear in analogues of the Weyl–Kac character formula for Lie algebras other than Kac–Moody algebras, such as the $\mathcal{W}$-algebras of conformal field theory. For some details we refer to chapter 7 of [Humphreys 1990] and to [Deodhar *et al.* 1982] and [Borho 1986].

*14.11 Characters and modular transformations

We have already mentioned (compare e.g. section 12.12) that in applications of untwisted affine Lie algebras in physics one is typically interested in the irreducible highest weight modules V_Λ at a fixed value $k^\vee$ of the level. One reason for this restriction is that the central element K plays the rôle of a 'ℂ-number' term in canonical commutation relations of quantum fields. Another special property of the modules at fixed level shows up when one studies the so-called *modified characters*

$$\tilde{\chi}_\Lambda := e^{-s_\Lambda \delta} \chi_\Lambda \qquad (14.50)$$

of irreducible highest weight modules. Here $\delta = (0,0,1)$ is the imaginary root (12.32) of the affine algebra $\mathfrak{g}$, and s_Λ, called the *modular anomaly* of V_Λ, is the number

$$s_\Lambda := \frac{1}{(\bar\theta,\bar\theta)} \left(\frac{(\Lambda+\rho,\Lambda+\rho)}{k^\vee+g^\vee} - \frac{(\rho,\rho)}{g^\vee} \right). \qquad (14.51)$$

An important theorem, which can be obtained with the help of the Weyl–Kac character formula, states that the modified characters $\tilde{\chi}_\Lambda$, with Λ taking values in the set of integrable $\mathfrak{g}$-weights at fixed level $k^\vee$, span a (finite-dimensional) unitary module of $\mathrm{SL}(2,\mathbb{Z})$, the group of 2×2-matrices with integral entries and determinant one. Roughly speaking, this can be traced back to the fact that the Weyl group of an affine Lie algebra is the semidirect product (10.44) of a lattice of translations and the Weyl group $\overline{W}$ of the horizontal subalgebra. The sum in the character formula (14.47) can therefore be written as a sum over a lattice and a sum over $\overline{W}$. As it turns out, the first sum yields so-called *theta functions*, which are known to be modular forms.

To describe the action of $\mathrm{SL}(2,\mathbb{Z})$ on this module, it proves to be convenient to employ the notation

$$\lambda \,\hat{=}\, (\zeta, t, \tau) := \frac{(\bar\theta,\bar\theta)}{4\pi\mathrm{i}} \, (\bar\lambda, k, n) \qquad (14.52)$$

for $\mathfrak{g}$-weights instead of the more familiar notation $\lambda = (\bar\lambda, k, n)$ that was e.g. used in equation (13.39). With this convention, the action of $\gamma \in \mathrm{SL}(2,\mathbb{Z})$ on the characters is defined as the transformation

$$\tilde{\chi}_\Lambda((\zeta,t,\tau)) \;\mapsto\; \gamma \bullet \tilde{\chi}_\Lambda((\zeta,t,\tau)) := \tilde{\chi}_\Lambda((\zeta',t',\tau')) \qquad (14.53)$$

with

$$(\zeta',t',\tau') := \left(\frac{\zeta}{c\tau+d}, t - \frac{c(\zeta,\zeta)}{2(c\tau+d)}, \frac{a\tau+b}{c\tau+d} \right) \quad \text{for} \quad \gamma = \begin{pmatrix} a & b \\ c & d \end{pmatrix}. \qquad (14.54)$$

The group obtained by dividing out the normal subgroup $\{\pm\mathbf{1}\}$ from $\mathrm{SL}(2,\mathbb{Z})$ is called the modular group and denoted by $\mathrm{PSL}(2,\mathbb{Z})$; the mapping (14.53) with $\gamma \in \mathrm{PSL}(2,\mathbb{Z})$ is called a modular transformation of the character $\tilde{\chi}_\Lambda$. The modular group can be defined as the group generated by two elements S and T modulo the relations $S^2 = \mathbf{1} = (ST)^3$. In matrix notation, these generators are

$$S = \begin{pmatrix} 0 & -1 \\ 1 & 0 \end{pmatrix}, \qquad T = \begin{pmatrix} 1 & 1 \\ 0 & 1 \end{pmatrix}. \qquad (14.55)$$

In particular, on the so-called *specialized* (or Virasoro specialized) characters $\tilde{\chi}_\Lambda(\tau) \equiv \tilde{\chi}_\Lambda((0,0,\tau))$, the modular transformations are generated by $\tau \mapsto -1/\tau$ and $\tau \mapsto \tau + 1$.

By insertion of the Weyl–Kac character formula, one can compute the explicit form of the matrices which represent the generators S and T on the module that is spanned by the modified characters at fixed level. For T one obtains immediately

$$(R(T))_{\Lambda,\Lambda'} = e^{2\pi\mathrm{i}\,s_\Lambda}\, \delta_{\Lambda,\Lambda'}. \qquad (14.56)$$

For S the calculation is more complicated; one finds

$$(R(S))_{\Lambda,\Lambda'} = (-\mathrm{i})^{(d-r)/2}\, |L_\mathrm{w}/L^\vee|^{-1/2} (k^\vee + g^\vee)^{-r/2}$$
$$\cdot \sum_{\bar{w}\in\overline{W}} (\mathrm{sign}\,\bar{w}) \exp\left[-\frac{4\pi\mathrm{i}}{(\bar\theta,\bar\theta)} \frac{1}{k^\vee+g^\vee} (\bar{w}(\bar\Lambda+\bar\rho), \bar\Lambda'+\bar\rho) \right]. \qquad (14.57)$$

This result, which we already mentioned in exercise 10.6, is known as the *Kac–Peterson formula* for the modular S-matrix. We have pointed out above that the modular group $\text{PSL}(2,\mathbb{Z})$ is represented on the characters only projectively, i.e. on the characters one has a representation only of the twofold covering $\text{SL}(2,\mathbb{Z})$. If one lifts the identity element of $\text{PSL}(2,\mathbb{Z})$ to $\text{SL}(2,\mathbb{Z})$, one must always carefully check whether it is represented by $\mathbf{1}$ or by C, where $C_{\Lambda,\Lambda'} = \delta_{\Lambda',\Lambda^+}$. For instance, the matrix $R(S)$ as given in (14.57) satisfies $(R(S)R(T))^3 = C$, while the complex conjugate matrix $\overline{R(S)}$ (which has a relative minus sign in the exponent) obeys the relation $(\overline{R(S)}R(T))^3 = \mathbf{1}$.

The Weyl–Kac character formula and the modular transformation prop-
erties of characters are the source of numerous combinatorial identities | *Information*
involving theta functions and the Dedekind eta function. Accordingly,
there are deep connections between the theory of affine Lie algebras and various other areas of mathematics such as number theory, the theory of modular forms, the theory of finite groups like the Monster group, and the theory of vertex operator algebras. For more information, see e.g. [Frenkel *et al.* 1988] and chapters 12 and 13 of [Kac 1990].

*14.12 Harish-Chandra theorem and higher Casimir operators

The quadratic Casimir operator $\mathcal{C}_2$ is an essential ingredient in the description of the set $B(\Lambda)$ (14.38), and hence for the proof of the Weyl–Kac character formula. This demonstrates that $\mathcal{C}_2$ is an extremely useful quantity. One is therefore interested also in generalizations of this operator, i.e. other distinguished elements of the center $\mathcal{Z}(\mathsf{U}(\mathfrak{g}))$ of $\mathsf{U}(\mathfrak{g})$. Just like $\mathcal{C}_2$, each $z \in \mathcal{Z}(\mathsf{U}(\mathfrak{g}))$ acts on any highest weight module as a multiple of the identity. To describe this multiple one defines for each highest weight module V_λ a function $\psi_\lambda\colon \mathcal{Z}(\mathsf{U}(\mathfrak{g})) \to \mathbb{C}$ through the action

$$R_\lambda(z)\, v_\lambda =: \psi_\lambda(z) \cdot v_\lambda \quad \text{for all } z \in \mathcal{Z}(\mathsf{U}(\mathfrak{g})) \tag{14.58}$$

of z on the highest weight vector v_λ. The function ψ_λ is called the *central character* determined by the highest weight λ; it must not be confused with the character χ_λ of the representation R_λ, which is defined as a trace over the module V_λ carrying the representation. Note that for (14.58) to make sense, the highest weight λ need not be dominant integral, nor even integral.

Two central characters ψ_λ and ψ_μ are equal if and only if λ and μ are related by

$$w(\lambda + \rho) = \mu + \rho \tag{14.59}$$

for an element w of the Weyl group W of $\mathfrak{g}$; this assertion is known as the *Harish-Chandra theorem* (for a proof, see e.g. chapter 23.3 of [Humphreys 1972]). If $\mathfrak{g}$ is a finite-dimensional semisimple or an affine Lie algebra, then for any integral weight μ there is precisely one dominant integral weight λ which obeys (14.59) for some $w \in W$. As a consequence, the Harish-Chandra theorem immediately implies that finite-dimensional highest weight modules of finite-dimensional semisimple Lie algebras, and also integrable highest weight modules of affine Lie algebras, are uniquely characterized by the central character of their highest weight.

If $\mathfrak{g}$ is a finite-dimensional semisimple Lie algebra, then a distinguished basis of the center $\mathcal{Z}(\mathsf{U}(\mathfrak{g}))$ of the universal enveloping algebra is provided by the so-called *higher order Casimir operators*. These are homogeneous polynomials

$$\mathcal{C}_n = d_{a_1 a_2 \ldots a_n}\, T^{a_1} T^{a_2} \cdots T^{a_n} \tag{14.60}$$

in the generators T^a of $\mathfrak{g}$, with $d_{a_1 a_2 \ldots a_n}$ suitable 'invariant tensors' of the adjoint representation (for more details about these tensors and the higher order Casimir operators see section 17.8). The order n of this polynomial is called the *order* of C_n.

Furthermore, it can be shown that $\mathcal{Z}(\mathsf{U}(\mathfrak{g}))$ is isomorphic to the algebra of polynomials over the base field F in $\mathrm{r} = \mathrm{rank}\,\mathfrak{g}$ variables. The number of algebraically independent Casimir operators of a simple Lie algebra $\mathfrak{g}$ is therefore equal to the rank r of $\mathfrak{g}$. For finite-dimensional simple Lie algebras, the orders of the Casimir operators that generate $\mathcal{Z}(\mathsf{U}(\mathfrak{g}))$ algebraically have been listed in table V; note e.g. that for D_r with even rank r, there are two distinct Casimir operators of order r. The numbers $m_i = n_i - 1$, with n_i, $i = 1, 2, \ldots, \mathrm{r}$, the orders of the independent Casimir operators, are called the *exponents* of $\mathfrak{g}$ (they appear e.g. in the formula (7.21) for the eigenvalues of the Cartan matrix).

In the case of $\mathfrak{sl}(2)$, the center $\mathcal{Z}(\mathsf{U}(\mathfrak{g}))$ is one-dimensional and spanned by the second order Casimir operator. The theorem of Harish-Chandra then reduces to the statement that one can characterize a finite-dimensional irreducible $\mathfrak{sl}(2)$-representation uniquely by the eigenvalue of the quadratic Casimir operator (i.e. in terms of the angular momentum operators L_i, by the eigenvalue of $\vec{L}^2$, the spin). Analogously, for general simple $\mathfrak{g}$ one could use the eigenvalues of all independent Casimir operators to characterize irreducible representations. However, since the Casimir eigenvalues are polynomials in the components of a weight, while the Dynkin labels are of course linear, in practice it is much more convenient to use Dynkin labels instead of Casimir eigenvalues for this purpose. On the other hand, in applications in physics the Casimir eigenvalues are often more directly linked to observable quantities.

Summary:

The universal enveloping algebra $\mathsf{U}(\mathfrak{g})$ of a Lie algebra $\mathfrak{g}$ is a quotient of the tensor algebra of $\mathfrak{g}$. Its basis is described by the Poincaré–Birkhoff–Witt theorem, and its center is generated algebraically by the Casimir operators.

Verma modules are obtained by acting with the enveloping algebra of the lowering operators on a highest weight vector with highest weight Λ. Any Verma module of a finite-dimensional simple or of an affine Lie algebra contains a unique maximal submodule. When quotienting out this submodule, one obtains the unique irreducible module of $\mathfrak{g}$ with highest weight Λ. The vectors in the maximal submodule are called null vectors.

Characters of Verma and irreducible modules can be computed with the Weyl–Kac character formula. This formula is based on the behavior of the characters with respect to Weyl group transformations and the fact that the quadratic Casimir operators belongs to the center of $\mathsf{U}(\mathfrak{g})$.

Keywords:

Universal enveloping algebra, quotient of an algebra by an ideal, Poincaré–Birkhoff–Witt theorem;
Verma module, (primitive) null vector, (quadratic) Casimir operator, Weyl–Kac character formula, Harish-Chandra theorem;
modular group, Kac–Peterson formula.

Exercises:

a) Show that if $\mathcal{I}$ is a two-sided ideal of an associative algebra $\mathfrak{A}$, then the quotient space $\mathfrak{A}/\mathcal{I}$ is again an algebra with an associative product. Is this still true if $\mathcal{I}$ is just a subalgebra?

b) Apply the same reasoning to Lie algebras. In particular, explain why by setting $[x + \mathfrak{k}, y + \mathfrak{k}] = [x, y] + \mathfrak{k}$ for all $x, y \in \mathfrak{g}/\mathfrak{k}$ one obtains a well-defined Lie bracket.

c) Prove the universal property of $\mathsf{U}(\mathfrak{g})$.

Exercise 14.1

Consider highest weight modules of $\mathfrak{sl}(2)$ for which the highest weight is not a positive integer. Is the Verma module in this case fully reducible? Is it irreducible?

Exercise 14.2

Compute the Killing form in the two bases $\{L_i\}$ and $\{L_0, L_\pm\}$ (compare equations (2.3) and (2.5)) of $\mathfrak{sl}(2)$. Explain why this suggests to consider the element (14.22) as a generalization of (14.21).

Exercise 14.3

Show that the 'length squared' of the vector $E_-^i v_\Lambda$ is given by $\Lambda^i (v_\Lambda \,|\, v_\Lambda)$ for any highest weight module with highest weight vector v_Λ.

Exercise 14.4

Derive the relation (14.30) between the Casimir eigenvalue in the adjoint module and the dual Coxeter number.
Hint: Use the relation between the dual Coxeter labels and the highest root, and the property (7.22) of the Weyl vector.

Exercise 14.5

a) Convince yourself that (14.13) implies that the character of the Verma module $\mathcal{V}_\Lambda$ is given by $\mathcal{X}_\Lambda = e^\Lambda \prod_{\alpha > 0} (1 + e^{-\alpha} + e^{-2\alpha} + \dots)^{\text{mult } \alpha}$.

b) Show that the only dominant integral weight in the set $B(\Lambda)$ (14.38) is the highest weight Λ.
Hint: Assume that $\mu \in B(\Lambda)$ is dominant integral, i.e. that all its Dynkin components are non-negative, and use that $|\Lambda + \rho|^2 - |\mu + \rho|^2$ has to vanish for all $\mu \in B(\Lambda)$.

Exercise 14.6

Work out the BGG resolution for the adjoint module of A_2.
Compare with the weight system (13.13) of the irreducible module.

Exercise 14.7

15

Tensor products of representations

15.1 Tensor products

As we have seen in section 5.3, whenever one is given two modules V and W of a Lie algebra $\mathfrak{g}$, $\mathfrak{g}$ can also be represented on the tensor product vector space $V \otimes W$, namely via

$$R_V \otimes R_W : \quad ((R_V \otimes R_W)(x))\,(v \otimes w) := \\ (R_V(x)\,v) \otimes w \,+\, v \otimes (R_W(x)\,w)\,. \tag{15.1}$$

This representation is called the *tensor product* of the $\mathfrak{g}$-representations R_V and R_W; analogously one speaks of the tensor product module $V \otimes W$ of the $\mathfrak{g}$-modules V and W. If $\mathcal{B}_V = \{v_i\}$ and $\mathcal{B}_W = \{w_j\}$ are (countable) bases of V and W, respectively, then a basis of $V \otimes W$ is provided by

$$\mathcal{B} = \{v_i \otimes w_j \mid v_i \in \mathcal{B}_V,\ w_j \in \mathcal{B}_W\}\,. \tag{15.2}$$

With this form of the basis it is manifest that $V \otimes W$ and $W \otimes V$ are different vector spaces, and hence also the modules are different. However, it turns out that the two modules are isomorphic, and hence

$$R_V \otimes R_W \;\cong\; R_W \otimes R_V\,. \tag{15.3}$$

In other words, the tensor product is commutative up to isomorphism. It is also associative up to isomorphism, i.e.

$$(R \otimes R') \otimes R'' \;\cong\; R \otimes (R' \otimes R'') \tag{15.4}$$

for all $\mathfrak{g}$-representations R, R', R''. We will study these isomorphisms in detail in chapter 16. The components of the isomorphism (15.4) are known as the $6j$-symbols of $\mathfrak{g}$. The reader should always keep in mind that when dealing with tensor products, it is quite natural to investigate properties that are valid only 'up to isomorphism'.

266

In the special case where $W = V$, it is natural to analyze the behavior of the tensor product with respect to the transposition $v \otimes w \mapsto w \otimes v$. As a vector space, $V \otimes V$ splits into the two invariant subspaces

$$V_{\mathrm{s}} := \{ v \otimes v' + v' \otimes v \mid v, v' \in V \}, \quad V_{\mathrm{a}} := \{ v \otimes v' - v' \otimes v \mid v, v' \in V \} \quad (15.5)$$

of symmetric respectively antisymmetric elements. By the linearity of the tensor product representation, it follows that these subspaces are actually again $\mathfrak{g}$-modules, so that one has the direct sum decompositions

$$R \otimes R = R_{\mathrm{s}} \oplus R_{\mathrm{a}}, \quad V \otimes V = V_{\mathrm{s}} \oplus V_{\mathrm{a}}, \quad (15.6)$$

with R_{s} and R_{a} representations of $\mathfrak{g}$ on the vector spaces V_{s} and V_{a}, respectively (see exercise 15.2). In physics, these two sub-representations (which are, however, not necessarily irreducible) of $R \otimes R$ are often referred to as the *symmetric* and *antisymmetric coupling* of R with itself. Analogously, one can decompose higher tensor powers $R^{\otimes l}$ of R, acting on the vector space $V^{\otimes l}$, into sub-representations according to the symmetry properties; this will be examined in more detail in chapter 19.

*15.2 Tensor products for other algebraic structures

As already mentioned in chapter 5, the concepts of representations and representation spaces (modules) can also be defined analogously for algebras which are not Lie algebras. A representation of the algebra $\mathfrak{A}$ is by definition a linear map $R \colon \mathfrak{A} \to \mathfrak{gl}(V)$ from $\mathfrak{A}$ to the general linear algebra of some vector space V which preserves the product of $\mathfrak{A}$, $R(x \circ y) = R(x) \circ R(y)$, i.e. is an $\mathfrak{A}$-homomorphism. However, unlike in the case of Lie algebras, there is generically no natural way of defining tensor products of representations.

A mathematical structure for which tensor product representations *can* be defined is the one of groups. If R_V and R_W are representations of a group G, then the tensor product representation $R_V \otimes R_W$ acts as

$$\big((R_V \otimes R_W)(\gamma) \big)(v \otimes w) = (R_V(\gamma)\,v) \otimes (R_W(\gamma)\,w) \quad (15.7)$$

for all $\gamma \in$ G. If G is a *Lie* group, then the formula (15.1) for the Lie algebra of G can be viewed as being the infinitesimal version of (15.7).

Now if R_V and R_W are representations of $\mathfrak{A}$ in $\mathfrak{gl}(V)$ and $\mathfrak{gl}(W)$, respectively, then the prescription $x \otimes y \mapsto R_V(x) \otimes R_W(y)$ defines a representation of $\mathfrak{A} \otimes \mathfrak{A}$ in $\mathfrak{gl}(V \otimes W)$. This is linear in the elements $x \otimes y$ of $\mathfrak{A} \otimes \mathfrak{A}$ (or in other words, bilinear in the elements of $\mathfrak{A}$), so that it might be tempting to expect that a tensor product could be defined analogously to (15.7), i.e. via $x \mapsto R(x) := R_V(x) \otimes R_W(x)$, which means that $R(x)$ acts as $R(x)(v \otimes w) = (R_V(x)\,v) \otimes (R_W(x)\,w)$; however, this map does *not* define a representation, because it is not linear in x. Another tentative definition which may come to mind is to mimic the structure of the formula (15.1), i.e. write

$$x \mapsto R(x) := R_V(x) \otimes id_W + id_V \otimes R_W(x), \quad (15.8)$$

with id_V and id_W the identity map on V and W, respectively, or more explicitly,

$(R(x))(v{\otimes}w) = (R_V(x)\,v){\otimes}w + v{\otimes}(R_W(x)\,w)$. But then it follows that

$$\begin{aligned}
R(x \circ y) &= R_V(x)R_V(y){\otimes}id_W + id_V{\otimes}R_W(x)R_W(y)\\
&\neq R(x)R(y) = R_V(x)R_V(y){\otimes}id_W + R_V(x){\otimes}R_W(y)\\
&\quad + R_V(y){\otimes}R_W(x) + id_V{\otimes}R_W(x)R_W(y)\,,
\end{aligned} \tag{15.9}$$

so that generically R as defined by (15.8) again does *not* provide an $\mathfrak{A}$-representation. In the particular case of Lie algebras, one obtains a representation because instead of ordinary products of the representation matrices $R(x)$ one has to consider the commutators $R(x)R(y) - R(y)R(x)$ in which the unwanted terms cancel.

In order for the definition of a tensor product to make sense, an algebra must therefore allow for some additional structure. When putting this requirement in a formal setting, one is led to the concept of *Hopf* algebras; this will be described in chapter 22.

15.3 Highest weight modules

Because of the summation on the right hand side of (15.1), upon forming the tensor product the weights of the modules of a finite-dimensional simple (or an affine) Lie algebra $\mathfrak{g}$ add up. As a consequence, the weight system of the tensor product $V{\otimes}V'$ of the $\mathfrak{g}$-modules V and V' consists of all weights of the form $\lambda + \lambda'$, with λ and λ' weights of V and of V', respectively. The multiplicity of these weights is

$$\text{mult}_{V{\otimes}V'}(\lambda + \lambda') = \sum_{\substack{\mu,\mu'\\ \mu+\mu'=\lambda+\lambda'}} \text{mult}_V(\mu)\cdot\text{mult}_{V'}(\mu')\,. \tag{15.10}$$

This description of the weight system of tensor products is in fact still valid for any Lie algebra $\mathfrak{g}$ which possesses a triangular decomposition. As already mentioned in chapter 5, these properties of the tensor product are the origin of the term 'additive quantum number' that is commonly used in physics; in contrast, according to the formula (15.7) groups of symmetries give rise to multiplicative quantum numbers.

For finite-dimensional highest weight modules $V = V_\Lambda$ and $V' = V_{\Lambda'}$ of a (semi-)simple Lie algebra, the tensor product $V_\Lambda{\otimes}V_{\Lambda'}$ is again finite-dimensional. This implies (compare chapter 5) that it is completely reducible into irreducible modules, i.e. that there is a decomposition

$$V_\Lambda \otimes V_{\Lambda'} \cong \bigoplus_i \mathcal{L}_{\Lambda\Lambda'}^{\Lambda_i}\, V_{\Lambda_i} \tag{15.11}$$

of the tensor product into a direct sum of its irreducible submodules V_{Λ_i}. In (15.11) the summation is meant in the sense that for any weight Λ_i the module V_{Λ_i} appears at most once in the sum, i.e. eventual multiplicities larger than one are taken care of by the coefficients $\mathcal{L}_{\Lambda\Lambda'}^{\Lambda_i}$, which are thus non-negative integers. These integers are called the *tensor product coefficients* or *multiplicities*, or the *Littlewood–Richardson coefficients*, of $\mathfrak{g}$ (the term Littlewood–Richardson coefficient is sometimes reserved for the case of $\mathfrak{g} = A_r$).

Let us stress that when studying the tensor product coefficients, one is usually only interested in isomorphism classes of modules. This is indicated by the use of the symbol '$\cong$' in the decomposition (15.11). When one makes this isomorphism in (15.11) explicit, one is lead to the concept of Clebsch–Gordan coefficients, see chapter 16.

The formula (15.10) tells us in particular that the sum $\Lambda + \Lambda'$ of the two highest weights is a non-degenerate weight of $V_\Lambda \otimes V_{\Lambda'}$. Therefore the tensor product $V_\Lambda \otimes V_{\Lambda'}$ contains the module $V_{\Lambda + \Lambda'}$ as an irreducible submodule, with multiplicity one. Furthermore, a quite non-trivial property of the tensor product of simple Lie algebras is that the singlet, i.e. the one-dimensional module with highest weight $\Lambda = 0$, occurs in the decomposition of $V_\Lambda \otimes V_{\Lambda'}$ if and only if the modules V_Λ and $V_{\Lambda'}$ are each others conjugates, i.e. $\Lambda' = \Lambda^+$, and in this case its multiplicity is equal to one (this can be proven with the help of the orthogonality properties of the Lie group characters of irreducible highest weight modules, see section 21.10); thus we have

$$\mathcal{L}_{\Lambda\Lambda'}^{\ 0} = \delta_{\Lambda',\Lambda^+} \, . \tag{15.12}$$

Furthermore, the commutativity property (15.3) translates to $\mathcal{L}_{\Lambda\Lambda'}^{\ \Lambda''} = \mathcal{L}_{\Lambda'\Lambda}^{\ \Lambda''}$ and the associativity (15.4) to $\sum_i \mathcal{L}_{\Lambda\Lambda'}^{\ \Lambda_i} \mathcal{L}_{\Lambda_i\Lambda''}^{\ \Lambda_j} = \sum_i \mathcal{L}_{\Lambda\Lambda_i}^{\ \Lambda_j} \mathcal{L}_{\Lambda'\Lambda''}^{\ \Lambda_i}$.

For $\mathfrak{g} = A_r$, and also to some extent for the other classical algebras, combinatorial formulæ for the tensor product coefficients can be obtained with the help of Young tableaux (see chapter 19), whereas for exceptional $\mathfrak{g}$ no general formula is known. With the Young tableaux method one can obtain tensor product decompositions by hand as long as the representations are not too large. However, nowadays fast algorithms are available on the computer so that it is no longer necessary to resort to this or to similar methods.

15.4 Tensor products for $\mathfrak{sl}(2)$

In order to write out the decomposition (15.11) explicitly one may in principle proceed by brute force. That is, one starts with the whole set $\mathcal{S} = \{\lambda + \lambda'\}$ of weights (with proper multiplicities) and picks out of it any maximal weight Λ_i, i.e. any weight such that no other weight $\mu \in \mathcal{S}$ satisfies $\Lambda_i \leq \mu$. Then one removes all weights of V_{Λ_i} (obtained, say, by the Freudenthal recursion formula (13.15)) from $\mathcal{S}$; next one picks a maximal weight $\Lambda_{i'}$ from the remaining weights, removes all weights of $V_{\Lambda_{i'}}$, and so on, until no weights are left over. (In the first round there is a unique maximal weight, of multiplicity 1, namely $\Lambda + \Lambda'$.)

While this prescription is easily implemented on a computer, even for modestly complicated tensor products it would require an enormous amount of memory and computer time, so that nobody should consider this method seriously. There is however one specific case, namely $\mathfrak{g} = \mathfrak{sl}(2)$, in which the structure of the modules is so simple that the prescription can even be directly worked out by hand. The weights of the $\mathfrak{sl}(2)$-module V_Λ are $\Lambda - 2n$ for $n = 0, 1, \ldots, \Lambda$, each with multiplicity one. Then the weights of $V_\Lambda \otimes V_{\Lambda'}$ are $\lambda = \lambda_n := \Lambda + \Lambda' - 2n$ for $n = 0, 1, \ldots, \Lambda + \Lambda'$, with multiplicity $\mathrm{mult}_{V_\Lambda \otimes V_{\Lambda'}}(\lambda_n) = 1 + \min\{n, \Lambda + \Lambda' - n\}$. This is illustrated in figure (15.13) for $\Lambda = 7$ and $\Lambda' = 4$.

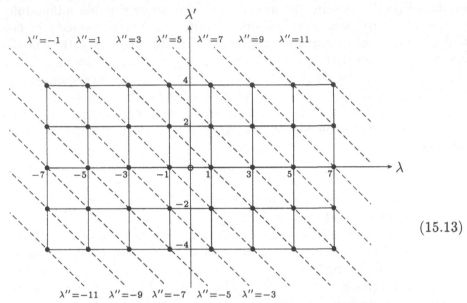

(15.13)

In this picture the x-axis gives the value of λ and the y-axis the value of λ'. Each dot stands for a basis vector in the tensor product $V_\Lambda \otimes V_{\Lambda'}$. Vectors with the same eigenvalue M of $R_{\Lambda \times \Lambda'}(J_0)$ lie on lines that intersect the x-axis at an angle of 135 degrees. For $-|\Lambda - \Lambda'| \leq \lambda'' \leq |\Lambda - \Lambda'|$ the dimension of the weight space to the weight λ'' is constant. This shows that $|\Lambda - \Lambda'| \leq \Lambda'' \leq \Lambda + \Lambda'$. We can also read off that when $\Lambda + \Lambda'$ is even, then only even weights Λ'' appear, and when $\Lambda + \Lambda'$ is odd, then Λ'' must be odd as well. We thus arrive at the formula

$$\mathcal{L}_{\Lambda \Lambda'}^{\Lambda''} = \begin{cases} 1 & \text{for} \quad |\Lambda - \Lambda'| \leq \Lambda'' \leq \Lambda + \Lambda' \\ & \quad \text{and} \quad \Lambda + \Lambda' + \Lambda'' \in 2\mathbb{Z}, \\ 0 & \text{else}. \end{cases} \qquad (15.14)$$

The corresponding tensor product decomposition

$$V_\Lambda \otimes V_{\Lambda'} = V_{|\Lambda-\Lambda'|} \oplus V_{|\Lambda-\Lambda'|+2} \oplus \cdots \oplus V_{\Lambda+\Lambda'-2} \oplus V_{\Lambda+\Lambda'} \qquad (15.15)$$

is often called the Clebsch–Gordan series of $\mathfrak{sl}(2)$ tensor products.

If none of the Littlewood–Richardson coefficients of a tensor product is larger than unity, then the product is said to be *simply reducible*. The decomposition (15.15) shows that all tensor products of irreducible $\mathfrak{sl}(2)$-modules are of this type. For other simple Lie algebras, also non-simply reducible tensor products appear.

When describing the 'coupling' of orbital angular momentum $\vec{L}$ and spin $\vec{S}$ of an electron in the hydrogen atom, physicists usually work with the operators $\vec{L}_{\text{tot}} := \vec{L} + \vec{S}$ for the total angular momentum and $\vec{L} \cdot \vec{S} \equiv \frac{1}{2}(\vec{L}_{\text{tot}}^2 - \vec{L}^2 - \vec{S}^2)$. In the explicit formulation in terms of representation matrices, these operators are described as follows. One is investigating the tensor product of two modules V_{j_1} and V_{j_2} of $\mathfrak{su}(2)$. The first module describes the degrees of freedom of orbital angular momentum, hence j_1 is an integer, while V_{j_2} describes the spin of the electron, i.e. $j_2 = \frac{1}{2}$. The possible states of the electron are described by the tensor product of these two modules. The operator L_i then stands as a short hand for $R_{j_1}(J_i) \otimes \mathbf{1}_{1/2}$, the operator S_i for $\mathbf{1}_{j_1} \otimes R_{1/2}(J_i)$, and $\vec{L} \cdot \vec{S}$ for $\frac{1}{2}[R_{j_1 \times 1/2}(\vec{J}^2) - R_{j_1}(\vec{J}^2) \otimes \mathbf{1}_{1/2} - \mathbf{1}_{j_1} \otimes R_{1/2}(\vec{J}^2)]$.

15.5 Conjugacy classes

A specific feature of the Clebsch–Gordan series (15.15) of $\mathfrak{sl}(2)$ is that $\mathcal{L}_{\Lambda\Lambda'}^{\Lambda''}$ vanishes unless $\Lambda + \Lambda' + \Lambda'' \in 2\mathbb{Z}$. In terms of the spin $j = \frac{1}{2}\Lambda$ this means that three spins j, j', j'' can be 'coupled' only if their sum is an integer, i.e. the tensor product of finite-dimensional irreducible representations contains either only integral spin or only half-integral spin representations. Put differently, the number

$$c_\Lambda := \Lambda \bmod 2 \qquad (15.16)$$

is additively conserved under forming tensor products, i.e. with respect to the tensor product the family of finite-dimensional highest weight modules of $\mathfrak{sl}(2)$ possesses a $\mathbb{Z}_2$-gradation.

The number $c_\Lambda \in \{0, 1\}$, taken modulo 2, is called the *conjugacy class* of the $\mathfrak{sl}(2)$-weight Λ. In order to extend this concept to other simple Lie algebras $\mathfrak{g}$, it is helpful to regard c_Λ as an element of an abelian group, namely of $\mathbb{Z}_2 \equiv \mathbb{Z} \bmod 2\mathbb{Z}$. Thus we would like to find some abelian group and associate to any $\mathfrak{g}$-weight Λ a group element c_Λ, its conjugacy class, such that

$$\mathcal{L}_{\Lambda\Lambda'}^{\Lambda''} \neq 0 \Rightarrow c_\Lambda + c_{\Lambda'} = c_{\Lambda''} \qquad (15.17)$$

(since the group is abelian, the group multiplication is written additively).

Such a group is indeed available. We already know that the root lattice L is a sublattice of the weight lattice L_{w}. Now any lattice constitutes an abelian group, where the group product is vector addition of lattice elements; hence the set of cosets L_{w}/L is an abelian group as well. It turns out that this group is finite; the number $I_c := |L_{\text{w}}/L|$ of conjugacy classes of $\mathfrak{g}$-weights is called the index of connection of $\mathfrak{g}$. (The index of connection of simple Lie algebras has been listed in table V.) We associate to any integral $\mathfrak{g}$-weight Λ as its conjugacy class just the equivalence class of Λ in L_w/L. That this definition of the conjugacy classes satisfies (15.17) follows from the observation that any two weights of the tensor product module differ by an element of the root lattice, so that $\Lambda + \Lambda' - \Lambda''$ must be a root lattice element in order for $\mathcal{L}_{\Lambda\Lambda'}^{\Lambda''}$ to be non-zero.

The group L_w/L is isomorphic to the center $\mathcal{Z}$ of the universal covering group G that has $\mathfrak{g}$ as its Lie algebra, which in turn is related to the symmetries of the ordinary and extended Dynkin diagrams via the relation (11.33). According to section 11.5, this group is $\mathcal{Z} = \mathbb{Z}_{r+1}$ for $\mathfrak{g} = A_r$, $\mathcal{Z} = \mathbb{Z}_3$ for E_6, $\mathcal{Z} = \mathbb{Z}_2$ for $\mathfrak{g} = B_r$, C_r, and E_7, and it is trivial for E_8, F_4 and G_2; for $\mathfrak{g} = D_r$ the group structure depends on whether the rank is even or odd, namely $\mathcal{Z} = \mathbb{Z}_2 \times \mathbb{Z}_2$ for r even and $\mathcal{Z} = \mathbb{Z}_4$ for r odd. For $\mathfrak{g} = A_2 = \mathfrak{sl}(3)$ where $\mathcal{Z} = \mathbb{Z}_3$, the conjugacy class is called the *triality* of an $\mathfrak{sl}(3)$-module.

The conjugacy class of a $\mathfrak{g}$-weight Λ can be expressed in terms of the Dynkin labels of Λ. For $\mathfrak{g} = D_r$ the conjugacy class has two components, $c_\Lambda = (c_\Lambda^{(1)}; c_\Lambda^{(2)})$, with

$$c_\Lambda^{(1)} = \Lambda^{r-1} + \Lambda^r \bmod 2,$$

$$c_\Lambda^{(2)} = \begin{cases} \sum_{j=1}^{r/2} \Lambda^{2j-1} \bmod 2 & \text{for } r \text{ even}, \\ 2\sum_{j=1}^{(r-1)/2} \Lambda^{2j-1} + (\Lambda^{r-1} - \Lambda^r) \bmod 4 & \text{for } r \text{ odd}. \end{cases} \tag{15.18}$$

If the rank r is even, the conjugacy class is characterized by two integers which are defined modulo 2, i.e. by an element of the discrete abelian group $\mathbb{Z}_2 \times \mathbb{Z}_2$. For odd rank, the conjugacy class is characterized by the group $\mathbb{Z}_4$, because $c_\Lambda^{(1)} = 2c_\Lambda^{(2)} \bmod 2$, so that $c_\Lambda^{(1)}$ and $c_\Lambda^{(2)}$ are not independent. The reason why we list $c_\Lambda^{(1)}$ separately in both cases is that this quantity distinguishes between the so-called *tensor* representations (with $c_\Lambda^{(1)} = 0$) and *spinor* representations (with $c_\Lambda^{(1)} = 1$). Tensor representations can be obtained from tensor products of the defining representation, while the construction of spinor representations is more complicated, compare chapter 20.

For the other simple algebras with non-trivial grading one has

$$c_\Lambda = \sum_{j=1}^{r} j \Lambda^j \bmod r + 1 \qquad \text{for} \quad \mathfrak{g} = A_r,$$

$$c_\Lambda = \Lambda^r \bmod 2 \qquad \text{for} \quad \mathfrak{g} = B_r,$$

$$c_\Lambda = \sum_{j=1}^{[r/2]} \Lambda^{2j-1} \bmod 2 \qquad \text{for} \quad \mathfrak{g} = C_r, \tag{15.19}$$

$$c_\Lambda = \Lambda^1 - \Lambda^2 + \Lambda^4 - \Lambda^5 \bmod 3 \quad \text{for} \quad \mathfrak{g} = E_6,$$

$$c_\Lambda = \Lambda^4 + \Lambda^6 + \Lambda^7 \bmod 2 \qquad \text{for} \quad \mathfrak{g} = E_7.$$

The property (15.17) of conjugacy classes constitutes a selection rule for tensor products which generalizes the 'spin selection rule' of $\mathfrak{sl}(2)$. In particular, the highest weight modules with weights in the zero conjugacy class $c_\Lambda = 0$ close among themselves under the tensor product. This family of modules always contains of course the singlet (i.e. $\Lambda = 0$), but also the adjoint module (i.e. $\Lambda = \theta$, which in particular lies on the root lattice). Since the singlet is contained in the tensor product of any module with its conjugate module, the conjugacy classes of conjugate modules must add up to zero, and hence $c_{\Lambda^+} = -c_\Lambda$.

15.6 Sum rules

Various sum rules are valid for tensor product decompositions (15.11) of representations of simple Lie algebras. The most important of these is the *character sum rule*

$$\chi_\Lambda \cdot \chi_{\Lambda'} = \sum_i \mathcal{L}_{\Lambda\Lambda'}^{\Lambda_i} \chi_{\Lambda_i}. \tag{15.20}$$

This relation is obtained by combining the properties (13.34) and (13.35) of the characters of direct sums and of tensor products with the decomposition (15.11). By evaluating the characters at the zero weight, (15.20) implies the dimension sum rule

$$d_\Lambda \cdot d_{\Lambda'} = \sum_i \mathcal{L}_{\Lambda\Lambda'}^{\Lambda_i} d_{\Lambda_i}, \tag{15.21}$$

where d_Λ denotes the dimension of the module V_Λ.

Another sum rule applies to the Dynkin index $I_\Lambda \equiv I_{V_\Lambda}$ that was introduced in section 14.9. The second order Dynkin index of the direct sum of irreducible modules is the sum of the Dynkin indices of the irreducible components (this follows e.g. from the formula (14.35) for I_Λ), while for tensor products one has $I_{\Lambda \times \Lambda'} \equiv I_{V_\Lambda \otimes V_{\Lambda'}} = I_\Lambda d_{\Lambda'} + I_{\Lambda'} d_\Lambda$. Thus

$$I_\Lambda d_{\Lambda'} + I_{\Lambda'} d_\Lambda = \sum_i \mathcal{L}_{\Lambda\Lambda'}^{\Lambda_i} I_{\Lambda_i}. \tag{15.22}$$

While in the general case the computation of tensor product decompositions by hand is tedious, for low-dimensional modules many tensor products can often be uniquely determined by combining the sum rules (15.21) and (15.22) with other generic properties of tensor products such as the conjugacy class selection rules, the decomposition (15.6), and the knowledge about the appearance of the singlet.

15.7 The Racah–Speiser algorithm

The character sum rule (15.20) gives rise to an algorithm for determining the tensor product coefficients of simple Lie algebras which is very efficient and can be easily implemented on a computer. This result is known as Weyl's *method of characters* or as the *Racah–Speiser algorithm*. This algorithm amounts to implementing the formula

$$\mathcal{L}_{\Lambda\Lambda'}^{\Lambda_i} = \sum_{w \in W}{}' (\text{sign}(w)) \, \text{mult}_{\Lambda'}(w(\Lambda_i + \rho) - \rho - \Lambda). \qquad (15.23)$$

for the tensor product multiplicities. Here W is the Weyl group of $\mathfrak{g}$; the sum $\sum'$ is only over those Weyl group elements for which $w(\Lambda_i + \rho) - \rho - \Lambda$ is a weight of the module $V_{\Lambda'}$; alternatively, the sum extends over the whole Weyl group W if one uses the convention that $\text{mult}_\Lambda(\nu) = 0$ whenever ν is not a weight of V_Λ.

To derive the result (15.23) one starts by expressing the characters χ_Λ and χ_{Λ_i} in (15.20) through the Weyl character formula (13.30). Using the fact that the denominator of that formula does not depend on the highest weight of the representation in question, one obtains

$$\sum_{w \in W} (\text{sign}(w)) \, \exp(w(\Lambda + \rho)) \cdot \chi_{\Lambda'}$$
$$= \sum_{w \in W} (\text{sign}(w)) \sum_i \mathcal{L}_{\Lambda\Lambda'}^{\Lambda_i} \, \exp(w(\Lambda_i + \rho)). \qquad (15.24)$$

Here for notational convenience we write $\exp(\lambda)$ for the formal exponential e^λ. The next step is to combine equation (15.24) with the identity $\chi_{\Lambda'} = \sum_{\lambda'} \text{mult}_{\Lambda'}(\lambda') \, e^{\lambda'}$ and use the fact that the multiplicities of the weight system of $V_{\Lambda'}$ are invariant under the Weyl group. This leads to the equation

$$\sum_{w \in W} \sum_{\lambda'} (\text{sign}(w)) \, \text{mult}_{\Lambda'}(\lambda') \exp(w(\Lambda + \rho + \lambda'))$$
$$= \sum_{w \in W} (\text{sign}(w)) \sum_i \mathcal{L}_{\Lambda\Lambda'}^{\Lambda_i} \, \exp(w(\Lambda_i + \rho)). \qquad (15.25)$$

Provided that we were allowed to forget about the sum over the Weyl group, a comparison of both sides of this equation would immediately lead to the conclusion that $\mathcal{L}_{\Lambda\Lambda'}^{\Lambda_i}$ is given by the multiplicity

$$\text{mult}_{\Lambda'}(\Lambda_i - \Lambda). \qquad (15.26)$$

To take also care of the presence of the summation over the Weyl group when comparing both sides, we implement the facts that the formal exponentials are independent and that the Weyl group W permutes the Weyl chambers. On the right hand side of (15.25), we take the term of the sum corresponding to the identity element of W. Then the argument of the formal exponential is a weight in the interior of the fundamental Weyl chamber, and we have to identify those terms of the right hand side which possess the same property.

If all weights of the form

$$\mu := \Lambda + \rho + \lambda', \qquad (15.27)$$

with λ' any weight of $V_{\Lambda'}$, are contained in the fundamental Weyl chamber, then also on the left hand side of the equation only the term with the identity element of the Weyl group can give a contribution lying in the fundamental chamber; as a consequence, one is in fact allowed to remove the summation over the Weyl group. In this case, (15.26) is indeed the correct result for the Littlewood–Richardson coefficients. Generically, however, it will also happen that some weight μ lies in a chamber other than the fundamental one. In this case the fact that the Weyl group permutes the Weyl chambers freely implies that there exists an element w of the Weyl group such that $w(\mu)$ lies in the fundamental chamber. For the left hand side we therefore obtain a contribution

$$\text{sign}(w)\exp(w(\Lambda + \rho + \lambda')) \cdot \text{mult}_{\Lambda'}(\lambda'),\tag{15.28}$$

which has to be compared to a contribution to the right hand side coming from the weight $\Lambda_i = w(\Lambda + \rho + \lambda') - \rho$ in the fundamental Weyl chamber, i.e. we have

$$\lambda' = w^{-1}(\Lambda_i + \rho) - \rho - \Lambda.\tag{15.29}$$

Summarizing these results, we arrive at the formula (15.23).

Let us now describe how the formula (15.23) can be implemented on a computer. To this end we first note that as an input for calculating a tensor product decomposition with (15.23), one needs the whole weight system $\{\lambda'\}$ of the module $V_{\Lambda'}$, but only the highest weight Λ of V_Λ. Further, because of the commutativity (15.3), without loss of generality one can always assume that $V_{\Lambda'}$ is a 'simpler' module (e.g. has smaller dimension) than V_Λ.

The formula (15.23) is already very explicit. It becomes even more transparent when it is translated into the following recipe, the Racah–Speiser algorithm. Given the highest weight Λ of V_Λ and the weight system $\{\lambda'\}$ of $V_{\Lambda'}$, consider the weights $\{\mu\} \equiv \{\lambda' + \Lambda + \rho\}$, where λ' belongs to the weight system of $V_{\Lambda'}$. Let us assume that the weight $\lambda' + \Lambda + \rho$ is not on the boundary of any Weyl chamber. Then there is a unique Weyl group element w such that $w(\lambda' + \Lambda + \rho)$ is inside the fundamental Weyl chamber. In this case we have a contribution of $\text{sign}(w) \cdot \text{mult}_{\Lambda'}(\lambda')$ to the Littlewood–Richardson coefficient $\mathcal{L}_{\Lambda\Lambda'}^{\Lambda_i}$ with $\Lambda_i = w(\Lambda + \rho + \lambda') - \rho$. The contributions for all weights λ' in the weight system of $V_{\Lambda'}$ have to be summed. In case $\lambda' + \Lambda + \rho$ lies on the boundary of some Weyl chamber, there is no Weyl group element that maps it to the interior of the fundamental Weyl chamber; therefore the weight λ' does not contribute to any Littlewood–Richardson coefficient.

As already described in section 10.3 (compare equation (10.23)), the unique Weyl group element $w \equiv w_\mu$ that maps a given weight μ (not lying on the boundary of any Weyl chamber) into the fundamental chamber can be determined as a product of fundamental Weyl reflections $w_{(i)}$ (i.e. reflections with respect to the hyperplane perpendicular to the ith simple root $\alpha^{(i)}$) as follows. First, one considers instead of μ the Weyl-transformed weight $\mu_1 := w_1(\mu)$, where w_1 is the fundamental reflection $w_1 := w_{(j_1)}$, with $j_1 \in \{1, 2, ..., r\}$ the smallest integer such that $\mu^{j_1} < 0$. Next one takes instead of μ_1 the weight $\mu_2 := w_2(\mu_1)$ with $w_2 := w_{(j_2)}$, with j_2 the smallest integer such that $(\mu_1)^{j_2} < 0$. Iterating this procedure, one ends up with a weight $\mu_n = w_n(\mu_{n-1}) = ... = w_n \circ w_{n-1} \circ \cdots \circ w_1(\mu)$ which obeys $(\mu_n)^j > 0$ for all $j = 1, 2, ..., r$ (none of the weights μ_i can have a Dynkin label equal to 0, because

otherwise μ_i, and hence also μ, would lie on the boundary of some Weyl chamber).
Then

$$w := w_n \circ \cdots \circ w_2 \circ w_1 \tag{15.30}$$

is the unique Weyl group element which one was looking for; in particular, the sign of w
is $\text{sign}(w) = (-1)^n$. The presentation of an element $w \in W$ as a product of fundamental
reflections is however not unique. The algorithm just described provides one specific
presentation of this type, which need not necessarily be reduced in the sense that the
number of fundamental reflections is minimal, i.e. the integer n appearing in (15.30)
may be larger than the length ℓ of the Weyl group element w; but as $\text{sign}(w)$ is fixed,
n must differ from ℓ by a multiple of 2.

15.8 Examples

Let us see how the recipe described above works in a non-trivial exam-
ple, namely the product of the adjoint module of $\mathfrak{g} = A_2 \cong \mathfrak{sl}(3)$ with
itself. The weights of this module have been displayed in (13.13) (see
also (3.30)). Adding to them the highest weight $\Lambda = \theta = (1,1)$, one obtains the weights $\mu - \rho = (2,2)$, $(3,0)$, $(0,3)$, twice $(1,1)$, $(0,0)$, $(2,-1)$,
and $(-1,2)$. The first six of these weights lie in the closure of the fundamental Weyl chamber and therefore need not be transformed. Also,
while the last two weights lie outside the fundamental Weyl chamber, the
weights $\mu = (3,0)$ and $(0,3)$ that are obtained by adding the Weyl vector
$\rho = (1,1)$ are mapped to themselves by the action of $w_{(1)}$ and $w_{(2)}$, respectively; hence they do not contribute to the tensor product coefficients.
The tensor product decomposition then reads

$$V_{(1,1)} \otimes V_{(1,1)} = V_{(2,2)} \oplus V_{(3,0)} \oplus V_{(0,3)} \oplus 2\,V_{(1,1)} \oplus V_{(0,0)}\,. \tag{15.31}$$

In physics it is common to denote irreducible modules by their dimension
(if several distinct irreducible modules have the same dimension, they
must be distinguished by additional notation; e.g. one denotes the conjugate of a module 'n' by '$\overline{n}$'). In this notation, the tensor product (15.31)
is written as $8 \times 8 = 27 + 10 + \overline{10} + 2 \cdot 8 + 1$.

As a slightly more complicated example, consider, still for $\mathfrak{g} = A_2$, the tensor product
between the adjoint module and the six-dimensional module $R_{(2,0)}$. Adding the highest
weight $(2,0)$ to the weights of the adjoint module, one obtains five weights that lie in the
closure of the fundamental Weyl chamber, namely $\mu - \rho = (3,1)$, $(1,2)$, $(0,1)$, and twice
$(2,0)$, as well as the two weights $\mu - \rho = (4,-1)$ and $(1,-1)$ outside the fundamental
chamber for which μ is mapped onto itself by $w_{(2)}$, and in addition the weight $\mu - \rho = (3,-2)$ for which $w_{(2)}(\mu) - \rho = w_{(2)}((4,-1)) - (1,1) = (2,0)$. Thus the weights $(4,-1)$
and $(1,-1)$ can be discarded, while the weight $(3,-2)$ cancels the contribution of one
of the two weights $(2,0)$. The resulting tensor product decomposition is thus

$$V_{(1,1)} \otimes V_{(2,0)} = V_{(3,1)} \oplus V_{(1,2)} \oplus V_{(0,1)} \oplus V_{(2,0)}\,, \tag{15.32}$$

or, in terms of dimensionalities, $8 \times 6 = 24 + 15 + \overline{3} + 6$.

The manipulations involved in these two examples are visualized in the following picture. (The weights μ which are mapped to themselves (and therefore do not contribute to the tensor product) are indicated by dots, while the corners of the shifted hexagons which are not marked by a dot stand for weights that need not be mapped at all, since they are already in the interior of the fundamental Weyl chamber.)

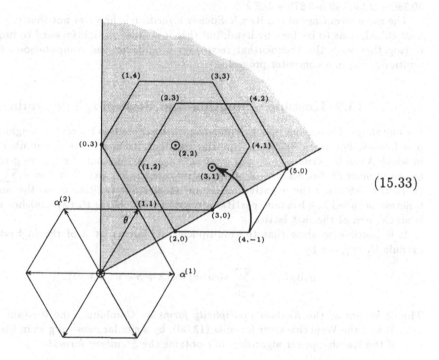

$$(15.33)$$

To see that the Racah-Speiser algorithm is still easily applicable to tensor products which already may seem quite involved, let us also study two tensor products of highest weight modules of the exceptional simple Lie algebra E_8. First, take the tensor product of the adjoint module $R_\theta = R_{\Lambda_{(1)}}$ with itself. Among the 248 weights λ of R_θ, there are eleven weights such that $\mu - \rho = \lambda + \theta$ lies in the closure of the fundamental Weyl chamber, namely $\mu - \rho = (2, 0, 0, 0, 0, 0, 0, 0)$, $(0, 1, 0, 0, 0, 0, 0, 0)$, $(0, 0, 0, 0, 0, 0, 1, 0)$, the zero weight, and eight times the weight $\theta = (1, 0, 0, 0, 0, 0, 0, 0)$. For 230 of the remaining weights $\mu - \rho$, at least one Dynkin label of $\mu + \rho$ vanishes so that $\mu + \rho$ is mapped to itself by a fundamental Weyl reflection; hence none of these weights contributes. There are seven other weights left which have to be analyzed further. These weights can be written as $\mu - \rho = \theta - \alpha^{(j)}$ for $j = 2, 3, \ldots, 8$; they obey $w_{(j)}(\theta - \alpha^{(j)} + \rho) - \rho = \theta$, and hence their contributions cancel those of all but one of the weights $\mu - \rho = \theta$. Thus one arrives at the decomposition

$$V_\theta \otimes V_\theta = V_{2\Lambda_{(1)}} \oplus V_{\Lambda_{(1)}} \oplus V_{\Lambda_{(2)}} \oplus V_{\Lambda_{(7)}} \oplus V_0 . \qquad (15.34)$$

The respective dimensionalities read $248 \times 248 = 27\,000 + 248 + 30\,380 + 3\,875 + 1$.

Finally, take the largest of these modules, $V_{\Lambda_{(2)}}$, and consider its tensor product with the adjoint. Now fifteen of the weights $\mu - \rho = \lambda + \Lambda_{(2)}$, with λ a weight of the adjoint, lie in the closure of the fundamental Weyl chamber, including in particular eight times the weight $\Lambda_{(2)}$. Of the remaining weights, only the seven weights $\mu - \rho = \theta - \Lambda_{(2)}$

contribute, each of which cancels the contribution of one of the weights $\Lambda_{(2)}$. One then obtains the decomposition

$$V_\theta \otimes V_{\Lambda_{(2)}} = V_{\Lambda_{(1)}+\Lambda_{(2)}} \oplus V_{\Lambda_{(1)}+\Lambda_{(7)}} \oplus V_{2\Lambda_{(1)}} \oplus V_{\Lambda_{(1)}} \oplus V_{\Lambda_{(2)}} \oplus V_{\Lambda_{(3)}} \oplus V_{\Lambda_{(7)}} \oplus V_{\Lambda_{(8)}}.$$

In terms of dimensions, this reads $248 \times 30\,380 = 4\,096\,000 + 779\,247 + 27\,000 + 248 + 30\,380 + 2\,450\,240 + 3\,875 + 147\,250$.

The main advantage of the Racah–Speiser algorithm is however not that it allows for such calculations to be done by hand, but that it is quite straightforward to implement it (together with the Freudenthal recursion formula for the computation of weight multiplicities) in a computer program.

*15.9 Kostant's function and Steinberg's formula

For any simple Lie algebra $\mathfrak{g}$, one defines the *Kostant function* P_K on the weight space of $\mathfrak{g}$ as follows. For any $\mathfrak{g}$-weight λ, the value $P_K(\lambda)$ is declared to be the number of ways in which λ can be written as a non-negative linear combination of negative $\mathfrak{g}$-roots, i.e. as the number of distinct sets of non-negative integers ℓ_α such that $\lambda = -\sum_{\alpha<0} \ell_\alpha \alpha$. (If in the definition the negative roots are replaced by positive ones, the analogous function is called the Kostant partition function.) Of course $P_K(\lambda)$ vanishes unless λ is an element of the root lattice.

It is possible to show that the multiplicity of the weight λ of the highest weight module V_Λ is given by

$$\mathrm{mult}_\Lambda(\lambda) = \sum_{w \in W} \mathrm{sign}(w)\, P_K(\lambda + \rho - w(\Lambda + \rho)). \tag{15.35}$$

This is known as the *Kostant multiplicity formula*. Combining the Kostant formula (15.35) and the Weyl character formula (13.30), by a similar reasoning as in the derivation of the Racah–Speiser algorithm one obtains the *Steinberg formula*

$$\mathcal{L}_{\Lambda\Lambda'}^{\Lambda_i} = \sum_{w,w' \in W} \mathrm{sign}(ww')\, P_K(\Lambda_i + 2\rho - w(\Lambda + \rho) - w'(\Lambda' + \rho)) \tag{15.36}$$

for tensor product multiplicities.

The formulæ (15.35) and (15.36) look quite suggestive, as they express the multiplicities $\mathrm{mult}_\Lambda(\lambda)$ and $\mathcal{L}_{\Lambda\Lambda'}^{\Lambda_i}$ rather directly through the weights of interest. But an obvious disadvantage is the (double) summation over the Weyl group W which is problematic for large W. For practical purposes (and in particular for computer implementation) the Freudenthal formula for the multiplicities of weights and the Racah–Speiser algorithm for the Littlewood–Richardson coefficients are therefore much more convenient.

*15.10 Affine Lie algebras

The description of tensor products does not generalize from finite-dimensional simple to affine Lie algebras as nicely as many other aspects of the representation theory. The reason is that any non-trivial module of an affine algebra is infinite-dimensional, which in general leads to infinite multiplicities in tensor product decompositions. To handle such infinite multiplicities, one must keep track of the grade at which an irreducible module arises in the tensor product; at any fixed grade the multiplicity is still finite. (Taking these finite multiplicities as the coefficients of a power series, one obtains a function which can be interpreted as the character of another Lie algebra, which contains the Virasoro algebra.)

Note that the weights add up when forming tensor products, and hence (compare e.g. the formula (13.51)) in particular so do the levels. Thus any irreducible module contained in the tensor product of modules which have level $k_1^\vee$ and $k_2^\vee$, respectively, has level $k^\vee = k_1^\vee + k_2^\vee$. In certain applications of affine Lie algebras in physics, such as in two-dimensional conformal field theory, a different product of irreducible highest weight modules is more interesting than the tensor product. This 'fusion product', to be described in section 22.13, preserves the value of the level. The analogues of the tensor product coefficients for the fusion product can be obtained by a direct generalization of the Racah–Speiser formula (15.23) to the affine case, which will be given in equation (22.54).

Summary:

The tensor product of finite-dimensional highest weight modules of simple Lie algebras is completely reducible into irreducible modules. The tensor product coefficients of this decomposition are conveniently calculated via the Racah–Speiser formula (15.23). For $\mathfrak{sl}(2)$ and for small modules of other simple Lie algebras, tensor product decompositions can also be obtained with the help of conjugacy class selection rules and sum rules for the dimension and second order Dynkin index.

Keywords:

Tensor product of representations and modules, (anti-) symmetric coupling, additive quantum number, tensor product multiplicity, Littlewood–Richardson coefficient, Clebsch–Gordan series, simply reducible tensor product, conjugacy class selection rule, character and dimension sum rule.

Exercises:

Compute the decomposition of the tensor product of the triplet 3 times the anti-triplet $\overline{3}$ of $\mathfrak{sl}(3)$ (compare chapter 3) with the following methods:

Exercise 15.1

a) the 'brute force' method described before (15.14);

b) by making use of the dimension sum rule (15.21) and other general properties of tensor products;

c) with the Racah–Speiser algorithm.

Perform the same manipulations for more complicated tensor products of $\mathfrak{sl}(3)$-representations.

Verify that the tensor product of a representation with itself can be decomposed as described in (15.6).

Exercise 15.2

The tensor product of the two-dimensional irreducible module $(1) \equiv V_{\Lambda_{(1)}}$ of $\mathfrak{sl}(2)$ with itself decomposes as $(1) \otimes (1) = (0) \oplus (2)$ into irreducible modules. Write down the corresponding representation matrices for the generators H and $E_{\pm}$ of $\mathfrak{sl}(2)$.

Exercise 15.3

What is the conjugacy class of an irreducible module whose weight system contains the weight $\lambda = 0$?
Is this still true for reducible modules with the same property?

Exercise 15.4

The Lie algebra B_r and C_r are *dual* to each other, i.e. their Cartan matrices are each others transpose. For some people (including the authors) it is difficult to remember which Cartan matrix belongs to which algebra. Convince yourself that by (15.19) this information can be recovered from the requirement that the r th simple root of B_r must be in the zero conjugacy class.

Exercise 15.5

Prove the formula for the Dynkin index of a tensor product that is used to obtain (15.22).
Hint: Recall the definition of the Dynkin index in terms of a trace of representation matrices and insert the formula (15.1).

Exercise 15.6

Design a computer program which performs tensor product decompositions with the help of the Freudenthal recursion formula and the Racah–Speiser algorithm. (Make sure that only the weight system of the 'smaller' factor of the product needs to be calculated.)
Technical remark: In order to avoid restrictions on the accessible tensor products, it is favorable to use a programming language which allows for dynamical memory allocation and can handle arbitrarily large integers.
Try to quantify the gain in computer time and memory of such a program as compared to a program based on the brute force method that was described before (15.14).

Exercise 15.7

16
Clebsch–Gordan coefficients
and tensor operators

16.1 Isomorphisms involving tensor products of modules

In the previous chapter we have encountered several isomorphisms involving tensor products of representations, in particular the decomposition (15.11) of the tensor product of two modules and the associativity (15.3) and commutativity (15.4) of the tensor product. In this chapter, we describe what these isomorphisms look like in a concrete basis, i.e. provide the changes of the basis that are involved. The coefficients of these basis changes make their appearance in various applications.

The explicit formulæ in this chapter are all presented for the Lie algebra $\mathfrak{sl}(2)$, or equivalently, for its compact real form $\mathfrak{su}(2)$ (the generalization of the concepts introduced in this chapter to other simple Lie algebras is in principle straightforward, but it is also rather tedious). This case is most interesting for practical purposes, since $\mathfrak{sl}(2)$ describes e.g. angular momentum in non-relativistic quantum mechanics. The main simplification in the case of $\mathfrak{sl}(2)$ stems from the two closely related facts that for $\mathfrak{sl}(2)$ all tensor products are simply reducible (i.e. all Littlewood–Richardson coefficients are either zero or one) and that the weights of irreducible representations are non-degenerate. These features imply that the freedom in choosing the relevant bases just consists of phase factors, whereas in the general case it is described by unitary matrices.

16.2 Clebsch–Gordan coefficients and $3j$-symbols

As we restrict our attention to $\mathfrak{sl}(2)$, and as a major application is in the description of angular momentum in quantum mechanics, we will use the language of spins instead of weights and the bra-ket notation that has been described at the end of chapter 5. Thus as generators of $\mathfrak{sl}(2)$ we take J_0 and $J_\pm$ with Lie brackets $[J_0, J_\pm] = \pm J_\pm$ and $[J_+, J_-] = 2J_0$, we often suppress the symbols for the representations in which the generators are taken, and the basis vectors of the irreducible module with spin j are chosen as $|j, m\rangle$ with $m = j, j - 1, \dots, -j + 1, -j$, which satisfy

$$
\begin{aligned}
J_0 |j, m\rangle &= m \cdot |j, m\rangle \\
J_\pm |j, m\rangle &= \sqrt{j(j + 1) - m(m \pm 1)} \cdot |j, m\pm 1\rangle,
\end{aligned}
\tag{16.1}
$$

as in (5.42). One can check (see exercise 16.1) that the vectors $|j, m\rangle$ form a complete and orthonormal set. Furthermore, the quadratic Casimir operator acts as

$$
\vec{J}^2 |j, m\rangle = j(j + 1) \cdot |j, m\rangle,
\tag{16.2}
$$

i.e. has eigenvalue $j(j + 1)$ on this module.

Let us now consider the tensor product $V_{j_1} \otimes V_{j_2}$ of two irreducible modules of $\mathfrak{sl}(2)$ with spins j_1 and j_2. A natural basis of this tensor product is given by the $(2j_1 + 1)(2j_2 + 1)$ vectors

$$
|j_1\, j_2, m_1\, m_2\rangle := |j_1, m_1\rangle \otimes |j_2, m_2\rangle.
\tag{16.3}
$$

These basis vectors of the tensor product are eigenvectors of the operators $R_{j_1}(J_0) \otimes \mathbb{1}_{j_2}$ and $\mathbb{1}_{j_1} \otimes R_{j_2}(J_0)$ to the eigenvalues m_1 and m_2, respectively, and of $R_{j_1}(\vec{J}^2) \otimes \mathbb{1}_{j_2}$ and $\mathbb{1}_{j_1} \otimes R_{j_2}(\vec{J}^2)$ to the eigenvalues $j_1(j_1 + 1)$ and $j_2(j_2 + 1)$, respectively. (Here for clarity we did indicate the representation in which J_0 is to be taken; $\mathbb{1}_{j_1} \equiv R_{j_1}(\mathbb{1})$ is the representation matrix of the unit $\mathbb{1}$ of the universal enveloping algebra $\mathsf{U}(\mathfrak{sl}(2))$ in the spin j_1 representation, i.e. the $(2j_1 + 1) \times (2j_1 + 1)$ unit matrix.) However, what we are really after, are the eigenvectors of J_0 evaluated in the tensor product representation, i.e. the eigenvectors of the operator

$$
R_{j_1 \times j_2}(J_0) := (R_{j_1} \otimes R_{j_2})(J_0) = R_{j_1}(J_0) \otimes \mathbb{1}_{j_2} \oplus \mathbb{1}_{j_1} \otimes R_{j_2}(J_0)
\tag{16.4}
$$

in the irreducible modules into which the tensor product can be decomposed. As we are working with modules of $\mathfrak{sl}(2)$, we can characterize these irreducible components uniquely by their spin J, or equivalently, by the eigenvalue $J(J + 1)$ of the quadratic Casimir operator. Accordingly we consider a second basis $|J, M\rangle \equiv |j_1\, j_2; J, M\rangle$ of the tensor product $V_{j_1} \otimes V_{j_2}$ which is characterized by

$$
\begin{aligned}
R_{j_1 \times j_2}(J_0) |j_1\, j_2; J, M\rangle &= M\, |j_1\, j_2; J, M\rangle \\
R_{j_1 \times j_2}(\vec{J}^2) |j_1\, j_2; J, M\rangle &= J(J + 1)\, |j_1\, j_2; J, M\rangle.
\end{aligned}
\tag{16.5}
$$

The vectors $|j_1\, j_2, m_1\, m_2\rangle$, with m_1 and m_2 ranging over the values $m_1 = -j_1, -j_1 + 1, \dots, j_1$ and $m_2 = -j_2, -j_2 + 1, \dots, j_2$, form a complete system in $V_{j_1} \otimes V_{j_2}$, and the same is true for the vectors $|j_1\, j_2; J, M\rangle$ with $J = |j_1 - j_2|, |j_1 - j_2| + 1, \dots, j_1 + j_2$ and $M = -J, -J + 1, \dots, J$; in fact they constitute orthonormal bases of the tensor product module. Therefore there is a unitary transformation connecting these two sets of vectors, which in the bra-ket notation is written as

$$|j_1 j_2; J, M\rangle = \sum_{m_1, m_2} |j_1 j_2, m_1 m_2\rangle \langle j_1 j_2, m_1 m_2 | J, M\rangle. \qquad (16.6)$$

The coefficients

$$\langle j_1 j_2, m_1 m_2 | J, M\rangle \equiv \langle j_1 j_2, m_1 m_2 | j_1 j_2; J, M\rangle \qquad (16.7)$$

appearing in this transformation are called *Clebsch–Gordan coefficients*. The phases of these coefficients are not yet determined unambiguously by the requirement that the transformation must be unitary. This stems from the fact that an eigenvector equation (together with a required normalization) does not provide a natural basis for its solutions. If the eigenspace is one-dimensional, the remaining freedom is a phase; in higher dimensions, the freedom is described by unitary transformations. (Notice however, that by equation (16.1) we already fix the relative phases within each irreducible representation.) Hence to fix this freedom, we prescribe the relative phases in the two basis sets of eigenvectors by requiring that the numbers $\langle j_1 j_2, j_1 j_1 - J | j_1 j_2; J, J\rangle$ are real and positive for all possible values of J. It turns out that with this convention *all* Clebsch–Gordan coefficients are real.

Moreover, the Clebsch–Gordan coefficients vanish unless certain selection rules are fulfilled. Namely, first of all, it is of course necessary that the module with spin J occurs in the decomposition of the tensor product of the two modules with spins j_1 and j_2; as seen in the previous chapter, this implies that

$$|j_1 - j_2| \le J \le j_1 + j_2 \qquad (16.8)$$

and

$$J - j_1 - j_2 \in \mathbb{Z}. \qquad (16.9)$$

Second, when applied to an eigenstate of the operators $R_{j_1}(J_0)$, $R_{j_2}(J_0)$ and $R_{j_1 \times j_2}(J_0)$, the definition of the tensor product representation as the direct sum (16.4) implies that

$$m_1 + m_2 = M. \qquad (16.10)$$

This is a necessary condition on the eigenvalues of J_0 for the Clebsch–Gordan coefficient to be non-zero, i.e. constitutes a selection rule.

Further properties of the Clebsch–Gordan coefficients follow from the fact that the two systems $|j_1 j_2; J, M\rangle$ and $|j_1 j_2, m_1 m_2\rangle$, with J and M, and m_1 and m_2, respectively, in the range mentioned above, are complete and orthonormal. Hence the basis transformation connecting these two bases is unitary. In the bra-ket notation this is expressed as

$$\delta_{J,J'}\delta_{M,M'} = \langle J, M \mid J', M' \rangle$$

$$= \sum_{m_1=-j_1}^{j_1} \sum_{m_2=-j_2}^{j_2} \langle j_1 j_2, m_1 m_2 \mid J, M \rangle \langle j_1 j_2, m_1 m_2 \mid J', M' \rangle$$

$$\delta_{m_1,m_1'}\delta_{m_2,m_2'} = \langle j_1 j_2, m_1 m_2 \mid j_1 j_2, m_1' m_2' \rangle$$

$$= \sum_{J=|j_1-j_2|}^{j_1+j_2} \sum_{M=-J}^{J} \langle j_1 j_2, m_1 m_2 \mid J, M \rangle \langle j_1 j_2, m_1' m_2' \mid J, M \rangle .$$

$$(16.11)$$

The Clebsch–Gordan coefficients also possess a number of symmetries. To exhibit them most clearly, it is convenient to consider rescaled coefficients, the so-called Wigner's $3j$-*symbols*

$$\begin{pmatrix} j_1 & j_2 & J \\ m_1 & m_2 & -M \end{pmatrix} := \frac{(-1)^{j_1-j_2+M}}{\sqrt{2J+1}} \langle j_1 j_2, m_1 m_2 \mid J, M \rangle . \qquad (16.12)$$

The $3j$-symbols defined by (16.12) are invariant under cyclic permutation of the three columns, while under exchange of any two columns they change by a sign factor $(-1)^{j_1+j_2+J}$. The same factor $(-1)^{j_1+j_2+J}$ arises when the signs of m_1, m_2 and M are flipped simultaneously.

16.3 Calculation of Clebsch–Gordan coefficients

To illustrate how Clebsch–Gordan coefficients are computed in practice, let us have a look at the simplest non-trivial example, the tensor product of two two-dimensional representations, i.e. $j_1 = j_2 = \frac{1}{2}$. As can be seen from the following picture, which is a special case of the figure (15.13), J then can only take the two values 0 and 1.

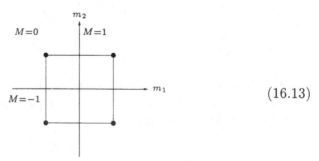

$$(16.13)$$

The four basis vectors of the tensor product module are therefore linear combinations of the three basis vectors of the spin-1 module and the single basis vector of the spin-0 module.

In the tensor product module there is a unique (up to a phase) vector with spin component $M = 1$, namely

$$|1,1\rangle = |\tfrac{1}{2},\tfrac{1}{2}\rangle \otimes |\tfrac{1}{2},\tfrac{1}{2}\rangle , \qquad (16.14)$$

corresponding to the dot in the upper left corner in (16.13). Similarly, the dot in the lower left corner corresponds to the unique vector

$$|1, -1\rangle = |\tfrac{1}{2}, -\tfrac{1}{2}\rangle \otimes |\tfrac{1}{2}, -\tfrac{1}{2}\rangle \tag{16.15}$$

with $M = -1$. On the other hand, the two vectors $|\tfrac{1}{2}, -\tfrac{1}{2}\rangle \otimes |\tfrac{1}{2}, \tfrac{1}{2}\rangle$ and $|\tfrac{1}{2}, \tfrac{1}{2}\rangle \otimes |\tfrac{1}{2}, -\tfrac{1}{2}\rangle$, which have eigenvalue $M = 0$, must be linear combinations of the two vectors $|1, 0\rangle$ and $|0, 0\rangle$. To determine these combinations, one notes that $\sqrt{2} \cdot |1, 0\rangle$ can be obtained by applying the operator

$$R_{j_1 \times j_2}(J_-) \equiv R_{j_1}(J_-) \otimes \mathbf{1}_{j_2} \oplus \mathbf{1}_{j_1} \otimes R_{j_2}(J_-) \tag{16.16}$$

to the vector $|1, 1\rangle$:

$$
\begin{aligned}
\sqrt{2}\,|1, 0\rangle &= R_{j_1 \times j_2}(J_-)|1, 1\rangle \\
&= R_{j_1}(J_-)|\tfrac{1}{2}, \tfrac{1}{2}\rangle \otimes |\tfrac{1}{2}, \tfrac{1}{2}\rangle + |\tfrac{1}{2}, \tfrac{1}{2}\rangle \otimes R_{j_2}(J_-)|\tfrac{1}{2}, \tfrac{1}{2}\rangle \\
&= |\tfrac{1}{2}, -\tfrac{1}{2}\rangle \otimes |\tfrac{1}{2}, \tfrac{1}{2}\rangle + |\tfrac{1}{2}, \tfrac{1}{2}\rangle \otimes |\tfrac{1}{2}, -\tfrac{1}{2}\rangle .
\end{aligned} \tag{16.17}
$$

The vector $|0, 0\rangle$ must be orthogonal to this vector. This requirement determines it uniquely (up to a phase, which depends on the chosen convention):

$$\sqrt{2}\,|0, 0\rangle = |\tfrac{1}{2}, -\tfrac{1}{2}\rangle \otimes |\tfrac{1}{2}, \tfrac{1}{2}\rangle - |\tfrac{1}{2}, \tfrac{1}{2}\rangle \otimes |\tfrac{1}{2}, -\tfrac{1}{2}\rangle . \tag{16.18}$$

Thus we have

$$
\begin{aligned}
\langle \tfrac{1}{2}\tfrac{1}{2}, \tfrac{1}{2}\tfrac{1}{2}\,|\,1, 1\rangle &= \langle \tfrac{1}{2}\tfrac{1}{2}, -\tfrac{1}{2}-\tfrac{1}{2}\,|\,1, -1\rangle = 1 \\
\langle \tfrac{1}{2}\tfrac{1}{2}, \pm\tfrac{1}{2} \mp\tfrac{1}{2}\,|\,1, 0\rangle &= \tfrac{1}{\sqrt{2}}, \qquad \langle \tfrac{1}{2}\tfrac{1}{2}, \pm\tfrac{1}{2} \mp\tfrac{1}{2}\,|\,0, 0\rangle = \pm\tfrac{1}{\sqrt{2}},
\end{aligned} \tag{16.19}
$$

while all other Clebsch–Gordan coefficients are zero.

The general procedure for computing Clebsch–Gordan coefficients is just a straightforward generalization of this simple calculation. First, the highest weight vector of the irreducible module of the tensor product with maximal spin $J = j_1 + j_2$ is equal to the product of the two highest weight vectors $|j_1, j_1\rangle$ and $|j_2, j_2\rangle$ of the factors of the tensor product. Any other vector in the same irreducible module can be obtained by applying $R_{j_1 \times j_2}(J_-)$ repeatedly to this vector. The highest weight vector in the irreducible module with $J = j_1 + j_2 - 1$ is then (up to a phase) fixed by the requirement that it is orthogonal to $|J, J-1\rangle$ in the two-dimensional subspace with eigenvalue $M = J - 1$. Then again, all vectors in the irreducible module with spin $J = j_1 + j_2 - 1$ are found by application of $R_{j_1 \times j_2}(J_-)$ to the highest weight vector. The highest weight vector of the irreducible representation with $J = j_1 + j_2 - 2$ is then in the orthogonal complement of $|J, J-2\rangle$ and $|J-1, J-2\rangle$ in the three-dimensional space of vectors with eigenvalue $M = J - 2$. Iterating this procedure, all Clebsch–Gordan coefficients are constructed.

Let us also mention that the computer package Mathematica has built-in functions for Clebsch–Gordan coefficients ('ClebschGordan') and 3j-symbols ('ThreeJSymbol'), and also for the 6j-symbols ('SixJSymbol') that will be introduced below.

Choosing the bases of the irreducible modules in the standard manner as above, the Clebsch–Gordan coefficients are typically algebraic numbers; on the other hand, the coefficient $\langle j_1\,j_2, j_1\,j_2\,|\,j_1 + j_2, j_1 + j_2\rangle$ is always equal to one, i.e. in particular an integer. Now the integrality property of the structure constants in a Chevalley basis can be regarded as a property of the representation matrices of the adjoint representation. It is therefore not surprising that by a judicious choice of bases this property can be achieved for every finite-dimensional irreducible representation. Analogously, it is in fact possible to choose specific bases for the irreducible modules in such a way that with respect to these bases *all* Clebsch–Gordan coefficients are integers [Smith 1965]. However, these bases are no longer orthonormal, so that for all practical purposes they are inconvenient.

16.4 Formulæ for Clebsch–Gordan coefficients

We now present a few useful relations for Clebsch–Gordan coefficients. To start with, we evaluate the matrix element $\langle j_1 j_2, m_1 m_2 \mid R_{j_1 \times j_2}(J_\pm) \mid J, M \rangle$ of the operator $R_{j_1 \times j_2}(J_\pm)$ in two different ways, namely by considering the action of the operator either on the 'bra' vector $\langle \cdot \mid$ to the left, or on the 'ket' vector $\mid \cdot \rangle$ to the right. This leads to the recursion relation

$$\sqrt{J(J+1) - M(M \pm 1)}\, \langle j_1 j_2, m_1 m_2 \mid J, M \rangle$$

$$= \sqrt{j_1(j_1 + 1) - m_1(m_1 \pm 1)}\, \langle j_1 j_2, m_1 \pm 1\, m_2 \mid J, M \pm 1 \rangle \qquad (16.20)$$

$$+ \sqrt{j_2(j_2 + 1) - m_2(m_2 \pm 1)}\, \langle j_1 j_2, m_1 m_2 \pm 1 \mid J, M \pm 1 \rangle$$

for the $\mathfrak{sl}(2)$ Clebsch–Gordan coefficients (see exercise 16.2). This can be employed to obtain fully explicit formulæ for the Clebsch–Gordan coefficients or, equivalently, the $3j$-symbols. The result for the $3j$-symbols, known as the *Racah formula*, reads

$$\begin{pmatrix} j & k & l \\ m & n & p \end{pmatrix} = (-1)^{j-k-p} [\Delta(j, k, l)\, (j + m)!\, (j - m)!\, (k + n)!\, (k - n)!$$

$$\cdot\, (l + p)!\, (l - p)!]^{1/2} \cdot \sum_q (-1)^q\, [q!\, (j - q - m)!\, (k - q + n)!$$

$$\cdot\, (j + k - l - q)!\, (l - k + q + m)!\, (l - j + q - n)!]^{-1},$$

where

$$\Delta(j, k, l) := \frac{(j+k-l)!\,(k+l-j)!\,(l+j-k)!}{(j+k+l+1)!}, \qquad (16.21)$$

and where the selection rules require that $m + n + p = 0$ and $|j - k| \le l \le j + k$. The sum over q in the Racah formula runs over all integers for which the corresponding expressions make sense; the number of non-vanishing terms in this sum is the minimum of the nine numbers $j \pm m$, $k \pm n$, $l \pm p$, $j + k - l$, $k + l - j$, and $l + j - k$. Obviously the recursion relations (16.20) are much more transparent and by far easier to use than the closed formula that follows from it.

All the considerations presented so far generalize in a straightforward manner to any other simple Lie algebra. However, technically this is clearly rather involved, due to the presence of higher multiplicities of the weights of irreducible modules and of the tensor product decompositions. The latter imply in particular that generically the Clebsch–Gordan coefficients carry an additional degeneracy index, and that the freedom in describing the Clebsch–Gordan coefficients is expressed by unitary matrices rather than just phases. Specific Clebsch–Gordan coefficients have been computed for various Lie algebras, and tables can be found in the literature, see e.g. [Chen *et al.* 1987]. However, unlike the Racah formula in the case of $\mathfrak{sl}(2)$, no explicit general formulæ are known for those Clebsch–Gordan coefficients.

16.5 Intertwiner spaces

One can interpret the Clebsch–Gordan coefficients also from a slightly different point of view. Namely, one may analyze what the intertwiners from an irreducible module V_{Λ_i} that appears in the tensor product $V_\Lambda \otimes V_{\Lambda'}$ of two highest weight modules to that tensor product module look like. Recall that Schur's Lemma (see chapter 5) states that the space of intertwiners of two *irreducible* modules is one-dimensional if the modules are isomorphic, and zero-dimensional otherwise. The tensor product decomposition (15.11) then tells us that the Littlewood–Richardson coefficient $\mathcal{L}_{\Lambda\Lambda'}{}^{\Lambda_i}$ is equal to the dimension of the space $\mathcal{V}_{\Lambda\Lambda'}{}^{\Lambda_i} = \mathrm{span}\left\{\left(\begin{smallmatrix} & \Lambda_i & \\ \Lambda & & \Lambda'\end{smallmatrix}\right)\right\}$ of intertwiners $\left(\begin{smallmatrix} & \Lambda_i & \\ \Lambda & & \Lambda'\end{smallmatrix}\right)$ from V_{Λ_i} to $V_\Lambda\otimes V_{\Lambda'}$. According to equation (16.6), the Clebsch–Gordan coefficients for fixed J are therefore nothing else than the matrix elements of the intertwiners $\left(\begin{smallmatrix} & \Lambda_i & \\ \Lambda & & \Lambda'\end{smallmatrix}\right) \equiv \left(\begin{smallmatrix} & 2J & \\ 2j_1 & & 2j_2\end{smallmatrix}\right)$ for a definite choice of the bases of all modules. Symbolically, one can depict these intertwiners, respectively the associated Clebsch–Gordan coefficients, by trivalent graphs:

$$
\begin{array}{c}
\uparrow \Lambda \\[4pt]
\bullet\, \alpha \\[4pt]
\Lambda_1 \nearrow \quad \nwarrow \Lambda_2
\end{array}
\tag{16.22}
$$

In order to cover tensor products of arbitrary irreducible representations of finite-dimensional simple Lie algebras, in this figure we use labels Λ, Λ' etc. to denote highest weights, and we have labelled the vertex by a degeneracy index α, which is present if the intertwiner space has dimension larger than 1. Furthermore, the arrows distinguish between the representations whose tensor product is formed and the irreducible subrepresentation that is contained in that tensor product. In the $\mathfrak{sl}(2)$ case the multiplicity index and the arrows are not needed.

Later on we will deal with tensor products of intertwiner spaces, and as a consequence we will combine several trivalent graphs of the type (16.22) to larger graphs. When doing so, we will encounter vertices with one incoming and two outgoing lines, which is opposite to the orientation of the lines in (16.22). This is consistent provided one introduces the convention that inverting the orientation of an arrow amounts to considering the conjugate module, i.e.

$$\underset{\Lambda}{\xrightarrow{\hspace{1.5cm}}} \quad = \quad \underset{\Lambda^+}{\xleftarrow{\hspace{1.5cm}}} \tag{16.23}$$

(That this prescription works is related to the fact that whenever a representation R_1 is contained in the tensor product $R_2 \otimes R_3$, then the conjugate representation R_2^+ is contained in $R_1^+ \otimes R_3$.)

In elementary particle physics, the representation theoretic statement that $\mathcal{V}_{\Lambda\Lambda'}{}^{\Lambda_i}$ is non-zero manifests itself in the fact that two particles carrying the representations R_Λ and $R_{\Lambda'}$ of a symmetry algebra can interact to form a particle carrying the representation R_{Λ_i}. Graphs like (16.22), or also the intertwiners themselves, are therefore often called 'three-point couplings'.

In terms of the intertwiner spaces, the isomorphism $V_\Lambda \otimes V_{\Lambda'} \cong V_{\Lambda'} \otimes V_\Lambda$ which describes the commutativity of the tensor product gives rise to an isomorphism

$$\mathsf{R}: \quad \mathcal{V}_{\Lambda\Lambda'}{}^{\Lambda_i} \to \mathcal{V}_{\Lambda'\Lambda}{}^{\Lambda_i} \tag{16.24}$$

of intertwiner spaces. This isomorphism squares to the identity, $\mathsf{R}^2 = id$. Choosing bases of $\mathcal{V}_{\Lambda\Lambda'}{}^{\Lambda_i}$ and $\mathcal{V}_{\Lambda'\Lambda}{}^{\Lambda_i}$ such that the matrix describing this map is diagonal, each diagonal entry must then be $+1$ or -1. This sign indicates whether a certain coupling is 'symmetric' or 'anti-symmetric', i.e. whether the relevant irreducible module appears in the symmetric part V_{s} or antisymmetric part V_{a} of the tensor product (compare the decomposition (15.6)). In the case of $\mathfrak{sl}(2)$, the sign is given by $(-1)^{j_1 + j_2 + J}$ (because of the spin selection rule for the tensor product, the exponent is always an integer).

16.6 Racah coefficients and $6j$-symbols

Let us now turn to the isomorphism $(V_\Lambda \otimes V_{\Lambda'}) \otimes V_{\Lambda''} \cong V_\Lambda \otimes (V_{\Lambda'} \otimes V_{\Lambda''})$ for the tensor product of three modules. This corresponds to an isomorphism

$$\mathsf{F}: \quad \sum_i \mathcal{V}_{\Lambda\Lambda'}{}^{\Lambda_i} \otimes \mathcal{V}_{\Lambda_i\Lambda''}{}^{\Lambda_j} \to \sum_i \mathcal{V}_{\Lambda\Lambda_i}{}^{\Lambda_j} \otimes \mathcal{V}_{\Lambda'\Lambda''}{}^{\Lambda_i} \tag{16.25}$$

of intertwiner spaces. It is worth mentioning that the tensor product of the underlying vector spaces is associative, so that as long as one is only interested in the vector space structure it is not necessary to write brackets. In contrast, according to (16.25) the tensor product of Lie algebra modules is associative only up to unitary equivalence (a tensor product with this property is often referred to as being 'quasi-associative'.) Also note that already for $\mathfrak{sl}(2)$ in the decomposition of the tensor product (16.25) (which is then $(2j_1 + 1) \cdot (2j_2 + 1) \cdot (2j_3 + 1)$-dimensional) into irreducible modules, irreducible modules typically appear with multiplicity larger than one.

A natural orthonormal basis for the tensor product module on the left hand side of (16.25) is obtained by first coupling the first two modules to a module of some spin k using the relevant Clebsch–Gordan coefficients, and then coupling this spin-k module with the third module:

$$|(j_1\, j_2)\, k\, j_3\, ;\, J, M\rangle := \sum_{m_1, m_2, m_3, m} |j_1\, j_2\, j_3, m_1\, m_2\, m_3\rangle$$
$$\cdot\, \langle j_1\, j_2, m_1\, m_2\, |\, k, m\rangle\, \langle k\, j_3, m\, m_3\, |\, J, M\rangle .$$
(16.26)

The analogous basis for the right hand side of (16.25) is obtained when one starts by coupling the second and third irreducible module; this leads to the vectors

$$|j_1\, (j_2\, j_3)\, k\, ;\, J, M\rangle := \sum_{m_1, m_2, m_3, m} |j_1\, j_2\, j_3, m_1\, m_2\, m_3\rangle$$
$$\cdot\, \langle j_2\, j_3, m_2\, m_3\, |\, k, m\rangle\, \langle j_1\, k, m_1\, m\, |\, J, M\rangle .$$
(16.27)

These two bases are related by a unitary basis transformation with coefficients of the form

$$\langle j_1\, (j_2\, j_3)\, k\, ;\, J, M\, |\, (j_1\, j_2)\, k'\, j_3\, ;\, J', M'\rangle =: \delta_{J, J'} \delta_{M, M'}$$
$$\cdot\, (-1)^{j_1 + j_2 + j_3 + J}\, \sqrt{(2k+1)(2k'+1)}\, \begin{Bmatrix} j_1 & j_2 & k' \\ j_3 & J & k \end{Bmatrix} .$$
(16.28)

Similarly as in (16.12), here a conventional factor has been split off on the right hand side such as to define a new set of symbols, the so-called Wigner's $6j$-symbols $\begin{Bmatrix} j_1 & j_2 & k \\ j_3 & J & l \end{Bmatrix}$.

In short, the $6j$-symbols describe how two different orderings of coupling modules are related. In terms of the pictorial description (16.22) of the couplings, $\begin{Bmatrix} \Lambda_1 & \Lambda_2 & \mu' \\ \Lambda_3 & \Lambda & \mu \end{Bmatrix}$ thus describes the mapping

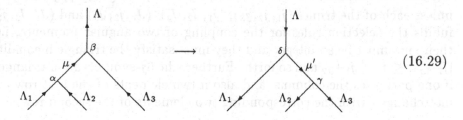
(16.29)

(the multiplicity labels α, β, γ, δ attached to the vertices, as well as the arrows, are included for completeness; in the $\mathfrak{sl}(2)$ case they are absent). A more suggestive diagrammatic description is obtained in terms of the dual diagrams, in which one replaces vertices by faces and all edges by lines perpendicular to them. Doing so, in place of (16.29) one obtains the

picture

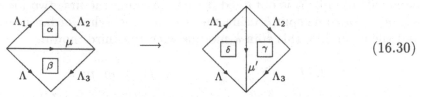

$$\tag{16.30}$$

Note that the multiplicity labels are now attached to the faces rather than to the vertices of the diagrams. Furthermore, one may now glue the diagrams on the two sides to each other by identifying the respective lines which carry identical labels; thereby one obtains a tetrahedron. This tetrahedron can be viewed as the pictorial representation of the $6j$-symbol; projected to the plane, it looks like

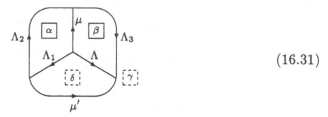

$$\tag{16.31}$$

Here three triangular faces of the tetrahedron are displayed directly, while the fourth face – the one labelled by the multiplicity index γ – is represented by the rest of the plane. (Thus one recovers the tetrahedron by a one-point compactification of the plane, e.g. using stereographic projection.) Also, depending on the ordering of the edges of a face, the face is oriented positively or negatively; this is indicated in the diagram by enclosing the multiplicity index in a box with solid and dashed lines, respectively.

The $6j$-symbols have again a number of nice properties. First, they are all real. Second, there is the selection rule that $\left\{ \begin{smallmatrix} j_1 & j_2 & j_3 \\ J_1 & J_2 & J_3 \end{smallmatrix} \right\}$ vanishes unless each of the triples (j_1, j_2, j_3), (j_1, J_2, J_3), (J_1, j_2, J_3) and (J_1, J_2, j_3) fulfills the selection rules for the coupling of two angular momenta, i.e. their sum must be an integer and they must satisfy the triangle inequality $|j_1 - j_2| \le j_3 \le j_1 + j_2$ and so forth. Further, the $6j$-symbols are unchanged if one permutes the columns, and also if two elements of the first row are interchanged with the corresponding two elements of the second row, e.g.

$$\left\{ \begin{matrix} j_1 & j_2 & j_3 \\ J_1 & J_2 & J_3 \end{matrix} \right\} = \left\{ \begin{matrix} J_1 & J_2 & j_3 \\ j_1 & j_2 & J_3 \end{matrix} \right\} . \tag{16.32}$$

These symmetries form the group $\mathcal{S}_4$; they are called the *tetrahedral* symmetries of the $6j$-symbols. For general finite-dimensional simple Lie algebras one must also take into account the multiplicity labels of the faces of the tetrahedra, which result in a matrix structure of the '6Λ'-symbols according to $\left\{ \begin{smallmatrix} \Lambda_1 & \Lambda_2 & \mu \\ \Lambda_3 & \Lambda & \mu' \end{smallmatrix} \right\}_{\alpha\beta}^{\gamma\delta}$, as well as the orientation of their edges. Nevertheless, the 6Λ-symbols again possess an analogous S_4-symmetry; the explicit form of the S_4-transformations then involves also the multiplicity labels. A peculiarity of the $\mathfrak{sl}(2)$ case is the presence of additional symmetries; they are of the type

$$\left\{ \begin{matrix} k & j_2 & j_3 \\ l & j_3 & j_4 \end{matrix} \right\} = \left\{ \begin{matrix} k & (j_1+j_2+j_3-j_4)/2 & (j_1+j_2-j_3+j_4)/2 \\ l & (j_1-j_2+j_3+j_4)/2 & (-j_1+j_2+j_3+j_4)/2 \end{matrix} \right\} \tag{16.33}$$

and form another S_3 group which commutes with the tetrahedral S_4-symmetry.

$6j$-symbols and certain generalizations thereof are used as building blocks in the Ponzano–Regge approach to three-dimensional quantum gravity and in three-dimensional topological field theories, which give rise to state-sum invariants of three-manifolds; see e.g. [Chung *et al.* 1994] and chapter VI of [Turaev 1994]. In two-dimensional conformal field theory analogues of the $6j$-symbol, known as fusing matrices (compare (16.40) below), encode information on the analytic properties of correlation functions. Racah coefficients also appear as the result of certain angular integrations that arise in intermediate steps in the calculation of certain multi-loop Feynman diagrams [Nickel 1978].

Information

Using the relations (16.27) and (16.28) one can relate the $6j$-symbols to the Clebsch–Gordan coefficients. One finds

$$\sum_{\substack{M_1,M_2,M_3,\\m_1,m_2}} (-1)^\Delta \begin{pmatrix} J_1 & J_2 & j_3 \\ M_1 & -M_2 & m_3 \end{pmatrix} \begin{pmatrix} J_2 & J_3 & j_1 \\ M_2 & -M_3 & m_1 \end{pmatrix} \begin{pmatrix} J_3 & J_1 & j_2 \\ M_3 & -M_1 & m_2 \end{pmatrix} \begin{pmatrix} j_1 & j_2 & j_3' \\ m_1 & m_2 & m_3' \end{pmatrix}$$

$$= \frac{1}{2j_3+1} \left\{ \begin{matrix} j_1 & j_2 & j_3 \\ J_1 & J_2 & J_3 \end{matrix} \right\} \delta_{j_3,j_3'} \delta_{m_3,m_3'} \tag{16.34}$$

with $\Delta = J_1 + J_2 + J_3 + M_1 + M_2 + M_3$. This in turn can be used to derive analogues of the orthogonality relations as well as a closed formula for the $6j$-symbols.

Finally, we remark that, using analogous isomorphisms for the tensor product of four or more irreducible representations, it is possible to define $9j$-symbols etc. The generalization of these concepts for algebras other than $\mathfrak{sl}(2)$ is again obvious, but rather involved, and we refrain from presenting it here.

16.7 The pentagon and hexagon identities

The isomorphisms R (16.24) and F (16.25) between (tensor products of) spaces of intertwiners can be employed to construct also similar isomorphisms for intertwiners between more complicated multiple tensor products. It turns out that in fact arbitrary tensor products can be treated without introducing any further independent isomorphisms. Now isomorphic multiple tensor products can typically be related by different combinations of the operations R (exchange of two 'neighboring' tensor factors) and F (change of the bracketing of a triple tensor product). As a consequence, these two isomorphisms satisfy various compatibility relations.

   It can be shown that all these compatibility conditions can be reduced to only three independent equations, a so-called *pentagon identity* and two *hexagon identities*. The former arises when describing isomorphisms between the four-fold tensor products $U \otimes (V \otimes (W \otimes X))$ and $((U \otimes V) \otimes W) \otimes X$, and the latter from isomorphisms between the triple tensor products $U \otimes (V \otimes W)$ to $(W \otimes U) \otimes V$, and from $(U \otimes V) \otimes W$ to $V \otimes (W \otimes U)$, respectively. The equality of the respective isomorphisms can be displayed conveniently in terms of commuting diagrams (compare section 5.3 for an instruction on how to read such diagrams). The names 'pentagon' and 'hexagon' arise from the shape of these diagrams: The pentagon diagram looks like

$$
\begin{array}{ccc}
& \boxed{U \otimes ((V \otimes W) \otimes X)} & \longrightarrow \boxed{(U \otimes (V \otimes W)) \otimes X} \\
\boxed{U \otimes (V \otimes (W \otimes X))} & & \downarrow \\
& \boxed{(U \otimes V) \otimes (W \otimes X)} & \longrightarrow \boxed{((U \otimes V) \otimes W) \otimes X}
\end{array}
\tag{16.35}
$$

while the first of the hexagon diagrams is

$$
\begin{array}{ccccc}
\boxed{U \otimes (V \otimes W)} & \longrightarrow & \boxed{U \otimes (W \otimes V)} & \longrightarrow & \boxed{(U \otimes W) \otimes V} \\
\downarrow & & & & \downarrow \\
\boxed{(U \otimes V) \otimes W} & \longrightarrow & \boxed{W \otimes (U \otimes V)} & \longrightarrow & \boxed{(W \otimes U) \otimes V}
\end{array}
\tag{16.36}
$$

   These statements apply to the maps R and F for arbitrary finite-dimensional simple Lie algebras. Except for $\mathfrak{sl}(2)$ the explicit form of the hexagon and pentagon identities is only rarely needed, and hence we do not display the identities here (they can e.g. be found in [Moore and Seiberg 1990] and in [Fuchs *et al.* 1995]).

   We do, however, mention that the pentagon identity can be presented geometrically as equating two alternative decompositions of a pentahedron into two or three tetrahedra respectively. Namely, one side of the pentagon identity consists of a product of two 6Λ-symbols, with a summation over a common multiplicity index. In the geometric description, each 6Λ-symbol is represented by a tetrahedron, and the summation is represented by gluing the two tetrahedra together along a common face f_0; this gluing procedure results in a pentahedron, which is displayed as the upper figure of (16.37)).

   On the other hand, this pentahedron can be decomposed into three tetrahedra representing the three terms on the other side of the pentagon identity, as depicted in the lower figure of (16.37). To describe this decomposition, we insert an additional edge e that links the two vertices v_+ and v_- of the pentahedron which do not belong to the face f_0. For any of the three edges e_j $(j = 1, 2, 3)$ that form the boundary of f_0, we then obtain a tetrahedron T_j; the edges of T_j are e, e_j and the edges joining the endpoints of e_j to $v_{\pm}$. The three tetrahedra T_i make up the pentahedron, and the summations in the pentagon equation correspond to the edge e and again to the faces along which the tetrahedra T_i are glued.

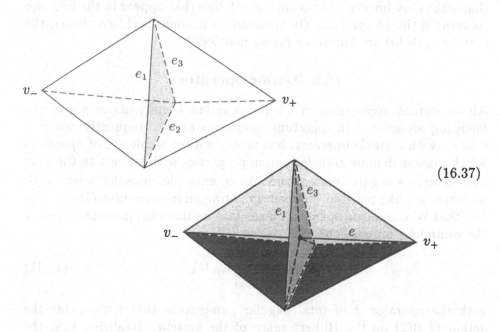

$$(16.37)$$

Let us now concentrate on the case of $\mathfrak{sl}(2)$. In this case the pentagon relation is known as the *Biedenharn sum rule* and reads

$$\sum_l (-1)^\Delta (2l+1) \left\{ \begin{matrix} j_1 & j_2 & l \\ j_3 & j_4 & k_1 \end{matrix} \right\} \left\{ \begin{matrix} j_3 & j_4 & l \\ j_5 & j_6 & k_2 \end{matrix} \right\} \left\{ \begin{matrix} j_5 & j_6 & l \\ j_2 & j_1 & k_3 \end{matrix} \right\}$$
$$= \left\{ \begin{matrix} k_1 & k_2 & k_3 \\ j_5 & j_1 & j_4 \end{matrix} \right\} \left\{ \begin{matrix} k_1 & k_2 & k_3 \\ j_6 & j_2 & j_3 \end{matrix} \right\} , \qquad (16.38)$$

with $\Delta = j_1 + j_2 + j_3 + j_4 + j_5 + j_6 + k_1 + k_2 + k_3 + l$. Also, the two hexagon identities become identical; they yield the *Racah–Elliott relation*

$$\sum_l (-1)^{k_1+k_2+l} (2l+1) \left\{ \begin{matrix} j_1 & j_2 & l \\ j_3 & j_4 & k_1 \end{matrix} \right\} \left\{ \begin{matrix} j_3 & j_4 & l \\ j_2 & j_1 & k_2 \end{matrix} \right\} = \left\{ \begin{matrix} j_1 & j_4 & k_1 \\ j_2 & j_3 & k_2 \end{matrix} \right\} ; \quad (16.39)$$

here already the explicit values $(-1)^{j_1+j_2+J}$ of R: $\mathcal{V}_{\Lambda\Lambda'}{}^{\Lambda_i} \rightarrow \mathcal{V}_{\Lambda'\Lambda}{}^{\Lambda_i}$ have been inserted.

With an appropriate choice of bases of the relevant modules, the $6j$-symbols also satisfy a certain completeness condition. This relation is best expressed by regarding the $6j$-symbols as matrices with the matrix indices given by the labels in the third column, according to

$$\left\{ \begin{matrix} j_1 & j_2 & k \\ j_3 & J & k' \end{matrix} \right\} = \mathsf{F}_{kk'} \left[\begin{matrix} j_1 & j_2 \\ J & j_3 \end{matrix} \right] . \qquad (16.40)$$

In this description, the completeness condition just says that each of the matrices $\mathsf{F} \left[\begin{matrix} j_1 & j_2 \\ J & j_3 \end{matrix} \right]$ is unitary (as a matrix with indices k and k'). However, the pentagon and hexagon identities can*not* be written as relations between matrix products of these matrices; also, for Lie algebras other than $\mathfrak{sl}(2)$ they involve summations over labels that appear in the first two columns of the 6Λ-symbols. (In applications in conformal field theory, the matrices (16.40) are known as *fusing* matrices.)

16.8 Tensor operators

An important application of Clebsch–Gordan coefficients arises in the following situation. In quantum mechanics one is frequently not just dealing with a single operator, but rather with a whole set of operators which possess definite transformation properties with respect to the $\mathfrak{sl}(2)$ that describes angular momentum. As an example, consider a *vector* $\vec{W}$ of operators, like position, momentum, or angular momentum itself. The fact that $\vec{W}$ is a vector operator means that its three components W_i obey the commutation relations

$$[J_i, W_j] = \mathrm{i} \sum_{k=1}^{3} \epsilon_{ijk} W_k \tag{16.41}$$

with the operator $\vec{J}$ of total angular momentum that implements the action of $\mathfrak{sl}(2)$ on the Hilbert space of the system. Realizing that the numbers ϵ_{ijk} are just the entries of the representation matrices $R_1(J_i)$ of the irreducible representation of $\mathfrak{sl}(2)$ with spin one, this can be rewritten as

$$[J_i, W_j] = \sum_{k=1}^{3} (R_1(J_i))_{jk} W_k \,. \tag{16.42}$$

This observation motivates the following generalization. Any collection $\mathcal{O} = \{\mathcal{O}_p\}$ of operators for which the commutator with the $\mathfrak{sl}(2)$-generators J_i is described by an $\mathfrak{sl}(2)$-representation R, in the sense that

$$[J_i, \mathcal{O}_p] = \sum_{q=1}^{\dim R} (R(J_i))_{pq}\, \mathcal{O}_q \,, \tag{16.43}$$

is called a *tensor operator*. In case the representation R is irreducible, say of spin k, $\mathcal{O}$ is called an *irreducible tensor operator* of *rank k*. For an irreducible tensor operator $\mathcal{O}^{(k)}$ of rank k one can find linear combinations $T_q^{(k)}$ of the individual operators $\mathcal{O}_p^{(k)}$, with $q = -k, -k+1, \ldots, k$, such that

$$\begin{aligned}
[J_\pm, T_q^{(k)}] &= \sqrt{k(k+1) - q(q \pm 1)} \cdot T_{q \pm 1}^{(k)} \,, \\
[J_0, T_q^{(k)}] &= q \cdot T_q^{(k)}
\end{aligned} \tag{16.44}$$

analogously to (16.1). The $T_q^{(k)}$ are called *standard components* of $\mathcal{O}^{(k)}$.

The vector operators correspond to irreducible tensor operators of rank $k = 1$. The tensor product of two vector operators W and W' produces a tensor operator, which however is not irreducible. Rather, the operators $W_i \otimes W'_j$ transform according to the representation $R_1 \otimes R_1 \cong R_2 \oplus R_1 \oplus R_0$ of $\mathfrak{sl}(2)$. The irreducible tensor operators contained in the tensor product are given by the 'trace' $W_1 \otimes W'_1 + W_2 \otimes W'_2 + W_3 \otimes W'_3$ which transforms as a singlet, the symmetric part, and the antisymmetric part which transforms like a vector (see exercise 16.3).

The relations (16.44) imply severe restrictions on the matrix elements

$$\langle \tau, JM \,|\, T_q^{(k)} \,|\, \tau', J'M' \rangle \tag{16.45}$$

of an irreducible tensor operator $\mathcal{O}^{(k)}$. To describe this in more detail, we decompose the Hilbert space of the physical system into irreducible modules of the $\mathfrak{sl}(2)$ spanned by J_0 and $J_\pm$. As an irreducible module may occur more than once, we have to include in the labelling of Hilbert space vectors a degeneracy index τ in addition to J and M, resulting in the notation $|\tau, JM\rangle$ employed in (16.45). (For example, in the description of the hydrogen atom, τ stands for the principal quantum number n.) Let us now compute the matrix element (16.45). We have

$$
\begin{aligned}
q \,\langle \tau, JM \,|\, T_q^{(k)} \,|\, \tau', J'M' \rangle &= \langle \tau, JM \,|\, [J_0, T_q^{(k)}] \,|\, \tau', J'M' \rangle \\
&= (M - M') \,\langle \tau, JM \,|\, T_q^{(k)} \,|\, \tau', J'M' \rangle,
\end{aligned} \tag{16.46}
$$

where in the first line we inserted (16.44), while in the second line we expressed the commutator as $[J_0, T_q^{(k)}] = J_0 T_q^{(k)} - T_q^{(k)} J_0$ and applied the J_0 appearing in the first term to the bra- and the one in the second term to the ket-vector. According to (16.46) the matrix element (16.45) can only be non-zero if the selection rule

$$M - M' = q \tag{16.47}$$

is satisfied.

At this place a few more general remarks on selection rules are in order. Selection rules force certain matrix elements to be zero. Whether the converse is true, i.e. whether all matrix elements allowed by the selection rules that are present in a theory are in fact non-zero, depends on the physical system under consideration. Actually, as already remarked in chapter 1, in physics one often does not deal with a single theory, but rather with a family of theories which are parametrized by the values of certain 'coupling constant' parameters. In this situation, the fact that matrix elements that are in principle allowed by the selection rules vanish typically indicates that one is considering the theory at a rather special point in parameter space. If these special values of the parameters are realized in a physical system, one may ask why nature has happened to choose just this particular point in parameter space, i.e. why it 'adjusts' the coupling constants in such a special manner; problems of this type are called *fine tuning problems*. In case matrix elements vanish for all values in the parameter space, one tries to identify a symmetry which explains this behavior. (This way e.g. the so-called *R*-parity has been discovered in supersymmetric theories; compare also the remarks in section 1.7.) When one perturbs such a theory and the perturbation respects the symmetry, it follows that those matrix elements must also vanish in the perturbed theory. This observation, which is important in many applications of perturbation theory, is rephrased in physicists' language by saying that these matrix elements are 'protected' by a symmetry. Also, the property of a system that only those matrix elements vanish that are forbidden by some selection rule associated to a symmetry, is sometimes called the *naturality* property.

16.9 The Wigner–Eckart theorem

The defining properties (16.44) of an irreducible tensor operator $\mathcal{O}^{(k)}$ allow to make a much stronger statement than (16.47), the so-called *Wigner–Eckart theorem*. This theorem asserts that there exists a set of numbers $\langle \tau, J \, \| \, T^{(k)} \, \| \, \tau', J' \rangle$ (called the *reduced* matrix elements of $\mathcal{O}^{(k)}$) which depend neither on q nor on M or M', such that

$$\langle \tau, J\, M \, | \, T_q^{(k)} \, | \, \tau', J'\, M' \rangle = \tfrac{(-1)^{2k}}{\sqrt{2J+1}} \, \langle \tau, J \, \| \, T^{(k)} \, \| \, \tau', J' \rangle \, \langle J'\, k, M'\, q \, | \, J, M \rangle$$

$$= (-1)^{J-M} \langle \tau, J \, \| \, T^{(k)} \, \| \, \tau', J' \rangle \begin{pmatrix} J & k & J' \\ -M & q & M' \end{pmatrix} . \qquad (16.48)$$

Thus all matrix elements are known once the reduced matrix elements have been calculated. Hence the Wigner–Eckart theorem reduces drastically the number of matrix elements that have to be computed separately.

To derive the Wigner–Eckart theorem, we consider the $(2k+1)(2J'+1)$ vectors $T_q^{(k)} | \tau', J' M' \rangle$ for $q = -k, \ldots, k$, $M' = -J', \ldots, J'$ and fixed value of τ', and study for all J'' the (not necessarily linearly independent) vectors

$$| \tau', J''\, M'' \rangle_T := \sum_{M', q} T_q^{(k)} | \tau', J'\, M' \rangle \, \langle J'\, k, M'\, q \, | \, J'', M'' \rangle . \qquad (16.49)$$

By the orthogonality relations (16.11) of the Clebsch–Gordan coefficients, this relation can be inverted to

$$T_q^{(k)} | \tau', J'\, M' \rangle = \sum_{J'', M''} | \tau', J'', M'' \rangle_T \, \langle J'\, k, M'\, q \, | \, J'', M'' \rangle . \qquad (16.50)$$

Now the transformation properties (16.44) of a tensor operator and the recursion relations (16.20) for the Clebsch–Gordan coefficients imply that the identities

$$J_\pm \, | \tau', J''\, M'' \rangle_T = \sqrt{J''(J''+1) - M''(M''\pm 1)} \cdot | \tau', J''\, M''\pm 1 \rangle_T ,$$

$$J_z \, | \tau', J''\, M'' \rangle_T = M'' \cdot | \tau', J''\, M'' \rangle_T \qquad\qquad (16.51)$$

hold (see exercise 16.5). Hence, unless they are all zero, the vectors $|\tau', J'' M''\rangle_T$ with fixed value of τ'' and J'' all have the same length and span an irreducible module of the $\mathfrak{sl}(2)$ of angular momentum. Therefore the matrix elements $\langle \tau, J M \,|\, \tau', J'' M'' \rangle_T$ are all zero unless $J = J''$ and $M = M''$, and, if they are non-zero, they are independent of $M = M''$. Using (16.50), we can now write the matrix element we are after as

$$\langle \tau, J M \,|\, T_q^{(k)} \,|\, \tau', J' M' \rangle = \sum_{J'', M''} \langle \tau, J M \,|\, \tau', J'' M'' \rangle_T \, \langle J' k, M' q \,|\, J'', M'' \rangle. \quad (16.52)$$

Fixing an arbitrary normalization of the reduced matrix elements, this implies the Wigner-Eckart theorem (16.48).

We also remark that in this formulation the selection rule (16.47) that we have derived with the help of (16.44) is directly reproduced by the selection rule (16.10) for the Clebsch-Gordan coefficients. Another selection rule for the matrix element (16.45) is implied by the selection rule (16.8) for the Clebsch-Gordan coefficients that follows from the triangle inequality: The matrix element vanishes unless $|J' - k| \le J \le J' + k$.

A typical situation in which the Wigner-Eckart theorem can be applied is the following. Any Hamiltonian $H(\varphi, \vartheta; \nu)$ which depends on the spherical coordinates φ and ϑ and possibly some other coordinates or parameters ν can be expanded in terms of spherical harmonics, i.e. in a series of irreducible tensor operators. Usually one also considers the action of the parity operator P, which satisfies $P^2 = 1$, and writes any irreducible tensor operator as the sum of two eigenoperators of P. The resulting decomposition is called the *multipole expansion*.
The component in the spin-k irreducible representation with parity $(-1)^k$ is called the electrical 2^k-pole moment $E^{(k)}$, and the component of parity $(-1)^{k+1}$ is called the magnetic 2^k-pole moment $M^{(k)}$. In applications the most important moments are the electric and magnetic (dipole) moments $E^{(1)}$ and $M^{(1)}$ and the electric quadrupole moment $E^{(2)}$. The Wigner-Eckart theorem provides selection rules for the matrix elements of the various moments. For example, only those matrix elements of $E^{(k)}$ and $M^{(k)}$ between states of spin J and J' can be non-vanishing for which $|J - J'| \le k \le J + J'$. In particular for dipole moments one has $J = J' \pm 1$.

—————— Information

Summary:

We have analyzed various isomorphisms between modules explicitly. The decomposition of the tensor product of two modules is described by the Clebsch-Gordan coefficients or, equivalently, by the $3j$-symbols. The associativity of the tensor product of three modules is described by the Racah coefficients or the $6j$-symbols. The $6j$-symbols satisfy the pentagon and hexagon identities.
The Wigner-Eckart theorem expresses the matrix elements of irreducible tensor operators as products of reduced matrix elements and Clebsch–Gordan coefficients.

Keywords:

Clebsch–Gordan coefficient, 3j-symbol, Racah coefficient, 6j-symbol, tetrahedral symmetry, pentagon and hexagon relations; (irreducible) tensor operator, selection rule, Wigner–Eckart theorem, reduced matrix element.

Exercises:

Express the vectors $|j, m\rangle$ in terms of the eigenvectors $v_{\Lambda,\lambda}$ with $\Lambda = 2j$ and $\lambda = 2m$, where $v_{\Lambda,\lambda} = (J_-)^{(\Lambda-\lambda)/2} v_{\Lambda,\Lambda}$. Check that the vectors given in (16.1) are in fact orthonormal.

Exercise 16.1

Prove the recursion relations (16.26).

Exercise 16.2

Determine the standard components of the tensor product of two vector operators.

Exercise 16.3

Find selection rules for the matrix elements of a scalar operator S, i.e. an operator which satisfies $[S, J_i] = 0$.
Compare with the Wigner–Eckart theorem.

Exercise 16.4

Use the transformation properties (16.44) of a tensor operator and the recursion relations (16.20) to derive the formula (16.51).

Exercise 16.5

Consider the quantum mechanical hydrogen atom, including the spin contribution $\vec{S}$ to the total angular momentum $\vec{J}$ of the electron. Use the Wigner–Eckart theorem to show that the two matrix elements $\langle \nu, J\, M \,|\, \vec{L} + 2\vec{S} \,|\, \nu', J'\, M' \rangle$ and $\langle \nu, J\, M \,|\, \vec{J} \,|\, \nu', J'\, M' \rangle$ are proportional.

(When $J = J'$, the proportionality factor g_J is called the Landé factor; it plays an important rôle in the description of the Zeeman effect. Namely, for homogeneous external magnetic field of field strength B the shift of the energy eigenvalue for the state with 'magnetic quantum number' M is $-M g_J \mu_B B$, where $\mu_B = e\hbar/2mc$ is Bohr's magneton.)

Exercise 16.6

17
Invariant tensors

17.1 Intertwiners and invariant tensors

It is often convenient to have information about tensor products which goes beyond the mere knowledge of the tensor product multiplicities $\mathcal{L}_{\Lambda\Lambda'}^{\Lambda_i}$ that were introduced in (15.11), but which nevertheless does not require the knowledge of the numerical values of Clebsch–Gordan coefficients. For instance, one is often interested in 'extracting the singlet contribution' to quantities that are obtained from multiple tensor products, i.e. isolate those elements of the tensor product which transform in the trivial one-dimensional representation. Such problems can be addressed with the help of intertwiners between singlets and tensor product representations, which are also known as invariant tensors.

A typical situation where this is relevant is Lagrangian field theory where the Lagrangian is formed by combining non-singlet fundamental fields. To demand that the algebra $\mathfrak{g}$ which organizes the fundamental field into multiplets is a symmetry of the theory amounts in particular to the requirement that the Lagrangian is a singlet with respect to $\mathfrak{g}$. Another example is provided by gauge theories. The gauge principle states that gauge transformations are pure redundancies in the description of physical states. Hence gauge transformations must not affect observable quantities; in other words, all observables must be gauge invariant, i.e. singlets with respect to gauge transformations.

In the physics literature, by a *tensor* of a Lie algebra $\mathfrak{g}$ one means any object which, roughly speaking, 'transforms as a $\mathfrak{g}$-representation R', or more precisely, on which the elements of the simple Lie algebra $\mathfrak{g}$ are defined to act in a specific finite-dimensional $\mathfrak{g}$-representation R. (In this chapter, we will consider only finite-dimensional simple Lie algebras, and only finite-dimensional representations.) In applications, this representation typically arises as the tensor product of finite-dimensional irreducible representations. One frequently fixes a basis $\mathcal{B} = \{v_{(i)}\}$ of the module V and describes any tensor by its coordinates with respect to the basis induced by $\mathcal{B}$ on the tensor product under consideration. The simplest example of a tensor is thus given by the components v^i ($i = 1, 2, \ldots, \dim V$) of a vector $v = \sum_i v^i v_{(i)}$ in an irreducible $\mathfrak{g}$-module V with respect to the basis $\mathcal{B}$. Further, given this basis, the elements $\tilde{v}$ of the module V^+ conjugate to V are conveniently described by their components with respect to the dual basis $\mathcal{B}^\star = \{\tilde{v}^{(i)}\}$ (defined by $\tilde{v}^{(i)}(v_{(j)}) = \delta^i_j$, compare e.g. (4.2)), i.e. by components $\tilde{v}_i$ with a lower index. The upper and lower indices are then often referred to as *covariant* and *contravariant* indices, respectively. The representation matrices act on the components of v and $\tilde{v}$ as

$$
\begin{aligned}
v^i &\mapsto (v')^i = \sum_j (R(x))^i{}_j\, v^j\,, \\
\tilde{v}_i &\mapsto (\tilde{v}')_i = \sum_j (R^+(x))_i{}^j\, \tilde{v}_j = -\sum_j \tilde{v}_j\, (R(x)^{\mathrm{t}})^j{}_i\,.
\end{aligned}
\tag{17.1}
$$

Let us denote the singlet of $\mathfrak{g}$, i.e. the trivial one-dimensional $\mathfrak{g}$-module, by V_0. An *invariant tensor* t for the tensor product module $W := V^{\otimes m} \otimes (V^+)^{\otimes n}$ is by definition an intertwiner between V_0 and the $\mathfrak{g}$-module W; roughly, an invariant tensor thus describes a possible way to 'form a singlet' out of irreducible modules. By abuse of terminology, one often does not mention the tensor product W explicitly, but simply speaks about an invariant tensor 'for V'. We have already encountered a module with an invariant tensor before, namely the adjoint module and the Killing form on it as defined in equation (6.12) (compare exercise 17.1).

In formulæ, the defining property of t reads

$$
0 = t\, R_0(x) = R_W(x)\, t
\tag{17.2}
$$

for all $x \in \mathfrak{g}$, where the first equality expresses the fact that in the singlet representation all elements x of $\mathfrak{g}$ are represented by zero, $R_0(x) = 0$, and the second equality is the defining intertwining property of t. Furthermore, with the conventions of (17.1), the invariant tensor t can be described by its coordinates with respect to a pair of dual bases in V and V^+, i.e. by a collection $\{t^{i_1 i_2 \ldots i_m}{}_{j_1 j_2 \ldots j_n}\}$ of numbers, where each of the m upper and n lower indices runs from 1 to $\dim V$. Equation (17.2) then translates to the following identity for these coordinates:

$$
\begin{aligned}
0 = \sum_\ell \big[&(R(T^a))^{i_1}{}_\ell\, t^{\ell\, i_2 \ldots i_m}{}_{j_1 \ldots j_n} + \ldots + (R(T^a))^{i_m}{}_\ell\, t^{i_1 \ldots i_{m-1}\ell}{}_{j_1 \ldots j_n} \\
&+ (R^+(T^a))_{j_1}{}^\ell\, t^{i_1 \ldots i_m}{}_{\ell\, j_2 \ldots j_n} + \ldots + (R^+(T^a))_{j_n}{}^\ell\, t^{i_1 \ldots i_m}{}_{j_1 \ldots j_{n-1}\ell} \big]
\end{aligned}
\tag{17.3}
$$

holds for all generators T^a of $\mathfrak{g}$. In this formulation the qualification 'invariant' which refers to the singlet property means more specifically that the transformations (17.1) leave the formal homogeneous polynomial $\sum_{i_1,\ldots,i_m,j_1,\ldots,j_n} t^{i_1 i_2 \ldots i_m}_{ j_1 j_2 \ldots j_n} \tilde{v}_{i_1} \otimes \cdots \otimes \tilde{v}_{i_n} \otimes v^{j_1} \otimes \cdots \otimes v^{j_n}$ invariant.

For brevity, one also refers to the components $t^{i_1 i_2 \ldots i_m}_{ j_1 j_2 \ldots j_n}$ as (invariant) tensors. Now if $t^{i_1 i_2 \ldots i_m}_{ j_1 j_2 \ldots j_n}$ is an invariant tensor, then so are the combinations $t^{\ldots i_p \ldots i_q \ldots}_{ j_1 j_2 \ldots j_n} \pm t^{\ldots i_q \ldots i_p \ldots}_{ j_1 j_2 \ldots j_n}$ that are obtained by symmetrizing or antisymmetrizing with respect to any pair of (say) upper indices. As a consequence, one can without loss of generality assume that all tensors one is dealing with only change by a definite factor of ± 1 upon exchanging any two like indices. Thus in agreement with the observations of chapter 15 (see equation (15.6)), any irreducible representation in a tensor product, and hence in particular the singlet describing invariant tensors, has definite symmetry properties when the bases of the irreducible modules are chosen appropriately.

Any finite-dimensional $\mathfrak{g}$-module V is unitary, and hence it can be identified with its conjugate module via the hermitian product $(\cdot | \cdot)$: To any $\tilde{w} \in V^+$ one can associate a $w \in V$ such that $\tilde{w}(v) = (w | v)$ for all $v \in V$. As a consequence, for any finite-dimensional module V the Kronecker symbol δ is an invariant tensor for $V \otimes V^+$. For the tensor δ, the requirement (17.3) just reduces to the defining equation (5.16) for the representation matrices of the conjugate representation.

Invariant tensors can also be defined for an arbitrary tensor product $V_{\Lambda_1} \otimes V_{\Lambda_2} \otimes \ldots \otimes V_{\Lambda_p}$; such tensors obey a relation analogous to (17.3):

$$\sum_{\ell=1}^{\dim V_{\Lambda_1}} (R_{\Lambda_1}(T^a))^{i_1}_{\ell}\, t^{\ell\, i_2 \ldots i_p} + \ldots + \sum_{\ell=1}^{\dim V_{\Lambda_p}} (R(T^a))^{i_p}_{\ell}\, t^{i_1 \ldots i_{p-1} \ell} = 0\,. \quad (17.4)$$

The notion of an invariant tensor has the following generalization. For any module V of an arbitrary Lie algebra $\mathfrak{g}$, one defines the space $[V]_{\mathfrak{g}}$ of *co-invariants* of V as the largest quotient of V on which $\mathfrak{g}$ acts trivially. The space $[V]_{\mathfrak{g}}$ can be described as the quotient of V by the subspace that is spanned by all those vectors $v \in V$ which can be obtained by acting with some element $x \in \mathfrak{g}$ on some vector w of V, $v = xw$. One can check that for an irreducible module V of a finite-dimensional simple Lie algebra the space of co-invariants has dimension zero, except when V is the trivial module. For a tensor product of irreducible modules, the prescription therefore just selects the singlets.

17.2 Orthogonal and symplectic modules

In the case of a self-conjugate irreducible representation, the notion of an invariant tensor has the following application. If the irreducible module V is self-conjugate, i.e. if the conjugate module V^+ is isomorphic to V, then there must exist an intertwiner between V and V^+, or equivalently an intertwiner between $V \otimes V$ and the singlet, i.e. there is an invariant tensor e for $V \otimes V$, and analogously also for $V^+ \otimes V^+$. In terms of components, this means that for any vector $v \in V$ there is a linear relation between the components of v and those of the associated dual element $\tilde{v} \in V^+$, and vice versa. These are provided by an invariant tensor e with two lower indices such that $\tilde{v}_i = \sum_j e_{ij} v^j$, and by an analogous tensor with two upper indices (which for simplicity we also denote by e) such that $v^i = \sum_j e^{ij} \tilde{v}_j$. It follows that the tensor $\sum_j e_{ij} e^{jk}$ is an invariant tensor of $V \otimes V^+$. But this tensor product contains the singlet just once, and the unique invariant tensor for it is given by the Kronecker delta; hence we can fix the phase freedom that is still present in the definition of e_{ij} and e^{jk} in such a way that these tensors are inverse to each other: $\sum_j e_{ij} e^{jk} = \delta_i^{\ k}$.

If the tensor e is symmetric, $e_{ij} = e_{ji}$, then the representation R is referred to as an *orthogonal* representation, while if e is antisymmetric, $e_{ij} = -e_{ji}$, then R is called a *symplectic* representation. This terminology stems from the following fact. The invariant tensor on $V \otimes V$ supplies an invariant scalar product on V, or an invariant symplectic form, respectively. As a consequence, $\mathfrak{g}$ can be embedded into the Lie algebra $\mathfrak{so}(V)$ and $\mathfrak{sp}(V)$, respectively (compare also section 18.4).

For self-conjugate modules, one can raise and lower indices by contracting with the tensor e, so that one can concentrate on tensors with one type of indices only. This is even possible for non-selfconjugate modules, provided that one is willing to pay the price of changing the number of indices; this is achieved by contraction with the completely antisymmetric Levi–Civita tensors $\epsilon^{i_1 i_2 \ldots i_d}$ and $\epsilon_{i_1 i_2 \ldots i_d}$, with $d = \dim V$, to raise and lower indices. (The value of $\epsilon^{i_1 i_2 \ldots i_d}$ is by definition $+1$ if $(i_1, i_2, \ldots, i_d)$ is an even permutation of $(1, 2, \ldots, d)$, -1 if it is an odd permutation, and zero else.)

Apart from most of the highest weight modules of A_r, D_{2n+1} and E_6, all irreducible highest weight modules of the simple Lie algebras are self-conjugate (see e.g. table XII for the information which of the basic modules of simple Lie algebras are orthogonal or symplectic). Also, the adjoint representation of any semisimple Lie algebra is orthogonal; the relevant symmetric two-index tensor is provided by the Killing form. Because of the existence of a non-degenerate antisymmetric two-index tensor, a symplectic module is necessarily even-dimensional. Also recall from chapter 13 (compare (13.16)) that the lowest weight of a module V_Λ with highest weight Λ is given by $\lambda_{\min} = -\Lambda^+$; thus for self-conjugate modules we have $\lambda_{\min} = -\Lambda$. If V_Λ is orthogonal, then the depth of $\lambda_{\min}$ (i.e. the number of times that a simple root must be subtracted from Λ to obtain $\lambda_{\min}$) is an even integer, while for symplectic V_Λ the depth of $\lambda_{\min}$ is odd.

17.3 Primitive invariants

In the previous subsection we just considered invariant tensors of a fixed tensor product; we now turn our attention to relations between invariant tensors for *different* tensor products of a module V and its conjugate module. Sums and products of invariant tensors are again invariant tensors, and the same is true for the tensor

$$\hat{t}^{i_1 i_2 \dots i_m}_{\ j_1 j_2 \dots j_n} = \sum_{\ell} t^{i_1 \dots i_p \, \ell \, i_{p+1} \dots i_m}_{\ j_1 \dots j_q \, \ell \, j_{q+1} \dots j_m} \tag{17.5}$$

that is obtained by *contracting* a single invariant tensor, i.e. by summation over a pair of upper and lower indices. Combining these operations, it follows e.g. that the tensor

$$u^{i_1 i_2 \dots i_m}_{\ j_1 j_2 \dots j_m} = \sum_{\ell} t^{i_1 \dots i_p \, \ell \, i_{p+1} \dots i_m} \, t'_{j_1 \dots j_q \, \ell \, j_{q+1} \dots j_m} \tag{17.6}$$

obtained by contracting the product of two invariant tensors $t^{i_1 \dots i_m}$ and $t'_{j_1 \dots j_m}$ is again an invariant tensor. As a consequence, one can obtain all invariant tensors via summations, multiplications and contractions from a small number of algebraically independent invariant tensors, the so-called *primitive invariants* which have the minimal possible numbers of indices. Also, for any invariant tensor t the tensor

$$\tilde{t}^{i_1 i_2 \dots i_m}_{\ j_1 j_2 \dots j_m} := t^{i_1 i_2 \dots i_m}_{\ j_1 j_2 \dots j_n} - \frac{1}{d} \, \delta^{i_p}_{j_q} \sum_{\ell} t^{i_1 \dots i_p \, \ell \, i_{p+1} \dots i_m}_{\ j_1 \dots j_q \, \ell \, j_{q+1} \dots j_m} \tag{17.7}$$

is traceless, i.e. satisfies $\sum_{\ell} \tilde{t}^{\dots \ell \dots}_{\dots \ell \dots} = 0$. Primitive invariants with more than two indices can therefore be chosen as totally traceless tensors, i.e. such that summation over any pair of upper and lower indices yields zero.

For tensor products of the defining modules that were listed in table XII and their conjugate modules, a complete set of primitive invariants is known for all simple Lie algebras except for E_8. For any simple $\mathfrak{g}$, this set contains the Kronecker symbol δ^i_j, as well as the Levi–Civita tensor $\epsilon^{i_1 i_2 \dots i_d}$ (together with the analogous tensor $\epsilon_{i_1 i_2 \dots i_d}$ with lower indices). In the case of A_r these are already all primitive invariants of the defining representation (this implies e.g. that any contraction of Levi–Civita tensors can be expressed through products of Kronecker symbols, compare exercise 17.2). For the other simple Lie algebras there are further primitive invariants; these are listed in table (17.8).

$\mathfrak{g}$	invariants
B_r, D_r	δ^{ij}
C_r	f^{ij}
E_6	d^{ijk}
E_7	$f^{ij}, \ d^{ijkl}$
E_8	$\delta^{ab}, \ f^{abc}, \ t^{abcd} \dots$
F_4	$\delta^{ij}, \ d^{ijk}$
G_2	$\delta^{ij}, \ f^{ijk}$

$$(17.8)$$

All invariant tensors in table (17.8) are either completely antisymmetric or completely symmetric. We have denoted the antisymmetric tensors generically by $f^{i_1\cdots i_n}$ and the symmetric ones by $d^{i_1\cdots i_n}$, and we also already implemented the fact that the quadratic symmetric invariants can, by an appropriate choice of basis, be taken to be δ^{ij}. It can be shown that in the case of E_8 there exists (at least) one primitive invariant $t^{abcd\cdots}$ which is not yet known explicitly (it is known that it must be higher than quartic, and there are indications that there is a symmetric eight-index primitive invariant). This lack of knowledge is related to the fact that for E_8 the lowest-dimensional module is the adjoint; in the table the latter property is indicated by the use of indices $a, b, \ldots$ rather than $i, j, \ldots$.

Above we have assumed that bases are chosen such that e.g. the symmetric two-index invariants are just Kronecker symbols. However, for most calculations it is in fact not necessary to know the numerical values of the components of the invariants. In particular, primitive invariants can be defined without reference to a basis, namely by prescribing their symmetry properties as well as basis-independent algebraic relations such as tracelessness.

The knowledge of a set of primitive invariants is indispensable whenever one needs to parametrize any kind of objects which correspond to invariant tensors in a *unique* manner.

Information

Compare e.g. [Minard 1983] for a description of polarization effects in nonlinear optics, [Kephart and Vaughn 1983] and [Wybourne 1980] for a discussion of potentials for Higgs scalar fields in Lagrangian field theory, and [ter Haar Romeny *et al.* 1991] for an application to image processing.

17.4 Adjoint tensors of $\mathfrak{sl}(n)$

Besides the defining representation, the adjoint representation of a simple Lie algebra is often the representation of most direct interest. We therefore now turn to the study of invariant tensors for the adjoint module.

We have already seen that the Killing form is a two-index invariant tensor for the adjoint module. A three-index primitive invariant tensor for the adjoint module of any simple Lie algebra $\mathfrak{g}$ is provided by the structure constants of $\mathfrak{g}$. A peculiarity of the algebras $A_{n-1} = \mathfrak{sl}(n)$ with $n \geq 3$ is that they also possess another three-index primitive invariant, denoted by $d^{ab}{}_c$, which is symmetric in a and b. This has its origin in the fact that together with the unit matrix $\mathbb{1}$, the generators in the defining representation of $\mathfrak{sl}(n)$ span the space of all complex $n \times n$-matrices. Indeed, one can write

$$R_\Lambda(T^a)\, R_\Lambda(T^b) = \gamma\kappa^{ab}\,\mathbb{1} + \sum_{c=1}^{n^2-1} (\mathrm{i}d^{ab}{}_c + f^{ab}{}_c)\, R_\Lambda(T^c)\,. \qquad (17.9)$$

The coefficients γ, d and f in the expansion (17.9) are real in any basis where the representation matrices $R_\Lambda(T^a)$ are anti-hermitian. Moreover, in such a basis the invariant d is traceless, i.e. satisfies $\sum_a d^{ab}{}_a = 0$.

For the rest of this chapter we now choose the basis with respect to which the components v^a in the adjoint module are taken in such a way that the Killing form κ^{ab} of the (complex) Lie algebra $\mathfrak{g}$ is proportional to δ^{ab} and that the representation matrices are anti-hermitian. In such a basis the structure constants $f^{abc} = \sum_d f^{ab}{}_d \kappa^{cd}$ with three upper indices are completely antisymmetric (compare exercise 8.3), the d-invariant d^{abc} with three upper indices is completely symmetric, and it is not necessary to distinguish between upper and lower indices. For composite invariants obtained from combinations of f, d and δ one then has the following relations.

- With the normalization $\mathrm{tr}(R_\Lambda(T^a)R_\Lambda(T^b)) = -\frac{1}{2}\delta^{ab}$ of the generators, f and d satisfy

$$\sum_{a,b} f^{abc} f^{abd} = n\,\delta^{cd}\,, \qquad \sum_{a,b} d^{abc} d^{abd} = \left(n - \tfrac{4}{n}\right)\delta^{cd}\,. \qquad (17.10)$$

- In addition to the Jacobi identity $\sum_e (f^{ade} f^{ebc} + f^{bde} f^{eca} + f^{cde} f^{eab}) = 0$ for f there is the similar formula

$$\sum_e \left(f^{ade} d^{ebc} + f^{bde} d^{eca} + f^{cde} d^{eab}\right) = 0\,. \qquad (17.11)$$

- A contraction of two f tensors can be expressed through δ and d tensors as

$$\sum_e f^{abe} f^{cde} = \frac{2}{n}\left(\delta^{ac}\delta^{bd} - \delta^{bc}\delta^{ad}\right) + \sum_e (d^{ace} d^{bde} - d^{ade} d^{bce})\,. \qquad (17.12)$$

- For triple products the contraction rules

$$\begin{aligned}
\sum_{a,b,c} f^{adb} f^{bec} f^{cfa} &= -\tfrac{1}{2}\, n\, f^{def}\,, \\
\sum_{a,b,c} d^{adb} f^{bec} f^{cfa} &= -\tfrac{1}{2}\, n\, d^{def}\,, \\
\sum_{a,b,c} d^{adb} d^{bec} f^{cfa} &= -\tfrac{1}{2}\left(n - \tfrac{4}{n}\right) f^{def}\,, \\
\sum_{a,b,c} d^{adb} d^{bec} d^{cfa} &= -\tfrac{1}{2}\left(n - \tfrac{12}{n}\right) d^{def}
\end{aligned} \qquad (17.13)$$

hold.

- For $n = 3$, the identity (17.12) is supplemented by

$$\begin{aligned}
\sum_e d^{abe} d^{cde} = \tfrac{1}{3}\big(&\delta^{ac}\delta^{bd} + \delta^{bc}\delta^{ad} - \delta^{ab}\delta^{cd} \\
&+ \textstyle\sum_e (f^{ace} f^{bde} + f^{ade} f^{bce})\big)\,.
\end{aligned} \qquad (17.14)$$

17.5 Tensors for spinor modules

According to chapter 13 the basic modules, from which all finite-dimensional modules of a simple Lie algebra can be obtained as tensor products, are just the defining modules, except for B_r and D_r, where they are the spinor modules of dimension 2^r and 2^{r-1}, respectively. Apart from the tensors for the adjoint and for the defining modules, the invariant tensors for the spinor modules are therefore the most interesting ones.

Recall that D_r has two inequivalent spinor modules. For the purposes of this section it is convenient to consider the direct sum of these modules; for brevity we refer to this reducible module as 'the' spinor of D_r. When doing so, the invariant tensors of the spinor can be described simultaneously for B_r and D_r. To derive these invariant tensors, one considers apart from the trivial primitive invariant tensor δ^α_β also an invariant tensor with index structure $(\gamma^i)^\alpha_\beta$. Here i is an index corresponding to the defining module, while α, β refer to the spinor module; thus i runs from 1 to n, with $n = 2r + 1$ for B_r and $n = 2r$ for D_r, while α, β run from 1 to 2^r. Thus γ is an invariant tensor on the tensor product of the defining module with the spinor module and the conjugate of the spinor module.

The tensor $(\gamma^i)^\alpha_\beta$ satisfies

$$\sum_{\alpha'} \left((\gamma^i)^\alpha_{\alpha'} (\gamma^j)^{\alpha'}_\beta + (\gamma^j)^\alpha_{\alpha'} (\gamma^i)^{\alpha'}_\beta \right) = 2\delta^{ij}\delta^\alpha_\beta , \qquad (17.15)$$

or expressed as matrices acting on the spinor module,

$$\{\gamma^i, \gamma^j\} := \gamma^i\gamma^j + \gamma^j\gamma^i = 2\delta^{ij}\,\mathbb{1} . \qquad (17.16)$$

(Here we consider the complex Lie algebra; when dealing with real forms, δ^{ij} gets replaced by a diagonal matrix η^{ij} of appropriate signature.) Matrices satisfying identities of this type are called *gamma matrices* or *Dirac matrices*; they furnish a representation of a so-called Clifford algebra (see section 20.6).

Let us now consider contractions of these tensors. We will use the matrix notation introduced in (17.16); thus products of the γ^i will be matrix products, i.e. contraction of the relevant spinor indices α, β, ... is understood. We set $\Gamma_{(0)} := \mathbb{1}$ and define for any integer p with $1 \leq p \leq n$ and any integers $i_\ell \in \{1, 2, ..., n\}$ with $i_1 < i_2 < \cdots < i_p$ the quantities

$$\Gamma^{i_1 i_2 \dots i_p}_{(p)} := i^{[p/2]}\, \gamma^{i_1}\gamma^{i_2}\cdots\gamma^{i_p} ; \qquad (17.17)$$

here $[p/2]$ is the integral part of $p/2$. Note that because of (17.16), for arbitrary integers $i_\ell \in \{1, 2, ..., n\}$ which are all distinct, the product $\gamma^{i_1}\gamma^{i_2}\cdots\gamma^{i_p}$ equals $\pm\Gamma^{i_1 i_2 \dots i_p}_{(p)}$, where the sign factor is the sign of the permutation $\sigma \in \mathcal{S}_p$ for which $\sigma(i_1) < \sigma(i_2) < \cdots < \sigma(i_p)$.

To describe the properties of the tensors (17.17), we now have to distinguish between B_r and D_r. In the case of D_r, the indices i_p run from 1 to 2r. One can show that the 2^{2r} matrices (17.17) with $p \in \{0, 1, \dots, 2r\}$ are linearly independent and hence form a basis of the complex $2^r \times 2^r$ - matrices. Also, except for $\Gamma_{(0)} = \mathbb{1}$ they are traceless, and the square of each of these matrices is the unit matrix. Moreover, together with their negatives, these matrices form a finite group; as a consequence there exists a choice of basis of the spinor module such that they are all unitary. (These remarks also apply to $\mathfrak{so}(4)$, which however is rather special in other respects because it is not simple. The case $n = 4$ is particularly interesting in physics, because it is the dimension of space-time. The real form $\mathfrak{so}(3, 1)$ is the Lie algebra of the Lorentz group; for details about its spinor representations, see section 20.7.)

For B_r, the indices i_p run from 1 to 2r + 1. One finds that now already the matrices $\Gamma_{(2q)}^{i_1 i_2 \dots i_{2q}}$ ($q \in \{0, 1, \dots, r\}$) with even value of the subscript are 2^{2r} in number, and indeed they provide a basis of the complex $2^r \times 2^r$ - matrices. Further, the set of all matrices $\Gamma_{(p)}$ (with both even and odd p) together with their negatives form a group of 2^{r+2} elements when the rank r is even; if r is odd, this remains true when the matrices $\pm\Gamma_{(2r+1)}$ are multiplied by i.

Since the matrices $\Gamma_{(q)}^{i_1 i_2 \dots i_q}$ (with q even when n is odd) form a basis, in particular the representation matrices $R_s(T^a)$ of the spinor representation can be expressed as linear combinations of these matrices. It turns out that in the orthonormal basis of $\mathfrak{so}(n)$ in which the generators T^a are given by antisymmetric matrices T^{ij} with $i < j$, one has in fact

$$R_s(T^{ij}) = \tfrac{1}{4} \Gamma_{(2)}^{ij}. \tag{17.18}$$

Another property of the matrices (17.17) which is important in various applications is the following. One finds that, given any two $2^r \times 2^r$ -matrices M and N, the relation

$$M_\beta^\alpha N_\delta^\gamma = 2^{-r} \sum_{p=0}^{n} \left[\sum_{i_1, i_2, \dots, i_p} (\Gamma_{(p)}^{i_1 i_2 \dots i_p})_\delta^\alpha (N\, \Gamma_{(p)}^{i_1 i_2 \dots i_p} M)_\beta^\gamma \right] \tag{17.19}$$

holds when n is even, and for n odd a similar formula is valid with the summation restricted to even p. One may apply this result in particular to the tensor product of two matrices of the type $\Gamma_{(q)}$; the formulæ one then obtains are known as *Fierz identities* (for details, see e.g. [Case 1955]).

Fierz identities are a common tool in Lagrangian field theories involving fermions, which carry a spinor representation of $\mathfrak{so}(n)$. Let us present the explicit form of (17.19) for $n = 4$ (euclidean) space-time dimensions, taking $M = N = \mathbb{1}$ for simplicity. Defining $\gamma_5 := i\gamma^1\gamma^2\gamma^3\gamma^4$, one obtains

$$
\begin{aligned}
\delta^\alpha_\beta \delta^\gamma_\delta = \tfrac{1}{4}\Big[\delta^\alpha_\delta \delta^\gamma_\beta &+ \sum_{i=1}^{4}(\gamma^i)^\alpha_\delta (\gamma^i)^\gamma_\beta - \tfrac{1}{2}\sum_{\substack{i,j=1\\i\neq j}}^{4}(\gamma^i\gamma^j)^\alpha_\delta (\gamma^i\gamma^j)^\gamma_\beta \\
&+ \sum_{i=1}^{4}(\gamma_5\gamma^i)^\alpha_\delta (\gamma_5\gamma^i)^\gamma_\beta - (\gamma_5)^\alpha_\delta (\gamma_5)^\gamma_\beta \Big].
\end{aligned}
\tag{17.20}
$$

The five terms in this sum are usually referred to as the scalar, vector, tensor, pseudovector and pseudoscalar term, respectively.

17.6 Singlets in tensor products

In principle, all tensor product coefficients $\mathcal{L}_{\Lambda\Lambda'}{}^{\Lambda''}$ can be deduced from the enumeration of linearly independent invariant tensors with the appropriate index structure. For example, since together with the unit matrix the representation matrices of the defining module V of $A_{n-1} = \mathfrak{sl}(n)$ span the algebra of all $n \times n$-matrices, one can write

$$
\tilde{v}_i \otimes v^j = \delta^j_i S + \sum_a (R(T^a))^j_i A^a
\tag{17.21}
$$

for the tensor product of V with its conjugate module. Applying the transformation (17.1) to both sides of this equation and using that both $(R(T^a))^j_i$ (compare exercise 17.3) and δ^j_i are invariant tensors, it follows that S and A^a must transform according to the singlet and adjoint representation, respectively. Hence one reads off that the irreducible components of $V \otimes V^+$ are the singlet and adjoint modules.

In matrix language, this can be understood as follows. Via the adjoint action, the algebra A_r acts on the n^2-dimensional vector space of $n \times n$-matrices. The decomposition just mentioned corresponds to the decomposition of a matrix into its trace part and a traceless part; each of these parts constitutes a separate irreducible representation of $\mathfrak{sl}(n)$.

In practice, this method of evaluating tensor product decompositions is rather inefficient, because it is in general quite difficult to find all independent invariants. Nevertheless there are a few circumstances where such considerations can be applied profitably. For instance, for $n \geq 3$ the number of adjoint modules in the tensor product of two adjoints of $\mathfrak{sl}(n)$ is equal to two because there are two linearly independent three-index tensors f^{abc} and d^{abc}, and from the symmetry properties of these tensors we learn that the two independent couplings of three adjoints can be characterized as being totally symmetric or totally antisymmetric.

In Yang–Mills theories, the gauge bosons carry the adjoint representation of the structure group. For example in QCD, the gauge theory of strong interactions, the structure group is SU(3), and the gauge bosons

Information

(gluons) carry the eight-dimensional representation. The combinatorics of invariant tensors of the adjoint representation then governs the allowed possibilities for coupling three gluons together. According to the result just mentioned, there are two inequivalent three-gluon couplings.

Another example is provided by the determination of the number of singlets in a tensor product. Of course, in the tensor product of two irreducible modules there is precisely one singlet if the two modules are conjugate to each other, and otherwise there is no singlet at all. But for multiple tensor products the situation can be more interesting.

As an illustration, let us determine the number of singlets in the product of four adjoint modules of $\mathfrak{sl}(n)$. With the help of the identities (17.10) – (17.13) one can check that for $n \geq 4$ there are nine linearly independent invariants with the index structure of $v^a \otimes v^b \otimes v^c \otimes v^d$, namely (say)

$$
\begin{array}{ccc}
\delta^{ab}\delta^{cd}, & \delta^{ac}\delta^{bd}, & \delta^{bc}\delta^{ad}, \\[2mm]
\sum_e f^{abe}f^{cde}, & \sum_e f^{ace}f^{bde}, & \sum_e d^{abe}d^{cde}, \\[2mm]
\sum_e d^{abe}f^{cde}, & \sum_e d^{ace}f^{bde}, & \sum_e d^{bde}d^{ace}.
\end{array}
\qquad (17.22)
$$

Hence the tensor product $R_\theta \otimes R_\theta \otimes R_\theta \otimes R_\theta$ contains the singlet with multiplicity 9.

For $n = 3$ there is the additional relation (17.14) so that only eight of the nine invariants (17.22) are linearly independent, while for $n = 2$ one has $d^{abc} \equiv 0$ as well as $f^{abc} \propto \epsilon^{abc}$ so that there are only three linearly independent invariants.

17.7 Projection operators

To describe the singlets in the four-fold tensor product $(V_{\Lambda_1} \otimes V_{\Lambda_2}) \otimes (V_{\Lambda_3} \otimes V_{\Lambda_4})$ in a concrete manner, we use the Littlewood–Richardson coefficients to express it as

$$
(V_{\Lambda_1} \otimes V_{\Lambda_2}) \otimes (V_{\Lambda_3} \otimes V_{\Lambda_4}) \cong \sum_{\Lambda, \Lambda'} \mathcal{L}_{\Lambda_1 \Lambda_2}{}^{\Lambda} \, \mathcal{L}_{\Lambda_3 \Lambda_4}{}^{\Lambda'} \, V_\Lambda \otimes V_{\Lambda'} .
\qquad (17.23)
$$

In this summation a singlet arises precisely for each term for which Λ' is the conjugate of Λ. Hence the invariant tensors of the four-fold tensor product are in one-to-one correspondence with the pairs of irreducible submodules V_Λ of $V_{\Lambda_1} \otimes V_{\Lambda_2}$ and V_{Λ^+} of $V_{\Lambda_3} \otimes V_{\Lambda_4}$. Pictorially, we can present this as in (17.24) below: A non-vanishing coupling of Λ_1 and Λ_2 to Λ is represented by a trivalent vertex just like we already did in figure (16.22); this is the left vertex in (17.24). Also recall the convention that Λ^+ is depicted by the same line as Λ, but with the direction of the arrow reversed; thus the right vertex of (17.24) describes the coupling of Λ_3 and Λ_4 to Λ^+.

$$(17.24)$$

Each allowed label Λ of the inner line of the figure corresponds to coupling the four external lines via the vectors of the module V_Λ respectively $V_{\Lambda+}$ as 'intermediate states'. To each irreducible module of intermediate states and each allowed combination of multiplicity labels α, β there is associated a specific invariant tensor for the tensor product $V_{\Lambda_1} \otimes V_{\Lambda_2} \otimes V_{\Lambda_3} \otimes V_{\Lambda_4}$. With such an invariant tensor one selects among all intermediate states in (17.24) those which belong to a definite irreducible module; these tensors are therefore called *projection operators*. (Quantum mechanically the left and right parts of the diagram (17.24) can be interpreted as in- and out-states, respectively. The projection operator is then a mapping from in- to out-states; the matrix elements of this operator are the components of the invariant tensor.) As a particularly important example, consider the singlets in the tensor product $V_\Lambda \otimes V_{\Lambda+} \otimes V_\Lambda \otimes V_{\Lambda+}$ of two defining and two conjugate defining modules of $\mathfrak{sl}(n)$, with components $v^i \otimes \tilde{v}_j \otimes v^k \otimes \tilde{v}_l$. It follows from (17.21) that this tensor product contains two singlets, with the associated invariant tensors given by $\delta^i_j \delta^k_l$ and

$$(T_\Lambda)^{ik}_{jl} := \sum_{a,b} \kappa_{ab} \, R_\Lambda(T^a)^i{}_j \, R_\Lambda(T^b)^k{}_l. \tag{17.25}$$

The tensor $\delta^i_j \delta^k_l$ corresponds to having the singlet as intermediate state, while the presence of the tensor κ_{ab} in (17.25) shows that $(T_\Lambda)^{ik}_{jl}$ describes the 'coupling through the adjoint' as module of intermediate states. Both the singlet and the adjoint appear also as intermediate states in the tensor product of two adjoints of each other finite-dimensional simple Lie algebra, but typically this tensor product also contains further irreducible modules.

In QCD, the gluons carry the adjoint representation (compare the previous section), while the quarks carry the defining representation of $\mathfrak{su}(3)$. Thus the diagram (17.24) can in this context be regarded as describing the scattering of two quarks by exchanging a gauge boson (with time now running in the vertical direction, and the arrows on the lines interpreted properly). The tensor (17.25) then corresponds to the exchange of a gluon, and is therefore called the gluon projection operator for quark-quark scattering, while $\delta^i_j \delta^k_l$ corresponds to exchanging a singlet gauge boson, such as the photon.

$\overline{\text{Information}}$

By comparison with table (17.8), one learns that the projection tensor $(T_\Lambda)^{ik}_{jl}$ is not a primitive invariant. For instance in the case of A_r it must clearly be a linear combination of the composite invariants $\delta^i_j \delta^k_l$ and $\delta^i_l \delta^k_j$. These decompositions in terms of the primitive invariant tensors of the list (17.8) are shown in table XIV. (Here the overall normalization is determined by a specific choice of the normalization of the representation matrices; also, the primitive invariant tensors are normalized such that $\sum_{i_1,\dots,i_n} t_{ij_1\dots j_n} t^{kj_1\dots j_n} = \delta^k_i$.)

In determining relations like those in table XIV, it is often necessary to expand an invariant tensor with a fixed index structure with respect to a given basis of such

Table XIV. *Decomposition of projection tensors into invariant tensors*

$\mathfrak{g}$	Λ	expansion of $(T_\Lambda)^{ik}_{jl}$
A_r	$\Lambda_{(1)}$	$\frac{1}{2}\left(\frac{1}{r+1}\,\delta^i{}_j\delta^k{}_l - \delta^i{}_l\delta^k{}_j\right)$
B_r	$\Lambda_{(1)}$	$\frac{1}{2}\left(\delta^{ik}\delta^{jl} - \delta^{il}\delta^{jk}\right)$
C_r	$\Lambda_{(1)}$	$-\frac{1}{4}\left(\delta^i{}_l\delta^k{}_j + f^{ik}f_{jl}\right)$
D_r	$\Lambda_{(1)}$	$\frac{1}{2}\left(\delta^{ik}\delta^{jl} - \delta^{il}\delta^{jk}\right)$
E_6	$\Lambda_{(1)}$	$-\frac{1}{6}\,\delta^i{}_j\delta^k{}_l - \frac{1}{2}\,\delta^i{}_l\delta^k{}_j + 5\,d^{ikm}d_{jlm}$
E_7	$\Lambda_{(6)}$	$-\frac{1}{4}\left(\delta^i{}_l\delta^k{}_j + f^{ik}f_{jl} + d^{ikmn}f_{mj}f_{nl}\right)$
F_4	$\Lambda_{(4)}$	$\frac{1}{3}\left(\delta^{ik}\delta^{jl} - \delta^{il}\delta^{jk}\right) + \frac{7}{3}\left(d^{ikm}d^{jl}{}_m - d^{ilm}d^{jk}{}_m\right)$
G_2	$\Lambda_{(2)}$	$\frac{1}{2}\left(\delta^{ik}\delta^{jl} - \delta^{il}\delta^{jk}\right) - f^{ijm}f^{kl}{}_m$

invariants. For tensors whose indices all correspond to real representations, this can be done as follows. If $\{(t_A)^{i_1 i_2 \dots}_{j_1 j_2 \dots}\}$ with $A = 1, 2, \dots, M$ is the basis of invariants, the expansion coefficients τ^A of a tensor

$$t = \sum_{A=1}^{M} \tau^A t_A \tag{17.26}$$

can be written as

$$\tau^A = \sum_{B=1}^{M} (G^{-1})^{AB}\, \mathrm{tr}(t_B\, t)\,, \tag{17.27}$$

where the trace 'tr' denotes summation over all tensor indices, and

$$G_{AB} := \mathrm{tr}(t_A\, t_B)\,. \tag{17.28}$$

In the case of non-real representations, one has to use similar formulæ involving also tensors for the conjugate representation, such that in particular the trace operation again makes sense.

Thus in order to determine the expansion coefficients τ^A one must compute the various traces appearing in (17.28) and (17.27) as well as invert the $M \times M$-matrix G.

17.8 Casimir operators

According to chapter 14 the Casimir operators of a simple Lie algebra $\mathfrak{g}$ are distinguished elements of the center of the universal enveloping algebra $\mathsf{U}(\mathfrak{g})$ of $\mathfrak{g}$; they are homogeneous polynomials in the generators of $\mathfrak{g}$, and they constitute a maximal set of algebraically independent elements of the center. Since the representation matrices in the adjoint representation are just the structure constants, it follows from equation (17.3) that the Casimir operators are invariant tensors of the adjoint representation. One can show that they are in fact in one-to-one correspondence with the set of tensors that generates the space of these invariant tensors. Namely, in order that a polynomial

$$P = c\mathbf{1} + \sum_a t_a T^a + \sum_{a,b} t_{ab} T^a T^b + \sum_{a,b,c} t_{abc} T^a T^b T^c + \dots \qquad (17.29)$$

in the generators T^a belongs to the center of $U(\mathfrak{g})$, it is sufficient that all coefficients $t_{a_1 a_2 \dots a_n}$ are invariant tensors of the adjoint representation. And conversely, when P is written in such a form that the tensors $t_{a_1 a_2 \dots a_n}$ are all completely symmetric (which is always possible), then this is also a necessary condition (see exercise 17.8). In short, all Casimir operators of $\mathfrak{g}$ are of the form

$$\mathcal{C}_n = \sum_{a_1, \dots, a_n = 1}^{d} d_{a_1 a_2 \dots a_n} \, T^{a_1} T^{a_2} \cdots T^{a_n} \,, \qquad (17.30)$$

with suitable symmetric invariant tensors $d_{a_1 a_2 \dots a_n}$. (Note that even though the Casimir operators are algebraically independent elements of $U(\mathfrak{g})$, the relevant tensors $d_{a_1 a_2 \dots a_n}$ need not necessarily be algebraically independent, i.e. need not all be primitive invariants.)

Now recall from chapter 14 that the quadratic Casimir operator is the element

$$\mathcal{C}_2 := \sum_{a,b=1}^{d} \kappa_{ab} \, T^a T^b \qquad (17.31)$$

of the universal enveloping algebra; thus the relevant invariant two-index tensor is the Killing form $\kappa_{ab} \propto \sum_{c,d} f^{ac}{}_d f^{bd}{}_c$. Likewise, other elements of the center of $U(\mathfrak{g})$ can be obtained by taking the tensors

$$\mathrm{tr}(\mathrm{ad}\,_{T^{a_1}} \circ \mathrm{ad}\,_{T^{a_2}} \circ \cdots \circ \mathrm{ad}\,_{T^{a_n}})$$

$$= \sum_{b_1, b_2, \dots, b_n = 1}^{d} f^{a_1 b_1}{}_{b_2} f^{a_2 b_2}{}_{b_3} \cdots f^{a_{n-1} b_{n-1}}{}_{b_n} f^{a_n b_n}{}_{b_1} \,. \qquad (17.32)$$

However, these elements are not all algebraically independent, and hence typically do not provide Casimir operators. For instance, in the case of $\mathfrak{sl}(n)$ the relations (17.10) and (17.13) can be used to show that the combination $\sum_{a,b,c,d,e,f} f_{ad}{}^e f_{be}{}^f f_{cf}{}^d \, T^a T^b T^c$ is just proportional to the quadratic Casimir operator.

On the other hand, for $\mathfrak{sl}(n)$ with $n \geq 3$ there is the independent invariant d^{abc}, and hence an independent third order Casimir operator

$$\mathcal{C}_3 \equiv \mathcal{C}_3^{(d)} := \sum_{a,b,c=1}^{d} d_{abc}\, T^a T^b T^c\,. \tag{17.33}$$

In fact, one can show that all the independent higher order Casimir operators of $\mathfrak{sl}(n)$ can be expressed through the three-index invariant d, namely as

$$\mathcal{C}_m := \sum_{\substack{a_1,a_2,\ldots,a_{m-3}, \\ b_1,b_2,\ldots,b_m}} d_{a_1 b_1 b_2}\, d_{a_1 a_2 b_3}\, d_{a_2 a_3 b_4} \cdots$$

$$\cdots d_{a_{m-2} a_{m-3} b_{m-2}}\, d_{a_{m-3} b_{m-1} b_m}\, T^{b_1} T^{b_2} \cdots T^{b_m} \tag{17.34}$$

for $m = 3, 4, \ldots, n$.

Let us also note that, as is clear from the formula $C_\Lambda = (\Lambda, \Lambda + 2\rho)$ (14.28), the quadratic Casimir eigenvalues C_Λ are rational numbers, but that typically they are not integers. In contrast, the second order Dynkin index I_Λ which is related to the quadratic Casimir eigenvalue by $I_\Lambda = d_\Lambda C_\Lambda / d$ (14.34), when divided by the length squared of the highest root, is a half integer for any dominant integral weight Λ; it is in fact an integer if the highest weight module is orthogonal (half integer values do occur for some symplectic or non-selfconjugate highest weight modules). Some explicit formulæ for the eigenvalues of higher order Casimir operators can be found in chapter 9 of [Barut and Rączka 1986], and also in [Micu 1964] and [Perelomov and Popov 1968].

Summary:

An invariant tensor of a $\mathfrak{g}$-module V is an intertwiner between some tensor power of V and the singlet. Invariant tensors are helpful in various situations where one wants to extract the 'singlet part' of an expression. Invariant tensors which correspond to 'intermediate states' in multiple tensor products are called projection operators. Casimir operators are characterized by invariant tensors of the adjoint module.

Keywords:

Intertwiner with the singlet, invariant tensor, primitive invariant, orthogonal and symplectic modules, Casimir operator;
Kronecker symbol, Levi–Civita tensor, gamma matrix, Fierz identities, gluon projection operator.

Exercises:

Check that the Killing form and the structure constants of a simple Lie algebra are invariant tensors in the sense of (17.3).

Exercise 17.1

Express the contraction $\sum_\ell \epsilon^{\ell i_2 \cdots i_d} \epsilon_{\ell j_2 \ldots j_d}$ of Levi–Civita tensors as a sum of products of Kronecker symbols.

Exercise 17.2

Write down the invariance property (17.4) for the matrix elements $(T^a)^i{}_j$ of the generators T^a in an arbitrary irreducible representation. Write the equation in matrix form and explain its meaning.

Exercise 17.3

Determine, for each simple Lie algebra $\mathfrak{g}$, the minimal tensor power of the defining module which contains a singlet.
In the $\mathfrak{su}(3)$ model of hadrons (compare chapter 3), baryons are regarded as bound states of three quarks. Could one also employ other simple Lie algebras to describe baryons in such a way? Would the answer to the latter question be different if the quarks were bosons rather than fermions?

Exercise 17.4

Determine a set of linearly independent fourth rank invariant tensors for the adjoint representation of E_8.
Identify the linear combinations of these tensors which correspond to the various irreducible components in the tensor product (15.34) of two adjoints.

Exercise 17.5

Find a set of linearly independent fifth rank invariant tensors for the defining representation of $\mathfrak{sl}(3)$.

Exercise 17.6

Use the Clifford algebra relation (17.15) and the properties of the matrices $\Gamma_{(p)}$ (17.17) to compute the traces of products of gamma matrices.
(Such traces are particularly interesting in the case of $\mathfrak{so}(3,1)$, because in Lagrangian quantum field theory they arise in the evaluation of Feynman graphs which involve fermions.)

Exercise 17.7

Show that for P as defined in (17.29), with totally symmetric $t_{a_1 a_2 \ldots a_n}$, to belong to the center of $U(\mathfrak{g})$, it is necessary and sufficient that the coefficients are invariant tensors of the adjoint representation.

Exercise 17.8

18
Subalgebras and branching rules

18.1 Subalgebras

In the analysis of the structure of highest weight modules in chapter 13 we have made heavy use of the decomposition of the weight system of a module of a simple Lie algebra $\mathfrak{g}$ into modules of its $\mathfrak{sl}(2)_\alpha$-subalgebras. This was both useful and easy because the irreducible modules of $\mathfrak{sl}(2)$ have a very simple structure. A more general, and more difficult, task is to decompose the modules of a simple Lie algebra $\mathfrak{g}$ with respect to an arbitrary Lie subalgebra $\mathfrak{h} \subset \mathfrak{g}$. (Recall from section 4.8 that in this context $\mathfrak{g}$ is called the ambient algebra.)

This problem arises quite often in applications, in particular whenever there is any kind of *symmetry breaking*. The term symmetry breaking refers to the situation that a system which possesses a certain symmetry is perturbed or deformed in such a way that in the perturbed system only part of the symmetry is still realized. (In practice, it is often the perturbed system that is physically relevant, while the system with the full symmetry describes an idealized situation which is not realized in nature; compare e.g. the $\mathfrak{sl}(3)$ symmetry of hadrons that was described in chapter 3.) If the full symmetry is described by a Lie algebra $\mathfrak{g}$ and the surviving symmetry by a subalgebra $\mathfrak{h}$ of $\mathfrak{g}$, one can of course use multiplets of $\mathfrak{h}$ to describe the physical states; however, one also would like to know how this is related to the description of the states in terms of multiplets of $\mathfrak{g}$. This is one of the questions we will address in this chapter.

To characterize a subalgebra $\mathfrak{h}$ of a Lie algebra $\mathfrak{g}$ it is usually not sufficient to know just to what abstract Lie algebras $\mathfrak{g}$ and $\mathfrak{h}$ are isomorphic; one must be careful to specify which subset of $\mathfrak{g}$ is to be identified as the subalgebra. Frequently it is even more convenient not to regard the subalgebra $\mathfrak{h}$ as just a subset of $\mathfrak{g}$, but rather to specify a definite embedding map, i.e. an injective Lie algebra homomorphism $\imath:\ \mathfrak{h}\to\mathfrak{g}$. Instead of $\mathfrak{h}\subset\mathfrak{g}$ we will therefore write an embedding as $\mathfrak{h}\overset{\imath}{\hookrightarrow}\mathfrak{g}$, or shortly, $\mathfrak{h}\hookrightarrow\mathfrak{g}$.

We have already encountered several important subalgebras of a finite-dimensional simple Lie algebra, e.g. the Cartan subalgebra and the Borel subalgebras in the triangular decomposition. Clearly, none of these algebras is simple, and the Borel subalgebras are not even reductive. Here we will be interested in reductive subalgebras, and mainly in simple or semi-simple subalgebras. When dealing with the problem to find simple Lie subalgebras of finite-dimensional simple Lie algebras, it is sufficient to consider only proper *maximal subalgebras* $\mathfrak{h}$, i.e. subalgebras such that there does not exist any 'intermediate' simple subalgebra $\mathfrak{k}$ obeying $\mathfrak{h}\hookrightarrow\mathfrak{k}\hookrightarrow\mathfrak{g}$. Non-maximal subalgebras can then be treated in a step-wise procedure, first considering maximal subalgebras $\mathfrak{h}$, then in turn the maximal subalgebras of these algebras $\mathfrak{h}$, and so on.

If all step operators of a reductive subalgebra $\mathfrak{h}$ of a simple Lie algebra $\mathfrak{g}$ are also step operators of $\mathfrak{g}$, then $\mathfrak{h}$ is called a *regular* subalgebra of $\mathfrak{g}$; otherwise $\mathfrak{h}$ is called a *special* subalgebra. More generally, subalgebras which are contained in some regular subalgebra are called *R-subalgebras*, while all others are called *S-subalgebras*. In order to classify all reductive subalgebras of simple Lie algebras, one thus has to look for the maximal regular subalgebras and for the maximal *S*-subalgebras. Also, one finds that every reductive subalgebra of a semisimple Lie algebra that does contain an abelian ideal is an *R*-subalgebra.

The root system and the basis of simple roots of the subalgebra $\mathfrak{h}\hookrightarrow\mathfrak{g}$ will be denoted by $\tilde{\Phi}$ and $\tilde{\Phi}_{\mathbf{s}}=\{\tilde{\alpha}^{(i)}\}$, respectively. According to one of the main properties of simple roots, the difference $\tilde{\alpha}^{(i)}-\tilde{\alpha}^{(j)}$ of two simple roots of the subalgebra is never a root: $\tilde{\alpha}^{(i)}-\tilde{\alpha}^{(j)}\notin\tilde{\Phi}$. If $\mathfrak{h}$ is a regular subalgebra, this immediately implies that $\tilde{\alpha}^{(i)}-\tilde{\alpha}^{(j)}\notin\Phi$ (the root system of $\mathfrak{g}$), because otherwise the element $E^{\tilde{\alpha}^{(i)}-\tilde{\alpha}^{(j)}}\propto[E^{\tilde{\alpha}^{(i)}},E^{-\tilde{\alpha}^{(j)}}]$ would not vanish and would also lie in $\mathfrak{h}$, in contradiction to $\tilde{\alpha}^{(i)}-\tilde{\alpha}^{(j)}\notin\tilde{\Phi}$. As a consequence, the regular subalgebras of $\mathfrak{g}$ are in one-to-one correspondence to those linearly independent subsets $\tilde{S}\subset\Phi$ which obey

$$\tilde{\alpha},\tilde{\beta}\in\tilde{S}\implies\tilde{\alpha}-\tilde{\beta}\notin\Phi. \tag{18.1}$$

Given such a subset, the generators $\{E^{\tilde{\alpha}},E^{-\tilde{\alpha}},H^{\tilde{\alpha}}\,|\,\tilde{\alpha}\in\tilde{S}\}$ generate the regular subalgebra upon taking Lie brackets, and $\tilde{\Phi}_{\mathbf{s}}=\tilde{S}$ is a basis of simple roots of $\mathfrak{h}$.

A rather common special embedding is the so-called *diagonal* embedding

$$\mathfrak{g}\hookrightarrow\mathfrak{g}\oplus\mathfrak{g}. \tag{18.2}$$

Given identical bases $\{T^a_{(1)}\}$ and $\{T^a_{(2)}\}$ of the two copies of $\mathfrak{g}$, the diagonal subalgebra is spanned by $\{T^a_{(+)}\}$ with $T^a_{(+)}:=T^a_{(1)}+T^a_{(2)}$.

18.2 The Dynkin index of an embedding

Using the embedding $\imath$ and the Killing form $\kappa_{\mathfrak{g}}$ of the ambient algebra $\mathfrak{g}$, we can define a bilinear form κ on the subalgebra $\mathfrak{h}$, namely $\kappa(x, y) := \kappa_{\mathfrak{g}}(\imath(x), \imath(y))$. Thus the Killing form of $\mathfrak{g}$ induces an invariant bilinear form on any semisimple subalgebra $\mathfrak{h}$ of $\mathfrak{g}$. Now if $\mathfrak{h}$ is simple, all invariant bilinear forms on $\mathfrak{h}$ are proportional, and hence in particular the induced bilinear form on $\mathfrak{h}$ is a numerical multiple of the Killing form of $\mathfrak{h}$. This constant of proportionality is (after dividing by a conventional normalization factor $c = I_{\mathrm{ad}}(\mathfrak{g})/I_{\mathrm{ad}}(\mathfrak{h})$) called the *Dynkin index* $I_{\mathfrak{h}\subset\mathfrak{g}}$ of the embedding $\mathfrak{h} \hookrightarrow \mathfrak{g}$. If the reductive subalgebra $\mathfrak{h}$ is not simple, then there is of course a separate Dynkin index for each simple ideal of $\mathfrak{h}$.

This discussion shows that

$$c\, I_{\mathfrak{h}\subset\mathfrak{g}} = \frac{\kappa_{\mathfrak{g}}(\imath(x), \imath(y))}{\kappa_{\mathfrak{h}}(x, y)} \tag{18.3}$$

for any two elements x, y of $\mathfrak{h}$ for which the denominator is non-zero. As the normalization of the Killing form is completely fixed by the length of the highest root θ, the Dynkin index can also be written as

$$I_{\mathfrak{h}\subset\mathfrak{g}} = \frac{(\theta(\mathfrak{g}), \theta(\mathfrak{g}))}{(\theta(\mathfrak{h}), \theta(\mathfrak{h}))}. \tag{18.4}$$

For regular subalgebras, the highest root $\theta(\mathfrak{h})$ is a root of $\mathfrak{g}$, so that according to the formula (18.4) the Dynkin index $I_{\mathfrak{h}\subset\mathfrak{g}}$ is equal to one for $\mathfrak{g}$ simply laced, either 1 or 2 for $\mathfrak{g}$ non-simply laced other than G_2, and either 1 or 3 for $\mathfrak{g} = G_2$. One can show that, more generally, the Dynkin index is an integer for *any* embedding $\mathfrak{h} \hookrightarrow \mathfrak{g}$ of simple Lie algebras.

There exists yet another possibility to characterize the Dynkin index of an embedding. Namely, if $R: \mathfrak{g} \to \mathfrak{gl}(V)$ is a representation of $\mathfrak{g}$ on a vector space V, then $R \circ \imath: \mathfrak{h} \to \mathfrak{gl}(V)$ is a representation of $\mathfrak{h}$ on V (compare equation (5.20) for the underlying general principle). According to the definition (14.33) of the second order Dynkin index I_R of a $\mathfrak{g}$-representation R one then has

$$\mathrm{tr}\left((R{\circ}\imath)(x)\,(R{\circ}\imath)(y)\right) = (I_{R\circ\imath}/I_{\mathrm{ad}}(\mathfrak{h}))\,\kappa_{\mathfrak{h}}(x, y)$$
$$\equiv \mathrm{tr}\left(R(\imath(x))\,R(\imath(y))\right) = (I_R/I_{\mathrm{ad}}(\mathfrak{g}))\,\kappa_{\mathfrak{g}}(\imath(x), \imath(y)). \tag{18.5}$$

(In general, the representation $R{\circ}\imath$ is reducible; the symbol $I_{R\circ\imath}$ stands for the sum of the Dynkin indices of all irreducible sub-representations of $R{\circ}\imath$.) It follows that for any $\mathfrak{g}$-representation R, the Dynkin index of an embedding $\mathfrak{h} \hookrightarrow \mathfrak{g}$ can be expressed as the ratio of the second order Dynkin indices of the representations R and $R{\circ}\imath$:

$$I_{\mathfrak{h}\subset\mathfrak{g}} = I_{R\circ\imath}/I_R. \tag{18.6}$$

The Dynkin index can be used to distinguish between various inequivalent embeddings of a subalgebra $\mathfrak{h}$ in $\mathfrak{g}$. For example, the table (18.14) below shows that E_8 contains three different S-subalgebras which are all isomorphic to A_1. If also non-maximal embeddings are considered, then there are, however, also some cases where several inequivalent embeddings of $\mathfrak{h}$ in $\mathfrak{g}$ possess the same Dynkin index.

18.3 Finding regular subalgebras

There exists a simple algorithm to find all sets $\tilde{S}$ of roots which satisfy the condition (18.1). Namely, up to isomorphism all *maximal regular semisimple* subalgebras are obtained by considering all subsets

$$\tilde{S} \subset \Phi_\mathbf{s} \cup \{-\theta\}\,, \tag{18.7}$$

where θ is the highest root of $\mathfrak{g}$. More precisely, for $\mathfrak{g} \neq A_r$ there exist no other maximal regular semisimple subalgebras besides the ones with simple root systems $\tilde{\Phi}_\mathbf{s} = \Phi_\mathbf{s} \cup \{-\theta\} \setminus \{\alpha^{(i)}\}$ for some $i \in \{1, 2, ..., r\}$; and conversely, up to very few exceptions each such choice does yield a maximal regular semisimple subalgebra. In contrast, for $\mathfrak{g} = A_r$ any subalgebra obtained with this prescription is just A_r itself. As a consequence, for A_r the relevant subalgebras are precisely the ones which have a simple root system $\tilde{\Phi}_\mathbf{s} = \Phi_\mathbf{s} \cup \{-\theta\} \setminus \{\alpha^{(i)}, \alpha^{(j)}\}$ with $i, j \in \{1, 2, ..., r\}$, $i \neq j$.

The exceptions just mentioned are encountered only for exceptional algebras, namely when removing the root $\alpha^{(3)}$ of F_4 or of E_7, or the roots $\alpha^{(3)}$, $\alpha^{(5)}$ or $\alpha^{(6)}$ of E_8 from $\Phi_\mathbf{s} \cup \{-\theta\}$. The chains of embeddings which show that these embeddings are not maximal are

$$
\begin{array}{llll}
F_4,\ \Phi_\mathbf{s} \cup \{-\theta\} \setminus \{\alpha^{(3)}\}\ : & A_3 \oplus A_1 \hookrightarrow & B_4 & \hookrightarrow F_4\,, \\
E_7,\ \Phi_\mathbf{s} \cup \{-\theta\} \setminus \{\alpha^{(3)}\}\ : & A_3 \oplus A_3 \oplus A_1 \hookrightarrow & D_6 \oplus A_1 & \hookrightarrow E_7\,, \\
E_8,\ \Phi_\mathbf{s} \cup \{-\theta\} \setminus \{\alpha^{(3)}\}\ : & A_3 \oplus D_5 \hookrightarrow & D_8 & \hookrightarrow E_8\,, \\
E_8,\ \Phi_\mathbf{s} \cup \{-\theta\} \setminus \{\alpha^{(5)}\}\ : & A_5 \oplus A_2 \oplus A_1 \hookrightarrow & E_6 \oplus A_2 & \hookrightarrow E_8\,, \\
E_8,\ \Phi_\mathbf{s} \cup \{-\theta\} \setminus \{\alpha^{(6)}\}\ : & A_7 \oplus A_1 \hookrightarrow & E_7 \oplus A_1 & \hookrightarrow E_8\,.
\end{array} \tag{18.8}
$$

That the prescription (18.7) always yields a regular subalgebra is easily understood. Namely, including minus the highest root into $\Phi_\mathbf{s}$ preserves the property that the difference of any two elements is not a root, and by removing one of the simple roots one then recovers a linearly independent set of roots. The rule (18.7) is also nicely illustrated in terms of Dynkin diagrams. As described in section 11.6, by adding a node corresponding to $-\theta$ to the Dynkin diagram of $\mathfrak{g}$, one obtains the *extended* Dynkin diagram of $\mathfrak{g}$, which is the same as the Dynkin diagram of the untwisted affine Lie algebra $X_r^{(1)}$ associated to $\mathfrak{g} = X_r$. For $\mathfrak{g} \neq A_r$, the Dynkin diagram of the subalgebra $\mathfrak{h}$ is then obtained from the extended Dynkin diagram of

$\mathfrak{g}$ by removing the node corresponding to a simple root $\alpha^{(i)}$. For $\mathfrak{g} = A_r$ the removal of a single node from the extended Dynkin diagram just gives back the original Dynkin diagram, which explains why one has to discard two simple roots in order to arrive at a (proper) maximal regular semisimple subalgebra. These subalgebras of A_r are either

$$\mathfrak{h} = A_{r-1} \quad \text{or} \quad \mathfrak{h} = A_{r'} \oplus A_{r-r'-1} \quad \text{with} \quad r' \in \{1, 2, \ldots, [\tfrac{r-1}{2}]\}. \quad (18.9)$$

There also exist non-semisimple maximal regular subalgebras. They are parabolic subalgebras and are algebraically generated by the set that is obtained from a Chevalley–Serre basis of $\mathfrak{g}$ when removing a single generator $E^{\alpha^{(i)}}$; they obviously contain the semisimple subalgebra that is obtained when removing in addition also the generators $E^{-\alpha^{(i)}}$ and $H^{\alpha^{(i)}}$. For instance, the subalgebras (18.9) of A_r are maximal only among the *semisimple* subalgebras, but not among all subalgebras of A_r; by removing the generators $E^{\alpha^{(i)}}$ and $E^{-\alpha^{(i)}}$ from a Chevalley basis, one gets the non-semisimple subalgebra $\mathfrak{h} \oplus \mathfrak{u}(1) \hookrightarrow A_r$, where $\mathfrak{h}$ is one of the semisimple subalgebras (18.9), namely the one corresponding to the choice $\tilde{\Phi}_s = \Phi_s \setminus \{\alpha^{(i)}\}$. Thus e.g. a maximal semisimple subalgebra of $\mathfrak{sl}(5)$ is $\mathfrak{sl}(3) \oplus \mathfrak{sl}(2)$, but this is contained in the maximal reductive subalgebra $\mathfrak{sl}(3) \oplus \mathfrak{sl}(2) \oplus \mathfrak{u}(1)$ of $\mathfrak{sl}(5)$.

18.4 *S-subalgebras*

To find also the maximal *S*-subalgebras, a different procedure must be followed. Let us consider an n-dimensional module V of $\mathfrak{g}$; the corresponding $\mathfrak{g}$-representation R is a homomorphism from $\mathfrak{g}$ to the general linear algebra $\mathfrak{gl}(V) \cong \mathfrak{gl}(n)$ (compare equation (5.3)). Using the fact that $\mathfrak{g}$ is simple, one can easily convince oneself that the image $R(\mathfrak{g})$ is already contained in $\mathfrak{sl}(n) \subset \mathfrak{gl}(n)$. Thus any n-dimensional representation of $\mathfrak{g}$ supplies us with an embedding of $\mathfrak{g}$ into $\mathfrak{sl}(n)$. As seen in section 17.2, in case the module is selfconjugate we can do even better: If the module is symplectic, we can embed it even in $\mathfrak{sp}(n)$, while if it is orthogonal, we can embed it into $\mathfrak{so}(n)$. These embeddings are special (unless they are a bijection), and up to very few exceptions they are maximal.

As an example, take the defining module of E_6; this is 27-dimensional and complex (compare e.g. table XII), and accordingly there is a maximal special embedding $E_6 \hookrightarrow A_{26} \cong \mathfrak{sl}(27)$. As an example for an orthogonal module, take the adjoint module of any simple $\mathfrak{g}$; this gives rise to the maximal special embedding

$$\mathfrak{g} \hookrightarrow \mathfrak{so}(\dim \mathfrak{g}). \quad (18.10)$$

The method also works for modules which are not irreducible, but then it does not yield maximal embeddings. For example, while the defining module $V = V_{\Lambda_{(1)}}$ of $\mathfrak{sl}(n) \cong A_{n-1}$, which is complex, corresponds to the trivial embedding $\mathfrak{sl}(n) \subseteq \mathfrak{sl}(n)$, the reducible module $V \oplus V^+ = V_{\Lambda_{(1)}} \oplus V_{\Lambda_{(n-1)}}$ is orthogonal, yielding the embedding $\mathfrak{sl}(n) \hookrightarrow \mathfrak{so}(2n)$. This embedding is however not maximal, but rather there is an intermediate maximal regular non-semisimple subalgebra

$$\mathfrak{sl}(n) \oplus \mathfrak{u}(1) \hookrightarrow \mathfrak{so}(2n). \tag{18.11}$$

All S-subalgebras obtained this way are simple. There are also a few non-simple S-subalgebras of the classical algebras; they can best be understood in terms of the realization of the relevant algebras by matrices. One finds the following non-simple embeddings (using the notation $\mathfrak{sl}(n)$ rather than A_{n-1}, etc., as is appropriate when dealing with the matrix realization):

$$\begin{aligned}
\mathfrak{sl}(m) \oplus \mathfrak{sl}(n) &\hookrightarrow \mathfrak{sl}(mn), \\
\mathfrak{so}(m) \oplus \mathfrak{so}(n) &\hookrightarrow \mathfrak{so}(mn), \\
\mathfrak{sp}(m) \oplus \mathfrak{sp}(n) &\hookrightarrow \mathfrak{so}(4mn), \\
\mathfrak{so}(m) \oplus \mathfrak{sp}(n) &\hookrightarrow \mathfrak{sp}(mn), \\
\mathfrak{so}(2m+1) \oplus \mathfrak{so}(2n+1) &\hookrightarrow \mathfrak{so}(2m+2n+2).
\end{aligned} \tag{18.12}$$

(The reader can work out the precise form of these embeddings in exercise 18.2.) The last type of embedding is a special case of the more general series

$$\mathfrak{so}(m) \oplus \mathfrak{so}(n) \hookrightarrow \mathfrak{so}(m+n), \tag{18.13}$$

which except for odd m and n describes a regular embedding.

The methods just described provide all maximal S-subalgebras of the classical Lie algebras. The maximal S-subalgebras of the exceptional simple Lie algebras require a case-by-case study. This yields the following result, where the labels in square brackets denote the Dynkin index of the embedding:

$\mathfrak{g}$	maximal S-subalgebras of $\mathfrak{g}$
E_6	$A_2^{[9]}$, $G_2^{[3]}$, $C_4^{[1]}$, $F_4^{[1]}$, $A_2^{[2]} \oplus G_2^{[1]}$
E_7	$A_1^{[231]}$, $A_1^{[399]}$, $A_2^{[21]}$, $A_1^{[15]} \oplus A_1^{[24]}$, $A_1^{[7]} \oplus G_2^{[2]}$, $A_1^{[3]} \oplus F_4^{[1]}$, $C_3^{[1]} \oplus G_2^{[1]}$
E_8	$A_1^{[520]}$, $A_1^{[760]}$, $A_1^{[1240]}$, $B_2^{[12]}$, $A_1^{[16]} \oplus A_2^{[6]}$, $F_4^{[1]} \oplus G_2^{[1]}$
F_4	$A_1^{[156]}$, $A_1^{[8]} \oplus G_2^{[1]}$
G_2	$A_1^{[28]}$

$$\tag{18.14}$$

18.5 Projection maps and defining representations

To learn more about embeddings of semisimple Lie algebras, we consider a Cartan subalgebra $\mathfrak{h}_\circ$ of a semisimple subalgebra $\mathfrak{h}$ of the finite-dimensional semisimple Lie algebra $\mathfrak{g}$. We know that $\mathfrak{h}_\circ$ consists entirely of semisimple elements of $\mathfrak{h}$. Now as we have learned in section 5.2, the decomposition $x = x_\mathrm{s} + x_\mathrm{n}$ of $x \in \mathfrak{g}$ into its semisimple and nilpotent parts induces the Jordan decomposition of $R(x)$ in *any* finite-dimensional representation R. In particular, if x_s is a semisimple element of $\mathfrak{g}$, then $R(x_\mathrm{s})$ is semisimple in any finite-dimensional representation of $\mathfrak{g}$. Now using the adjoint representation of $\mathfrak{g}$, the ambient Lie algebra $\mathfrak{g}$ can be viewed as a particular (reducible) $\mathfrak{h}$-module; it follows that the elements of $\mathfrak{h}_\circ$ are not only semisimple in $\mathfrak{h}$, but also in $\mathfrak{g}$. Thus $\mathfrak{h}_\circ$ is an abelian subalgebra of semisimple elements of $\mathfrak{g}$, and therefore contained in a maximal abelian subalgebra of semisimple elements, i.e. in a Cartan subalgebra $\mathfrak{g}_\circ$ of $\mathfrak{g}$:

$$\imath(\mathfrak{h}_\circ) \subseteq \mathfrak{g}_\circ. \tag{18.15}$$

If $\mathfrak{g}$ is simple, this can even be extended to the full triangular decompositions, i.e. we also have

$$\imath(\mathfrak{h}_\pm) \subseteq \mathfrak{g}_\pm. \tag{18.16}$$

To see this, we first note that the algebra $\mathfrak{h}_+$ that is spanned by all step operators for positive roots of $\mathfrak{h}$ is solvable, and is therefore contained in a Borel subalgebra $\mathfrak{b}$ of $\mathfrak{g}$. Now similarly as Cartan subalgebras, any two Borel subalgebras are conjugate under such inner automorphisms which leave the chosen Cartan subalgebra fixed. Hence we can use $\mathfrak{b}$ to define the distinction between positive and negative roots of $\mathfrak{g}$, i.e. take the triangular decomposition of $\mathfrak{g}$ in such a way that $\mathfrak{g}_\circ \oplus \mathfrak{g}_+ = \mathfrak{b}$. Then we have $\imath(\mathfrak{h}_+) \subseteq \mathfrak{g}_\circ \oplus \mathfrak{g}_+$, i.e. we can decompose any step operator y_+ corresponding to a positive root β of $\mathfrak{h}$ as $\imath(y_+) = x_+ + x_\circ$ in components $x_+ \in \mathfrak{g}_+$ and $x_\circ \in \mathfrak{g}_\circ$. Now since e_+ is a step operator of $\mathfrak{h}$, we must have $[y_\circ, y_+] = \beta(y_\circ) y_+$ for all $y_\circ$ in the Cartan subalgebra $\mathfrak{h}_\circ$ of $\mathfrak{h}$, and since the roots are non-zero functions, one can choose $y_\circ \in \mathfrak{h}_\circ$ such that $\alpha(y_\circ) \neq 0$. Comparison with the bracket $[\imath(y_\circ), x_+ + x_\circ]$ then shows that in fact $x_\circ = 0$, and hence $\imath(\mathfrak{h}_+) \subseteq \mathfrak{g}_+$. Having chosen $\mathfrak{g}_+$ in this manner, the reasoning for $\mathfrak{h}_-$ is somewhat more complicated, because now the triangular decomposition of $\mathfrak{g}$ is already fixed.

The injective map $\imath$ that embeds the Cartan subalgebra $\mathfrak{h}_\circ$ of $\mathfrak{h}$ into $\mathfrak{g}_\circ$ gives rise to a dual map $\imath^\star \colon \mathfrak{g}_\circ^\star \to \mathfrak{h}_\circ^\star$ of the weight spaces. The map $\imath^\star$ is surjective and a projection on the weight space of $\mathfrak{h}$. Thus the weights of $\mathfrak{h}$ can be regarded as projections of weights of $\mathfrak{g}$, i.e. there is a *projection matrix* P – a matrix with rank$\,\mathfrak{g}$ rows and rank$\,\mathfrak{h}$ columns – such that for any $\mathfrak{g}$-weight μ the associated $\mathfrak{h}$-weight is given by

$$\imath^\star(\mu) = P\,\mu. \tag{18.17}$$

In concrete applications, this projection between the weight spaces is often one of the most helpful tools. For instance, one can show that for simple

$\mathfrak{g}$ the embedding (18.16) of a step operator $E^{\alpha'}$ of $\mathfrak{h}$ is given by

$$\imath(E^{\alpha'}) = \sum_{\alpha \in \Gamma_{\alpha'}} c^{\alpha'}_{\alpha} E^{\alpha}, \qquad (18.18)$$

where $\Gamma_{\alpha'}$ is the subset

$$\Gamma_{\alpha'} := \{\alpha \in \Phi \,|\, P\alpha = \alpha'\} \qquad (18.19)$$

of the root system Φ of $\mathfrak{g}$, and where $c^{\alpha'}_{\alpha}$ are complex numbers which satisfy $c^{-\alpha'}_{-\alpha} = \bar{c}^{\alpha'}_{\alpha}$. The sets $\Gamma_{\alpha'}$ obey $\Gamma_{-\alpha'} = \{-\alpha \,|\, \alpha \in \Gamma_{\alpha'}\}$ as well as $\Gamma_{\alpha'} \cap \Gamma_{\beta'} = 0$ for all distinct roots α', β' of $\mathfrak{h}$.

For regular embeddings, the projection matrix has a particularly simple form. Namely, the Dynkin labels of the $\mathfrak{h}$-weights λ are just the Dynkin labels $\lambda^i = (\lambda, \alpha^{(i)\vee})$ with respect to those simple $\mathfrak{g}$-roots $\alpha^{(i)}$ which survive the inclusion (18.7) – i.e. precisely the corresponding Dynkin labels of the $\mathfrak{g}$-weights – together with the number $(\lambda, -\theta^\vee)$ which may be regarded as the Dynkin label with respect to $-\theta$. By expanding θ in the basis of simple $\mathfrak{g}$-roots, the latter can be expressed in terms of the Dynkin labels of the $\mathfrak{g}$-weights and the Coxeter labels a_i.

One can show that two embeddings $\imath_1, \imath_2 \colon \mathfrak{h} \hookrightarrow \mathfrak{g}$ are equivalent if and only if for any finite-dimensional (linear) representation R of $\mathfrak{g}$ the $\mathfrak{h}$-representations $R_1 = R \circ \imath_1$ and $R_2 = R \circ \imath_2$ are isomorphic. However, this characterization of an embedding by the collection of representations $R \circ \imath$ is to a large extent redundant. In fact, for all simple Lie algebras $\mathfrak{g}$ except D_r, E_7 and E_8, for equivalence of $\imath_1$ and $\imath_2$ it is already sufficient that

$$R_\circ \circ \imath_1 \cong R_\circ \circ \imath_2 \qquad (18.20)$$

for a single irreducible $\mathfrak{g}$-representation $R_\circ$, namely the lowest-dimensional (non-trivial) representation, i.e. the so-called defining representation (compare table XII in chapter 13). For the algebras D_r, E_7 and E_8 one needs instead two distinct irreducible representations, namely the vector and the spinor, i.e. the representations with highest weight $\Lambda_{(1)}$ and $\Lambda_{(r)}$, for D_r, those with highest weight $\Lambda_{(1)}$ (133-dimensional) and $\Lambda_{(6)}$ (56-dimensional) for E_7, and $\Lambda_{(1)}$ (248-dimensional) and $\Lambda_{(7)}$ (3 875-dimensional) for E_8.

In particular, for any embedding one can fix the projection matrix P by determining how the weights of these special representations are projected. Let us illustrate this by an embedding of $B_2 \cong \mathfrak{so}(5)$ into $A_5 \cong \mathfrak{sl}(6)$. There is an embedding for which the six weights of the defining module of A_5 are mapped to the weights of the defining module of B_2 plus once the zero weight. The corresponding projection matrix reads

$$P = \begin{pmatrix} 1 & 0 & 0 & 0 & 1 \\ 0 & 2 & 2 & 2 & 0 \end{pmatrix}. \qquad (18.21)$$

18.6 Branching rules

In the application to symmetry breaking, the classification of subalgebras tells us which symmetries can survive the breaking. However, typically one also wants to know how the states that in the original system form representation spaces of the larger symmetry get organized into modules of the surviving symmetry algebra. This information is given by the *branching rules*

$$V_\Lambda(\mathfrak{g}) \rightsquigarrow \bigoplus_j V_{\lambda_j}(\mathfrak{h}) \tag{18.22}$$

of the embedding $\mathfrak{h} \hookrightarrow \mathfrak{g}$, which express an irreducible highest weight module V_Λ of the ambient algebra $\mathfrak{g}$ as the direct sum of irreducible modules of the subalgebra $\mathfrak{h}$.

We have in fact already encountered specific examples of branching rules. Namely, the tensor product $R_V \otimes R_W$ of two $\mathfrak{g}$-representations is again a $\mathfrak{g}$-representation. On the other hand, it can also be regarded as a representation of the direct sum $\mathfrak{g} \oplus \mathfrak{g}$ of two copies of the ambient algebra $\mathfrak{g}$; the first copy of $\mathfrak{g}$ acts in the representation R_V on the module V while the second copy of $\mathfrak{g}$ acts trivially on this factor of the tensor product, and conversely for the second factor. As a consequence, the decomposition of the tensor product module $V \otimes W$ into irreducible modules can be understood as a branching rule for the diagonal embedding $\mathfrak{g} \hookrightarrow \mathfrak{g} \oplus \mathfrak{g}$.

For any branching rule, the characters obey the sum rule

$$\chi_\Lambda^{(\mathfrak{g})} = \sum_j \chi_{\lambda_j}^{(\mathfrak{h})}, \tag{18.23}$$

with the argument of the $\mathfrak{g}$-character restricted to the Cartan subalgebra of $\mathfrak{h}$. By evaluation on the zero weight, this yields the dimension sum rule

$$\sum_j \dim(V_{\lambda_j}(\mathfrak{h})) = \dim(V_\Lambda(\mathfrak{g})). \tag{18.24}$$

In addition, as a consequence of (18.6) there is a sum rule for the second order Dynkin index:

$$I_\Lambda(\mathfrak{g}) \cdot I_{\mathfrak{h} \subset \mathfrak{g}} = \sum_j I_{\lambda_j}(\mathfrak{h}). \tag{18.25}$$

To any orthogonal, symplectic or complex $\mathfrak{g}$-module V, respectively, there is associated a specific embedding of $\mathfrak{g}$ into $\tilde{\mathfrak{g}} = \mathfrak{so}(n)$, $\mathfrak{sp}(n)$ or $\mathfrak{sl}(n)$, respectively. For this embedding the branching rule of the defining module of $\tilde{\mathfrak{g}}$ contains precisely one term, namely the module V. Furthermore, the matrix which according to the considerations in section 5.6 is needed to characterize the classical matrix realization of $\tilde{\mathfrak{g}}$ is precisely the two-index invariant tensor which, as explained in section 17.2, characterizes V as orthogonal, symplectic or complex. More generally, whenever $\mathfrak{h}$ is an S-subalgebra of $\mathfrak{g} = A_r$, B_r or C_r, then it possesses an irreducible module of dimension r, $2r + 1$, and $2r$, respectively.

If rank $\mathfrak{h} =$ rank $\mathfrak{g}$, the projection of the weight space of $\mathfrak{g}$ to that of $\mathfrak{h}$ is in fact a bijection and we can identify both spaces. Now the step operators E^α of $\mathfrak{h}$ are characterized by the property that they are eigenvectors under the adjoint action of the Cartan subalgebra $\mathfrak{h}_\circ$ of $\mathfrak{h}$. Since $\mathfrak{h}_\circ$ coincides with the Cartan subalgebra $\mathfrak{g}_\circ$ of $\mathfrak{g}$, they are eigenvectors under the adjoint action of $\mathfrak{g}_\circ$ as well, and hence are also step operators of $\mathfrak{g}$. It follows that the roots (i.e. the weights of the adjoint module) of $\mathfrak{h}$ are among the roots of $\mathfrak{g}$, or in other words, that the embedding is a regular one. Now not any root of $\mathfrak{g}$ can also be a root of $\mathfrak{h}$, since otherwise the two algebras would coincide. Thus for a (proper) regular embedding the branching rule of the adjoint module reads

$$V_{\mathrm{ad}}(\mathfrak{g}) \;\rightsquigarrow\; V_{\mathrm{ad}}(\mathfrak{h}) \oplus \Big(\bigoplus_{\ell \in L} V_{\lambda_\ell} \Big), \qquad (18.26)$$

with a non-empty set L. More generally, for any regular embedding $\imath :$ $\mathfrak{h} \hookrightarrow \mathfrak{g}$ and any representation R of $\mathfrak{g}$, the representation $R \circ \imath$ of $\mathfrak{h}$ is reducible.

To compute branching rules, in addition to the weight system of the $\mathfrak{g}$-module one only has to know the projection matrix. Given the explicit form of the various weights of the reducible $\mathfrak{h}$-module, it is straightforward to group these weights into the weight systems of irreducible modules of $\mathfrak{h}$. Together with the highest $\mathfrak{g}$-weight, the projection matrix restricts in particular the possible conjugacy classes of these $\mathfrak{h}$-modules. (For many embeddings the conjugacy class with respect to $\mathfrak{h}$ is already uniquely fixed by the conjugacy class of the highest $\mathfrak{g}$-weight alone, so that all $\mathfrak{h}$-modules are in the same conjugacy class. As an example consider the diagonal subalgebra $\mathfrak{h} = \mathfrak{su}(2)$ of $\mathfrak{g} = \mathfrak{su}(2) \oplus \mathfrak{su}(2)$; in physicists' language, this corresponds to coupling two angular momenta to a total angular momentum. In this case, our statement reduces to the well-known fact that the tensor product of two integer spin or two half-integer spin modules can only contain integer spin modules of the diagonal subalgebra, while half-integer spin modules of $\mathfrak{h}$ are contained in tensor products of integer with half-integer spin modules.) However, there is no need to go into any details here, because it does not yield any further structural insight.

Several computer programs are available which allow for the computation of branching rules (compare the list in the Epilogue of the book). Also, various general formulæ for branching rules, in particular closed expressions for the case of classical Lie algebras of arbitrary rank, can be found in the literature (see e.g. [King 1975], [Yang and Wybourne 1986] and [Thoma and Sharp 1996]), and exhaustive tables of branching rules, obtained by computer, are available as well (e.g. the tables in [McKay and Patera 1981]).

According to the remarks after (18.20), one can specify any embedding of simple Lie algebras by the branching rules of the defining module, possibly (for D_r, E_7 and E_8) supplemented by the branching rule of a second low-dimensional module. For instance, there are two inequivalent embeddings of A_1 into A_2. They can be distinguished by the branching rule for the defining three-dimensional module of $A_2 \cong \mathfrak{sl}(3)$, which reads $(1,0) \leadsto 1 \oplus 0$ and $(1,0) \leadsto 2$, respectively. The corresponding branching rules for the adjoint module read $(1,1) \leadsto 2 \oplus 1 \oplus 1 \oplus 0$ and $(1,1) \leadsto 4 \oplus 2$, respectively; in terms of the weight diagrams, this corresponds to

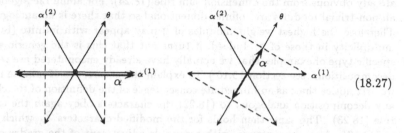

$$(18.27)$$

Thus in the first case, the embedded $\mathfrak{sl}(2)$ is the $\mathfrak{sl}(2)_\alpha$-subalgebra that corresponds to any of the $\mathfrak{sl}(3)$-roots α, while in the second case the embedding is a special one.

As another example, consider the embeddings of B_2 into A_5. One finds that there besides the embedding specified by (18.21) there is one other inequivalent embedding. The two embeddings are characterized by the branching of the 6-dimensional A_5-module with highest weight $(1,0,0,0,0)$ as follows:

$$(1,0,0,0,0) \;\leadsto\; (0,0) \oplus (1,0) \hat{=} 1 + 5\,,$$
$$(1,0,0,0,0) \;\leadsto\; 2\,(0,0) \oplus (0,1) \hat{=} 1 + 1 + 4\,. \tag{18.28}$$

Finally, as an embedding of a reductive subalgebra, consider the regular embedding

$$\mathfrak{sl}(3) \oplus \mathfrak{sl}(2) \oplus \mathfrak{u}(1) \;\hookrightarrow\; \mathfrak{sl}(5)\,. \tag{18.29}$$

Note that there is in fact a three-parameter family of such embeddings, because the direction of the $\mathfrak{u}(1)$-ideal is only restricted by the requirement that it must not lie in the Cartan subalgebra of $\mathfrak{sl}(2) \oplus \mathfrak{sl}(3)$. Fixing this freedom in such a way that the $\mathfrak{u}(1)$-generator is orthogonal (with respect to the Killing form of $\mathfrak{sl}(5)$) to $\mathfrak{sl}(2)$ and $\mathfrak{sl}(3)$, the branching rules (expressed in terms of dimensions of $\mathfrak{sl}(n)$-modules and charges of $\mathfrak{u}(1)$-modules) of the modules $V_{\Lambda_{(1)}}$ and $V_{\Lambda_{(2)}}$ of $\mathfrak{sl}(5)$ read

$$5 \;\leadsto\; (3;1;\tfrac{1}{3}) \oplus (1;2;-\tfrac{1}{2})\,, \quad 10 \;\leadsto\; (\bar{3};1;\tfrac{2}{3}) \oplus (3;2;-\tfrac{1}{6}) \oplus (1;1;-1)\,. \tag{18.30}$$

These branching rules are those relevant for the 'grand unification' in high energy physics, in which the Lie algebras $\mathfrak{su}(3)$ appearing in the gauge theory of strong interactions (QCD) and $\mathfrak{su}(2) \oplus \mathfrak{u}(1)$ of the electroweak interactions are embedded in an ambient $\mathfrak{su}(5)$ which is postulated to describe the symmetry of the theory at ultrahigh energies. (The additional gauge bosons that are present in the extended theory mediate e.g. proton decay). The two summands in the decomposition of the $\bar{5}$ of $\mathfrak{su}(5)$

Information

correspond to the anti-down-quark and to the electron and neutrino, respectively, while the three summands for the 10 of $\mathfrak{su}(5)$ correspond to the anti-up-quark, the up- and down-quarks, and the positron, respectively.

*18.7 Embeddings of affine Lie algebras

For embeddings $\mathfrak{h} \hookrightarrow \mathfrak{g}$ of finite-dimensional semisimple Lie algebras, the unitary highest weight modules of $\mathfrak{g}$ decompose into a finite number of unitary highest weight modules of $\mathfrak{h}$. One says that these embeddings are *finitely reducible*. This finite reducibility is already obvious from the dimension sum rule (18.24). For affine Lie algebras, however, all non-trivial modules are infinite-dimensional so that there is no analogue of (18.24). Therefore the highest weight modules of $\mathfrak{h}$ may appear with infinite (but countable) multiplicity in those of $\mathfrak{g}$. Indeed, it turns out that this is the generic situation. (A specific type of examples that we actually have already encountered are tensor product decompositions, see section 15.10.) To explain the occurrence of infinite multiplicities, we first notice that, as an immediate consequence of the definition of the characters, for any decomposition analogous to (18.22) the characters obey again the character sum rule (18.23). The same then holds for the modified characters $\tilde{\chi}_\Lambda$ which were defined in (14.50). As a consequence, with a suitable adjustment of the grades of the highest weight vectors we have

$$\tilde{\chi}_\Lambda^{(\mathfrak{g})} = \sum_i \tilde{\chi}_{\lambda_i}^{(\mathfrak{h})} \,, \tag{18.31}$$

where the summation generically contains an infinite number of terms.

So far our considerations were valid for any Lie subalgebra of an affine Lie algebra. In the rest of this section, we focus our attention on a particularly important class of subalgebras which can be described as follows. Any embedding $\bar{\mathfrak{h}} \hookrightarrow \bar{\mathfrak{g}}$ of finite-dimensional simple (and similarly for semisimple or reductive) Lie algebras determines in a natural and unique way an embedding $\mathfrak{h} = \bar{\mathfrak{h}}^{(1)} \hookrightarrow \mathfrak{g} = \bar{\mathfrak{g}}^{(1)}$ of untwisted affine algebras, namely by constructing $\mathfrak{h}$ and $\mathfrak{g}$ from the horizontal subalgebras $\bar{\mathfrak{h}}$ and $\bar{\mathfrak{g}}$, respectively, via the central extensions of their loop algebras. In fact, except for some very specific embeddings, such as subalgebras that arise as the fixed point algebras of certain automorphisms, the embeddings which have been analyzed in detail are all induced in this manner by an embedding of finite-dimensional Lie algebras.

The implications of the sum rule (18.31) can be investigated by studying the modified characters, specialized to $\tilde{\chi}_\Lambda((0, 0, \tau))$ as after equation (14.55), in the limit $\tau \to 0$. To evaluate this limit, one relates it, via the modular transformation $\tau \mapsto -1/\tau$, to the limit $\tau \to i\infty$. The result is

$$\lim_{\tau \to 0} \tilde{\chi}_\Lambda(\tau) = \tilde{D}_\Lambda \, e^{\pi i c / 12\tau}, \tag{18.32}$$

with

$$\tilde{D}_\Lambda = |L_w / L^\vee|^{-1/2} (k^\vee + g^\vee)^{-r/2} \prod_{\bar{\alpha} > 0} \left[2 \sin \left(\frac{2\pi}{(\bar{\theta}, \bar{\theta})} \frac{(\bar{\Lambda} + \bar{\rho}, \bar{\alpha})}{k^\vee + g^\vee} \right) \right] \tag{18.33}$$

and

$$c = \frac{k^\vee \dim \bar{\mathfrak{g}}}{k^\vee + g^\vee} \,. \tag{18.34}$$

Here $\bar{\mathfrak{g}}$ is the horizontal subalgebra of $\mathfrak{g}$, $g^\vee$ its dual Coxeter number and $\bar{\rho}$ its Weyl vector, and the product in (18.33) is over all positive $\bar{\mathfrak{g}}$-roots. The number c, called the conformal anomaly of $\mathfrak{g}$ at level $k^\vee$, is the eigenvalue of the central charge of the associated Virasoro algebra (compare section 12.12).

Implementing the asymptotic behavior (18.32) into the character sum rule, one learns that the summation over i in the decomposition (18.31) is finite if and only if the conformal anomalies of $\mathfrak{g}$ and $\mathfrak{h}$ are equal,

$$c^{(\mathfrak{g})} = c^{(\mathfrak{h})}. \tag{18.35}$$

(In particular, all modules of $\mathfrak{g}$ and $\mathfrak{h}$ have a fixed value of the level of $\mathfrak{g}$ and $\mathfrak{h}$, respectively.) An embedding fulfilling this condition is called a *conformal embedding*. Via the relation between embeddings of affine Lie algebras and those of their horizontal subalgebras, the classification of maximal subalgebras of simple Lie algebras also supplies a complete list of conformal embeddings. One finds e.g. that for a maximal conformal embedding the level of $\mathfrak{g}$ must be equal to one.

The construction of the affine embedding in terms of the embedding of the horizontal subalgebras also shows that the central extension K has to be the same for $\mathfrak{g}$ and $\mathfrak{h}$ or in other words that the eigenvalues k and h of the central generators of $\mathfrak{g}$ and $\mathfrak{h}$ must coincide. For the levels $h^\vee$ and $k^\vee$ this implies that their ratio is equal to the relative length squared of the highest roots of $\mathfrak{h}$ and $\mathfrak{g}$, i.e. to the Dynkin index (18.3) of the embedding $\mathfrak{h} \hookrightarrow \mathfrak{g}$,

$$h^\vee = I_{\mathfrak{h} \subset \mathfrak{g}} \, k^\vee. \tag{18.36}$$

In particular, for maximal conformal embeddings one has $k^\vee = 1$ and hence $h^\vee = I_{\mathfrak{h} \subset \mathfrak{g}}$.

In the case of conformal embeddings, the asymptotic form of the character sum rule (18.31) reduces to

$$\tilde{D}_\Lambda^{(\mathfrak{g})} = \sum_i \tilde{D}_{\Lambda_i}^{(\mathfrak{h})}, \tag{18.37}$$

where the sum is understood to count also multiplicities. This sum rule is quite similar to the dimension sum rule (18.24) for embeddings of semisimple Lie algebras. Accordingly, $\tilde{D}_\Lambda$ may be regarded as the relative 'size' of an affine module as compared to the module with highest weight $k^\vee \Lambda_{(0)}$. Indeed, in certain applications the quotient $\tilde{D}_\Lambda / \tilde{D}_{k^\vee \Lambda_{(0)}}$ plays the rôle of a 'quantum' version of the dimension, see section 22.13. For conjugate modules $\tilde{D}$ takes the same value: $\tilde{D}_{\Lambda^+} = \tilde{D}_\Lambda$. More generally, one has $\tilde{D}_{\omega(\Lambda)} = \tilde{D}_\Lambda$ for any automorphism ω that corresponds (compare section 11.5) to a symmetry of the Dynkin diagram of the affine Lie algebra $\mathfrak{g}$. For example, for $A_r^{(1)}$ at level one all integrable highest weights have the same value, $\tilde{D}_{\Lambda_{(i)}} = \tilde{D}_{\Lambda_{(0)}} = 1/\sqrt{r+1}$ for all $i = 1, 2, \ldots, r$.

For non-conformal embeddings the summation in (18.31) is infinite. However, it can be shown that

$$\tilde{\chi}_\Lambda^{(\mathfrak{g})}(\tau) = \sum_j b_{\Lambda;\Lambda_j}(\tau) \, \tilde{\chi}_{\Lambda_j}^{(\mathfrak{h})}(\tau), \tag{18.38}$$

where the sum is now over all integrable highest weights Λ_j of $\mathfrak{h}$ at a given level and hence is finite. In this formula the coefficients $b_{\Lambda;\Lambda_j}$ are no longer numbers, but are functions which depend non-trivially on the parameter τ; they are known as the *branching functions* of the embedding.

Unfortunately, branching functions are rather difficult to compute. Although for certain embeddings some more explicit formulæ are known, one can compute branching rules in full generality only by matching the states of a given $\mathfrak{g}$-module, grade by grade, by the states of the various possible $\mathfrak{h}$-modules, which in particular requires the complete knowledge of the null vector structure of the modules. But the numbers

a_Λ are typically irrational, so that the sum rule (18.37) fortunately implies that these numbers can already determine a branching rule to a large extent. In fact, at low levels of $\mathfrak{h}$ the knowledge of these numbers for all integrable highest weights Λ together with some information about the conjugacy classes of the relevant modules of the horizontal subalgebras is often sufficient to determine the branching rules of a conformal embedding completely. For more information on branching functions we refer to [Kac and Wakimoto 1988].

Summary:

The subalgebras of a simple or affine Lie algebra $\mathfrak{g}$ fall in two classes, regular and special subalgebras. For regular subalgebras, any step operator of the subalgebra is also a step operator of $\mathfrak{g}$. Subalgebras are characterized by the Dynkin index of the embedding. A convenient description is in terms of projection matrices.

For embeddings of affine Lie algebras, irreducible modules of $\mathfrak{g}$ are typically not finitely reducible as modules of the subalgebra. Only conformal embeddings are finitely reducible.

Keywords:

Subalgebra and ambient algebra, regular and special subalgebras, R- and S-subalgebras, Dynkin index of an embedding, extended Dynkin diagram, projection matrix, branching rule;
finite reducibility, conformal embedding, branching function.

Exercises:

Apply the prescription (18.7) for the construction of maximal regular semisimple subalgebras to all simple Lie algebras. In particular, deduce from the extended Dynkin diagrams that the following are maximal regular embeddings:

Exercise 18.1

$$A_2 \oplus A_2 \oplus A_2 \hookrightarrow E_6 \,, \quad A_7 \hookrightarrow E_7 \,, \quad D_8 \hookrightarrow E_8 \,. \qquad (18.39)$$

Derive the same result from the description of the root systems in an orthogonal basis (compare table VII).

Determine the precise form of the embeddings (18.12) by considering block-diagonal subgroups of transformations acting on tensor products or direct sums of suitable modules.

Exercise 18.2

Determine the projection matrix for the special embedding $A_3 \hookrightarrow B_3$ which has the branching rule $(1,0,0) \rightsquigarrow (0,1,0) \oplus (0,0,0)$.

Exercise 18.3

a) Check that the matrix P (18.21) is the projection matrix for the embedding of B_2 in A_5 that corresponds to the first of the branching rules (18.28).

b) Compute the projection matrix for the second of the branching rules (18.28).

c) In both cases, use the projection matrix to determine the branching rule of the adjoint module of A_5.

Exercise 18.4

What is the precise form of the $\mathfrak{u}(1)$ generator in the embedding (18.29) which results in the branching rules (18.30)?

Exercise 18.5

Find the branching rule of the 27-dimensional defining module of E_6 under the chain $A_2 \oplus A_1 \hookrightarrow A_4 \hookrightarrow D_5 \hookrightarrow E_6$ of (non-maximal) regular embeddings.

(In terms of Dynkin diagrams this chain of embeddings is visualized as follows:

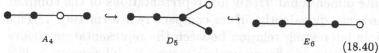

$$A_4 \qquad\qquad D_5 \qquad\qquad E_6 \qquad (18.40)$$

In high energy physics, $A_4 \cong \mathfrak{sl}(5)$ and $D_5 \cong \mathfrak{so}(10)$ have been investigated as candidates for the symmetries of grand unified theories. Because of $E_4 \equiv A_4$ and $E_5 \equiv D_5$, the chain (18.40) is in this context sometimes called 'E-series unification'.)

By referring to the matter content of the standard model (see e.g. chapter 22 of [Nachtmann 1990]), explain why the defining E_6-module is of special interest for particle physics.

Exercise 18.6

a) Check that the subalgebra $\hat{\mathfrak{g}}^{[N]}$ of a centrally extended loop algebra $\hat{\mathfrak{g}}$ that is spanned by $\{T^a_{Nm} \mid m \in \mathbb{Z}\} \cup \{K\}$ for some $N \in \mathbb{Z}_{>0}$ again satisfies the relations of a centrally extended loop algebra. ($\hat{\mathfrak{g}}^{[N]}$ is called a *winding* subalgebra of $\hat{\mathfrak{g}}$; why?)

b) Determine the canonical central element, the simple roots and the fundamental weights of $\hat{\mathfrak{g}}^{[N]}$. To which centrally extended loop algebra is $\hat{\mathfrak{g}}^{[N]}$ isomorphic?

c) Consider the analogous subalgebra of the Virasoro algebra. Find linear combinations of the generators L_{Nm} and C of the subalgebra which satisfy precisely the Virasoro bracket relations (12.58). Does the latter property extend to the semidirect sum of the Virasoro algebra and the untwisted affine Lie algebra that is associated to $\hat{\mathfrak{g}}^{[N]}$?

Exercise 18.7

Express the embedding matrix M, which describes the mapping $\imath(\mathfrak{h}_\circ)$ introduced in (18.15) as $\imath(h) = Mh$ for all $h \in \mathfrak{h}_\circ$, in terms of the projection matrix P and the quadratic form matrices of $\mathfrak{g}$ and $\mathfrak{h}$.

Exercise 18.8

19

Young tableaux and
the symmetric group

19.1 Young tableaux

In this chapter we introduce a pictorial way, so-called *Young tableaux*, to represent finite-dimensional irreducible representations of the complex Lie algebra $\mathfrak{sl}(n)$, or equivalently, of its compact real form $\mathfrak{su}(n)$. This will also reveal an interesting relation between the representation theory of the Lie algebra $\mathfrak{sl}(n)$ and the symmetric groups $\mathcal{S}_l$. The group $\mathcal{S}_l$ is a discrete group with $l!$ elements, which describes the permutations of a set of l objects.

To introduce Young tableaux, let us start in a somewhat more general setting and consider the l-fold tensor product $V^{\otimes l}$ of a unitary module V of some simple Lie algebra $\mathfrak{g}$. (We will not really use that $\mathfrak{g}$ is a simple Lie algebra; analogous considerations actually apply to linear representations of any other suitable algebraic structure.) In chapter 15 we have already seen that $V^{\otimes l}$ carries some unitary representation of $\mathfrak{g}$ (which typically is highly reducible). Now $V^{\otimes l}$ also carries a representation of $\mathcal{S}_l$. The action of this permutation group can be described as follows. Given a basis $\mathcal{B} = \{v_{(i)}\}$ of V, a basis of $V^{\otimes l}$ is provided by the n^l elements

$$v_{(i_1 i_2 \ldots i_l)} := v_{(i_1)} \otimes v_{(i_2)} \otimes \cdots \otimes v_{(i_l)}, \tag{19.1}$$

where each of the indices i_k runs independently from 1 to $n = \dim V$. An element σ of $\mathcal{S}_l$ then acts as follows: It maps the basis element (19.1) to the basis element $v_{(\sigma(i_1)\sigma(i_2)\ldots\sigma(i_l))}$. This action on the basis elements can be extended uniquely to a linear mapping of $V^{\otimes l}$.

Let us denote by $\tilde{W}_r$ the subspace of $V^{\otimes l}$ that consists of the direct sum of all modules of $\mathcal{S}_l$ that are isomorphic to a given irreducible $\mathcal{S}_l$-module W_r. This way we decompose $V^{\otimes l}$ as $V^{\otimes l} = \bigoplus_r \tilde{W}_r$ into modules of $\mathcal{S}_l$. As is not hard to see (see exercise 19.1), the action of $\mathfrak{g}$ and the action of $\mathcal{S}_l$ on $V^{\otimes l}$ commute. This implies that any $\tilde{W}_r$ in this decomposition is by itself already a module of $\mathfrak{g}$. (This generalizes the considerations in section 15.1, which in the present setting amount to using the action of $S_2 \equiv \mathbb{Z}_2$ to define symmetric and antisymmetric couplings in the tensor product of two identical modules.)

Now we specialize these general considerations to the case that $\mathfrak{g}$ is the Lie algebra $\mathfrak{sl}(n)$. We can then use two special properties of this algebra. First, any irreducible module of $\mathfrak{sl}(n)$ appears in some multiple tensor product of the defining n-dimensional module of $\mathfrak{sl}(n)$. Therefore we assume from now on that the module V of our interest is the defining module. The second special property of $\mathfrak{sl}(n)$ is that the modules $\tilde{W}_r$ as introduced above are in fact *irreducible* modules. We will see the reason for that in section 19.4. Conversely, one can also show that in the tensor product $V^{\otimes l}$ all irreducible representations of S_l occur (see exercise 19.2).

This means that any irreducible representation of $\mathfrak{sl}(n)$ can be characterized by two data: First, the power l of the tensor product $V^{\otimes l}$ in which the irreducible representation occurs; and second, its symmetry properties with respect to S_l. (However, this characterization is not quite unique. The totally antisymmetrized n-fold product of the defining module with itself is isomorphic to the trivial one-dimensional module; in terms of weight vectors, one has $v_{(i_1 i_2 \ldots i_n)} = \epsilon_{i_1 i_2 \ldots i_n} v$ with $\epsilon_{i_1 i_2 \ldots i_n}$ the Levi–Civita symbol.) Both data can be pictorially illustrated as follows. Each factor V of the tensor product is represented by a quadratic box $\square$. The representation of S_l can be specified by specifying which indices in the components with respect to the basis (19.1) are antisymmetrized and which are symmetrized.

For example, in agreement with the results of section 15.1, the tensor product of two defining modules of $\mathfrak{sl}(n)$ contains two irreducible submodules, the *symmetric tensor* module, with highest weight $2\Lambda_{(1)}$ and basis vectors $w_{\{ij\}} = w_{\{ji\}} \propto v_{(i)} \otimes v_{(j)} + v_{(j)} \otimes v_{(i)}$, and the *antisymmetric tensor*, with highest weight $\Lambda_{(2)}$ and basis vectors $w_{[ij]} = -w_{[ji]} \propto v_{(i)} \otimes v_{(j)} - v_{(j)} \otimes v_{(i)}$. These correspond to the two irreducible (one-dimensional) representations of $S_2 \equiv \mathbb{Z}_2$: The symmetric tensor to the trivial representation in which every permutation $\sigma \in S_l$ is represented by the number 1, and the antisymmetric tensor to the irreducible representation in which every permutation is represented by its sign.

In terms of the boxes $\square$, the symmetry properties with respect to pairs of indices are encoded by the following rule: Boxes representing indices that are to be symmetrized are glued together in horizontal rows, while boxes which stand for indices that are to be antisymmetrized are glued together in vertical columns. The figure obtained by this prescription is called the *Young tableau* or Young diagram of the $\mathfrak{sl}(n)$-, or of the S_l-module, respectively. (Note that the tensor $t_{i_1 i_2 \ldots i_l}$ that corresponds to some irreducible module in $V^{\otimes l}$ is symmetric in a pair of indices only if the corresponding boxes are in one and the same row of the tableau, and antisymmetric only if the boxes are in one and the same column. In all other pairs of indices, the tensor is neither symmetric nor antisymmetric.) Thus the Young tableau for the defining module is a single box, while the Young tableaux of the symmetric and antisymmetric tensor consist of two boxes, arranged in a single row and column, respectively, as displayed in figure (19.2).

$$\square \qquad\qquad \begin{array}{c}\square\\\square\end{array} \qquad\qquad \square\square \qquad\qquad (19.2)$$

$$\Lambda_{(1)} \qquad\qquad \Lambda_{(2)} \qquad\qquad 2\Lambda_{(1)}$$

The other fundamental modules, which have highest weights $\Lambda_{(j)}$, $j = 3, 4, \ldots, n-1$, can be obtained analogously: They are the j-fold antisymmetric tensor modules, and hence their Young tableaux consist of a single column with j boxes. Columns with n boxes describe the antisymmetrized n-fold product of V, which, as mentioned above, is isomorphic to the one-dimensional module. If such a column appears within a Young tableau with other columns, it can therefore be discarded. Columns with more than n boxes cannot occur, since antisymmetrizing more than n indices gives zero.

An arbitrary irreducible highest weight module of highest weight $\Lambda = \sum_{i=1}^{n-1} \Lambda^i \Lambda_{(i)}$ can be described in a way quite analogous to how we obtained the module $2\Lambda_{(1)}$ in the previous example, namely by taking, for each i, Λ^i copies of the i-fold antisymmetric tensor module and symmetrizing them. Hence the Young tableau has $\sum_i \Lambda^i$ columns, which we will always arrange such that the first Λ^{n-1} columns consist of $n-1$ boxes, the next Λ^{n-2} columns consisting of $n-2$ boxes, and so on. For example, the Young tableaux Y_θ of the adjoint module (with highest weight $\Lambda_{(1)} + \Lambda_{(n-1)}$) and Y_ρ of the irreducible module having the Weyl vector $\rho = \sum_{i=1}^{n-1} \Lambda_{(i)}$ as highest weight look as shown in figure (19.3).

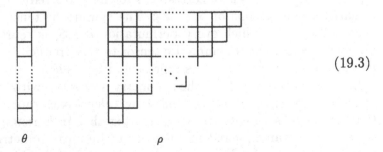

$$\theta \qquad\qquad\qquad \rho \qquad\qquad\qquad\qquad (19.3)$$

The total number of boxes of the Young tableau of $V_{\Lambda_{(j)}}$ is $\sum_{j=1}^{n-1} j\Lambda^j$. Taking this number modulo n, one obtains the conjugacy class of the module, as defined in (15.19).

19.2 Dimensions and quadratic Casimir eigenvalues

One might wonder whether other quantities that can be associated to a highest weight can also be easily expressed in terms of the Young tableaux. In this section we show that this is in fact possible for the dimension of an irreducible module and for its eigenvalue of the quadratic Casimir operator.

The dimension of an $\mathfrak{sl}(n)$-module is equal to the number of independent n-tuples $(i_1, i_2, ..., i_n)$ corresponding to the basis vectors $v_{(i_1 i_2 ... i_l)}$ with l indices $i_1, i_2, ..., i_l \in \{1, 2, ..., n\}$ that have the correct symmetry properties. A pair of indices i_p and i_q which is antisymmetrized can never have the same value of i_p and i_q; moreover, since interchanging two such indices just yields a minus sign, for counting purposes we can impose the restriction that $i_p < i_q$ for $1 \le p < q \le l$ on indices that are antisymmetrized. Indices that have to be symmetrized can take also identical values, and hence the prescription is that $i_p \le i_q$ for $1 \le p < q \le l$. As a consequence, the dimension of an $\mathfrak{sl}(n)$-module is equal to the number of distinct possibilities to fill the boxes of its Young tableau with integers $i_p \in \{1, 2, ..., n\}$, $p = 1, 2, ..., l$, obeying $i_p \le i_q$ when i_p appears to the left of i_q in the same row of the tableau, as well as $i_p < i_q$ when i_p appears above i_q in the same column of the tableau.

Let us illustrate this with a few simple examples. First consider the defining module whose Young tableau consists of a single box. The allowed fillings of the box are just

$$\boxed{1}, \boxed{2}, \boxed{3}, \cdots, \boxed{n} \tag{19.4}$$

One may regard the box with entry i as a pictorial realization of the basis vector $v_{(i)}$ of V; thus our counting merely reflects the fact that the defining module of $\mathfrak{sl}(n)$ is n-dimensional. Next take the symmetric and antisymmetric tensor modules. The allowed fillings of the boxes are then as given in figures (19.5) and (19.6), respectively:

$$\boxed{1\,1}, \boxed{1\,2}, \cdots, \boxed{1\,n}, \boxed{2\,2},$$
$$\boxed{2\,3}, \cdots, \boxed{2\,n}, \cdots, \boxed{3\,n}, \cdots, \boxed{n\,n} \tag{19.5}$$

and

$$\boxed{\begin{smallmatrix}1\\2\end{smallmatrix}}, \boxed{\begin{smallmatrix}1\\3\end{smallmatrix}}, \cdots, \boxed{\begin{smallmatrix}1\\n\end{smallmatrix}}, \boxed{\begin{smallmatrix}2\\3\end{smallmatrix}}, \cdots, \boxed{\begin{smallmatrix}2\\n\end{smallmatrix}}, \cdots, \boxed{\begin{smallmatrix}3\\n\end{smallmatrix}}, \cdots, \boxed{\begin{smallmatrix}n-1\\n\end{smallmatrix}} \tag{19.6}$$

One can think of these fillings to stand for the vectors

$$v_{(1)} \otimes v_{(1)}, \; v_{(1)} \otimes v_{(2)} + v_{(2)} \otimes v_{(1)}, \; \cdots, \; v_{(1)} \otimes v_{(n)} + v_{(n)} \otimes v_{(1)}, \; v_{(2)} \otimes v_{(2)},$$
$$v_{(2)} \otimes v_{(3)} + v_{(3)} \otimes v_{(2)}, \; \cdots, \; v_{(2)} \otimes v_{(n)} + v_{(n)} \otimes v_{(2)}, \; \cdots, \; v_{(n)} \otimes v_{(n)} \tag{19.7}$$

and

$$v_{(1)} \otimes v_{(2)} - v_{(2)} \otimes v_{(1)}, \ v_{(1)} \otimes v_{(3)} - v_{(3)} \otimes v_{(1)}, \ \ldots, \ v_{(1)} \otimes v_{(n)} - v_{(n)} \otimes v_{(1)},$$

$$v_{(2)} \otimes v_{(3)} - v_{(3)} \otimes v_{(2)}, \ \ldots, \ v_{(2)} \otimes v_{(n)} - v_{(n)} \otimes v_{(2)}, \ \ldots, \quad (19.8)$$

$$v_{(3)} \otimes v_{(n)} - v_{(n)} \otimes v_{(3)}, \ \ldots, \ v_{(n-1)} \otimes v_{(n)} - v_{(n)} \otimes v_{(n-1)},$$

respectively, of the tensor product $V \otimes V$. From the figures one reads off that the dimensionalities of these modules are $n(n+1)/2$ and $n(n-1)/2$, respectively.

Finally we display the possible fillings for the adjoint module $V_\theta = V_{\Lambda_{(1)} + \Lambda_{(2)}}$ of $\mathfrak{g} = \mathfrak{sl}(3)$, which show that this module is eight-dimensional (figure (19.9)).

$$\young(11,2)\,,\ \young(12,2)\,,\ \young(13,2)\,,\ \young(11,3)\,,\ \young(12,3)\,,\ \young(13,3)\,,\ \young(22,3)\,,\ \young(23,3) \qquad (19.9)$$

For more complicated Young tableaux the enumeration of all possible fillings becomes lengthy. However, it is possible to shortcut the calculation and express the dimensionality of an irreducible highest weight module V_Λ of $\mathfrak{sl}(n)$ rather directly in terms of the associated Young tableau Y_Λ, namely by the following so-called *hook formula*

$$d_\Lambda = \prod_{(i,j) \in Y_\Lambda} \frac{n - i + j}{h_{ij}}. \qquad (19.10)$$

Here the pairs (i,j) label the boxes of Y_Λ, with i numbering the rows from top to bottom, and j numbering the columns from left to right. The integer h_{ij}, the so-called hook length, is the number of boxes which belong to the hook that has (i,j) as upper left hand corner, i.e. of those boxes (i', j') for which $i' = i$, $j' \geq j$ or $i' \geq i$, $j' = j$.

In terms of the lengths a_j of the jth column and b_i of the ith row, the hook length is

$$h_{ij} = a_j + b_i - i - j + 1. \qquad (19.11)$$

The number of columns of length i of Y_Λ is just the Dynkin label Λ^i, so that the lengths of the rows are

$$b_i = \sum_{j=i}^{n-1} \Lambda^j, \qquad (19.12)$$

while the lengths of the columns may be expressed as $a_i = \max\{j \mid \sum_{k=j}^{n-1} \Lambda^k \geq i\}$.

When combining the latter formulæ, one observes that the numerator and denominator of the hook formula contain a huge number of common factors. Cancelling (most of) them between numerator and denominator, one obtains

$$d_\Lambda = \prod_{i=1}^{n-1}(\Lambda^i + 1) \cdot \prod_{i=1}^{n-2}\left(\tfrac{1}{2}(\Lambda^i + \Lambda^{i+1}) + 1\right)$$

$$\cdot \prod_{i=1}^{n-3}\left(\tfrac{1}{3}(\Lambda^i + \Lambda^{i+1} + \Lambda^{i+2}) + 1\right) \cdots \prod_{i=1}^{2}\left(\tfrac{1}{n-2}(\Lambda^i + \Lambda^{i+1} + \ldots \right. \tag{19.13}$$

$$\left. \ldots + \Lambda^{i+n-3}) + 1\right) \cdot \left(\tfrac{1}{n-1}(\Lambda^1 + \Lambda^2 + \ldots + \Lambda^{n-1}) + 1\right).$$

This result for the dimension of V_Λ is in turn nothing but the Weyl dimension formula (13.32) for the case of $\mathfrak{sl}(n)$ (see exercise 19.4).

Quadratic Casimir eigenvalues $C_\Lambda = (\Lambda, \Lambda + 2\rho)$ can be computed in a similar manner. Consider a Young tableau with n_{row} rows of length $b_1, b_2, \ldots, b_{n_{\mathrm{row}}}$ and n_{col} columns of length $a_1, a_2, \ldots, a_{n_{\mathrm{col}}}$, and denote the total number of boxes of the tableau by l. The quadratic Casimir eigenvalue of the $\mathfrak{sl}(n)$-module V_Λ described by this Young tableau is given by

$$C_\Lambda = \tfrac{(\theta,\theta)}{2}\left[l\,(n - l/n) + \sum_{i=1}^{n_{\mathrm{row}}}(b_i)^2 - \sum_{i=1}^{n_{\mathrm{col}}}(a_i)^2\right]. \tag{19.14}$$

This relation can e.g. be obtained by relating the integers a_i and b_j to the components of the highest weight Λ in an orthogonal basis of the weight space. (More details, and analogous formulæ for other Lie algebras, can be found in the appendix of [Pilch and Schellekens 1984].)

19.3 Tensor products

Another application of Young tableaux is that they give rise to a simple recipe for the decomposition of the tensor product of two irreducible modules into irreducible modules. It should be emphasized that this recipe is only convenient as long as one does calculations by hand which involve only low-dimensional modules. In general – in particular when implemented in a computer program – the Racah–Speiser algorithm that was described in section 15.7 is far superior to the prescription that we are presenting now.

Suppose that we want to determine into which irreducible modules the tensor product of two given irreducible modules V_{Λ_1} and V_{Λ_2} of $\mathfrak{sl}(n)$ decomposes. From the first section of this chapter we know that we can realize both irreducible modules as subspaces with definite behavior under permutations in two appropriately chosen tensor products of the defining module: $V_{\Lambda_i} \subset V^{\otimes l_i}$. Therefore the tensor product $V_{\Lambda_1} \otimes V_{\Lambda_2}$ is contained in the $l_1 + l_2$-fold tensor product of the defining module V. In other words, the Young tableaux corresponding to the irreducible modules in the tensor product are all composed out of $l_1 + l_2$ boxes.

The $l_1 + l_2$-fold tensor product of the defining module contains, however, far more irreducible modules than those which appear in the tensor product $V_{\Lambda_1} \otimes V_{\Lambda_2}$. To obtain only those which occur in this tensor product we have to keep the symmetry properties among the first l_1 and among the last l_2 indices in the same way as they behave for V_{Λ_1} and V_{Λ_2}. This results in the following prescription. The Young tableaux of the irreducible submodules in $V_{\Lambda_1} \otimes V_{\Lambda_2}$ are obtained by gluing the boxes of (say) Y_{Λ_1} to the diagram Y_{Λ_2} in all possible ways which are compatible with the required symmetry properties (and also such that the top left corner of Y_{Λ_2} remains the top left corner of the new tableau). Also, whenever a column of length n is produced in this procedure, then according to our general remarks it stands for the trivial module so that, if it occurs in some larger Young tableau, it can be discarded.

Let us consider as an example the tensor product of two adjoint modules of $\mathfrak{sl}(3)$. We mark the three boxes of the Young tableau for one of the modules by the letters a, b and c such that a and b correspond to symmetrized indices, and a and c to anti-symmetrized ones. These three boxes now have to be glued to the other Young tableau such that a is neither in the same column as b (because symmetrized indices must not be antisymmetrized) nor in the same row as c (because antisymmetrized indices must not be symmetrized). In addition, a must come 'before' b and c in the natural ordering of boxes, i.e. proceeding row after row from top to bottom, and from left to right within each row; in particular, if a and b are in the same row, then a must be to the left of b, and if a and c are in the same column, then a has to be above c. This recipe leaves six possibilities of gluing the boxes together, leading to the decomposition shown in the figure below. In that figure the filled dot in the last line stands for the trivial one-dimensional module (the singlet). Also, when proceeding to the second equality the labels a, b and c have been suppressed, and the irrelevant columns of length three have been discarded.

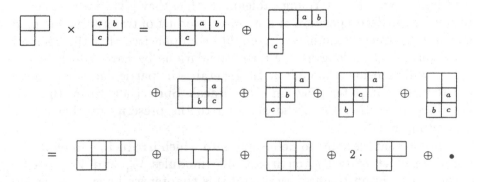

Already this simple example shows that in order to obtain the correct multiplicities one has to be rather careful in keeping track of the relevant symmetry properties. For higher-dimensional modules and higher-rank algebras, this makes the use of Young tableaux inconvenient.

*19.4 Young tableaux for other Lie algebras

One can use Young tableaux also to describe irreducible modules of other Lie algebras and Lie groups. Such a calculus has been worked out for all classical Lie algebras. However, in applications it quickly gets rather complicated, and it is also rather difficult to computerize. We therefore focus on the basic ideas. The reader who really wants to get familiar with the formalism can find the details in [Koike and Terada 1987] or chapter 10 of [Hamermesh 1962].

Let us first sketch what happens if one uses Young tableaux not to describe the modules of the Lie *algebra* $\mathfrak{sl}(n)$, or $\mathfrak{su}(n)$, respectively, but of some closely related Lie *groups*. Starting with the defining n-dimensional module V of the general linear group $\mathrm{GL}(n)$, the formalism sketched in the first section of this chapter carries through, and again Young tableaux are in one-to-one correspondence with irreducible highest weight modules. The crucial property that one has to verify to this end is that the subspaces of $V^{\otimes l}$ whose symmetry properties are described by some definite Young tableau are still irreducible as modules of $\mathrm{GL}(n)$. One can check that this is indeed true for the group $\mathrm{GL}(n)$, i.e. the group of all invertible complex $n \times n$-matrices, and also for its real counterpart $\mathrm{GL}(n, \mathbb{R})$, for which the matrix elements are restricted to be real. Moreover, it remains true for the special linear groups $\mathrm{SL}(n)$ and $\mathrm{SL}(n, \mathbb{R})$, for which in addition the determinant of the matrix is required to be one, and finally also for the unitary group $\mathrm{U}(n)$ and the special unitary group $\mathrm{SU}(n)$.

There is however one rule which has to be modified: In the Lie algebra case a column with n boxes could be removed from the Young tableau without affecting the result. In contrast, in the case of groups the presence of such columns is irrelevant only if the matrices of the group have determinant one. Therefore this rule must be dropped for all groups that are not unimodular, so that in these cases Young tableaux which differ by the presence of columns of length n describe different irreducible modules.

The extension of this calculus to the other classical groups $\mathrm{SO}(n)$ and $\mathrm{SP}(n)$ is based on the following observation. Via the defining n-dimensional module, these groups can be embedded into $\mathrm{SU}(n)$. Hence any module of $\mathrm{SU}(n)$, and therefore any Young tableau, provides a module of these groups. The question then is whether the irreducible module of $\mathrm{SU}(n)$ is still irreducible as a module of the subgroup.

To investigate this problem, let W be some irreducible module of $\mathrm{SU}(n)$ contained in $V^{\otimes l}$. This means in particular that each entry of a representation matrix $R_W(T^a)$ of W in any basis is a homogeneous polynomial $P(M) = P_{IJ}(M)$ of degree l in the entries $M \equiv M_{ij}$ of the representation matrices of the defining representation R_V. Now W is a fully reducible module of the classical subgroup G we are interested in. If it is not irreducible, we can find a basis in which the elements of the subgroup are all represented by block-diagonal matrices. In this basis those matrix elements $P_{IJ}(M)$ which do not belong to one of the blocks have to vanish when a is restricted to the subgroup. It follows that $P_{IJ}(M)$ is an invariant tensor for $V^{\otimes l}$; hence there must exist an invariant tensor for the l-fold tensor product of the defining module of the subgroup (compare the discussion in chapter 17 after equation (17.3)). Thus the fact that Young tableaux describe *irreducible* modules of $\mathrm{SU}(n)$ reflects the absence of non-trivial primitive invariant tensors for $\mathfrak{sl}(n)$. For the other classical algebras there do exist non-trivial invariant tensors, and this makes the use of Young tableaux considerably more involved.

For $\mathrm{SO}(n)$ the primitive invariant tensor of the defining representation is given by the trace. Therefore one must restrict oneself to the completely traceless tensors in W in order to arrive at irreducible modules. We do not want to present the resulting calculus here, but rather refer the interested reader to chapter 10.5, 10.6 of [Hamermesh 1962]. The representations of SO(n) are now described by Young tableaux with n rows, whereas the rank of these groups is considerably smaller, namely $n/2$ for n even and $(n-1)/2$ for n odd. In fact, the information contained in these diagrams is highly redundant and can be reduced such that one only has to deal with Young tableaux which have at most as many rows as the rank is. The price one then pays is that the tracelessness condition does not single out any more a definite irreducible module. Rather an irreducible SO(n)-module and its conjugate are then described by one and the same Young tableau (these two modules combine into a single irreducible module of the group O(n) which has two connected components, one of which is isomorphic to SO(n)).

For the symplectic group the situation is quite similar. This time the relevant contraction is not given by the trace, but rather one must use a two-index antisymmetric tensor. In all other respects the theory closely parallels that for the orthogonal groups; for details see chapter 10.8, 10.9 of [Hamermesh 1962].

*19.5 Brauer–Weyl theory

Implicit in the derivation of the Young tableau formalism is a basic algebraic fact which can be employed to develop similar techniques also in various other situations. Namely, two different structures act on the space $V^{\otimes l}$, and these actions commute with each other. In the discussion above, the two structures were the Lie algebra $\mathfrak{sl}(n)$ and the permutation group S_l. To treat these two structures in a more symmetric manner, we replace $\mathfrak{sl}(n)$ by its universal enveloping algebra $\mathsf{U}(\mathfrak{sl}(n))$ and the symmetric group by its group algebra $\mathbb{C}S_l$; we are then dealing with two associative algebras which act linearly on $V^{\otimes l}$.

The *group algebra* $F\mathrm{G}$ of some finite group G (over the complex numbers $F = \mathbb{C}$, or more generally over any field F) is defined as follows. The $|G|$-dimensional vector space $F^{|G|}$ over F, with $|G|$ the number of elements of G, has a standard basis $\mathcal{B}$ whose elements we label by the group elements, i.e. $\mathcal{B} = \{e_\gamma \,|\, \gamma \in G\}$. On the vector space $F^{|G|}$ one can define a product by setting

$$e_\gamma \circ e_{\gamma'} := e_{\gamma\gamma'} \,. \tag{19.15}$$

One can show that this turns $F^{|G|}$ into an associative algebra. This algebra is called the group algebra $F\mathrm{G}$ of G. It can also be shown that group algebras are semisimple.

Also, linear representations of G extend to linear representations of the algebra $F\mathrm{G}$. It follows that the $\mathfrak{sl}(n)$-module $V^{\otimes l}$ carries a linear representation of $\mathbb{C}S_l$, just like it carries a linear representation of $\mathsf{U}(\mathfrak{sl}(n))$. Now both $\mathsf{U}(\mathfrak{sl}(n))$ and $\mathbb{C}S_l$ are subalgebras of the ambient algebra $\mathfrak{gl}(V^{\otimes l})$. One can show that in fact

$$\mathsf{U}(\mathfrak{sl}(n)) = \{x \in \mathfrak{gl}(V^{\otimes l}) \,|\, [x, \mathbb{C}S_l] = 0\} \,,$$
$$\mathbb{C}S_l = \{x \in \mathfrak{gl}(V^{\otimes l}) \,|\, [x, \mathsf{U}(\mathfrak{sl}(n)) = 0]\} \,, \tag{19.16}$$

i.e. they are each others' centralizers in $\mathfrak{gl}(V^{\otimes l})$. Instead of centralizer one also uses the term *commutant* and calls each of the subalgebras the *commutator algebra* of the other. The identities (19.16) can actually be regarded as the origin of the intimate relationship between the representations of $\mathfrak{sl}(n)$ and those of S_l.

Analogous relations arise whenever identities like (19.16) hold for two subalgebras of an associative algebra. When the ambient algebra is again of the type $\mathfrak{gl}(V^{\otimes l})$ and one of the subalgebras is the enveloping algebra $U(\mathfrak{g})$ of the relevant finite-dimensional simple Lie algebra $\mathfrak{g}$, this observation amounts to another approach to the representation theory of $\mathfrak{g}$ which is known as *Brauer-Weyl theory* (or, in the case of $\mathfrak{sl}(n)$, as *Frobenius-Schur duality*). The centralizer of $U(\mathfrak{g})$ is then larger than the group algebra $\mathbb{C}S_l$; the additional generators correspond to the presence of additional invariant tensors for $\mathfrak{gl}(V^{\otimes l})$. This approach is, of course, especially powerful when the centralizer of $U(\mathfrak{g})$ has a nice representation theory, as is the case for $\mathfrak{g} = \mathfrak{sl}(n)$. Indeed, for any other choice of $\mathfrak{g}$ the representation theory of the centralizer is much more complicated; for details, see e.g. chapters IV, V, VI of [Weyl 1939].

Summary:

Young tableaux characterize in a pictoral way the finite-dimensional irreducible modules of $\mathfrak{sl}(n)$, respectively $\mathfrak{su}(n)$. Several characteristic quantities, like the dimension or the Dynkin index, can be computed in this formalism. There is also an algorithm to decompose tensor products of irreducible modules. Young tableaux can be generalized to other classical groups. The mathematical structure behind these algorithms is Brauer-Weyl-Frobenius-Schur duality.

Keywords:

Young tableau, hook formula, symmetrizing and antisymmetrizing, symmetric group;
group algebra, Brauer-Weyl theory, Frobenius-Schur duality.

Exercises:

Show that the action of $\mathfrak{g}$ and the action of S_l on $V^{\otimes l}$ commute.

Exercise 19.1

Show that in the tensor product $V^{\otimes l}$ all irreducible modules of S_l occur.
Hint: Start from some suitable tensor in $V^{\otimes l}$ and apply appropriate (anti-)symmetrizations.

Exercise 19.2

Write down explicitly the linear combinations of the basis vectors $v_{(i_1 i_2 i_3)}$ of $V \otimes V \otimes V$ that correspond to the fillings of figure (19.9).

Exercise 19.3

Compute the root system of the Lie algebra $\mathfrak{sl}(n)$. Use the result to check that the relation (19.13) reproduces the Weyl dimension formula (13.32).

Exercise 19.4

20

Spinors, Clifford algebras, and supersymmetry

20.1 Spinor representations

In this chapter we investigate in some more detail the representations of the orthogonal simple Lie algebras $\mathfrak{so}(n)$. We start by recalling a few basic facts about conjugacy classes of irreducible representations of these algebras, compare equation (15.18). Let us first assume that $n = 2r + 1$ is odd. Then there are two conjugacy classes, which for highest weight $\Lambda = \sum_{i=1}^r \Lambda^i \Lambda_{(i)}$ are characterized by $\Lambda^r \bmod 2$. If Λ^r is even, the representation is called a *tensor* representation, otherwise it is called a *spinor* representation. In the case of n even, $n = 2r$, the conjugacy classes are classified by an element of $\mathbb{Z}_2 \times \mathbb{Z}_2$ if r is even, while for odd r they are described by an element of the cyclic group $\mathbb{Z}_4$. However, for our present purposes we do not need the full information and only recall that in any case there is a $\mathbb{Z}_2$-valued conjugacy class $c_\Lambda = \Lambda^{r-1} - \Lambda^r \bmod 2$. Again it distinguishes between spinors (c_Λ odd) and tensors (c_Λ even).

Recall from chapter 19 that all finite-dimensional representations of the algebras A_n can be obtained by taking an appropriate tensor power of the defining $(n + 1)$-dimensional representation. The same cannot be true for $\mathfrak{so}(n)$: The defining n-dimensional representation has highest weight $\Lambda = \Lambda_{(1)}$ and is therefore a tensor. Because of the conjugacy class selection rule (15.17), any tensor power of the defining representation can only contain tensor representations; this is, of course, also the origin of the name 'tensor representation'.

In this chapter, we will provide an explicit construction for some spinor representations. Our main tool will be a finite-dimensional associative algebra, the Clifford algebra, which is also interesting in its own right. However, before we consider this algebra in more detail, we introduce a realization of $\mathfrak{so}(n)$ in terms of antisymmetric $n \times n$-matrices, which is different from the one described in section 5.6. As generators we choose

$$s_{ij} := \mathcal{E}_{i,j} - \mathcal{E}_{j,i} \tag{20.1}$$

with $i < j$, where the matrices $\mathcal{E}_{i,j}$ are the matrix units defined in equation (5.28), i.e. $(\mathcal{E}_{i,j})_{kl} = \delta_{ik}\delta_{jl}$. The matrices (20.1) span the Lie algebra $\mathfrak{so}(n)$, with commutation relations

$$[s_{ij}, s_{kl}] = \delta_{jk}s_{il} + \delta_{il}s_{jk} - \delta_{jl}s_{ik} - \delta_{ik}s_{jl} . \tag{20.2}$$

The generators of a Cartan subalgebra in this basis are $s_{12}, s_{34}, \dots, s_{p,p+1}$, where $p = n$ for n even and $p = n - 1$ for n odd. (This realization of $\mathfrak{so}(n)$ in terms of $n \times n$-matrices is different from the one given in equation (5.32). The present realization and the one in (5.32) are unitarily equivalent, compare exercise 20.1. For reasons that will become apparent soon, in this chapter we prefer the realization in terms of antisymmetric matrices; the realization in equation (5.32) displays however more clearly the triangular decomposition of $\mathfrak{so}(n)$.)

20.2 Clifford algebras

We now construct the Clifford algebras. The basic ingredient in this construction is a finite-dimensional vector space V which is endowed with a non-degenerate symmetric bilinear form g. At this point we have to be careful and decide whether we work over the field $F = \mathbb{R}$ of real numbers or over the complex numbers $\mathbb{C}$. For $F = \mathbb{C}$ one can choose an orthogonal basis of V, i.e. a basis e_i which obeys $g(e_i, e_j) = \delta_{ij}$, while for $F = \mathbb{R}$ one can only find a pseudo-orthogonal basis, i.e. $g(e_i, e_j) = \eta_i \delta_{ij}$, where the η_i are signs, $\eta_i = \pm 1$. In other words, for a fixed dimension all non-degenerate bilinear symmetric forms are equivalent in the complex case while for the real case they are classified by their signature: One says that a non-degenerate symmetric bilinear form g has signature p, q if there are p positive signs and q negative signs. Correspondingly, given the dimension n of V, we will find just a single complex Clifford algebra $\mathcal{C}(\mathbb{C}^n)$, while for $F = \mathbb{R}$ one finds different Clifford algebras $\mathcal{C}_{p,q}(\mathbb{R}^n)$ for different signatures of g. (This is analogous to the classification of real forms of a complex simple Lie algebra in terms of the signature of the Killing form, compare chapter 8.) The *Clifford algebra* is by definition the associative algebra that is freely generated by a unit element $\mathbf{1}$ and the basis elements e_i, $i = 1, 2, \dots, n$, modulo the anti-commutation relations

$$e_i e_j + e_j e_i = 2 g(e_i, e_j) \mathbf{1} . \tag{20.3}$$

(We do not introduce a special symbol for the associative product in the Clifford algebra, but rather write it by juxtaposition of the symbols.)

To examine the structure of the Clifford algebra defined by (20.3), let us as a first step count the number of its elements. The algebra is definitely generated by all finite products $e_{i_1} e_{i_2} \cdots e_{i_\ell}$ for all $\ell \in \mathbb{Z}_{\geq 0}$. Using the anti-commutation relations (20.3), we can furthermore assume that $i_1 \leq i_2 \leq \ldots \leq i_\ell$. Moreover, using (20.3) for $i = j$, we see that any e_i can appear in the product at most once. Hence the Clifford algebra is generated by all elements of the form $e_{i_1}^{m_1} e_{i_2}^{m_2} \cdots e_{i_n}^{m_n}$, where all exponents m_q are either 0 or 1. One can check that these elements are in fact linearly independent, which implies that the dimension of any Clifford algebra of an n-dimensional space is 2^n.

This construction of the Clifford algebra can be rephrased in a slightly more fancy language: Starting from a vector space V with a non-degenerate bilinear symmetric form g, we consider the *tensor algebra* of V, i.e. the direct sum of all tensor powers $V^{\otimes m}$ ($m \in \mathbb{Z}_{\geq 0}$) of V endowed with the tensor multiplication as a product. (Compare the analogous construction (14.1) for arbitrary vector spaces.) By counting the tensor power, this infinite-dimensional algebra comes with a $\mathbb{Z}_{\geq 0}$-gradation. According to the quotienting procedure outlined in section 14.2, in this algebra we can implement the relations of the Clifford algebra by dividing out the two-sided ideal $\mathcal{I}$ that is spanned by all elements of the form $x \otimes y - 2g(x,y)\mathbf{1}$. Clearly, this ideal $\mathcal{I}$ violates the $\mathbb{Z}_{\geq 0}$-gradation of the tensor algebra. However, the $\mathbb{Z}_{\geq 0}$-gradation also induces a coarser gradation, namely the $\mathbb{Z}_2$-gradation that is obtained by taking the value of the $\mathbb{Z}_{\geq 0}$-gradation modulo two. The ideal $\mathcal{I}$ respects this gradation, and hence the induced $\mathbb{Z}_2$-gradation still makes sense in the Clifford algebra. Correspondingly we can also introduce the *even subalgebras* $\mathcal{C}^+(\mathbb{C}^n)$ and $\mathcal{C}_{p,q}^+(\mathbb{R}^n)$ of $\mathcal{C}(\mathbb{C}^n)$ and $\mathcal{C}_{p,q}(\mathbb{R}^n)$, respectively.

Let us now make contact to the Lie algebra $\mathfrak{so}(n+1)$. We consider the elements

$$ \mathcal{S}_{0,i} = -\mathcal{S}_{i,0} := e_i \qquad \text{for} \quad i = 1, 2, \ldots, n \qquad (20.4) $$

and

$$ \mathcal{S}_{i,j} = -\mathcal{S}_{j,i} := [e_i, e_j] \qquad \text{for} \quad i, j = 1, 2, \ldots, n, \; i \neq j \qquad (20.5) $$

of the Clifford algebra. They are linearly independent and span an $n(n+1)/2$-dimensional subspace. By direct computation one checks (see exercise 20.2) that these elements span in fact the Lie algebra $\mathfrak{so}(n+1)$, with the Lie bracket (17.18) given by the commutator with respect to the associative product of the Clifford algebra (compare also equation (17.18)). Thus we can identify $\mathfrak{so}(n+1)$ as a subspace of the Clifford algebra.

20.3 Representation theory of Clifford algebras

Let us now turn to the representation theory of complex Clifford algebras $\mathcal{C}(\mathbb{C}^n)$. We start with $n = 1$. The algebra $\mathcal{C}(\mathbb{C})$ is two-dimensional and is generated by the unit element $\mathbf{1}$ and the element e_1 which obeys $(e_1)^2 = 1$, hence it is isomorphic to the group algebra of the discrete group $\mathbb{Z}_2$. Using the elements $(\mathbf{1} \pm e_1)/2$ as a basis of this algebra, it follows that it is isomorphic to the direct sum of two copies of $\mathbb{C}$:

$$\mathcal{C}(\mathbb{C}^1) \cong \mathbb{C} \oplus \mathbb{C}. \tag{20.6}$$

The discrete group $\mathbb{Z}_2$ (and hence also its group algebra) has two irreducible representations: The trivial one representing both elements by $1 \in \mathbb{C}$, and the faithful representation in which the two elements are represented by $+1$ and -1, respectively. These two representations are in one-to-one correspondence with the two summands in (20.6).

The Clifford algebra in two dimensions is four-dimensional. It can be realized in terms of Pauli matrices: We set $e_1 := \sigma_1$ and $e_2 := \sigma_2$. Apart from the unit matrix, the algebra then also contains $e_1 e_2 = i\sigma_3$. We therefore see that $\mathcal{C}(\mathbb{C}^2)$ is isomorphic to the algebra $\mathcal{M}_2(\mathbb{C})$ of all complex 2×2-matrices:

$$\mathcal{C}(\mathbb{C}^2) \cong \mathcal{M}_2(\mathbb{C}). \tag{20.7}$$

Now any 'full matrix algebra' $\mathcal{M}_m(\mathbb{C})$ is a simple associative algebra. Thus $\mathcal{C}(\mathbb{C}^2)$ is a simple associative algebra, and as a consequence the two-dimensional representation is the only irreducible representation.

Before describing the general Clifford algebra, let us look at the case of $\mathcal{C}(\mathbb{C}^3)$ which is $2^3 = 8$-dimensional. It is easy to check that the Pauli matrices form an irreducible representation of $\mathcal{C}(\mathbb{C}^3)$ by $R_1(e_i) = \sigma_i$ which, however, is not faithful (see exercise 20.3). Another representation of $\mathcal{C}(\mathbb{C}^3)$ is given by $R_1(e_i) = -\sigma_i$; the reader can show in exercise 20.3 that these two representations are inequivalent. In fact, $\mathcal{C}(\mathbb{C}^3)$ is isomorphic to the direct sum of two matrix algebras of 2×2-matrices:

$$\mathcal{C}(\mathbb{C}^3) \cong \mathcal{M}_2(\mathbb{C}) \oplus \mathcal{M}_2(\mathbb{C}). \tag{20.8}$$

The two representations we described exhaust all inequivalent irreducible representations.

These results generalize to arbitrary dimensions n as follows. One can show that the recurrence relation

$$\mathcal{C}(\mathbb{C}^{n+2}) \cong \mathcal{C}(\mathbb{C}^n) \otimes \mathcal{C}(\mathbb{C}^2) \tag{20.9}$$

holds. When combined with our previous results, this relation implies that the Clifford algebra is isomorphic to a matrix algebra when n is even, while for n odd it is isomorphic to the direct sum of two matrix algebras:

$$\mathcal{C}(\mathbb{C}^n) \cong \begin{cases} \mathcal{M}_{2^{n/2}}(\mathbb{C}) & \text{for } n \in 2\mathbb{Z}, \\ \mathcal{M}_{2^{(n-1)/2}}(\mathbb{C}) \oplus \mathcal{M}_{2^{(n-1)/2}}(\mathbb{C}) & \text{for } n \in 2\mathbb{Z}+1. \end{cases} \tag{20.10}$$

Clifford algebras are therefore semisimple associative algebras, and hence their representations are fully reducible. (The notion of semisimple associative algebras must not be confused with that of semisimple *Lie* algebras.) For even n there is, up to isomorphism, just a single irreducible representation which is $2^{n/2}$-dimensional, and any representation of $\mathcal{C}(\mathbb{C}^n)$ is faithful. For odd n there are two inequivalent irreducible representations of dimension $2^{(n-1)/2}$; in this case a representation of $\mathcal{C}(\mathbb{C}^n)$ is faithful if and only if it contains at least one copy of both irreducible representations. Finally we mention the following isomorphism for the even subalgebra:

$$\mathcal{C}^+(\mathbb{C}^n) \cong \mathcal{C}(\mathbb{C}^{n-1}). \tag{20.11}$$

20.4 Representations of orthogonal algebras and groups

We are now in a position to apply our results for Clifford algebras to construct representations of the orthogonal Lie algebras. Since the Clifford algebra $\mathcal{C}(\mathbb{C}^{n-1})$ contains the Lie algebra $\mathfrak{so}(n)$ as a subalgebra, any irreducible representation of $\mathcal{C}(\mathbb{C}^{n-1})$ gives rise to a representation of $\mathfrak{so}(n)$. Moreover, since $\mathfrak{so}(n)$ contains all the elements e_i whose products span $\mathcal{C}(\mathbb{C}^{n-1})$, any irreducible representation of $\mathcal{C}(\mathbb{C}^{n-1})$ will also be irreducible as a representation of $\mathfrak{so}(n)$.

We consider first the case that $n-1$ is odd. Then $\mathcal{C}(\mathbb{C}^{n-1})$ has two inequivalent irreducible representations of dimension $2^{n/2-1}$, which give rise to two irreducible representations of $\mathfrak{so}(n)$. Already the dimension of these irreducible representations suggests that they are the two inequivalent spinor representations, with highest weights $\Lambda_{(n/2-1)}$ and $\Lambda_{(n/2)}$; that this is indeed the case can be checked explicitly by computing the weights. When $n-1$ is even, there is a single irreducible representation of dimension $2^{(n-1)/2}$. Similar calculations as in the previous case show that this gives rise to the spinor representation of $\mathfrak{so}(n)$ with highest weight $\Lambda_{((n-1)/2)}$.

These remarks concern the construction of spinor representations of Lie *algebras*. Let us also comment on the construction of spinor representations of orthogonal Lie *groups*. The Lie groups $SO(n)$ are not simply connected (compare section 9.7); we will not obtain representations of $SO(n)$, but rather of its universal covering group. In fact, as a by-product we will construct that group.

We concentrate on the case that $n = 2r + 1$ is odd. We start by choosing a 2^r-dimensional irreducible representation R_s of the Clifford algebra $C(\mathbb{C}^{n-1})$, and denote the representation matrices of the generators e_i by ρ_i: $\rho_i := R_s(e_i)$. Next we choose some element γ in the Lie group $SO(n)$. We write $T \equiv R_{\Lambda_{(1)}}(\gamma)$ for the representation matrix of this element in the defining n-dimensional representation of $SO(n)$ and denote the entries of this matrix by T_j^i. The matrix T is orthogonal, $T^t T = \mathbb{1}$, and it has determinant one. The first property implies that along with the matrices ρ_i also the matrices

$$\rho_i' := \sum_{j=1}^n T_i^j \rho_j \qquad (20.12)$$

form a representation of the Clifford algebra (see exercise 20.5). But for n odd there is up to isomorphism only a single irreducible representation of the Clifford algebra $C(\mathbb{C}^{n-1})$, and hence these two representations must be equivalent. In other words, for any $\gamma \in SO(n)$ there is an invertible $2^r \times 2^r$-matrix $M(\gamma)$ which intertwines the two representations, i.e.

$$\rho_i' = M(\gamma)\, \rho_i\, M(\gamma)^{-1}, \qquad (20.13)$$

where matrix multiplication is understood. Now we remember that, since the representation R_s is irreducible, Schur's Lemma implies that the inter-twiner $M(\gamma)$ is uniquely determined up to a scalar multiple. Furthermore, the matrix product $M(\gamma_1)M(\gamma_2)$ implements the action of the element $\gamma_1 \gamma_2 \in SO(n)$ on the matrices ρ_i. This implies that

$$M(\gamma_1 \gamma_2) = \omega(\gamma_1, \gamma_2)\, M(\gamma_1)\, M(\gamma_2) \qquad (20.14)$$

with ω a scalar depending on γ_1 and γ_2. As we have learned in section 12.2, the relation (20.14) is the defining relation for a *ray* or *projective* representation of $SO(n)$.

The multipliers $\omega(\gamma_1, \gamma_2)$ are rather inconvenient for practical purposes, and we would like to get rid of them. According to the discussion in section 12.2 they depend on the specific choice of the intertwiners $M(\gamma)$; we are free to multiply $M(\gamma)$ with a scalar, in a way compatible with the group product. One can show that in the present case this freedom allows to fix ω to either -1 or 1. It is *not* possible to get also rid of these signs, and hence the matrices M form a genuine ray representation of $SO(n)$. But as described in section 12.2 this ray representation can be lifted to an ordinary representation of an extended group $\widehat{G}$, which in the case of interest is the universal covering group $Spin(n)$ of $SO(n)$. That ω takes values in $\{\pm 1\}$, i.e. in a $\mathbb{Z}_2$-group, reflects the fact that $Spin(n)$ is a *two*fold covering of $SO(n)$ (compare also section 9.7).

When $n = 2r$ is even, the construction of the covering group is technically a bit more involved, owing to the fact that $C(\mathbb{C}^{n-1})$ has then two inequivalent irreducible representations. The construction yields the two inequivalent 2^{r-1}-dimensional spinor representations of the universal covering group Spin(n) of SO(n). The Spin(n)-representation by the matrices M is unitary; in fact, it is reducible and can be decomposed into the spinor and conjugate spinor representation. (However, as noted in section 19.4, this reducible representation is irreducible as a representation of the non-connected Lie group O(n).)

*20.5 SU(2) versus SO(3) revisited

It is instructive to sketch the relation between the construction of the universal covering of SO(n) in the previous section and the results for SO(3) we mentioned in section 9.6. To any vector $\vec{x} = (x_i)$ of the three-dimensional *real* vector space carrying the defining representation of SO(3) we can associate the traceless hermitian matrix

$$\sum_{i=1}^{3} x_i \sigma^i = \begin{pmatrix} x_3 & x_1 - ix_2 \\ x_1 + ix_2 & -x_3 \end{pmatrix}, \tag{20.15}$$

where σ^i are the Pauli matrices. The determinant of this matrix is

$$\det\left(x_i \sigma^i\right) = -|\vec{x}|^2, \tag{20.16}$$

where $|\vec{x}|$ stands for the usual euclidean norm of $\vec{x} \in \mathbb{R}^3$ (here and in the rest of this section we use the Einstein summation convention).

Consider now the map

$$x_i \sigma^i \mapsto U x_i \sigma^i U^\dagger, \tag{20.17}$$

where U is a unitary 2×2-matrix. The matrix on the right hand side of (20.17) is again hermitian and traceless, and can therefore be expanded in terms of Pauli matrices with real coefficients x_i': $U x_i \sigma^i U^\dagger = x_i' \sigma^i$. Moreover, the relation between x and x' is linear and the matrix on the right hand side has the same determinant $|\vec{x}|^2$ as the original one. As a consequence there is a matrix $T \in$ O(3) such that

$$U x_i \sigma^i U^\dagger = T(U)_i^j x_j \sigma^i. \tag{20.18}$$

This way we have realized the relation (20.12) with complex unitary matrices. Since we can multiply the matrix U with an arbitrary complex phase without changing this relation, we can in fact assume that U has determinant 1, i.e. $U \in$ SU(2). The Lie group SU(2) is connected, so that U can be continuously connected to the identity element $\mathbf{1}_{2\times 2}$ of SU(2). For the latter, however, the associated matrix T is the identity as well, $T(\mathbf{1}_{2\times 2}) = \mathbf{1}_{3\times 3}$. Therefore also any other $T(U)$ appearing in (20.18) lies in the component of the identity element of O(3), i.e. we have $T \in$ SO(3).

This way we constructed a map which sends $U \in$ SU(2) to $T(U) \in$ SO(3); this map is in fact a group homomorphism. In the present formulation it is obvious that U and $-U$ are mapped to the same matrix T. Therefore the kernel of the group homomorphism is $\{\pm\mathbf{1}\}$, i.e. the center of SU(2). This completes the proof that SU(2) is isomorphic to the universal covering group of SO(3): SU(2) $\equiv$ Spin(3).

20.6 Zoology of spinors

In this section we comment on some further aspects of spinors which are particularly relevant for certain applications in physics. We will consider a *real* vector space V of real dimension n and a symmetric bilinear form η on it, which without loss of generality we take to be diagonal, and which has signature (p, q) (i.e. has p positive and q negative eigenvalues) with $p + q = n$. The case $n = 4$ with Lorentz signature $(3, 1)$ is of most direct physical interest. However, in various developments in modern physics other dimensions and signatures play an equally important rôle, and hence in this section we will allow for arbitrary values of p and q.

> **Information**
>
> There have been attempts, starting with the work of Kaluza and Klein in the 1920s, to employ additional dimensions of space-time in order to describe internal degrees of freedom of elementary particles. In the simplest scenario, this amounts to considering instead of space-time $\mathcal{M}$ the product manifold $\mathcal{M} \times K$. Usually K is assumed to be compact; then the extra dimensions are said to be *compactified*. Such compactifications play an important rôle in the construction of supergravity and of superstring models. However, what really matters for physics are the internal degrees of freedom. Correspondingly, in more advanced treatments one starts with an abstract conformal field theory construction. Later on one then arrives eventually at an interpretation in terms of compactified spaces, but there are also perfectly acceptable models which do not possess any geometrical interpretation in terms of a product $\mathcal{M} \times K$.

The algebra we want to realize is again a Clifford algebra, with generators e_i and relations $e_i e_j + e_j e_i = 2\eta^{ij}$. In physics, one usually works directly with a matrix realization $\gamma^i = \rho(e_i)$ of the generators, so that the relations read

$$\gamma^i \gamma^j + \gamma^j \gamma^i = 2\eta^{ij} \,. \tag{20.19}$$

The matrices γ^i are called *Dirac* or simply *gamma* matrices; they are complex $2^{[n/2]} \times 2^{[n/2]}$-matrices, where the square brackets denote the integral part of a rational number. (Gamma matrices also play the rôle of invariant tensors of the relevant orthogonal Lie algebra, compare section 17.5.)

To give one specific realization of gamma matrices, we consider the case that $n = p + 1$ is even and the signature is $(p, 1)$. In this case, the realization is unique up to equivalence; one can choose

$$\gamma^0 := i \begin{pmatrix} \mathbb{1} & 0 \\ 0 & -\mathbb{1} \end{pmatrix}, \qquad \gamma^i := \begin{pmatrix} 0 & \Sigma^i \\ \Sigma^i & 0 \end{pmatrix} \text{ for } i = 1, 2, \dots, p \,. \tag{20.20}$$

Here Σ^i are hermitian $2^{(p-1)/2} \times 2^{(p-1)/2}$-matrices that obey appropriate anti-commutation relations. In $n = 4$ dimensions they can be chosen to be the Pauli matrices: $\Sigma^i = \sigma^i$ for $i = 1, 2, 3$. The matrix γ^0 is anti-hermitian while all other γ^i are hermitian. This is closely related to the fact that in quantum theory one employs hermitian conjugation to implement the operation of time reversal (compare section 1.6). In physics hermiticity properties therefore play an important rôle. However, complex conjugation is not a $\mathbb{C}$-linear operation, so that for applications in physics we cannot simply consider *complex* Clifford algebras and spinors. Another way to arrive at this conclusion is to realize that complex Clifford algebras are insensitive to the signature of the quadratic form η, which typically plays the rôle of the signature of space-time and hence is of considerable physical importance.

Nonetheless, we can use the results of the complex case as a starting point, similarly as one can approach the analysis of real Lie algebras by first studying complex Lie algebras and then examining their real forms. We start by associating to the real vector space V its complexification $V_{\mathbb{C}} \cong V \oplus iV$ (compare section 4.3). This is a complex vector space, which is endowed with an additional structure that an ordinary complex vector space does not have, namely a conjugation which is a real automorphism of order two interchanging $x + iy$ and $x - iy$ for all $x, y \in V$.

From the complexification $V_{\mathbb{C}}$ of V and the complex-bilinear extension $g := \eta_{\mathbb{C}}$ of the bilinear form η we can construct a complex Clifford algebra $\mathcal{C}(V_{\mathbb{C}}, g)$. When considered as *real* associative algebras, the algebra $\mathcal{C}(V_{\mathbb{C}}, g)$ and the complexification $\mathcal{C}(V, \eta)_{\mathbb{C}}$ of the algebra $\mathcal{C}(V, \eta)$ are isomorphic:

$$\mathcal{C}(V_{\mathbb{C}}, g) \; \cong \; \mathcal{C}(V, \eta)_{\mathbb{C}} \; \equiv \; \mathcal{C}(V, \eta) \otimes \mathbb{C}. \qquad (20.21)$$

The conjugation on $V_{\mathbb{C}}$ induces a conjugation on $\mathcal{C}(V, \eta)_{\mathbb{C}}$, i.e. an automorphism of order two of $V_{\mathbb{C}}$ as a real algebra; the subspace of $\mathcal{C}(V, \eta)_{\mathbb{C}}$ that is invariant under this automorphism is a subalgebra, called the *real subalgebra*. Thus just as in the case of real forms of complex Lie algebras, the structure of real Clifford algebras can be described in terms of their complexification together with a suitable conjugation on this complexification. However, we warn the reader that when one represents $V^{\mathbb{C}}$ as a simple matrix algebra (or a direct sum of simple matrix algebras), this conjugation does not in general correspond to taking the complex conjugate of all matrix elements.

We are now in a position to describe the various possible types of spinors. First, a *Dirac spinor*, commonly denoted by ψ or ψ_{D}, is by definition an element of a complex vector space that carries an irreducible representation of the complexified Clifford algebra; concretely, a Dirac spinor can be described as a vector in $\mathbb{C}^{2^{[n/2]}}$. However, we are primarily interested in representations of the real subalgebra, so we need to know more about the structure of that algebra.

As already announced, we consider a bilinear form η with signature

(p, q) on a real vector space V of dimension $n = p + q$. We will assume that $p - q \neq 3$ or 7 mod 8; it can then be shown that there is an invertible matrix m such that for any element a of $\mathcal{C}(V, \eta)_{\mathbb{C}}$ the conjugate element a^* can be written as $a^* = ma^\#m^{-1}$, where $a^\#$ is obtained from a by complex conjugation of all elements in a matrix realization (the explicit form of m of course depends on the specific realization one chooses). The matrix m can be used to define a conjugation also on the space of spinors, by

$$\psi^c := \psi^* m, \tag{20.22}$$

where the spinor ψ^* is obtained from the spinor ψ by complex conjugation of all its components. (In contrast to the rest of the book, we do not use an overbar to indicate complex conjugation of spinors, because in the physics literature the notation $\overline{\psi}$ is usually reserved for what we call ψ^c.) In the remaining cases, i.e. $p - q = 3$ or 7 mod 8, n is odd and $\mathcal{C}(\mathbb{C}^n)$ consists of two simple ideals. In these cases the real subalgebra is isomorphic to the matrix algebra of *complex* matrices, and the conjugation which determines the real subalgebra is the permutation of the two ideals.

The complexification $\mathcal{C}(V, \eta)_{\mathbb{C}}$ certainly contains the real Clifford algebra $\mathcal{C}(V, \eta)$. Therefore each irreducible representation of the complex algebra branches into irreducible representations of the real subalgebra. One finds that Dirac spinors are reducible precisely in those cases where the real subalgebra is a simple real matrix algebra or a direct sum of simple real matrix algebras, which is true for $p - q \bmod 8 \in \{0, 1, 2\}$. In these cases one obtains an irreducible representation of the real Clifford algebra by imposing an additional condition on the Dirac spinors, the *Majorana condition*

$$\psi \overset{!}{=} \psi^c, \tag{20.23}$$

which reduces the number of degrees of freedom of a spinor by a factor of two. (Instead of (20.23), one can equivalently also require that ψ^c equals $-\psi$; this minus sign will then appear also in (20.35) below.) Spinors $\psi \equiv \psi_M$ which satisfy (20.23) are called *Majorana spinors*.

So far we have been concerned with the question of whether an irreducible representation of the complex Clifford algebra is still an irreducible representation of an embedded real Clifford algebra. Another interesting question for applications in physics is whether an irreducible representation of the complex Clifford algebra $\mathcal{C}(\mathbb{C}^n)$ is still irreducible under the even subalgebra $\mathcal{C}^+(\mathbb{C}^n)$. Let us investigate this question first in the case of even n. We introduce the two projectors

$$P_\pm := \tfrac{1}{2}\left(1 \pm \lambda\, e_1 e_2 \cdots e_n\right), \tag{20.24}$$

where the phase factor λ is chosen such that $(\lambda e_1 e_2 \cdots e_n)^2 = 1$, namely such that $\lambda^2 = (-1)^{n(n-1)/2+q}$. With their help a Dirac spinor ψ_D can be decomposed into two components $\psi_\pm := P_\pm \psi_D$ which both transform irreducibly under the action of the even subalgebra. Such a spinor $\psi \equiv \psi_W$ which transforms in an irreducible representation of the even subalgebra is usually termed a *Weyl spinor*.

Here we have been dealing with the *complex* even subalgebra. It can still happen that a Weyl spinor carries a representation which is reducible under the *real* even subalgebra, so that the Majorana and the Weyl conditions can be imposed simultaneously, leading to a so-called *Majorana–Weyl spinor*. Indeed, this is the case if and only if λ is real, which is true for signatures with $p - q = 0$ mod 8.

If n is odd, any irreducible representation of the complex Clifford algebra is also irreducible under its even subalgebra $\mathcal{C}^+(\mathbb{C}^n)$. However, it can be reducible under its *real* even subalgebra, which happens in fact for $p - q = 1$ mod 8. For $p - q = 7$ mod 8 the situation is slightly more involved: In this case irreducible representations of the complex Clifford algebra are irreducible both under the real subalgebra and the even subalgebra, but are nevertheless reducible under the real even subalgebra. One then has

$$\mathcal{C}^+_{p,q}(\mathbb{R}^n) \cong \mathcal{M}_{2^{(n-1)/2}}(\mathbb{R})\,. \tag{20.25}$$

(Unlike in the cases of ψ_M and ψ_W, no particular name has been invented for the latter type of spinors.)

20.7 Spinors of the Lorentz group

The spinors for the *Lorentz group* SO(3,1) play a key rôle in relativistic quantum field theory. In this specific case an approach which is somewhat different from the one used in the previous section is convenient, the so-called *Infeld–van der Waerden* or *two-component spinor* formalism. The starting point is the observation that the universal covering Lie group of SO(3,1)$^+$, the component of the Lorentz group that is connected to the identity element, is isomorphic to the group SL(2,$\mathbb{C}$). (Here it is in order to repeat from chapter 9 a word of caution: Even though the definition of the group SL(2,$\mathbb{C}$) involves the complex numbers, here we regard SL(2,$\mathbb{C}$) as a *real* Lie group.) This isomorphism SO(3,1)$^+ \cong$ SL(2,$\mathbb{C}$) follows in a way analogous to how it was shown above that SU(2) is isomorphic to the universal covering group of SO(3). Namely, one uses the one-to-one correspondence (see exercise 20.6)

$$x = (x_m) \longleftrightarrow x_m \sigma^m \equiv x^0 \mathbb{1} + \sum_{i=1}^{3} x_i \sigma^i = \begin{pmatrix} x^0 + x^3 & x^1 + ix^2 \\ x^1 - ix^2 & x^0 - x^3 \end{pmatrix} \tag{20.26}$$

between Lorentz vectors $x \equiv (x_m)_{m=0,1,2,3}$ and hermitian matrices of rank two. Here we adopt the convention to use italic letters $m, n, \ldots$ (which take values in $\{0, 1, 2, 3\}$) for the indices of Lorentz vectors and for the labelling of the different 'sigma matrices' $\sigma^m \equiv \{\sigma^0 = \mathbb{1}, \sigma^i\}$ (σ^i are the Pauli matrices). Such indices are raised or lowered by the Lorentz metric $\eta^{mn} = (-+++) \equiv \text{diag}(-1, 1, 1, 1)$.

Two-component spinors are by definition elements of a two-dimensional irreducible module of SL(2,$\mathbb{C}$). Such irreducible modules exist because (see exercise 20.8) the complexification of the Lie algebra of SL(2,$\mathbb{C}$) is isomorphic to the direct sum $A_1 \oplus A_1$. More precisely, there are two inequivalent two-dimensional irreducible modules, both of which are self-conjugate, with highest weights $(1, 0)$ and $(0, 1)$ in the Dynkin basis (i.e. spins $(\frac{1}{2}, 0)$ and $(0, \frac{1}{2})$), respectively. A convenient manner to distinguish between these two spinor modules is to label elements of the irreducible module with highest weight $(1, 0)$ by subscripts which are ordinary greek letters (so-called *undotted indices*), while the subscripts attached to elements of the module with highest weight $(0, 1)$ are taken to be greek letters with an additional dot (*dotted indices*). Vectors ψ_α of the former module then transform according to

$$\psi_\alpha \mapsto \psi'_\alpha = \sum_\beta M_\alpha{}^\beta \psi_\beta, \tag{20.27}$$

with M a matrix in SL(2,$\mathbb{C}$). Moreover, along with these matrices M also their complex conjugate matrices $\overline{M}$ form a representation of SL(2,$\mathbb{C}$), which is precisely the one with highest weight $(0, 1)$; hence conjugation maps undotted two-component spinors to dotted two-component spinors, and dotted spinors $\psi_{\dot\alpha}$ transform as

$$\psi_{\dot\alpha} \mapsto \psi'_{\dot\alpha} = \sum_{\dot\beta} \overline{M}_{\dot\alpha}{}^{\dot\beta} \psi_{\dot\beta}. \tag{20.28}$$

The matrices that were associated to Lorentz vectors in the formula (20.26) transform under Lorentz transformations according to

$$x^n \sigma_n \mapsto M x^m \sigma_m M^\dagger \tag{20.29}$$

(see exercise 20.6), and hence the sigma matrices σ^m must have the index structure $\sigma^m_{\alpha\dot\alpha}$. This means that the four-dimensional vector module of SO(3, 1) corresponds to the irreducible module with highest weight $(1, 1)$ of the complexified Lie algebra. The sigma matrices can thus be regarded as the coefficients of the basis transformation which makes this correspondence explicit.

From the isomorphism of the complexified Lorentz algebra to $A_1 \oplus A_1$ we learn that the spinor modules with dotted or undotted spinors are symplectic (cf. table XII). The corresponding two-index invariant tensor is the two-dimensional Levi–Civita tensor $\epsilon_{\alpha\beta}$, together with its inverse $\epsilon^{\alpha\beta}$. For explicit calculations, it is convenient to fix the over-all sign freedom that is present in the definition of these tensors by setting $\epsilon^{12} = \epsilon_{21} = 1$ and $\epsilon_{12} = \epsilon^{21} = -1$, and analogously for dotted indices. The fact that the modules are symplectic then means that the quantities

$$\psi^\alpha \phi_\alpha \equiv \epsilon^{\alpha\beta} \psi_\alpha \phi_\beta = -\phi^\alpha \psi_\alpha \qquad \text{and} \qquad \psi_{\dot\alpha} \phi^{\dot\alpha} = -\phi_{\dot\alpha} \psi^{\dot\alpha} \qquad (20.30)$$

are antisymmetric in ϕ and ψ and are scalars under SL(2,$\mathbb{C}$).

In principle, we could now sketch how to describe in this language also the connected components of the Lorentz group which are not connected to the identity. This involves, however, various phase choices so that many different conventions are possible, and we refrain from discussing them here. Rather, we prefer to make contact with the formalism developed in the previous section. To this end we define matrices $\bar\sigma^m$ by

$$(\bar\sigma^m)^{\dot\alpha\alpha} := \epsilon^{\dot\alpha\dot\beta} \epsilon^{\alpha\beta} \sigma^m_{\beta\dot\beta} . \qquad (20.31)$$

Together with the σ^m they can be used to obtain yet another realization of gamma matrices:

$$\gamma^m = \begin{pmatrix} 0 & \sigma^m \\ \bar\sigma^m & 0 \end{pmatrix} \qquad (20.32)$$

for $m = 0, 1, 2, 3$. In this basis we can associate to each undotted Weyl spinor ψ a dotted Weyl spinor $\tilde\psi$ by setting

$$\tilde\psi^{\dot\alpha} := \mathrm{i}\, (\bar\sigma_2)^{\dot\alpha\beta} \psi^*_\beta , \qquad (20.33)$$

where ' * ' stands for complex conjugation. (In the literature, the notation $\bar\psi$ instead of $\tilde\psi$ is common.) We can then construct a Dirac spinor ψ_D out of two Weyl spinors χ and ϕ:

$$\psi_\mathrm{D} = \begin{pmatrix} \chi_\alpha \\ \tilde\phi^{\dot\alpha} \end{pmatrix} . \qquad (20.34)$$

For Majorana spinors ψ_M the number of degrees of freedom is reduced by a factor of two by the condition (20.23); they can therefore be constructed from a single Weyl spinor χ:

$$\psi_\mathrm{M} = \begin{pmatrix} \chi_\alpha \\ \tilde\chi^{\dot\alpha} \end{pmatrix} . \qquad (20.35)$$

For applications in physics the following remark is very important. Above we have always treated the components ψ_α and $\phi_{\dot\alpha}$ of the vectors in spinor modules in the manner to which we are accustomed from previous chapters, i.e. as ordinary (complex or real) numbers. (For example, in (20.30) we used $\psi^\alpha \phi_\alpha = \phi_\alpha \psi^\alpha$, so that the bilinear form (ϕ, ψ) was antisymmetric, as expected for a symplectic module.) In the context of spinors of the Lorentz group one refers to these objects as *commuting spinors*. In contrast, the major application of spinors of the Lorentz group in physics is the description of elementary particles which are fermions.

According to the spin-statistics relation spinors must then be required to *anti-commute* rather than commute (and hence can no longer be regarded as ordinary numbers, but rather as so-called *Grassmann* numbers). This introduces an additional minus sign in the relation (20.30), so that for anti-commuting spinors the relation becomes

$$\psi^\alpha \phi_\alpha = +\phi^\alpha \psi_\alpha \quad \text{and} \quad \psi_{\dot\alpha} \phi^{\dot\alpha} = +\phi_{\dot\alpha} \psi^{\dot\alpha}. \tag{20.36}$$

Whenever one intends to make use of a relation involving Lorentz spinors that one has found in the literature, one should be careful to check whether that relation applies to commuting or anti-commuting spinors.

One might wonder whether spinors occur also for other physically interesting groups like e.g. the general linear group GL(n) which describes the effect of arbitrary changes of the coordinates of some n-dimensional manifold and accordingly plays an important rôle in general relativity. This group, just like the orthogonal groups, is not simply connected; however, any *linear* representation of its universal covering group is already also a representation of GL(n). This explains why all attempts to define spinors in general relativity using GL(n) were doomed to fail.

20.8 The Poincaré algebra. Contractions

The symmetries of an d-dimensional pseudo-euclidean space V with signature (p, q) include the rotations (or rather, pseudo-rotations, also called boosts) which we studied above. There are, however, further obvious symmetries of this space, namely the translations. Let us fix a basis $\{v_{(m)}\}$ of the vector space V and denote the generator of the translations along the basis vector $v_{(m)}$ by iP_m; the factor of $i = \sqrt{-1}$ is chosen because then in physical terms P_m is precisely the momentum operator. In a pseudo-euclidean space all translations commute,

$$[P_m, P_n] = 0, \tag{20.37}$$

and hence the momenta P_m span an d-dimensional abelian Lie algebra, with the Lie bracket given by the commutator.

On the other hand, just like the positions x^m the momenta P_m transform in the defining representation of $\mathfrak{so}(p, q)$, i.e. form a vector operator with respect to $\mathfrak{so}(p, q)$. More precisely, we can describe the generators of $\mathfrak{so}(p, q)$ by antisymmetric matrices $M_{mn} = -M_{nm}$, with $1 \leq m, n \leq d$, having the commutation relations (20.2), i.e.

$$[M_{kl}, M_{mn}] = \eta_{kn} M_{lm} + \eta_{lm} M_{kn} - \eta_{km} M_{ln} - \eta_{ln} M_{km}. \tag{20.38}$$

The commutators between the rotations M_{mn} and the momenta P_l read

$$[M_{mn}, P_l] = \eta_{nl} P_m - \eta_{ml} P_n. \tag{20.39}$$

The Lie algebra $\mathcal{P}$ that is spanned by the P_l and M_{mn}, with Lie brackets (20.37), (20.38) and (20.39), is called the *Poincaré algebra*. It is a semidirect sum of $\mathfrak{so}(p, q)$ and the abelian algebra spanned by the momenta.

From this structure it is apparent that the Poincaré algebra is not a semisimple nor even a reductive Lie algebra. Thus we cannot directly use the structure theory developed in chapter 6 and other chapters to study the Poincaré algebra. However, there exists a trick, the so-called *contraction* of Lie algebras, by which one can relate the Poincaré algebra and many other non-reductive Lie algebras to reductive Lie algebras, so that one can still exploit the structure theory of the latter. In the case of the Poincaré algebra in $d = 4$ dimensions with signature (3,1), which will be written as $(-+++)$, the relevant reductive (in fact, simple) complex Lie algebra is the algebra $\mathfrak{so}(5)$. To determine the relevant real forms, we note that the maximal compact subalgebra of $\mathfrak{so}(p,q)$ is $\mathfrak{so}(p,\mathbb{R})$, so that the Killing form of $\mathfrak{so}(p,q)$ has $p(p-1)/2$ negative and $q(2p+q-1)/2$ positive eigenvalues. Thus the algebra $\mathfrak{so}(5)$ has three inequivalent real forms: Besides the compact real form corresponding to the signature $(+++++)$ of η, there are real forms corresponding to signatures (4,1) and (3,2), i.e. (say) $(-++++)$ and $(-+++-)$, which are called the de Sitter algebra and anti-de Sitter algebra. We will denote the generators of both the de Sitter and the anti-de Sitter algebra by M_{ab} with $0 \le a,b \le 4$ (and ordered such that $\eta = (-+++\pm)$); then in both cases the subset of generators M_{mn} with $0 \le m,n \le 3$ spans a subalgebra which is isomorphic to the Lorentz algebra $\mathfrak{so}(3,1)$. We now consider the rescaled generators

$$P_m^\rho := \rho^{-1} M_{m4} \qquad (20.40)$$

with ρ a positive real number. They satisfy

$$[P_m^\rho, P_n^\rho] = \rho^{-2}[M_{m4}, M_{n4}], \qquad [M_{lm}, P_n^\rho] = \eta_{mn}P_l^\rho - \eta_{ln}P_m^\rho. \qquad (20.41)$$

In the limit $\rho \to \infty$, the brackets $[M_{lm}, P_n^\rho]$ do not change their form, while

$$[P_m^\rho, P_n^\rho] = \rho^{-2}[M_{m4}, M_{n4}] \to 0. \qquad (20.42)$$

Thus by comparison with the relations (20.39) and (20.37) we learn that the limit $\rho \to \infty$ precisely leads to the commutation relations of the Poincaré algebra.

Let us describe in more general terms what we have achieved. We have constructed a family of simple (more generally, reductive) Lie algebras, which are parametrized by the real number ρ. For finite ρ all these algebras are isomorphic. Furthermore, the structure constants approach a well-defined limit for $\rho \to \infty$. However, the Lie algebra obtained in that limit is no longer isomorphic to the Lie algebras that are obtained at finite values of ρ. Rather, we have arrived at the (non-reductive) Poincaré algebra. This method of establishing relations between non-isomorphic Lie algebras is called *contraction*. The contraction procedure is helpful whenever one knows how to deal with the (isomorphic) algebras at finite

ρ; a large part of this information can then be carried over to obtain results about the Lie algebra that is obtained in the limiting case, in our example the Poincaré algebra. Typical applications include the study of special functions associated to these algebras (cf. chapter 21) and the construction of 'quantum deformations' of the algebras (cf. chapter 22).

Another contraction that appears frequently is the contraction of $\mathfrak{so}(3,1)$ to the algebra $\mathfrak{iso}(3,1)$. This algebra, called the *euclidean* algebra and also denoted by $\mathfrak{e}(3)$, is the semidirect sum of the algebra $\mathfrak{so}(3)$ of rotations and the abelian algebra of translations in a three-dimensional space. One can in fact also recover $\mathfrak{so}(3,1)$ from $\mathfrak{iso}(3,1)$; this plays a rôle in the description of aberration phenomena in optics, see e.g. [Mondragón and Wolf 1986], [Atakishiyev *et al.* 1989] and [Wolf 1995].

Information

20.9 Supersymmetry

The *supersymmetry* algebra was discovered in the early 1970s in the search for the most general symmetries of a relativistic quantum field theory. It would go far beyond the scope of this book to explain in any detail what precisely a relativistic quantum field theory is. (The interested reader should consult e.g. [Itzykson and Zuber 1980], [Weinberg 1995] and / or chapters I and II of [Haag 1992].) The relevant result from the investigation of such theories that is needed here is the so-called *Coleman–Mandula theorem*. This theorem asserts that – under a number of plausible assumptions – the most general Lie algebra of symmetries of a (four-dimensional) relativistic quantum field theory has the form of a direct sum $\mathcal{P} \oplus \bar{\mathfrak{g}}$ of the Poincaré algebra $\mathcal{P}$ and some reductive compact real Lie algebra $\bar{\mathfrak{g}}$. The statement that the symmetry algebra is such a direct sum means that the transformations described by $\bar{\mathfrak{g}}$ and the Poincaré algebra transformations commute. In particular, the elements of $\bar{\mathfrak{g}}$ are Lorentz scalars, and hence the corresponding degrees of freedom of the field theory are insensitive to the structure of space-time; $\bar{\mathfrak{g}}$ is therefore called the algebra of *internal symmetries* of the theory.

One of the corner stones of the Coleman–Mandula theorem is the restriction (which at the time seemed rather innocent) to infinitesimal symmetries that form a Lie algebra. It has been a major insight in particle physics to realize that for many purposes this assumption is too restrictive. Indeed, examples of field theories could be constructed in which the symmetry structure is more comprehensive, forming a so-called *Lie superalgebra*. A Lie superalgebra is by definition a vector space $\mathfrak{g}$ which can be decomposed into two complementary subspaces $\mathfrak{g}_0$ and $\mathfrak{g}_1$,

$$\mathfrak{g} = \mathfrak{g}_0 \oplus \mathfrak{g}_1 , \tag{20.43}$$

and which is endowed with a bilinear product

$$\{\cdot,\cdot\] : \quad \mathfrak{g} \times \mathfrak{g} \to \mathfrak{g} \tag{20.44}$$

that possesses the following properties which generalize the defining properties of a Lie bracket.

• The decomposition (20.43) of $\mathfrak{g}$ constitutes a $\mathbb{Z}_2$-gradation with respect to the bilinear product $\{\cdot,\cdot\]$, i.e.

$$\begin{aligned} \{\mathfrak{g}_0,\mathfrak{g}_0\] \subseteq \mathfrak{g}_0 \,, &\qquad \{\mathfrak{g}_1,\mathfrak{g}_0\] \subseteq \mathfrak{g}_1 \,, \\ \{\mathfrak{g}_1,\mathfrak{g}_1\] \subseteq \mathfrak{g}_0 \,, &\qquad \{\mathfrak{g}_0,\mathfrak{g}_1\] \subseteq \mathfrak{g}_1 \,. \end{aligned} \tag{20.45}$$

Thus to any element $x_i \in \mathfrak{g}_i$ one assigns its grade $|x|_i \equiv \deg(x_i) = i$ which is $\mathbb{Z}_2$-valued, $\deg(x_i) \in \{0,1\}$. Accordingly, the subspaces $\mathfrak{g}_0$ and $\mathfrak{g}_1$ are called the *even* and *odd* subspace of $\mathfrak{g}$, respectively.

• The antisymmetry of the Lie bracket is replaced by the condition that

$$\{x,y\] = (-1)^{1+|x|\cdot|y|}\{y,x\] \tag{20.46}$$

for all homogeneous elements (i.e., elements of definite degree) $x,y \in \mathfrak{g}$. When $\{x,y\]$ is antisymmetric, one also uses the symbol $[x,y]$, and when it is symmetric, one also writes $\{x,y\}$. Furthermore, since in applications there is an underlying associative product (of matrices, or of operators on a Hilbert space), one usually refers to $[\cdot,\cdot]$ as the commutator and to $\{\cdot,\cdot\}$ as the anti-commutator.

• The Jacobi identity is generalized to the *super Jacobi identity*, which states that

$$(-1)^{|x||z|}\{x,\{y,z\]\] + (-1)^{|x||y|}\{y,\{z,x\]\] + (-1)^{|y||z|}\{z,\{x,y\]\] = 0 \tag{20.47}$$

for all homogeneous elements $x,y,z \in \mathfrak{g}$.

We have already touched this structure in section 4.11; there we have used the notation $[x,y]_s$ instead of $\{x,y\}$. Note that Lie superalgebras are *not* Lie algebras. (The terms *graded* or *$\mathbb{Z}_2$-graded* Lie algebra, which are sometimes used in place of Lie superalgebra, are therefore misleading.).

It follows immediately that the even subspace $\mathfrak{g}_0 \subset \mathfrak{g}$ is a Lie algebra, with the restriction of $\{\cdot,\cdot\]$ to $\mathfrak{g}_0$ as the Lie bracket; $\mathfrak{g}_0$ is called the *bosonic subalgebra* of $\mathfrak{g}$. Furthermore, $\mathfrak{g}_1$ (also called the *fermionic* subspace) carries a representation of $\mathfrak{g}_0$. As a consequence, various results from the structure theory of Lie algebras can be applied to the study of Lie superalgebras.

Let us now have a look at symmetries of relativistic quantum field theory which include supersymmetries. (To be able to employ the isomorphism of the complexified Lorentz algebra with $A_1 \oplus A_1$, we work with complex Lie algebras.) According to the Coleman–Mandula theorem the bosonic subalgebra $\mathfrak{g}_0$ is the direct sum $\mathcal{P} \oplus \bar{\mathfrak{g}}$ of the Poincaré algebra and a reductive Lie algebra. The fermionic subspace $\mathfrak{g}_1$ then carries in particular a representation of $A_1 \oplus A_1 \subset \mathcal{P}$. Now the elements of $\mathfrak{g}_1$ obey anti-commutation relations; because of the spin-statistics relation, in a relativistic quantum field theory they therefore must describe fermionic degrees of freedom, i.e. must have half-integral Lorentz spin. The anti-commutator of two elements of $\mathfrak{g}_1$ is in $\mathfrak{g}_0$ and transforms in the tensor product representation of the Lorentz algebra. Since $\mathfrak{g}_0$ contains only elements of spin 0 (the internal symmetries, which are Lorentz scalars) and 1 (the Lorentz algebra itself, or in other words its adjoint representation), it follows that $\mathfrak{g}_1$ can contain only operators of spin $\frac{1}{2}$. More concretely, there is a certain number of fermionic operators Q_α^A transforming in the irreducible representation with highest $A_1 \oplus A_1$-weight $(1,0)$, i.e. as undotted two-component spinors; these are accompanied by their conjugates $\bar{Q}_{\dot\alpha B}$ carrying the $(0,1)$-representation. (Recall that conjugation maps undotted two-component spinors to dotted two-component spinors.) If the number N of values which the indices A, B can take is larger than one, one speaks of *N-extended* supersymmetry.

By analyzing the further constraints that follow from the super Jacobi identity, one can now characterize the most general symmetry Lie superalgebra of a (four-dimensional) relativistic field theory. In this process it is also crucial to have in mind that $\mathfrak{g}_1$ has the Poincaré algebra as a subalgebra. The result is known as the *Haag–Lopuszanski–Sohnius* (or HLS) *theorem* (for a simple derivation we refer to the first chapter of [Wess and Bagger 1992]). The HLS theorem states that apart from the Poincaré algebra, whose Lie brackets were already displayed in the previous section, and the relations

$$[M_{mn}, Q_\alpha^A] = (R_{(1,0)}(M_{mn}))_\alpha^{\ \beta} Q_\beta^A \,,$$
$$[M_{mn}, \bar{Q}_{\dot\alpha,B}] = (R_{(0,1)}(M_{mn}))_{\dot\alpha}^{\ \dot\beta} \bar{Q}_{\dot\beta,B} \tag{20.48}$$

which tell us that the operators Q and $\bar{Q}$ are two-spinors, the relations

$$[P_m, Q_\alpha^A] = 0 = [P_m, \bar{Q}_{\dot\alpha,B}]\,, \qquad \{Q_\alpha^A, \bar{Q}_{\dot\beta,B}\} = 2\,\sigma_{\alpha\dot\beta}^m P_m \delta_B^A \tag{20.49}$$

are required to hold. (From the point of view of physics, the last of these relations is the most interesting new ingredient. It states that, loosely speaking, the *supercharges* Q_α^A and $\bar{Q}_{\dot\alpha,B}$ can be regarded as 'square roots' of the momentum operators P_m.) Finally, the anti-commutation relations of the supercharges read

$$\{Q_\alpha^A, Q_\beta^B\} = \epsilon_{\alpha\beta} X^{AB} \qquad \text{and} \qquad \{\bar{Q}_{\dot\alpha,A}, \bar{Q}_{\dot\beta,B}\} = \epsilon_{\dot\alpha\dot\beta} X_{AB}^+ \,, \tag{20.50}$$

where the operators X^{AB} and X_{AB}^+ are required to be central elements, i.e. to commute with all generators of the supersymmetry algebra. They are in fact linear combinations of the generators T^a of the reductive ideal

$\bar{\mathfrak{g}}$ of the bosonic subalgebra $\mathfrak{g}_0$:

$$X^{AB} = \xi_a^{AB} \, T^a \, . \tag{20.51}$$

The supercharges transform with respect to $\bar{\mathfrak{g}}$ according to

$$[Q_\alpha^A, T^a] = S^{a,A}{}_B \, Q_\alpha^B \, , \qquad [\bar{Q}_{\dot\alpha,A}, T^a] = -(S^+)^{a,B}{}_A \, \bar{Q}_{\dot\alpha,B} \, , \tag{20.52}$$

where $S^a \equiv R(T^a)$ and $(S^+)^a \equiv -R^+(T^a)$ are the representation matrices for two conjugate $\bar{\mathfrak{g}}$-representations R and R^+. Furthermore, one can derive the consistency condition

$$S^{a,A}{}_B \, \xi_b^{BC} = -\xi_b^{AB} \, (S^+)^{a,C}{}_B \tag{20.53}$$

on the coefficients ξ_a^{AB} that appear in (20.51). In other words, these quantities intertwine the $\bar{\mathfrak{g}}$-representations realized by S and S^+ on the two types of supercharges. This property constitutes a strong restriction on the existence of central terms.

* 20.10 Twistors

From the results of section 20.5 it is apparent that the conventional formalism for spinors in non-relativistic quantum mechanics relies heavily on the 'accidental' iso-morphism between SU(2) and the universal covering Lie group Spin(3) of SO(3). In contrast, according to section 9.7 the universal covering group Spin(n) of SO(n) is not any classical matrix Lie group, except when $n = 3, 4, 5, 6$. In the latter cases there are again accidental isomorphisms, namely Spin(3) $\cong$ SU(2), Spin(4) $\cong$ SU(2) × SU(2), Spin(5) $\cong$ USP(4) and Spin(6) $\cong$ SU(4). One might wonder whether they are useful in physics as well. In this section we focus our attention on an application of the isomorphism Spin(6) = SU(4).

As long as gravitational effects are irrelevant, the physical space-time is *Minkowski space* M, i.e. a four-dimensional real vector space with a metric of signature (3, 1). As a manifold, this space is isomorphic to its group of translations. However, also the transformation of physical quantities under the Lorentz group SO(3, 1) of rotations of M contains important information on the physics of the system. Therefore one wants to build in the rotation group in the formalism as well. This is achieved by regarding M as the homogeneous space of the Poincaré group modulo the Lorentz group. However, according to the previous section the Poincaré group is not semisimple, so that we cannot directly apply the tools developed in this book.

Apart from contractions, discussed in section 20.8, one other possible way out is to regard M as a coset of an even larger group, namely the group of all conformal transformations of M. (A conformal transformation preserves by definition the metric up to a factor whose value may depend on space-time.) The conformal group of d-dimensional Minkowski space is generated by d translations, a dilatation which scales $x \mapsto \lambda x$, and $d(d-1)/2$ rotations which span SO($d-1, 1$). One can show that M can be compactified, and that on the compact manifold $\overline{M}$ also d special conformal mappings act. (Special conformal transformations act as $x_j \mapsto c_j |x|^2 - 2\sum_k c_k x_j x_k$, with c a constant vector.) One can check that all these transformations generate the universal covering group of the non-compact group SO($d, 2$) (note that the conformal *group* is finite-dimensional even in $d = 2$ dimensions, where the conformal *algebra* (the Witt algebra respectively the Virasoro algebra, see section 12.12) is infinite-dimensional).

For four space-time dimensions, the conformal group is thus SO(4, 2) (for some details about this group, see e.g. [Mack and Salam 1969]). Recalling the isomorphism

$$\mathfrak{so}(4,2)_{\mathbb{C}} \cong \mathfrak{so}(6)_{\mathbb{C}} \cong \mathfrak{su}(4)_{\mathbb{C}} \tag{20.54}$$

of the associated complex Lie algebras, one is led to consider complexified compactified Minkowski space $\overline{M}_{\mathbb{C}}$ with an action of the group $G := SL(4, \mathbb{C})$. We can then describe $\overline{M}_{\mathbb{C}}$ as the coset space G/P_1, where P_1 is the parabolic subgroup of G which consists of matrices of the form

$$\begin{pmatrix} * & * & 0 & 0 \\ * & * & 0 & 0 \\ * & * & * & * \\ * & * & * & * \end{pmatrix}, \tag{20.55}$$

with $*$ standing for an arbitrary complex entry. It can be shown (see exercise 20.7) that G/P_1 is isomorphic to the space of complex two-dimensional subspaces of a four-dimensional complex vector space $\mathbb{C}^4$.

Next we consider also two other parabolic subgroups of G, namely the group P_2 consisting of all matrices of the form

$$\begin{pmatrix} * & 0 & 0 & 0 \\ * & * & * & * \\ * & * & * & * \\ * & * & * & * \end{pmatrix} \tag{20.56}$$

and $P_{12} := P_1 \cap P_2$. One can show that G/P_2 is the complex projective space $\mathbb{PC}^3$, compare exercise 20.7. The coset space G/P_{12} can be related to $\overline{M}_{\mathbb{C}}$ and $\mathbb{PC}^3$ by two natural projections:

$$\pi_1 : \quad G/P_{12} \to G/P_1 \equiv \overline{M}_{\mathbb{C}} \quad \text{and} \quad \pi_2 : \quad G/P_{12} \mapsto G/P_2 \equiv \mathbb{PC}^3 \,. \tag{20.57}$$

A closer analysis shows that these two projections form a so-called *double fibration*, i.e. the restriction of each of the maps π_i on a fiber $\pi_j^{-1}(x)$ of the other map is injective.

To make this construction somewhat more explicit we restrict ourselves to complexified Minkowski space, dropping the compactification. Then $M_{\mathbb{C}} \equiv \mathbb{C}^4$ is lifted by π_1 to the subset $M_{\mathbb{C}} \times \mathbb{PC}^1$ of G/P_{12}. The projection π_2 maps this to a subset $P := \mathbb{PC}^3 \setminus \mathbb{PC}^1$ of $\mathbb{PC}^3$, i.e. to the three-dimensional complex projective space with one complex line deleted. The space P is called the *projective twistor space*. Given a point x in Minkowski space and a two-vector v with components $v_{\dot\alpha}$, the explicit form of the projection π_1 is

$$\pi_1 : \quad (x, [v]) \mapsto x \,, \tag{20.58}$$

where $[v]$ denotes the ray containing v. To describe points in P, we introduce four-vectors, which we write as a pair $(u^\alpha, v_{\dot\alpha})$ of two-vectors. These four-vectors are called *twistors*; they can be regarded as vectors in the defining module of $G = SL(4, \mathbb{C})$, which is called the twistor space and denoted by T. If to any four-vector in Minkowski space we associate a 2×2-matrix as in (20.26), then π_2 can be written

$$\pi_2 : \quad (x, [v]) \mapsto [(u^\alpha = -ix^{\alpha\dot\alpha} v_{\dot\alpha}, v)] \,. \tag{20.59}$$

We read (20.58) and (20.59) also as follows: To any point x of Minkowski space we can associate the complex line $[(u^\alpha = -ix^{\alpha\dot\alpha} v_{\dot\alpha}, v)]$, a so-called twistor line in twistor space T. This way all objects in a theory defined on Minkowski space can be mapped to objects defined over twistor space, which are however trivial on any twistor line. Conversely, given a twistor line (u, v) we can associate to it all points in Minkowski space that satisfy $u^\alpha = -ix^{\alpha\dot\alpha}v_\alpha$; the set of these points forms a so-called β-plane. This is a two-dimensional linear subspace of Minkowski space whose tangent space consists only of lightlike vectors and which is anti-selfdual in the sense that the tensors constructed from any two linearly independent tangent vectors are anti-selfdual. Thus the transformation to twistor lines is one-to-one only for objects that are trivial on β-planes. In gauge theories e.g. the curvature has to vanish on any β-plane; correspondingly the twistor transform is particularly useful for the study of anti-selfdual gauge fields. Another application of this transform is in supersymmetric theories: Any construction formulated in this language easily carries over to the supersymmetric case by replacing groups by supergroups and Lie algebras by Lie superalgebras. For more information, we refer the reader to chapter 1 (in particular §3) of [Manin 1988].

Summary:

Clifford algebras allow for a construction of spinor representations of orthogonal Lie algebras and groups; the construction also provides a realization of the covering group $\mathrm{Spin}(n)$ of $\mathrm{SO}(n)$.

On Dirac spinors one can impose the Majorana and the Weyl condition; the details depend on the signature of the metric. In the case of the Lorentz group the two-component spinor formalism is convenient. The Poincaré algebra is not reductive, but via contraction it can be obtained as a limit of a family of simple Lie algebras.

The most general symmetry Lie superalgebra of a relativistic field theory is described by the HLS theorem. Its bosonic subalgebra is the direct sum of the Poincaré algebra and a reductive Lie algebra. The supercharges can be regarded as square roots of the momentum operators.

Keywords:

Tensor and spinor representations, Clifford algebra, anti-commutation relations, projective (ray) representation, twistor;
spinor; Dirac, Majorana, Weyl, and Majorana–Weyl spinors, two-component spinors;
Lorentz group, Grassmann number, Poincaré algebra, contraction of Lie algebras;
internal symmetry, Coleman–Mandula theorem, supersymmetry, Haag–Lopuszanski–Sohnius theorem, twistor.

Exercises:

Show that the realization (20.2) of the generators of $\mathfrak{so}(n)$ is unitarily equivalent to the one given in equation (5.32). What are the virtues of the respective realizations?

Hint: In terms of the elements γ of the Lie *group* $SO(n)$, the relation reads $\gamma_{(I)} = U\gamma_{(II)}U^{-1}$, where the matrices $\gamma_{(I)}$ corresponding to the choice (20.2) are orthogonal, $\gamma_{(I)}\gamma_{(I)}^t = \mathbf{1}$, while the matrices $\gamma_{(II)}$ corresponding to (5.32) satisfy (why?) $\gamma_{(II)}^t K\gamma_{(II)} = K$ with K as given in equation (5.26) (for n even, respectively as K' in (5.27) for n odd).

Exercise 20.1

Let e_i, $i = 1, 2, \ldots, n$, satisfy the Clifford algebra relations (20.3). Show that the elements

$$S_{0,i} := -\tfrac{i}{2}S_{0,i} = -\tfrac{i}{2}e_i \quad \text{and} \quad S_{j,k} := \tfrac{1}{4}S_{j,k} = \tfrac{1}{4}[e_j, e_k] \quad (20.60)$$

($i = 1, 2, \ldots, n$ and $1 \leq j < k \leq n$) span the Lie algebra $\mathfrak{so}(n+1)$.

Exercise 20.2

Compute the representation matrix for $e_1 e_2 e_3$ in each of the two irreducible representations of $\mathcal{C}(\mathbb{C}^3)$. Conclude that both representations are not faithful, and that the two representations are inequivalent.

Exercise 20.3

Use the definition (20.1) of the $\mathfrak{so}(n+1)$-generators (which correspond to the defining (vector) representation) to show that the representations of Clifford algebras provide the spinor representations of $\mathfrak{so}(n+1)$.

Exercise 20.4

Check that orthogonal transformations of the type (20.12) of the representation matrices of a complex Clifford algebra yield again a representation of the Clifford algebra. What is the analogue of this statement in the case of *real* Clifford algebras?

Exercise 20.5

Show that the universal covering group of $SO(3,1)^+$ is isomorphic to $SL(2,\mathbb{C})$.

Hint: Identify the determinant of the matrix (20.26) with the Lorentz norm of x. Then examine $Mx_m\sigma^m M^\dagger$ with $M \in SL(2,\mathbb{C})$.

Exercise 20.6

Verify that G/P_1, with P_1 as in (20.55), is isomorphic to the space of complex two-dimensional subspaces of a four-dimensional complex vector space $\mathbb{C}^4$, and that G/P_2 is isomorphic to the three-dimensional complex projective space $\mathbb{PC}^3$.

Exercise 20.7

Show that the complexification of the Lie algebra of the Lorentz group is isomorphic to the direct sum $A_1 \oplus A_1$.

Exercise 20.8

Hint: Consider the maximal compact subgroup isomorphic to SO(3), consisting of rotations of the three space coordinates. As generators for the non-compact subspace choose boosts (pseudo-rotations) $K_i := M_{0i}$ along the space coordinate axes. Compute the algebra spanned by the K_i and consider appropriate complex linear combinations with infinitesimal rotations.

Find the explicit form of the 'conjugate' sigma matrices $\bar{\sigma}^m$ defined in (20.31). Verify the relations

Exercise 20.9

$$\sigma^m \bar{\sigma}^n + \sigma^n \bar{\sigma}^m = -2\eta^{mn} \, \mathbf{1} \,, \qquad \mathrm{tr}(\sigma^m \bar{\sigma}^n) = -2\eta^{mn} \,,$$

$$\sum_m \sigma^m_{\alpha\dot{\alpha}} \bar{\sigma}_m^{\beta\dot{\beta}} = -2\delta^\beta_\alpha \delta^{\dot{\beta}}_{\dot{\alpha}} \,, \tag{20.61}$$

and compute $\bar{\sigma}^m \sigma^n + \bar{\sigma}^n \sigma^m$ and $\mathrm{tr}(\sigma^k \bar{\sigma}^l \sigma^m \bar{\sigma}^n)$.

Consider a 'Grassmann number' variable θ, which by definition satisfies $\theta^2 = 0$. By a function of θ one means a power series in θ with complex coefficients. The derivative $\frac{d}{d\theta}$ with respect to a Grassmann variable is defined by $\frac{d}{d\theta} 1 = 0$, $\frac{d}{d\theta} \theta = 1$, and an indefinite integral by

Exercise 20.10

$$\int d\theta = 0 \,, \qquad \int \theta \, d\theta = 1 \,. \tag{20.62}$$

a) Show that on functions of Grassmann variables, differentiation $\frac{d}{d\theta}$ and integration $\int d\theta$ give identical results.

b) How does partial integration work? How does $d\theta$ transform under a change $\theta \mapsto \xi\theta$ ($\xi \in \mathbb{C}$) of variables? What is the analogue of Dirac's delta function?

c) Extend differentiation and integration to functions of several Grassmann variables θ_i which satisfy $\theta_i^2 = 0$ and $\theta_j \theta_i = -\theta_i \theta_j$. Compute the integral of the exponential power series $\exp(-\sum_{i,j} \xi_{ij} \theta_i \theta_j)$ over all Grassmann variables, and compare the result with Gaussian integration over complex variables.

21

Representations on function spaces

21.1 Group actions on functions spaces

The results presented in this chapter are all based on a rather simple observation about the action of groups on function spaces. To arrive at this observation, let us consider a space X which carries some *action* of a group G. This means that for each $\gamma \in$ G there is a mapping $\xi \mapsto \gamma \cdot \xi$ which associates to any $\xi \in X$ another element $\gamma \cdot \xi \in X$, and that this mapping is compatible with the group multiplication, in the sense that $\gamma \cdot (\gamma' \cdot \xi) = (\gamma\gamma') \cdot \xi$ for all $\gamma, \gamma' \in$ G, and that $e \cdot \xi = \xi$ for the unit element e of G. (This is also called a *left* action of G on X; similarly one can consider *right* actions $\xi \mapsto \xi \cdot \gamma$ which obey $(\xi \cdot \gamma) \cdot \gamma' = \xi \cdot (\gamma\gamma')$.)

We do *not* assume here that X is a vector space (in the notation we indicate this by using greek letters for the elements of X); the action of G will therefore in general not be represented by linear maps – the very notion of a linear map does not make sense unless X is a vector space. Nevertheless one can associate to X in a natural manner a vector space, namely the space $\mathrm{Map}(X, \mathbb{C})$ of all maps from X to the complex numbers $\mathbb{C}$ (or to some other field F). By defining addition and scalar multiplication of such maps f pointwise, i.e. $(f_1 + f_2)(\xi) = f_1(\xi) + f_2(\xi)$ and $(zf)(\xi) = z f(\xi)$ for all $\xi \in X$ and all $z \in \mathbb{C}$, this space $\mathrm{Map}(X, \mathbb{C})$ has the structure of a vector space. If we allow for all maps from X to $\mathbb{C}$, this space is infinite-dimensional unless X consists of only finitely many points.

Moreover, the action of G on X induces a natural action of G on $\mathrm{Map}(X, \mathbb{C})$: For any $\gamma \in$ G and any function $f \in \mathrm{Map}(X, \mathbb{C})$ we can define a function $R(\gamma)f$ on X by setting

$$(R(\gamma)f)(\xi) := f(\xi \cdot \gamma) \tag{21.1}$$

363

for all $\xi \in X$. (Note that because of

$$(R(\gamma')\,[R(\gamma)\,f])(\xi) = [R(\gamma)\,f](\xi \cdot \gamma') = f(\xi \cdot (\gamma'\gamma)) = [R(\gamma'\gamma)f](\xi) \quad (21.2)$$

a right action of G on X induces a left action on the function space $\mathrm{Map}(X,\mathbb{C})$. Thus (21.1) is a left action of G on $\mathrm{Map}(X,\mathbb{C})$.) One can check (see exercise 21.2) that this construction defines in fact a representation R of G on the vector space $\mathrm{Map}(X,\mathbb{C})$, and that this representation is linear. Let us also remark that often it is reasonable to restrict the space of functions to some linear subspace, like e.g. the space of continuous, differentiable, smooth, or analytic functions, depending on what other structure the space X carries.

These considerations can still be generalized. Instead of the space of functions from X to $\mathbb{C}$ one can also consider the space $\mathrm{Map}(X,W)$ of maps from X to W, where W is a vector space that carries a representation R_W of G. The space $\mathrm{Map}(X,W)$ is again a vector space (see exercise 21.3). In this situation, we have to generalize the prescription (21.1) to

$$(R(\gamma)f)(\xi) := R_W(\gamma)\,f(\xi \cdot \gamma). \quad (21.3)$$

The reader should also check that this provides a linear representation of G on $\mathrm{Map}(X,W)$. The G-singlets contained in the G-module $\mathrm{Map}(X,W)$ are characterized by the property that $f(\xi \cdot \gamma) = (R(\gamma)f)(\xi)$; they are called *equivariant* functions.

In physics the simplest example (to be described later on) for a representation of this type appears in the treatment of angular momentum in spherical coordinates. Representations of this type are highly reducible, and the study of the reducibility properties leads to interesting results about the relevant function spaces. The examples we examine in this chapter are either spaces of functions defined on (subsets of) $\mathbb{C}^n$, or spaces of functions defined on the Lie group manifold.

21.2 Multiplier representations and Lie derivatives

We will now be more specific and assume that we are given a right action of a Lie group G on an open subset U of $\mathbb{C}^n$. (For the action of a *Lie* group one requires that $\xi \cdot \gamma$ is analytic in ξ and γ.) To be precise, we will only need an action for all group elements in a neighborhood of the unit element e of G. To emphasize this, one sometimes speaks of G as a *local* Lie group or local Lie transformation group. (Note, however, that the qualifier 'local' is then used in the same sense as in the term 'local one parameter group', compare chapter 1; hence it does not refer to a property of G, but of its action. Moreover, the elements for which the local action is defined do *not* constitute a subgroup of G (see exercise 21.4), so that this term is actually a misnomer.)

Given an analytic function f on U, for any element x of the Lie algebra $\mathfrak{g}$ of G we consider the operation

$$L_x(f) := \frac{\mathrm{d}}{\mathrm{d}t} \left(\mathrm{Exp}(tx)\, f \right)\Big|_{t=0}, \qquad (21.4)$$

with the function $\mathrm{Exp}(tx)\, f$ defined analogously to (21.1):

$$(\mathrm{Exp}(tx)\, f)(\xi) := f(\xi \cdot \mathrm{Exp}(tx)) \qquad (21.5)$$

for all $\xi \in U$ and t in a small interval around 0. The operation (21.4), which is called the *Lie derivative* of f in direction x, generalizes in a natural way the notion of a partial derivative. Namely, when U is the euclidean space $\mathbb{R}^n$ endowed with the action of the group of translations, and x_i the generator of translations in the direction of the element e_i of some orthonormal basis, then $L_{x_i} f$ is just the partial derivative $L_{x_i} f = \partial f / \partial \xi_i$. In particular, it follows from the definition that when expressed in terms of coordinates on U, the Lie derivative is a first order linear differential operator which is homogeneous in the sense that it is linear in the derivatives and does not contain a constant term. Indeed, we have already encountered Lie derivatives in chapter 9. Namely, according to the results in chapter 9 we can associate to any element A of the Lie algebra $\mathfrak{g}$ of a Lie group G a left-invariant vector field $A(\xi) = \sum_a A^a(\xi)\, \partial/\partial \xi^a$. Now vector fields are first order linear differential operators on the space of differentiable functions on the Lie group manifold G, and hence the relation (9.4) just tells us that these operators form a set of Lie derivatives for $\mathfrak{g}$. (Thus the notion of a Lie derivative in this chapter must not be confused with the one of section 9.2, which simply amounts to associating to two vector fields a third one.)

Upon commutation, the set of all Lie derivatives forms a Lie algebra which is a homomorphic image of the Lie algebra $\mathfrak{g}$ of G (this is known as Lie's second fundamental theorem). If this algebra is *iso*morphic to $\mathfrak{g}$, the transformation group G is said to act *effectively* (compare section 9.4). Conversely, any set of linearly independent homogeneous first order differential operators L^a on U which are closed under commutation can be 'integrated up' in the sense that the complex n-dimensional Lie algebra spanned by the L^a is the algebra of Lie derivatives of a (local) Lie transformation group acting effectively on U. Also, the Lie derivatives already determine the action of G completely. For instance, that the operation of partial differentiation commutes reflects the fact that the group of translations in euclidean space is abelian.

Instead of the Lie derivatives (21.4), one can more generally consider so-called *generalized Lie derivatives* $\mathcal{L}_x$. By these, one means operations of the form

$$\mathcal{L}_x(f) := \frac{\mathrm{d}}{\mathrm{d}t} \left(\mathcal{R}_\Lambda(\mathrm{Exp}(tx))\, f \right)\Big|_{t=0}, \qquad (21.6)$$

where, by the prescription

$$(\mathcal{R}_\Lambda(\gamma)\,f)(\xi) := \nu_\Lambda(\gamma,\xi)\,f(\xi\cdot\gamma) \tag{21.7}$$

for all $\gamma \in G$ and all $\xi \in U$, $\mathcal{R}_\Lambda$ maps the complex-valued functions on U which are analytic in a neighbourhood of $0 \in U$ to other functions of the same type. In the formula (21.7), Λ is a parameter labelling different possible choices of $\mathcal{R}_\Lambda$, and the multiplier ν_Λ is a complex-valued function analytic in γ and ξ which satisfies $\nu_\Lambda(e,\xi) = 1$ for all ξ as well as $\nu_\Lambda(\gamma\gamma',\xi) = \nu_\Lambda(\gamma',\xi)\nu_\Lambda(\gamma,\gamma'\cdot\xi)$ for all $\gamma, \gamma' \in G$. Note that the latter relation means that for fixed value of the second argument ξ the multiplier ν_Λ generically does not satisfy the representation property; also, this relation implies that

$$\mathcal{R}_\Lambda(\gamma\gamma')\,f = \mathcal{R}_\Lambda(\gamma)(\mathcal{R}_\Lambda(\gamma')\,f). \tag{21.8}$$

In short, up to the action on the argument ξ, the effect of $\mathcal{R}_\Lambda$ is to multiply f with the function ν_Λ; $\mathcal{R}_\Lambda$ is therefore referred to as a *multiplier representation* of G.

When expressed in terms of coordinates on U, any generalized Lie derivative is a (possibly inhomogeneous, i.e. also containing a constant term) first order differential operator. Similarly as for ordinary Lie derivatives, with respect to commutation the generalized Lie derivatives form a Lie algebra which is a homomorphic image of $\mathfrak{g}$. G is said to act effectively in the multiplier representation $\mathcal{R}_\Lambda$ if this Lie algebra is in fact isomorphic to $\mathfrak{g}$. One can show that any multiplier representation of G is completely determined by its generalized Lie derivatives. Conversely, any set of linearly independent first order differential operators which are defined and analytic in an open subset U of $\mathbb{C}^n$ and which are closed under commutation span a Lie algebra of suitable generalized Lie derivatives for some effective multiplier representation of a Lie group G.

Any multiplier representation can be interpreted as an ordinary linear representation on a larger space. To do so, one includes an additional coordinate ζ, so that $(\xi;\zeta) \in \mathbb{C}^{n+1}$, and defines the action of G on the larger space as $(\xi;\zeta) \mapsto (\xi\cdot\gamma; \ln[\nu(\gamma,\xi)]\zeta)$. One can show that *any* multiplier representation can be lifted in this manner to an ordinary representation on $\mathbb{C}^{n+1}$. Thus a multiplier representation of G is a projective representation of G on a function space, and just like according to section 12.2 any projective representation can be lifted to a genuine representation of the simply connected covering group $\tilde{G}$ of G, multiplier representations correspond to genuine representations on a different space.

In the description of quantum mechanical systems one typically deals with function spaces, or more precisely, with the space of functions (or sections in some line bundle) over the classical configuration space. The simple observation that the formula (21.1) provides a linear representation clarifies why symmetries of the configuration space –

which is typically not a linear space – nevertheless induce a linear action on the function space describing the associated quantum mechanical system. This explains in yet another guise why *linear* representations are that important in physics. Furthermore, when one is interested in the action of the symmetries on physical states which are rays in the function space, then one is naturally led to a description of the symmetries in terms of multiplier representations.

21.3 Multiplier representations of SL(2)

As mentioned above, in terms of coordinates generalized Lie derivatives are first order differential operators. The following is the prime example for such operators. Let z denote a complex variable (taking values in some neighborhood of $0 \in \mathbb{C}$) and Λ a non-negative integer, and consider the three inhomogeneous differential operators

$$\mathcal{L}_0 := \Lambda - 2z\tfrac{\mathrm{d}}{\mathrm{d}z}, \quad \mathcal{L}_+ := \tfrac{\mathrm{d}}{\mathrm{d}z}, \quad \mathcal{L}_- := \Lambda z - z^2\tfrac{\mathrm{d}}{\mathrm{d}z}. \tag{21.9}$$

The commutation relations of these differential operators just reproduce the Lie brackets of $\mathfrak{sl}(2)$; according to the results of the previous section these operators therefore form a set of generalized Lie derivatives for an effective multiplier representation of the matrix group SL(2, $\mathbb{C}$).

Now any element

$$\gamma = \begin{pmatrix} a & b \\ c & d \end{pmatrix} \tag{21.10}$$

($ad - bc = 1$) of SL(2, $\mathbb{C}$) in a small neighborhood of the identity can be written uniquely as

$$\gamma = \exp(\tfrac{b}{d}L_+) \, \exp(cd\,L_-) \, \exp(-\ln(d)\,L_0), \tag{21.11}$$

with L_0 and $L_\pm$ the defining 2×2 representation matrices of the $\mathfrak{sl}(2)$ Lie algebra (see exercise 21.5). To obtain the multiplier representation $\mathcal{R}_\Lambda$ of SL(2, $\mathbb{C}$) which corresponds to the generalized Lie derivatives (21.9), one compares the action of (21.9) on powers t^n with the formula (21.6) to obtain the multipliers for the case of the Lie algebra generators L_0 and $L_\pm$, and exponentiates the result by comparison with equation (21.11). The result is

$$(\mathcal{R}_\Lambda(\gamma)\,f)(z) = (cz + d)^\Lambda \, f\!\left(\tfrac{az+b}{cz+d}\right), \tag{21.12}$$

i.e. the multiplier function ν_Λ is given by $\nu_\Lambda(\gamma, z) = (cz + d)^\Lambda$, and the group action is $z \mapsto \gamma \cdot z = (az + b)/(cz + d)$, which can be interpreted as acting with the matrix (21.10) on a column vector with entries z and 1.

Next we consider the space of polynomials in z whose order is smaller or equal to Λ. Using the formula (21.12), one can check that this space is invariant under the action of the differential operators (21.9). A basis of this space is given by the power functions f_m defined by $f_m(z) = z^m$, with $m = 0, 1, \ldots, \Lambda$. The expansion coefficients of $\mathcal{R}_\Lambda(\gamma)z^n$ in this basis, i.e. the functions $R_\Lambda^{m,n}$ determined by

$$(\mathcal{R}_\Lambda(\gamma)f_n)(z) = \sum_{m=0}^{\Lambda} R_\Lambda^{m,n}(\gamma)\, z^m \,, \tag{21.13}$$

are called the *matrix elements* of the multiplier representation $\mathcal{R}_\Lambda$. From the explicit form of (21.13), i.e.

$$\sum_{m=0}^{\Lambda} R_\Lambda^{m,n}(\gamma)\, z^m = (az+b)^n(cz+d)^{\Lambda-n} \,, \tag{21.14}$$

one obtains an expression for the matrix elements in terms of hypergeometric polynomials, namely, for $m \leq n$,

$$R_\Lambda^{m,n}(\gamma) = \frac{a^m d^{\Lambda-n} b^{n-m}\, n!}{m!\,(n-m)!}\, {}_2F_1(-m, -\Lambda-n; n-m+1; \tfrac{bc}{ad}) \tag{21.15}$$

and a similar formula for $m > n$. (The hypergeometric function ${}_2F_1$ is the solution of the differential equation $t(1-t)\,f''(t) + [\delta - (\alpha + \beta + 1)t]\,f'(t) - \alpha\beta\,f(t) = 0$ which around $t = 0$ has the power series expansion ${}_2F_1(\alpha, \beta; \delta; t) = \sum_{n=0}^{\infty}[\Gamma(\alpha+n)\Gamma(\beta+n)\Gamma(\delta)/n!\Gamma(\alpha)\Gamma(\beta)\Gamma(\delta+n)]\,t^n$.)

In other words, the $\Lambda+1$-dimensional space of polynomials of order not exceeding Λ is a module of the group SL(2), and hence of its Lie algebra $\mathfrak{sl}(2)$, with the representation of the Lie algebra given by the differential operators (21.9). In fact, this module is irreducible, and hence (isomorphic to) the highest weight module V_Λ with highest weight Λ; the highest weight vector is the constant function $z^0 = 1$. Of course, this result is the reason why above we chose the symbol Λ to label the different multiplier representations.

When one gives up the restriction that Λ must be a non-negative integer, the differential operators (21.9) still define a representation of $\mathfrak{sl}(2)$, but now the multiplier function $\nu_\Lambda(\gamma, z) = (cz+d)^\Lambda$ must be interpreted as its power series expansion in z about $z = 0$, so that the previous considerations are valid only in an appropriate neighborhood of $z = 0$. Further, there is no longer a finite-dimensional subspace of polynomials in z which is invariant, and hence one is dealing with an infinite-dimensional module of $\mathfrak{sl}(2)$ (spanned, say, by the powers z^n with $n \in \mathbb{Z}_{\geq 0}$). The matrix elements are still given by formulæ like (21.15) (with factorials replaced by the corresponding Gamma functions if the argument is not an integer), but now the hypergeometric functions are no longer polynomials, in particular the expansion in powers of bc/ad has a finite radius of convergence. The property (21.8) of $\mathcal{R}_\Lambda$ provides an addition theorem for the hypergeometric functions.

The module so obtained is (isomorphic to) the Verma module with highest weight Λ. For $\Lambda \notin \mathbb{Z}_{\geq 0}$ this module is irreducible. For $\Lambda \in \mathbb{Z}_{\geq 0}$ it is highly reducible, and, as described at length in chapter 14, its irreducible quotient is obtained by 'setting to zero' the primitive null vectors. For $\mathfrak{sl}(2)$ there is precisely one primitive null vector, and in the present setting it is obviously given by the function $z^{\Lambda+1}$. Thus the irreducible module can be described as the space of all polynomials, modulo the relation $z^{\Lambda+1} = 0$; this is nothing but the space of polynomials of order at most Λ, which as seen above is indeed an irreducible module with highest weight Λ.

21.4 General function spaces

When generalizing the above observations from SL(2) to other Lie groups, one has to be aware of the following. If R is a representation of the Lie algebra $\mathfrak{g}$ on a (possibly infinite-dimensional) vector space V, one can try to define a corresponding representation of the Lie group G via the exponential mapping, i.e. write $\gamma \in$ G as $\gamma = \text{EXP}(x)$ with $x \in \mathfrak{g}$ and

$$R(\gamma)\,v = R(\text{EXP}(x))\,v = \exp(R(x))\,v \equiv \sum_{j=0}^{\infty} \tfrac{1}{j!}\,(R(x))^j\,v \qquad (21.16)$$

for $v \in V$, where exp is the exponential power series. In other words, one would like to relate the exponential mapping of the Lie group to an exponential of the endomorphism $R(x)$. Now the right hand side of (21.16) is an infinite linear combination of vectors. But this formal object may not be defined as an element of V, since in general an interpretation of the infinite sum as a limit of finite sums is not available (if the notion of a limit does exist in V, then V is called a topological vector space).

However, if the abstract vector space V can be 'realized' in the space $\mathcal{F}$ of all functions which are analytic in a neighborhood of some point, say 0, in $\mathbb{C}^m$, in the sense that there is an isomorphism φ from V to a subspace $\mathcal{V} = \varphi(V)$ of $\mathcal{F}$, then we can use φ to define a representation R^φ of $\mathfrak{g}$ on $\mathcal{V}$ as well: The induced representation R^φ acts as $R^\varphi(x) = \varphi \circ R(x) \circ \varphi^{-1}(x)$ for $x \in \mathfrak{g}$. If this induced representation on $\mathcal{V}$ is by first order differential operators which are analytic at the point 0, then the previous results imply that the series (21.16) does make sense. More specifically, in this situation there is a multiplier representation $\mathcal{R}_\Lambda$ of G such that

$$R(\text{EXP}(tx))\,v = \mathcal{R}_\Lambda(\text{EXP}(tx))\,v \qquad (21.17)$$

for all $x \in \mathfrak{g}$, all $v \in V$ and for sufficiently small t. Further, if $\{v_{(n)} \,|\, n \in \mathbb{Z}_{\geq 0}\}$ is a basis of V, then the matrix elements of $\mathcal{R}_\Lambda(\text{EXP}(tx))$ are determined by the expansions

$$\left(\mathcal{R}_\Lambda(\text{EXP}(tx))\right) \varphi(v_{(n)}) = \sum_{m=0}^{\infty} R_\Lambda^{m,n}(\text{EXP}(tx))\, \varphi(v_{(m)}) \qquad (21.18)$$

of the images of the basis functions under the isomorphism φ.

To be precise, in general the matrix elements obtained from the formula (21.18) not only depend on the representation R and on the chosen basis of V, but also on properties of the function space $\mathcal{V}$, in particular on the number m of variables on which these functions depend. However, one does obtain unique matrix elements which depend only on R and the basis $\{v_{(n)}\}$ of V if the smallest subspace of $\mathcal{F}$ that contains $\mathcal{V}$ and is invariant under $\mathcal{R}_\Lambda$ is suitably well-behaved, e.g. the images $\varphi(v_{(n)})$ must provide a topological basis (compare section 12.3) of $\mathcal{V}$. We will assume that this requirement is fulfilled. Under this assumption, the expansion (21.18) amounts to the relation

$$(\mathcal{R}_\Lambda(\gamma))\,\varphi(v_{(n)}) = \sum_{m=0}^{\infty} R_\Lambda^{m,n}(\gamma)\,\varphi(v_{(m)}), \qquad (21.19)$$

valid in a neighborhood of the identity element of G, which can be regarded as a generating function for the matrix elements $R_\Lambda^{m,n}(\gamma)$ and the basis functions $\varphi(v_{(n)})$. Moreover, the representation property implies that the matrix elements satisfy

$$\sum_{m=0}^{\infty} R_\Lambda^{l,m}(\gamma)\,R_\Lambda^{m,n}(\gamma') = R_\Lambda^{l,n}(\gamma\gamma'), \qquad (21.20)$$

and the generalized Lie derivatives associated to $\mathcal{R}_\Lambda$ provide differential recurrence relations

$$\mathcal{L}_x\,\varphi(v_{(m)}) = \varphi(R(x)\,v_{(m)}) \qquad (21.21)$$

for the functions $\varphi(v_{(n)})$.

For appropriate choices of the group G and a basis of V, the matrix elements $R_\Lambda^{m,n}(\gamma)$ turn out to be well-known examples of the 'special functions' of mathematical physics, and often the same is true for the basis functions $\varphi(v_{(n)})$. A lot of information about these special functions can therefore be obtained by working out the explicit form of the relations (21.19) – (21.21). In the following sections some particularly important examples are presented.

21.5 Spherical harmonics

As a simple, and at the same time important, example we consider the Lie group SO(3) of rotations of three-dimensional space. Its Lie algebra is the compact real form $\mathfrak{su}(2)$ of $\mathfrak{sl}(2,\mathbb{C})$, but for the present purposes we can also work with the complex Lie algebra $\mathfrak{sl}(2) \equiv \mathfrak{sl}(2,\mathbb{C})$. In the spirit of the foregoing discussion, we realize the generators of $\mathfrak{sl}(2)$ by angular momentum differential operators $L^a = \sum_{b,c=1}^{3} \epsilon_{abc}\, q_a\, \partial/\partial q_b$ as in equation (2.1). Further, instead of the cartesian coordinates q_a we use spherical coordinates r, ϑ, φ of three-dimensional space. Then the angular momentum operators read

$$\mathcal{L}_0 = -2\mathrm{i}\,\frac{\partial}{\partial\varphi}, \qquad \mathcal{L}_\pm = \mathrm{e}^{\pm\mathrm{i}\varphi}\left(\pm\frac{\partial}{\partial\vartheta} + \mathrm{i}\cot(\vartheta)\,\frac{\partial}{\partial\varphi}\right). \qquad (21.22)$$

These operators do not depend on the coordinate r any more; hence all our considerations will have their natural setting in the space of functions on the two-dimensional sphere $S^2 \subset \mathbb{R}^3$. Our aim is therefore to realize the irreducible representations of $\mathfrak{sl}(2)$ as the differential operators (21.22) on spaces of analytic functions of ϑ and φ with $0 \le \varphi < 2\pi$ and $0 \le \vartheta < \pi$.

In other words, we would like to find functions $Y_{j,m}(\vartheta, \varphi)$ such that (using the normalization conventions of (5.39) and (5.40) for the basis vectors of $\mathfrak{sl}(2)$-modules)

$$\mathcal{L}_0 Y_{j,m} = 2m Y_{j,m}, \quad \mathcal{L}_\pm Y_{j,m} = [(j \mp m)(j \pm m + 1)]^{1/2} Y_{j,m\pm 1}. \quad (21.23)$$

The explicit form of the differential equation that is implied by $\mathcal{L}_0$ reads $\partial Y_{j,m}/\partial\varphi = im Y_{j,m}$; this has the general solution

$$Y_{j,m}(\vartheta, \varphi) = P_{j,m}(\cos(\vartheta))\, e^{im\varphi} \qquad (21.24)$$

with some function $P_{j,m}$ in a single complex variable, which for convenience is taken as $x \equiv \cos(\vartheta)$. Inserting this result into $\mathcal{L}_+ Y_{j,j} = 0$, one arrives at the differential equation

$$\left(\tfrac{\mathrm{d}}{\mathrm{d}\vartheta} - j \cot(\vartheta)\right) P_{j,j}(\cos(\vartheta)) = 0, \qquad (21.25)$$

which is solved by

$$P_{j,j}(x) \propto (1 - x^2)^{j/2}. \qquad (21.26)$$

By employing the equality (21.23) for $\mathcal{L}_-$ as a recursion relation, the functions $P_{j,m}(\cos(\vartheta))$ with $m < j$ and $j - m \in \mathbb{Z}$ are completely determined by (21.26). The requirement $\mathcal{L}_- Y_{j,-j} = 0$ then implies that the parameter j must be a non-negative integer. In this case there are precisely $2j + 1$ functions $Y_{j,m}$ for fixed j, with $m = -j, -j + 1, \dots, j - 1, j$, and the relations (21.23) just express the fact that the vector space spanned by these functions is the irreducible highest weight module of $\mathfrak{sl}(2)$ with highest weight $\Lambda = 2j$. These are, of course, just the considerations we already made in section 2.3 in yet another guise.

The overall normalization of $P_{j,j}(\cos(\vartheta))$ is not determined by (21.26). When this freedom is fixed in such a way that the functions $Y_{j,m}(\vartheta, \varphi)$ are orthonormal with respect to the standard measure on the two-sphere S^2, i.e. satisfy

$$\int_0^{2\pi} \mathrm{d}\varphi \int_0^{\pi} \sin(\vartheta)\, \mathrm{d}\vartheta\, Y_{j,m}(\vartheta, \varphi)\, \overline{Y_{j',m'}}(\vartheta, \varphi) = \delta_{jj'} \delta_{mm'}, \qquad (21.27)$$

then one has

$$P_{j,m}(x) = (1 - x^2)^{m/2} \frac{\mathrm{d}^m}{\mathrm{d}x^m} \mathcal{P}_j(x) \qquad (21.28)$$

with

$$\mathcal{P}_j(x) = \frac{1}{2^j\, j!} \frac{\mathrm{d}^j}{\mathrm{d}x^j} (x^2 - 1)^j. \qquad (21.29)$$

With this normalization, the functions $Y_{j,m}(\vartheta, \varphi)$ are called the *spherical harmonics*, while the functions $P_{j,m}$ and $\mathcal{P}_j$ are known as the Legendre functions and Legendre polynomials, respectively.

While the identities (21.23) provide recursion relations for the functions $P_{j,m}(\cos(\vartheta))$, they do not give rise to a differential equation for any of these functions alone. In order to obtain such an equation, one needs an operator which acts as a multiple of the identity. An obvious candidate for such an object is the quadratic Casimir operator of $\mathfrak{sl}(2)$. Indeed, combining the general form of the quadratic Casimir with the expressions (21.22), one obtains the second order differential equation

$$\frac{1}{\sin(\vartheta)}\frac{\mathrm{d}}{\mathrm{d}\vartheta}\Big(\sin(\vartheta)\frac{\mathrm{d}}{\mathrm{d}\vartheta} + j(j+1) - \frac{m^2}{\sin^2(\vartheta)}\Big)P_{j,m}(\cos(\vartheta)) = 0\,. \qquad (21.30)$$

The quadratic Casimir operator (or analogously, in the case of general G, the set of independent Casimir operators) can also be used to identify the irreducible submodules of a function space as those subspaces on which it acts as a specific multiple of the identity. Geometrically, the quadratic Casimir operator is just the Laplacian with respect to the standard metric on S^2.

21.6 Hypergeometric functions

In chapter 9 we have defined the Lie algebra $\mathfrak{g}$ of a Lie group G using left-invariant vector fields on G. As already mentioned in section 4.7, a vector in a tangent space should be thought of as a derivation on the space of (germs of) functions on the manifold, i.e. as a first order differential operator. In view of this relationship it is not surprising that the associative algebra of differential operators on G that is invariant under translations is isomorphic to the universal enveloping algebra $\mathsf{U}(\mathfrak{g})$ of $\mathfrak{g}$. When restricting to specific classes of functions, one considers specific types of differential operators. Now in classical analysis, one usually characterizes the special functions of mathematical physics, such as hypergeometric functions, Bessel functions and the like, by the differential equations they obey. The above results therefore imply that various types of special functions can be analyzed in Lie algebraic terms. Put differently, the nature of various special functions can be understood in terms of an underlying symmetry, which is described by a Lie algebra.

Let us consider a few examples. We generalize both the form of the differential operators $\mathcal{L}_0$ and $\mathcal{L}_\pm$ and the Lie algebra $\mathfrak{sl}(2)$ in a specific manner. First, instead of (21.22) we make the ansatz

$$\mathcal{L}_0 = 2\frac{\mathrm{d}}{\mathrm{d}\eta}\,, \qquad \mathcal{L}_\pm = \mathrm{e}^{\pm\eta}\,\big(\pm\frac{\mathrm{d}}{\mathrm{d}\xi} + \Phi(\xi)\frac{\mathrm{d}}{\mathrm{d}\eta} + \Psi(\xi)\big)\,, \qquad (21.31)$$

where Φ and Ψ are functions to be determined later. Second, we generalize the $\mathfrak{sl}(2)$ commutation relations to

$$[\mathcal{L}_0, \mathcal{L}_\pm] = \pm 2\,\mathcal{L}_\pm\,, \qquad [\mathcal{L}_+, \mathcal{L}_-] = \beta\mathcal{L}_0 + \gamma\mathcal{L}_4\,,$$
$$[\mathcal{L}_0, \mathcal{L}_4] = 0 = [\mathcal{L}_\pm, \mathcal{L}_4] \qquad (21.32)$$

with complex constants β, γ. Thus the additional generator $\mathcal{L}_4$ introduced here is a central element; therefore we also require that it acts as multiplication by a constant c. The differential operators (21.31) obey all of the relations (21.32) automatically, except for $[\mathcal{L}_+, \mathcal{L}_-] = \beta \mathcal{L}_0 + \gamma \mathcal{L}_4$; the latter imposes the condition that

$$\Phi' - \Phi^2 = \beta, \qquad \Psi' - \Phi\Psi = \tfrac{1}{2}\gamma c, \tag{21.33}$$

where the prime indicates differentiation with respect to ξ. Given a solution to (21.33), one may use Φ and Ψ to realize the operators (21.31) on spaces of analytic functions F in the variables ξ and η.

The representation theory of the Lie algebra defined by the brackets (21.32) can be worked out analogously as for the $\mathfrak{sl}(2)$ case, leading to relations similar to (21.23). In particular, there is an eigenvalue equation for $\mathcal{L}_0$ by which F can be factorized as $F(\xi, \eta) = f_m(\xi)\, e^{m\eta}$ analogously to (21.24). The quadratic Casimir operator, which reads

$$\mathcal{C}_2 = \mathcal{L}_+ \mathcal{L}_- + \mathcal{L}_- \mathcal{L}_+ + \tfrac{1}{2}\beta \mathcal{L}_0 \mathcal{L}_0 + \gamma\, (\mathcal{L}_0 + 1)\mathcal{L}_4 \,, \tag{21.34}$$

then reduces to a second order differential operator acting on the functions $f(\xi) \equiv f_m(\xi)$.

When describing the solutions to (21.33), we actually need to consider only a few special values of the parameters β and γ. Namely, for $\beta \neq 0$ the Lie algebra with brackets (21.32) is nothing but a central extension of $\mathfrak{sl}(2)$, and therefore (since finite-dimensional simple Lie algebras do not possess non-trivial central extensions, see section 12.1) is isomorphic to $\mathfrak{sl}(2) \oplus \mathfrak{u}(1) \cong \mathfrak{gl}(2)$. As a consequence, for any $\beta \neq 0$ we obtain equivalent results as for the specific choice $\beta = 1$, $\gamma = 0$. Similarly, by suitable rescaling of the generators we can reduce the cases with $\beta = 0$ to either of the two special choices $\beta = 0$, $\gamma = 1$ or $\beta = 0$, $\gamma = 0$. In each of these three cases, the eigenvalue equation for the quadratic Casimir operator is the differential equation for a class of special functions, or more precisely, of products of these special functions (possibly expressed in terms of complicated arguments) and elementary functions such as powers and exponentials.

We will not display all these functions explicitly, but only remark the following. For $\beta = 1$, $\gamma = 0$ the differential equations (21.33) possesses two types of solutions; for the first type, the functions f are related to hypergeometric functions, and for the second to confluent hypergeometric functions. For $\beta = 0$, $\gamma = 1$ there are again two types of solutions of (21.33); the first leads to confluent hypergeometric functions and the second to parabolic cylinder functions. Finally, for $\beta = 0$, $\gamma = 0$ (21.33) possesses the following solutions. Either $\Phi(\xi) \equiv 0$ and $\Psi(x)$ is a constant; then $\mathcal{C}_2 \propto d^2/d\xi^2$ so that the functions $f(\xi)$ are simply exponentials. Or else, $\Phi(\xi) = (\xi + a)^{-1}\Psi(\xi) = b(\xi + a)^{-1}$ with complex constants a, b; then

$$\mathcal{C}_2 = -\frac{1}{\xi}\frac{\mathrm{d}}{\mathrm{d}\xi}\,\xi\,\frac{\mathrm{d}}{\mathrm{d}\xi} + \frac{m^2}{\xi^2}\,, \tag{21.35}$$

and the solution to the eigenvalue equation $C_2 f_m(\xi) = \lambda f_m(\xi)$ are the Bessel functions $J_{\pm m}(\sqrt{\lambda} \xi)$.

Similarly to the $\mathfrak{sl}(2)$ case, one can also describe the multiplier representations associated to the generalized Lie derivatives encountered above, and also generalized Lie derivatives and multiplier representations in terms of functions of a single variable, analogously to equation (21.9). But this is considerably more complicated and will not be pursued here.

Let us also mention that just like solutions of differential equations are related to Lie groups, the solutions of algebraic equations are related to certain discrete groups. The solutions of an algebraic equation with coefficients in the rational numbers $\mathbb{Q}$ lie in an algebraic number field, i.e. a finite extension of $\mathbb{Q}$. Certain permutations of these solutions induce automorphisms of this number field which leave $\mathbb{Q}$ fixed. The set of all such automorphisms forms a group, called the Galois group of the field extension. (A simple example is provided by the equation $x^2 + 1 = 0$ which has the two solutions $x = \pm \mathrm{i}$. The Galois group is then isomorphic to $\mathbb{Z}_2$; it consists of the identity map and complex conjugation.) Historically, the observations by Galois led to the introduction of the concept of a group into mathematics; later this structure was generalized to continuous groups, now known as Lie groups, by Sophus Lie.

Galois groups also play a rôle in Lie algebra theory, namely when one considers Lie algebras $\mathfrak{g}$ over a number field F that is not algebraically closed. In that case the problem arises that the characteristic polynomials of the ad-diagonalizable elements of $\mathfrak{g}$ (compare section 6.1) cannot be split into linear factors. However, there still exists a field $\overline{F}$, extending F, in which this can be done. One can then extend $\mathfrak{g}$ to a Lie algebra over $\overline{F}$, quite in the same way as we have introduced the complexification of a real Lie algebra. Provided that the field F (and hence also $\overline{F}$) has characteristic 0, the simple Lie algebras over $\overline{F}$ can be classified as in chapter 7, with precisely the same result. (Note that the structure constants in a Chevalley–Serre basis are all integral, and that the integers $\mathbb{Z}$ can be embedded as a subring into any field of characteristic zero.) In a next step, one considers the group of F-automorphisms of the simple Lie algebras over $\overline{F}$. The Galois group of the extension $\overline{F}/F$ acts naturally on this automorphism group; using this action on the automorphism group, one can finally classify all simple Lie algebras over F.

*21.7 Classification of Lie derivatives

So far we have concentrated on a few specific examples of Lie algebras and of realizations of their representations by generalized Lie derivatives $\mathcal{L}_x$. A much more ambitious task is to classify all the realizations of a given representation of a finite-dimensional simple Lie algebra $\mathfrak{g}$ by generalized Lie derivatives.

As the following simple argument shows, there will typically be many different realizations. Given a specific realization $\mathcal{L}_x$, we can use any function f belonging to the function space under consideration which is non-vanishing in the relevant neighborhood of 0 to construct a new realization as $\tilde{\mathcal{L}}_x := f \, \mathcal{L}_x \, f^{-1}$, where f^{-1} means that at each point one takes the inverse of the value of f at that point. The special functions obtained in this realization are just those of the original realization multiplied by f, and the multiplier functions of the associated multiplier representations are related by

$$f(\xi) \, \tilde{\nu}_\Lambda(\gamma, \xi) = f(\gamma \cdot \xi) \, \nu_\Lambda(\gamma, \xi) \,. \tag{21.36}$$

When studying the classification problem, all realizations that are related in this simple manner by a function f should be treated as equivalent. (This is a special case of the corresponding statement about general ray representations, compare equation (12.9).)

Using the freedom to multiply by a function as in (21.36), one can reduce the classification of generalized Lie derivatives to the classification of ordinary Lie derivatives (possibly at the price of considering functions on a space with one additional dimension, see the remarks at the end of section 21.2). The latter problem, which is equivalent to the determination of all possible ways in which the Lie group G can act on arbitrary manifolds, has been solved for arbitrary finite-dimensional complex Lie algebras in the case of one or two variables. To explain this solution, and to work out its implications for the explicit form of the generalized Lie derivatives would go beyond the scope of this book (for some details, see [Miller 1968]). Instead, we only mention two specific results which arise as by-products of this classification.

First, for the two-variable realizations of the Lie algebras defined in equation (21.32), one finds that the special ansatz made in (21.31) already supplies almost all inequivalent solutions. There do exist additional solutions, but none of them leads to any new type of special functions.

Second, among the solutions for $\mathfrak{g} = \mathfrak{sl}(2)$, there is in particular one which (in contrast to the spherical harmonics discussed above) also describes irreducible representations with half-integer spin $j = \Lambda/2$. To obtain this solution, one does not start from a formulation of $\mathfrak{sl}(2)$ in terms of angular momentum operators, but directly makes an ansatz of the form (21.31). The result for the generalized Lie derivatives is

$$\mathcal{L}_0 = -2\mathrm{i}\frac{\partial}{\partial\varphi}\,, \qquad \mathcal{L}_\pm = \mathrm{e}^{\pm\mathrm{i}\varphi}\left(\pm\frac{\partial}{\partial\vartheta} + \mathrm{i}\cot(\vartheta)\frac{\partial}{\partial\varphi} + \frac{\mathrm{i}\ell}{\sin(\vartheta)}\right), \qquad (21.37)$$

where ℓ is an arbitrary complex parameter. The function space on which these differential operators act is spanned by the functions

$$\mathcal{Y}_{j,m;\ell}(\vartheta,\varphi) \propto \mathrm{e}^{\mathrm{i}m\varphi} \cdot (\cos(\vartheta)+1)^{(m-\ell)/2}(\cos(\vartheta)-1)^{(m+\ell)/2}$$
$$\cdot {}_2F_1(j+m+1,m-j;m+\ell+1;1-\tfrac{1}{2}\cos(\vartheta)) \qquad (21.38)$$

with $m = j, j-1, \ldots$. This space is finite-dimensional if and only if both j is a half integer and the parameter ℓ is one of the numbers $-j, -j+1, \ldots, j$; in this case also $m \in \{-j, -j+1, \ldots, j\}$, and the function space is isomorphic to the irreducible highest weight module with highest weight $\Lambda = 2j$. If, in addition, j is an integer, then one may choose $\ell = 0$; doing so one recovers the generalized Lie derivatives (21.22) and, using the identity

$$\mathcal{P}_j(x) = {}_2F_1(j+1,-j;1;1-\tfrac{1}{2}(\cos(\vartheta)))\,, \qquad (21.39)$$

the Legendre polynomials $\mathcal{P}_j$ and hence the spherical harmonics (21.24). In contrast, for $j \in \mathbb{Z} + \frac{1}{2}$ the relevant hypergeometric function is no longer a polynomial.

21.8 The Haar measure

We now turn to a rather specific type of function, namely functions defined on the group manifold G itself. We have already seen in section 9.3 that any finite-dimensional Lie group G is also a (pseudo-)Riemannian manifold with an invariant metric on it. Now any (pseudo-)Riemannian manifold of dimension d also has a volume element, which inherits analyticity and translation invariance from the metric. In terms of the metric tensor m the line element is written as $\mathrm{d}s^2 = \sum_{i,j} m_{ij}\mathrm{d}t^i\mathrm{d}t^j$, and the volume element reads $|\det(m)|^{1/2} \, dt^1 dt^2 \cdots dt^d$. Up to a multiplicative constant, the volume element that is invariant under translations is unique. We will denote the volume element on a Lie group G by

$$d\mu_\gamma = \Omega(t) \, dt^1 \, dt^2 \cdots dt^d \,, \tag{21.40}$$

i.e. in particular the notation $d\mu_\gamma$ makes explicit reference to the group element $\gamma = \gamma(t)$ that is parametrized by the coordinates t^i. The function $\Omega = |\det(m)|^{1/2}$ is analytic in these coordinates.

The existence of a metric structure with associated volume element allows us to introduce the notion of integration on the group manifold G, like on any other (pseudo-) Riemannian manifold. The right-invariance of the metric implies that also the measure induced by the volume element is invariant, i.e. that $d\mu_{\gamma\gamma'} = d\mu_\gamma$. Equivalently,

$$\int_G d\mu_\gamma \, f(\gamma\gamma') = \int_G d\mu_\gamma \, f(\gamma) \tag{21.41}$$

for all $\gamma' \in G$ and for all integrable functions, i.e. all functions f for which the right hand side exists. The volume of a measurable subset $H \subseteq G$ can be written as

$$\mu(H) := \int_G d\mu_\gamma \, \theta_H(\gamma) = \int_H d\mu_\gamma \,. \tag{21.42}$$

Here θ_H is the characteristic function of H, defined by $\theta_H(\gamma) = 1$ for $\gamma \in H$ and $\theta_H(\gamma) = 0$ for $\gamma \notin H$; thus the volume of H, which generically can be infinite, is finite precisely if θ_H is an integrable function.

The volume $\mu(H)$ is referred to as the *Haar measure* of H; the term Haar measure is also often used for the volume element $d\mu_\gamma$ itself. If G is a *compact* Lie group, then the Haar measure is both left- and right-invariant, and it is also invariant under inversion, $d\mu_{\gamma^{-1}} = d\mu_\gamma$. The volume $\mu(G)$ of a compact Lie group is finite; one then conventionally normalizes $d\mu_\gamma$ such that $\mu(G) = 1$, so that it can be interpreted as a probability measure.

We also remark that an invariant measure can be defined on any finite-dimensional topological group. (A *topological* or *continuous* group is a group which is a topological space and for which multiplication and inverse are continuous maps; in particular, any two points of the topological space are connected by a group operation.) However, if the group is not a Lie group, the construction of the measure is considerably more involved; for details see e.g. chapter XIV in vol. III of [Reed and Simon 1972]. (Also, for a discrete group the integral is just the summation over its elements.)

Let us list the explicit form of the Haar measure for several cases of interest.

- The Haar measure on the (non-compact) Lie group G = ℝ is nothing but the ordinary Lebesgue measure dx.
- The Haar measure for U(1), the abelian group of complex numbers z of absolute value one, reads

$$\mathrm{d}\mu_\gamma = \tfrac{1}{2\pi}\,\mathrm{d}\varphi\,, \qquad (21.43)$$

where φ, ranging from 0 to 2π, is the standard variable on the unit circle, in terms of which $z = \mathrm{e}^{i\varphi}$.

Contrary to these examples, generically there does not exist a globally defined coordinate system on the manifold G. However, one can often find coordinates which behave sufficiently well, e.g. which are defined on all of G except for a subset of measure zero so that the Haar measure can still be described globally.

- The group SU(2) is topologically the three-sphere S^3. Therefore the Haar measure reads

$$\mathrm{d}\mu_\gamma = \tfrac{1}{16\pi^2}\,\sin(\vartheta)\mathrm{d}\vartheta\,\mathrm{d}\varphi\,\mathrm{d}\psi \qquad (21.44)$$

in spherical coordinates, which satisfy with $0 \le \vartheta \le \pi$, $0 \le \varphi \le 2\pi$, $0 \le \psi \le 4\pi$.

- Another possibility is to regard S^3 as embedded in $\mathbb{R}^4$; thus we parametrize the elements of SU(2) as $\gamma = \xi_0 \mathbb{1} + \sum_{i=1}^3 \xi_i \sigma^i$ with σ^i the Pauli matrices and $\vec{\xi}^2 \equiv \sum_{i=0}^3 \xi_i^2 = 1$. Then the integration is over $\mathbb{R}^4$ with

$$\mathrm{d}\mu_\gamma = \pi^{-2}\,\mathrm{d}^4\xi\,\delta(\vec{\xi}^2 - 1)\,. \qquad (21.45)$$

- The easiest way to treat SU(n) with $n > 2$ is to write the group elements $\gamma \in \mathrm{SU}(n)$ as $\gamma = \gamma'\beta$, where γ' is an element of SU($n-1$) (with SU($n-1$) embedded regularly in SU(n)) and β an element of the coset space SU(n)/SU($n-1$). By induction, this reduces the Haar measure on SU(n) to that of SU(2) and of the coset spaces SU(m)/SU($m-1$) with $m = 3, 4, \ldots, n$. The coset space space SU(m)/SU($m-1$), in turn, is nothing but the sphere S^{2m-1}, and the relevant measure is just the standard measure on S^{2m-1}, which is analogous to (21.44) respectively, when S^{2m-1} is embedded in $\mathbb{R}^{2m}$, to (21.45).

- The Haar measure on GL($n, \mathbb{R}$) is

$$\mathrm{d}\mu_\gamma = |\det\gamma|^{-n}\prod_{i,j=1}^{n}\mathrm{d}\gamma_{ij}\,, \qquad (21.46)$$

where γ_{ij}, $i, j = 1, 2, \ldots, n$, are the matrix entries of $\gamma \in \mathrm{GL}(n, \mathbb{R})$.

- For SL(2, $\mathbb{C}$), parametrized as $\gamma = \left(\begin{smallmatrix} a & b \\ c & d \end{smallmatrix}\right)$, the measure reads

$$\mathrm{d}\mu_\gamma = |d|^{-2}\,\mathrm{d}b\,\mathrm{d}\bar{b}\,\mathrm{d}c\,\mathrm{d}\bar{c}\,\mathrm{d}d\,\mathrm{d}\bar{d}\,. \qquad (21.47)$$

As an application of the Haar measure, we mention the so-called *Weyl unitarity trick*, which consists of the following construction. Given an arbitrary finite-dimensional representation $R\colon \mathrm{G} \to \mathrm{GL}(V)$ of a *compact* Lie group G by endomorphisms of a Hilbert space V, one can obtain an associated *unitary* representation by replacing the given inner product $(v\,|\,w)$ on V by its group-averaged counterpart

$$\langle v\,|\,w\rangle := \int_{\mathrm{G}} \mathrm{d}\mu_\gamma\,(R(\gamma)v\,|\,R(\gamma)w)\,; \tag{21.48}$$

this new inner product is invariant under the action of G. Indeed, when the operation of forming adjoint operators is defined with respect to the new inner product $\langle v\,|\,w\rangle$, all representing endomorphisms $R(\gamma)$ are unitary. Thus the G-module V is unitarizable. Notice that for this result one needs the property that the Lie group is compact; this is one of the reasons why for compact Lie groups the representation theory is so much easier than in the non-compact case.

21.9 The Peter–Weyl theorem

The matrix elements of any representation $R_\Lambda(\gamma)$ can be interpreted as functions on the group manifold G. If G is compact and the representation R_Λ is finite-dimensional, then it is possible to show that every continuous function on G can be uniformly approximated by linear combinations of the matrix elements $R_\Lambda^{m,n}$, where Λ runs over the labels of all inequivalent finite-dimensional irreducible modules of G, i.e. over all dominant integral highest weights if G is semisimple. In other words, the set of these matrix elements forms a topological basis of the space of continuous functions on the group manifold. These statements are known as the fundamental approximation theorem or the *Peter–Weyl theorem*. Since the continuous functions are dense in the space

$$L^2(\mathrm{G}) := \{f\colon \mathrm{G} \to \mathbb{C} \mid \int_{\mathrm{G}} \mathrm{d}\mu_\gamma |f(\gamma)|^2 < \infty\} \tag{21.49}$$

of functions on the group manifold which are square integrable with respect to the Haar measure, the Peter–Weyl theorem applies to this larger function space as well.

Furthermore, it turns out that the basis elements $R_\Lambda^{m,n}$ are orthogonal with respect to the Haar measure, and they become orthonormal when multiplied by the square root of $d_\Lambda = \dim(V_\Lambda)$:

$$\int_{\mathrm{G}} \mathrm{d}\mu_\gamma\, R_\Lambda^{m,n}(\gamma)\, R_{\Lambda'}^{m',n'}(\gamma) = (d_\Lambda)^{-1}\,\delta_{\Lambda\Lambda'}\delta_{mm'}\delta_{nn'}\,. \tag{21.50}$$

When completed with respect to the scalar product

$$(f_1, f_2) := \int_G d\mu_\gamma \, f_1(\gamma) \, \overline{f_2(\gamma)}, \qquad (21.51)$$

$L^2(G)$ becomes a Hilbert space. The Peter–Weyl theorem then states that for finite-dimensional Lie groups the span of the matrix elements of the inequivalent finite-dimensional representations is dense in the Hilbert space $L^2(G)$. In the case of semisimple Lie groups, an additional simplification comes from the fact that all finite-dimensional modules can be obtained as irreducible components of tensorial powers of the small set of basic modules which were listed in section 13.6. Namely, for these Lie groups, already the space generated by the matrix elements of a finite set of representations is dense in $L^2(G)$. However, this finite set contains more irreducible representations than just those listed in section 13.6. For instance, as seen in chapter 19, any irreducible representation of A_r is contained in some tensor power of the defining representation with highest weight $\Lambda_{(1)}$. Nevertheless the space generated by the matrix elements of this single representation is not yet dense in $L^2(G)$. Rather, one must take the space generated by the matrix elements of all fundamental irreducible representations, with highest weights $\Lambda_{(i)}$, $i = 1, 2, \ldots, r$. For example, the tensor product $R_{\Lambda_{(1)}} \otimes R_{\Lambda_{(1)}}$ contains both $R_{2\Lambda_{(1)}}$ and $R_{\Lambda_{(2)}}$, but the subspace of $L^2(G)$ generated by the matrix elements of $R_{\Lambda_{(1)}}$ contains only certain linear combinations of the matrix elements of the two irreducible representations in the tensor product.

Next we observe that any Lie group acts on itself by (left or) right multiplication. Correspondingly, there is a representation $\mathcal{R}_{\text{reg}}$ of G on suitable functions $f \colon G \to \mathbb{C}$, given by

$$(\mathcal{R}_{\text{reg}}(\gamma)f)(\gamma') = f(\gamma'\gamma). \qquad (21.52)$$

This is called the (right) *regular representation* of G. When G is compact, then the relevant function space is $L^2(G)$, and the regular representation is unitary and fully reducible. The decomposition of $\mathcal{R}_{\text{reg}}$ into irreducible representations corresponds to decomposing $L^2(G)$ into a Hilbert space direct sum of the finite-dimensional irreducible modules of $\mathfrak{g}$:

$$L^2(G) \cong \bigoplus_\Lambda d_\Lambda V_\Lambda. \qquad (21.53)$$

Here the summation is over all inequivalent finite-dimensional irreducible modules of G, and the factors d_Λ indicate that each irreducible module V_Λ appears $d_\Lambda = \dim(V_\Lambda)$ times. In particular, any finite-dimensional irreducible module of $\mathfrak{g}$ is isomorphic to some invariant subspace of $L^2(G)$. (There is also a generalization of this result, by which one can obtain representations of G, so-called *induced* representations, from representations of a closed subgroup of G. For a description of this construction see e.g.

chapters 16 – 19 of [Barut and Rączka 1986], [Coleman 1968] and chapter 13 of [Kirillov 1975].)

Furthermore, the right and the left action of G on itself commute with each other. Considering both actions simultaneously, the result (21.53) can therefore be refined to a decomposition in terms of the combined left and right action, namely

$$L^2(G) \cong \bigoplus_\Lambda (V_{\Lambda+} \otimes V_\Lambda), \tag{21.54}$$

where $V_{\Lambda+}$ is the representation conjugate to V_Λ.

21.10 Group characters and class functions

Similarly as for the characters of Lie algebras, the traces of representation matrices of a group G are called the *characters* of G. More explicitly, the character ch_Λ of a finite-dimensional representation R_Λ is the map from G to $\mathbb{C}$ which is defined by

$$ch_\Lambda(\gamma) := \operatorname{tr}(R_\Lambda(\gamma)) \equiv \sum_{m=1}^{d_\Lambda} R_\Lambda^{m,m}(\gamma) \tag{21.55}$$

for all $\gamma \in G$. The character ch_Λ depends only on the isomorphism class of the representation R_Λ, not on its specific realization. Furthermore, because of the cyclic invariance of the trace, any character $f = ch_\Lambda$ obeys

$$f(\gamma\gamma'\gamma^{-1}) = f(\gamma') \tag{21.56}$$

for all $\gamma, \gamma' \in G$. A continuous function f which obeys the relation (21.56) is called a *class function* or *central function* of the group G, and the space of square integrable class functions is frequently denoted by $L^2(G)^G$.

Thus by the cyclic invariance of the trace, all group characters are class functions. Moreover, the group characters inherit the orthogonality property

$$\int_G d\mu_\gamma \, ch_\Lambda(\gamma) \overline{ch_{\Lambda'}(\gamma)} = \delta_{\Lambda,\Lambda'} \tag{21.57}$$

from the matrix elements.

If G is a compact Lie group, then according to the Peter–Weyl theorem any class function f can be uniformly approximated by linear combinations f_i of the matrix elements of finite-dimensional irreducible modules. Using the invariance (21.56), it follows that f is uniformly approximated by the group averages

$$\hat{f}_i(\gamma') := \int_G d\mu_\gamma\, f_i(\gamma\gamma'\gamma^{-1}) \tag{21.58}$$

of the f_i as well. Now combining the orthogonality relations (21.50) of the matrix elements $R_\Lambda^{m,n}$ and the representation properties $R_\Lambda^{m,n}(\gamma\gamma'\gamma^{-1}) = \sum_{k,l} R_\Lambda^{m,k}(\gamma)R_\Lambda^{k,l}(\gamma')R_\Lambda^{l,n}(\gamma^{-1})$ and $R_\Lambda^{l,n}(\gamma^{-1}) = R_\Lambda^{n,l}(\gamma)$, it follows that

$$\int_G d\mu_\gamma\, R_\Lambda^{m,n}(\gamma\gamma'\gamma^{-1}) = \delta_{mn}\, d_\Lambda^{-1}\, ch_\Lambda(\gamma'). \tag{21.59}$$

When inserting this result into the right hand side of equation (21.58), it follows that the functions $\hat{f}_i$ are actually linear combinations of the characters ch_Λ. In short:

> Any class function on a compact Lie group G can be uniformly approximated by linear combinations of characters of finite-dimensional irreducible G-modules.

If the Lie group G is semisimple, then any $\gamma \in G$ can be written as $\gamma = \gamma'\gamma_\circ(\gamma')^{-1}$ with $\gamma_\circ$ an element of a maximal torus $G_\circ$ of G, i.e. of a subgroup $G_\circ \subset G$ whose Lie algebra is the Cartan subalgebra $\mathfrak{g}_\circ$ of the Lie algebra $\mathfrak{g}$ of G. In terms of the representation matrices, this simply means that each matrix can be diagonalized by conjugation with a group element, compare section 10.6. Class functions, and hence in particular characters, of semisimple Lie groups are therefore completely determined by their values on the maximal torus. This can be used to reduce the integral $\int_G d\mu_\gamma\, f(\gamma)$ of a class function f to an integral over the maximal torus. More precisely, the Haar measure μ of a compact connected simple Lie group G can be expressed in terms of the Haar measure $\hat{\mu}$ of a maximal torus $G_\circ$ of G and a G-invariant measure ν on the coset space $G/G_\circ$ as follows. Provided that all measures μ, $\hat{\mu}$ and ν are normalized to volume 1, for any continuous function f on G the formula

$$\int_G d\mu_\gamma\, f(\gamma) = \frac{1}{|W(\mathfrak{g})|} \int_{G_\circ} d\hat{\mu}_{\gamma_\circ} \int_{G/G_\circ} d\nu\, f(\gamma\gamma_\circ\gamma^{-1})\, J(\gamma_\circ) \tag{21.60}$$

holds, where the function J on the maximal torus $G_\circ$ is given by

$$J(\text{Exp}(h)) := \Big| \prod_{\alpha\in\Phi_+} \big(e^{\alpha(h)/2} - e^{-\alpha(h)/2}\big) \Big|^2. \tag{21.61}$$

This result is known as *Weyl's integration formula*. (It can be shown that, although for given $\gamma_\circ \in G_\circ$ there can be several elements h of the

Cartan subalgebra $\mathfrak{g}_\circ \subset \mathfrak{g}$ such that $\gamma_\circ = \mathrm{EXP}(h)$, the function J is single-valued on $G_\circ$.) The formula (21.60) is particularly easy to apply to class functions, for which it takes the form

$$\int_G \mathrm{d}\mu_\gamma \, f(\gamma) = \frac{1}{|W(\mathfrak{g})|} \int_{G_\circ} \mathrm{d}\hat{\mu}_{\gamma_\circ} \, f(\gamma_\circ) \, J(\gamma_\circ) \,. \qquad (21.62)$$

Let us also note that the relation between group and Lie algebra characters is roughly as follows. According to (21.55), we have $ch_\Lambda(\gamma) = \mathrm{tr}(R_\Lambda(\gamma))$, while from section 13.5 we know that $\mathcal{X}_\Lambda(h) = \mathrm{tr}(\exp[R_\Lambda(h)])$ for h in a Cartan subalgebra of $\mathfrak{g}$. Now if $\gamma_\circ$ is an element of a maximal torus $G_\circ$, then $\gamma_\circ = \mathrm{EXP}(h)$ with $h \in \mathfrak{g}_\circ$, the Cartan subalgebra associated to $G_\circ$. Combining this information, one obtains the formula $R_\Lambda(\gamma_\circ) = R_\Lambda(\mathrm{EXP}(h)) = \exp[R_\Lambda(h)]$, from which one concludes that $\mathcal{X}_\Lambda(h) = ch_\Lambda(\mathrm{EXP}(h))$. However, this simple recipe in general does not work globally on $G_\circ$, because according to the remarks in section 9.4 a single exponential may not be sufficient to obtain all of $G_\circ$ from $\mathfrak{g}_\circ$.

It is worth mentioning that integrals over groups which appear in applications can often be performed without knowing the explicit form of $\mathrm{d}\mu_\gamma$, by simply making judicious use of the invariance properties of the measure and of the orthogonality relations (21.50). For instance, one has $ch_{\Lambda^+}(\gamma) = ch_\Lambda(\gamma^{-1}) = \overline{ch_\Lambda(\gamma)}$, so that the identity (21.57) tells us that $\int_G \mathrm{d}\mu_\gamma \, ch_\Lambda(\gamma) \, ch_{\Lambda'}(\gamma) = \delta_{\Lambda',\Lambda^+}$. Also, by taking $\Lambda' = 0$ and allowing also for reducible finite-dimensional G-representations R, one finds that the integral $\int_G \mathrm{d}\mu_\gamma \, ch_R(\gamma)$ is just the number of singlets in R. Together, this shows in particular that the singlet appears, with multiplicity 1, in the tensor product $R_\Lambda \otimes R_{\Lambda'}$ precisely if $\Lambda' = \Lambda^+$. (For further examples see e.g. chapter 8 of [Creutz 1983].)

21.11 Fourier analysis

Together with the orthogonality with respect to the Haar measure, the approximation property of the matrix elements allows for a generalization of Fourier transformation to arbitrary group manifolds. Namely, according to the Peter–Weyl theorem we can approximate any function f on a compact Lie group G as

$$f(\gamma) = \sum_\Lambda \sum_{m,n=1}^{d_\Lambda} \tilde{f}(\Lambda,m,n) \, \overline{R_\Lambda^{n,m}(\gamma)} \qquad (21.63)$$

with suitable coefficients $\tilde{f}(\Lambda,m,n)$. Moreover, if f is integrable with respect to the Haar measure, then the sum over Λ on the right hand side indeed converges to f (the convergence is uniform on G if f is sufficiently smooth), and these numbers can be interpreted as the coefficients of the

Fourier transform of f, i.e. are given by

$$\tilde{f}(\Lambda, m, n) = \int_G d\mu_\gamma \, f(\gamma) \, (d_\Lambda)^{1/2} \, R_\Lambda^{m,n}(\gamma) \,. \tag{21.64}$$

In short, one can represent f as a *Fourier series*. For class functions, the relations (21.64) and (21.63) simplify to formulæ containing only the group characters rather than general matrix elements:

$$f(\gamma) = \sum_\Lambda \tilde{f}(\Lambda) \, \overline{ch_\Lambda(\gamma)} \,, \qquad \tilde{f}(\Lambda) = (d_\Lambda)^{-1} \int_G d\mu_\gamma \, f(\gamma) \, ch_\Lambda(\gamma) \,. \tag{21.65}$$

When applied to the compact abelian Lie group $(S^1)^D \equiv (\mathrm{U}(1))^D$, this prescription is just the standard Fourier transform in D dimensions for functions in a hypercube with periodic boundary conditions.

Similarly, the Fourier transform of functions on D-dimensional euclidean space is obtained when our general prescription is applied to the non-compact abelian group $\mathbb{R}^D$. An analogous representation of f as a Fourier *integral* exists for all non-compact Lie groups, and more generally, for any locally compact topological group. Here we use the following terminology. A topological group, or more generally, a topological space, is called locally compact if each of its points possesses a neighborhood with compact closure. Any Lie group is locally compact, but there also exist important topological groups which are not locally compact, such as the group of unitary operators on an infinite-dimensional Hilbert space.

Summary:

The action of a group on a space X induces a linear representation of the group on the space of functions on X, and also on the space of mappings from X to any vector space W. These representations can be used to describe various special functions of mathematical physics.

According to the Peter–Weyl theorem, the span of the matrix elements of all inequivalent finite-dimensional representations of a compact simple Lie group G is dense in the Hilbert space of square integrable (with respect to the Haar measure) functions on G. In particular, any class function on G can be approximated by the characters of these representations. These results allow for a generalization of Fourier analysis on the group manifold G.

Keywords:

(Right or left) action of a group, (generalized) Lie derivative, multiplier representation, spherical harmonics;
Haar measure, Peter–Weyl theorem, regular representation;
group character, class function, Fourier transform.

Exercises:

Is the action $\xi \mapsto \gamma^{-1} \cdot \xi$ of a group G on a space X a left or a right action?

Exercise 21.1

Verify that R as defined by equation (21.1) is a linear representation of the group G.

Exercise 21.2

Show that the space $\mathrm{Map}(X, W)$ of maps from a space V to a G-module W is a vector space, and that with the action (21.3) of G it becomes a G-module.

Exercise 21.3

Consider a connected topological group G and a neighborhood M of the identity element of G. Show that the subgroup of G that consists of all products of elements of M is all of G.
Hint: Without loss of generality one can assume that $M^{-1} = M$ (otherwise replace M by $M \cap M^{-1}$). Consider the subgroup $G_0 := \bigcup_{n=0}^{\infty} M^n$ of G.

Exercise 21.4

a) Verify the expression (21.11) of an element $\gamma \in \mathrm{SL}(2, \mathbb{C})$ in terms of the matrices

Exercise 21.5

$$L_0 = \begin{pmatrix} 1 & 0 \\ 0 & -1 \end{pmatrix}, \quad L_+ = \begin{pmatrix} 0 & 1 \\ 0 & 0 \end{pmatrix}, \quad L_- = \begin{pmatrix} 0 & 0 \\ 1 & 0 \end{pmatrix}. \quad (21.66)$$

b) Derive the formula (21.12) for the action of $\gamma \in \mathrm{SL}(2, \mathbb{C})$ by first computing the multipliers for L_0 and $L_{\pm}$ and then exponentiating with the help of (21.11).

Use the relation

Exercise 21.6

$$q_1 = x = r \, \cos(\vartheta) \, \cos(\varphi), \quad q_2 = y = r \, \cos(\vartheta) \, \sin(\varphi),$$
$$q_3 = z = r \, \sin(\vartheta) \quad (21.67)$$

between cartesian and spherical coordinates to derive the differential operators (21.22) from the formula (2.1).
Check that the commutation relations of these operators reproduce the Lie brackets of $\mathfrak{sl}(2)$.

Compute the normalization of the spherical harmonic $Y_{j,j}$ that is needed in order to satisfy (21.27).

Express the quadratic Casimir operator of $\mathfrak{sl}(2)$ as the differential operator which leads to (21.30).

Exercise 21.7

Check that the quadratic Casimir operator defined by (21.34) indeed commutes with $\mathcal{L}_0$ and $\mathcal{L}_\pm$.

Exercise 21.8

Explain why the Haar measure of a compact Lie group is both left- and right-invariant and invariant under inversion.

Exercise 21.9

Consider the Lie group of upper triangular $n \times n$-matrices $\gamma = (\gamma_{i,j})$, $\gamma_{i,j} = 0$ for $i < j$.

Exercise 21.10

$$d\mu_r = d\gamma_{11}\, d\gamma_{12} \cdots d\gamma_{22} \cdots d\gamma_{nn}\, |\gamma_{11}(\gamma_{22})^2 \cdots (\gamma_{nn})^n|^{-1},$$

$$d\mu_l = d\gamma_{11}\, d\gamma_{12} \cdots d\gamma_{22} \cdots d\gamma_{nn}\, |(\gamma_{11})^n(\gamma_{22})^{n-1} \cdots \gamma_{nn}|^{-1} \tag{21.68}$$

are right- and left-invariant, respectively.

Explain the occurrence of the conjugate module in the formula (21.54). Hint: Determine the relation analogous to (21.52) for the left regular representation of a group (compare exercise 21.1).

Exercise 21.11

Show that the regular representation $\mathcal{R}_{\mathrm{reg}}$ *absorbs* any other unitary representation, i.e. that the tensor product of any unitary representation with $\mathcal{R}_{\mathrm{reg}}$ is a multiple of $\mathcal{R}_{\mathrm{reg}}$.

Show that any two absorbing representations are proportional.

Exercise 21.12

22

Hopf algebras and representation rings

22.1 Co-products

In this chapter we will formalize certain ideas that were already implicit in various constructions used earlier in this book. This will lead us to concepts that play a key rôle in recent developments in theoretical physics and mathematics. Some of these constructions require a more abstract treatment than most of the other structures we had been dealing with. Nonetheless, we will see that these structures arise in a rather natural way in several applications, and that they are even indispensable if one wants to deal with the symmetries of certain quantum systems.

The starting point is the observation of section 15.2 that given representations R_V and R_W of an algebra $\mathfrak{A}$ in the spaces $\mathfrak{gl}(V)$ and $\mathfrak{gl}(W)$, respectively, one obtains a representation of the tensor product $\mathfrak{A} \otimes \mathfrak{A}$ in $\mathfrak{gl}(V \otimes W)$ through the prescription

$$x \otimes y \mapsto R_V(x) \otimes R_W(y) \tag{22.1}$$

for all $x, y \in \mathfrak{A}$, which is extended linearly to all of $\mathfrak{A} \otimes \mathfrak{A}$. (To verify the representation property, recall that $\mathfrak{A} \otimes \mathfrak{A}$ is an algebra over the same field F as $\mathfrak{A}$. The product of the algebra $\mathfrak{A} \otimes \mathfrak{A}$ is the map from $(\mathfrak{A} \otimes \mathfrak{A}) \times (\mathfrak{A} \otimes \mathfrak{A})$ to $\mathfrak{A} \otimes \mathfrak{A}$ that acts as

$$(x \otimes y) \otimes (u \otimes v) \mapsto (x \circ u) \otimes (y \circ v), \tag{22.2}$$

where '$\circ$' denotes the product of $\mathfrak{A}$.) Now as already mentioned in section 15.2, generically it is nevertheless not possible to construct a tensor product representation of $\mathfrak{A}$, i.e. a representation of $\mathfrak{A}$ itself in $\mathfrak{gl}(V \otimes W)$. What is needed as a prerequisite for obtaining such an $\mathfrak{A}$-representation is a prescription of how to 'share out' elements of $\mathfrak{A}$, or more precisely, how to associate the elements of $\mathfrak{A} \otimes \mathfrak{A}$ in a linear manner to the elements of $\mathfrak{A}$. In other words, there must exist a linear mapping

$$\Delta : \quad \mathfrak{A} \to \mathfrak{A} \otimes \mathfrak{A} . \tag{22.3}$$

The direction of such a linear map Δ is just the opposite of what one is used from a product, which is a map from the Kronecker product $\mathfrak{A} \times \mathfrak{A}$ to $\mathfrak{A}$; correspondingly, Δ is called a 'dual product' or *co-product* of the algebra $\mathfrak{A}$. Given a co-product Δ, we make the following ansatz for the tensor product representation:

$$R_{V \otimes W} : \quad x \in \mathfrak{A} \mapsto (R_V \otimes R_W)(\Delta(x)) \in \mathfrak{A} \otimes \mathfrak{A} . \tag{22.4}$$

When we write out $\Delta(x)$ like

$$\Delta(x) = \sum_p x_p^{(1)} \otimes x_p^{(2)} \tag{22.5}$$

as a sum of tensor products of suitable elements $x_p^{(1)}$ and $x_p^{(2)}$ of $\mathfrak{A}$, the formula (22.4) reads more explicitly

$$x \mapsto \sum_p (R_V(x_p^{(1)}) \otimes R_W(x_p^{(2)})) . \tag{22.6}$$

In order that $R_{V \otimes W}$ as defined by (22.4) obeys the representation property $R(x \circ y) = R(x)R(y)$, it is necessary that

$$\Delta(x) \circ \Delta(y) = \Delta(x \circ y) \tag{22.7}$$

for all $x, y \in \mathfrak{A}$. (Here on both sides the symbol ' $\circ$ ' stands for multiplication; however, while on the right hand side this is just the product of $\mathfrak{A}$, on the left hand side it means the induced product of $\mathfrak{A} \otimes \mathfrak{A}$.) In terms of the decomposition (22.5), the left hand side of (22.7) reads

$$\Delta(x) \circ \Delta(y) = \sum_{p,q} (x_p^{(1)} \circ y_q^{(1)}) \otimes (x_p^{(2)} \circ y_q^{(2)}) . \tag{22.8}$$

In short, the requirement (22.7) says that the co-product must respect the algebra structure of $\mathfrak{A}$, or more abstractly, that it is an algebra homomorphism from A to $\mathfrak{A} \otimes \mathfrak{A}$; a co-product with this property is called a *compatible co-product* of $\mathfrak{A}$. The situation can thus be summarized as follows: If V and W are $\mathfrak{A}$-modules with respect to the F-algebra structure of $\mathfrak{A}$, then their tensor product $V \otimes W$ over the field F is naturally an $\mathfrak{A} \otimes \mathfrak{A}$-module, and by composition with a compatible co-product this tensor product constitutes an $\mathfrak{A}$-module.

When the algebra $\mathfrak{A}$ possesses a unit element e, then equation (22.7) tells us that

$$\Delta(x) \circ \Delta(e) = \Delta(x) = \Delta(e) \circ \Delta(x) . \tag{22.9}$$

By considering (22.9) for $x = e$, it follows in particular that $\Delta(e)$ is a projector i.e. idempotent, $(\Delta(e))^2 = \Delta(e)$. A natural solution to this requirement is

$$\Delta(e) = e \otimes e. \tag{22.10}$$

A co-product satisfying (22.10) is called a *unital* co-product.

22.2 Co-algebras and bi-algebras

In the previous section we have considered a vector space $\mathfrak{A}$ with a product and a compatible co-multiplication as an algebra endowed with some additional structure Δ. This fits well with the spirit of this book, in which the basic ingredients are Lie algebras and associative algebras. However, one may also turn things around and regard the co-product as the primary operation and the product M as a supplementary structure. Formalizing this idea, one arrives at the notion of a *co-algebra*; by definition, this is a vector space $\mathfrak{A}$ together with a linear map $\Delta : A \to \mathfrak{A} \otimes \mathfrak{A}$. From this point of view, the compatibility equation (22.7) says that the product M: $x \otimes y \mapsto x \diamond y$ is a co-algebra homomorphism.

The product M is a map M: $\mathfrak{A} \times \mathfrak{A} \to \mathfrak{A}$, while for the co-product Δ the arrow is reversed, $\Delta : \mathfrak{A} \to \mathfrak{A} \otimes \mathfrak{A}$. Nevertheless, the previous remarks suggest that the two maps M and Δ can be treated on a rather similar footing. Indeed, typically the properties of co-algebras can be deduced from those of corresponding algebras by simply 'reversing all arrows'. Therefore it is often not necessary to study co-algebras in their own right.

When considering triple tensor products of $\mathfrak{A}$-representations, it follows from (22.4) that

$$\begin{aligned}
R_{U \otimes (V \otimes W)}(x) &= (R_U \otimes R_{V \otimes W})(\Delta(x)) \\
&= (R_U \otimes (R_V \otimes R_W))((id \times \Delta) \circ \Delta(x))
\end{aligned} \tag{22.11}$$

and

$$\begin{aligned}
R_{(U \otimes V) \otimes W}(x) &= (R_{U \otimes V} \otimes R_W)(\Delta(x)) \\
&= ((R_U \otimes R_V) \otimes R_W)((\Delta \times id) \circ \Delta(x)).
\end{aligned} \tag{22.12}$$

Note that when expressing the co-product in terms of the decomposition (22.5), we have

$$\begin{aligned}
(id \times \Delta) \circ \Delta(x) &= (id \times \Delta)(\textstyle\sum_p x_p^{(1)} \otimes x_p^{(2)}) \\
&= \sum_{p,q} x_p^{(1)} \otimes (x_p^{(2)})_q^{(1)} \otimes (x_p^{(2)})_q^{(2)},
\end{aligned} \tag{22.13}$$

and analogously for $(\Delta \times id) \circ \Delta$. Requiring the tensor product of $\mathfrak{A}$-representations to be associative, i.e. that $R_{(U \otimes V) \otimes W} = R_{U \otimes (V \otimes W)}$, comparison of (22.11) and (22.12) leads to

$$(id \times \Delta) \circ \Delta = (\Delta \times id) \circ \Delta. \tag{22.14}$$

The relation (22.14) says that when one 'shares out' a second time, it does not matter which of the two tensor factors obtained in the first step is shared out; hence it is the analogue of the associativity of an ordinary product, and accordingly a co-product with the property (22.14) is called *co-associative*. Thus the 'dual' analogue of an associative algebra is a co-associative co-algebra. An analogue of a *unital* algebra can also be obtained in a natural manner, but to do so we must first formalize what the existence of a unit element $e \in \mathfrak{A}$ means. To this end we observe that the presence of a unit element distinguishes a one-dimensional subspace of $\mathfrak{A}$, namely the one that is spanned by elements proportional to e. By introducing the *unit map* $\mathrm{E}: F \to \mathfrak{A}$, defined to act as $\mathrm{E}(\xi) := \xi e$ for all $\xi \in F$, we can identify this subspace with the number field over which $\mathfrak{A}$ is defined. The properties of the unit element e can then be rephrased by saying that the map $(\mathrm{E} \times id): F \otimes \mathfrak{A} \to \mathfrak{A} \otimes \mathfrak{A}$, when composed with the multiplication M, reduces (up to scalar multiplication) to the identity map on $\mathfrak{A}$, and that the same is true for $id \times \mathrm{E}$:

$$\mathrm{M} \circ (\mathrm{E} \times id) = id = \mathrm{M} \circ (id \times \mathrm{E}). \qquad (22.15)$$

Analogously, for co-algebras there is a *co-unit map*

$$\varepsilon: \quad \mathfrak{A} \to F, \qquad (22.16)$$

which is characterized by the property that the composition of the co-product with $\varepsilon \times id$ or $id \times \varepsilon$ reduces to the identity map, i.e. that

$$(\varepsilon \times id) \circ \Delta = id = (id \times \varepsilon) \circ \Delta. \qquad (22.17)$$

When a vector space is endowed with all the structures listed above, one speaks of a bi-algebra. More precisely, a vector space $\mathfrak{A}$ is called a *bi-algebra* if, first, it carries both the structure of a unital associative algebra and that of a co-algebra with co-unit and co-associative co-product; and second, if these structures are compatible in the sense that the relations (22.7) and (22.10) hold and that the co-unit is an algebra homomorphism as well, i.e.

$$\varepsilon(x \diamond y) = \varepsilon(x)\, \varepsilon(y) \qquad (22.18)$$

for all $x, y \in \mathfrak{A}$. In this context, the compatibility relation (22.10) is often called the *connecting axiom* of the bi-algebra.

22.3 Hopf algebras

In any bi-algebra $\mathfrak{A}$ one can form two natural endomorphisms of $\mathfrak{A}$, namely the composition $\mathrm{M} \circ \Delta: \mathfrak{A} \to \mathfrak{A} \otimes \mathfrak{A} \to \mathfrak{A}$ of the co-product with the product, and the composition $\mathrm{E} \circ \varepsilon: \mathfrak{A} \to F \to \mathfrak{A}$ of the co-unit with the unit map. There are no reasons to expect that these are closely related. But in fact it turns out that often one obtains identical endomorphisms of $\mathfrak{A}$ as soon as one of the tensor factors at the intermediate stage of the mapping $\text{M} \circ \Delta$ is 'twisted' by a linear map

$$\text{S}: \quad \mathfrak{A} \to \mathfrak{A}, \tag{22.19}$$

i.e. then we have

$$\text{M} \circ (\text{S} \times id) \circ \Delta = \text{E} \circ \varepsilon = \text{M} \circ (id \times \text{S}) \circ \Delta. \tag{22.20}$$

A linear map S which satisfies these relations is called an antipodal map or *antipode* of $\mathfrak{A}$, and a bi-algebra which also possesses an antipode is called a *Hopf algebra*.

It follows from the defining properties of the various maps that the antipode of a Hopf algebra is not an (algebra) homomorphism, but rather an anti-homomorphism (see exercise 22.2), i.e. we have

$$\text{S}(x \circ y) = \text{S}(y) \circ \text{S}(x) \tag{22.21}$$

for all $x, y \in \mathfrak{A}$ (a similar statement applies to the compatibility of S with the co-product).

Besides the construction of tensor product representations, there is another situation where co-products arise in a simple manner, namely when one considers the dual vector space $\mathfrak{A}^*$ of a (finite-dimensional) associative algebra $\mathfrak{A}$, i.e. the space of linear maps φ from $\mathfrak{A}$ to the base field. The idea is to find out what structure is induced on $\mathfrak{A}^*$ by the fact that $\mathfrak{A}$ carries the structure of an algebra. To this end first recall the elementary fact from linear algebra that, given a linear map f from a vector space V to a vector space W, the *dual map* f^* of f is the linear map from W^* to V^* that is described by the transpose of the matrix which describes f. In the situation of our interest it is crucial to keep in mind that the arrows are 'reversed'; this principle provides us with a natural prescription for the co-product $\Delta^*(\varphi)$: For any element of the space $\mathfrak{A}^*$ we set

$$(\Delta^*(\varphi))(x \otimes y) := \varphi(x \circ y) \tag{22.22}$$

for all $x, y \in \mathfrak{A}$. Similarly, if $\mathfrak{A}$ has a unit e, then there is a natural co-unit ε^* on $\mathfrak{A}^*$, namely

$$\varepsilon^*(\varphi) := \varphi(e). \tag{22.23}$$

Analogously, if $\mathfrak{A}$ is a co-algebra, then its dual $\mathfrak{A}^*$ inherits an algebra structure, where the unit $e^* \in \mathfrak{A}^*$ is just the co-unit function and the multiplication is given by

$$(\varphi_1 \circ^* \varphi_2)(x) := (\varphi_1 \times \varphi_2)(\Delta(x)). \tag{22.24}$$

Finally, if $\mathfrak{A}$ is a Hopf algebra with antipode S, then $\mathfrak{A}^*$ is a Hopf algebra as well, with co-algebra and algebra structure as defined above, and with antipode s^* acting as

$$(s^*(\varphi))(x) := \varphi(s(x)).\qquad(22.25)$$

If $\mathfrak{A}$ is infinite-dimensional, one may still use this construction. Generically, though, complications will arise which are related to the fact that the double dual $(\mathfrak{A}^*)^*$ of an infinite-dimensional vector space need not be isomorphic to $\mathfrak{A}$. These can, however, be circumvented by restriction to a suitable subspace of $\mathfrak{A}^*$.

22.4 Universal enveloping algebras

Thus far in this chapter we have essentially introduced various new concepts at a rather abstract level, and it is time to discuss their manifestation in a more concrete setting. A natural example is provided by the universal enveloping algebras of semisimple Lie algebras over $\mathbb{C}$, which is in fact our motivation to study Hopf algebras in this book. (Recall from chapter 14 that the representations of the universal enveloping algebra $U(\mathfrak{g})$ are just the representations of $\mathfrak{g}$.) We recall that according to the formula (15.1) the tensor product of representations of a semisimple Lie algebra $\mathfrak{g}$ acts as

$$(R_{V\otimes W}(x))(v\otimes w) = (R_V(x)\,v)\otimes w + v\otimes(R_W(x)\,w)\qquad(22.26)$$

for all $x\in\mathfrak{g}$ and all $v\in V$, $w\in W$. This can indeed be rewritten in the form (22.4), namely with the map Δ given by

$$\Delta(x) = x\otimes e + e\otimes x.\qquad(22.27)$$

This simple form of Δ neatly reflects what physicists express as the 'additivity of quantum numbers' for a symmetry that is described by a Lie algebra. The presence of the unit element e in the formula (22.27) also shows that this relation should not be considered as an equation in the Lie algebra $\mathfrak{g}$ itself, but rather in the universal enveloping algebra $U(\mathfrak{g})$ of $\mathfrak{g}$. More precisely, (22.27) defines a co-product of $U(\mathfrak{g})$ for those elements which lie already in $\mathfrak{g}\subset U(\mathfrak{g})$. One can show (see exercise 22.1) that when combined with the connecting axiom (22.7) and $\Delta(e) = e\otimes e$, this already determines Δ uniquely on all of $U(\mathfrak{g})$. (Also note that on a generic element of the universal enveloping algebra $U(\mathfrak{g})$ the co-product is typically of a more complicated form than (22.27); that formula is valid *precisely* for elements which are already in $\mathfrak{g}$.)

One can also define a co-unit ε and an antipode s on $U(\mathfrak{g})$, namely by

$$\begin{aligned}\varepsilon(e) &= e\,, & \varepsilon(x) &= 0 \quad\text{ for all } x\in\mathfrak{g}\,,\\ s(e) &= e\,, & s(x) &= -x \quad\text{ for all } x\in\mathfrak{g}\,.\end{aligned}\qquad(22.28)$$

Together with the (anti-) homomorphism requirements (22.18) and (22.21), respectively, this prescription again determines these maps uniquely on all of $U(\mathfrak{g})$.

The fact that the co-product is a homomorphism from the algebra $U(\mathfrak{g})$ to the tensor product $U(\mathfrak{g}) \otimes U(\mathfrak{g})$ also explains why the decomposition of a tensor product representation into irreducible representations is described by the branching rules of the diagonal subalgebra $\mathfrak{g}$ in $\mathfrak{g} \oplus \mathfrak{g}$, compare the corresponding remarks in section 18.6.

The co-product appears also naturally in a different situation, namely when one deals with tensor operators, which were introduced in equation (16.43) in section 16.8. Let us define generalized tensor operators or *covariant* operators $\mathcal{O}^i$, where i labels the basis vectors of some representation $R_{\mathcal{O}}$ of a Hopf algebra $\mathfrak{A}$. When R_T is an irreducible representation, then we require that the relation

$$x \, \mathcal{O}^i = \sum_{j=1}^{\dim R_{\mathcal{O}}} \sum_p (R_{\mathcal{O}}(x_p^{(1)}))^i{}_j \mathcal{O}^j \, x_p^{(2)} \,, \tag{22.29}$$

holds, where the notation (22.18) for the co-product is used. For enveloping algebras $\mathfrak{A} = U(\mathfrak{g})$ and elements $x \in \mathfrak{g}$, for which the co-product takes the form (22.27), this reduces to the definition (16.43). It turns out [Mack and Schomerus 1993] that when one wants to generalize the notion of covariant operators to symmetries more general than Lie algebras (or their universal enveloping algebras), (22.29) is the right definition to use.

22.5 Quasi-triangular Hopf algebras

The co-product (22.27) of $U(\mathfrak{g})$ has the property that it remains unchanged when the order of the two tensor factors is reversed. A co-product with this property is said to be *co-commutative;* abstractly, this property of the co-product is written as

$$\pi_{12} \circ \Delta = \Delta \,, \tag{22.30}$$

where π_{12} is the transposition map on $U(\mathfrak{g}) \otimes U(\mathfrak{g})$, i.e. $\pi_{12}(x \otimes y) = y \otimes x$. In case that for some Hopf algebra $\mathfrak{A}$ the property (22.30) does not hold, the map $\pi_{12} \circ \Delta$ is again a co-associative co-product, known as the *opposite co-product.* Furthermore, provided that the antipodal map is invertible, together with the same product, unit and co-unit and the inverse antipode it yields another Hopf algebra structure for $\mathfrak{A}$. (For any commutative or co-commutative Hopf algebra, the antipode obeys $s \circ s = id$. But for arbitrary Hopf algebras this is no longer true, and there even are infinite-dimensional Hopf algebras whose antipode does not possess an inverse mapping.)

Given any two $\mathfrak{A}$-representations R_V and R_W, the permutation map constitutes an intertwiner between the representations $R_{V \otimes W}$ and $R_{W \otimes V}$. Indeed, because of the description (22.4) of the tensor product in terms of the co-product Δ (and using the notation (22.5)), the latter statement amounts to the equality

$$\sum_p (R_V(x_p^{(1)}))(v) \otimes (R_W(x_p^{(2)}))(w) = \sum_p (R_V(x_p^{(2)}))(v) \otimes (R_W(x_p^{(1)}))(w)$$

of the representation matrices for all $x \in \mathfrak{A}$ and all $v \in V$, $w \in W$; this relation is indeed implied by the co-commutativity (22.30). Conversely, if for *each* pair R_V and R_W of $\mathfrak{A}$-representations the permutation map is an intertwiner between $R_{V \otimes W}$ and $R_{W \otimes V}$, then it is natural to require that this reflects a property of the algebra $\mathfrak{A}$ itself, and again because of (22.4) this property of $\mathfrak{A}$ is precisely the co-commutativity of the co-product.

Now for generic Hopf algebras the permutation map does *not* constitute an intertwiner between $R_{V \otimes W}$ and $R_{W \otimes V}$ for all pairs of $\mathfrak{A}$-representations. However, it often happens that the two representations $R_{V \otimes W}$ and $R_{W \otimes V}$ are still isomorphic for arbitrary R_V and R_W, and hence there still exist intertwiners. In short, the tensor product of $\mathfrak{A}$-representations is no longer commutative, but still *quasi*-commutative, i.e. commutative up to isomorphism. Put differently, for any $x \in \mathfrak{A}$ the representation matrices $R_{V \otimes W}(x)$ and $R_{W \otimes V}(x)$ are related by conjugation. Again it is natural to demand that this behavior of representations should follow from a similar property of the underlying co-product, namely that Δ and the opposite co-product are conjugate to each other as well.

More explicitly, in this situation the equality (22.30) gets replaced by the somewhat weaker condition that there exists an invertible element $\mathcal{R}$ of the algebra $\mathfrak{A} \otimes \mathfrak{A}$ such that

$$\pi_{12} \circ \Delta(x) = \mathcal{R} \circ \Delta(x) \circ \mathcal{R}^{-1} \tag{22.31}$$

for all $x \in \mathfrak{A}$. Invertibility of $\mathcal{R}$ means that an inverse $\mathcal{R}^{-1}$ satisfying $\mathcal{R} \circ \mathcal{R}^{-1} = e \otimes e = \mathcal{R}^{-1} \circ \mathcal{R}$ exists. Note that the existence of invertible elements (also called units, see section 4.2) in an algebra is by no means guaranteed.

Provided that $\mathcal{R}$ also satisfies a few further technical consistency requirements, a Hopf algebra with the property (22.31) is called *quasi-co-commutative* or *quasi-triangular,* and $\mathcal{R}$ is called the R-element or *universal* R-*matrix* of the Hopf algebra. The qualification 'universal' expresses the fact that the intertwiner between the representations $R_{V \otimes W}$ and $R_{W \otimes V}$ depends on the representations R_V and R_W only very weakly: It is just the representation matrix for one single element $\mathcal{R}$ of the algebra $\mathfrak{A} \otimes \mathfrak{A}$. Among the finite-dimensional Hopf algebras, the quasi-triangular ones are singled out by the fact that, via the so-called quantum double construction, any finite-dimensional Hopf algebra can be naturally embedded in a suitable finite-dimensional quasi-triangular Hopf algebra.

22.6 Deformed enveloping algebras

An important class of examples for Hopf algebras with quasi-triangular (but not co-commutative) co-product is provided by a generalization of the universal enveloping algebras $U(\mathfrak{g})$, namely the so-called *deformed enveloping algebras* $U_q(\mathfrak{g})$ of simple (or affine) Lie algebras. The vector space structure of $U_q(\mathfrak{g})$ is the same as for the ordinary enveloping algebra $U(\mathfrak{g})$, but the prescription (22.27) for the co-product is generalized to

$$\Delta(H^i) = H^i \otimes e + e \otimes H^i,$$

$$\Delta(E^i_\pm) = E^i_\pm \otimes q^{H^i/4} + q^{-H^i/4} \otimes E^i_\pm. \tag{22.32}$$

This formula requires some explanation. The $U_q(\mathfrak{g})$-elements H^i and $E^i_\pm$ are the generators of a Chevalley basis of $\mathfrak{g}$. Further, $q^{\xi H^j}$ stands for elements of $U_q(\mathfrak{g})$ which with respect to multiplication behave precisely as exponentials do; before trying to make this more precise, we study a few consequences. First, for $q = 1$ (22.32) reduces to the $U(\mathfrak{g})$-formula (22.27); the $U_q(\mathfrak{g})$-co-product is therefore called a deformation or *q-deformation* of the co-product of $U(\mathfrak{g})$, and q the *deformation parameter*. Also, the co-product of the Cartan subalgebra generators H^i does not get deformed, so that the H^i still give rise to additive quantum numbers.

Second, given the co-product (22.32), the validity of the connecting axiom (22.18) requires that the multiplication must be modified as well. More precisely, the commutators involving the Cartan subalgebra generators H^i can be kept, but the relations $[E^i_+, E^j_-] = \delta_{i,j} H^i$ of the Chevalley basis must be changed to

$$[E^i_+, E^j_-] = \delta_{i,j} \frac{q^{H^i/2} - q^{-H^i/2}}{q^{1/2} - q^{-1/2}}, \tag{22.33}$$

and the Serre relations are modified by the insertion of functions of the deformation parameter q, too.

To give a more detailed explanation of the rôle of the quantities $q^{\xi H^j}$, it is worth recalling that, strictly speaking, in order for the formal object $\Delta(x)$ as defined by (22.5) to be an element of the algebra $\mathfrak{A} \otimes \mathfrak{A}$, it must be a *finite* linear combination of basis elements of $\mathfrak{A} \otimes \mathfrak{A}$. Nevertheless one often also considers mappings (22.5) for which one deals with infinite linear combinations. Such a mapping it is still called a co-product of $\mathfrak{A}$, provided that it can be interpreted as a linear map to a suitable completion of $\mathfrak{A} \otimes \mathfrak{A}$. Now when expressions like $q^{\xi H^j}$ appearing in (22.32) and (22.33) are interpreted in terms of exponential power series, i.e. as $q^{\xi H^i} \equiv \exp(\xi \ln q \cdot H^i)$, then they are infinite series in the H^i, and accordingly the co-product (22.32) of $U_q(\mathfrak{g})$ provides an example for the situation just described. In this approach the deformation parameter q is considered as a complex number. The problem can be circumvented if instead one regards expressions like $q^{H^i/4}$ as independent new generators which are required to obey the same multiplication rules as exponentials; in this case one deals only with finite linear combinations. Yet another possibility is to interpret q as a *formal variable* or indeterminate in the sense that under ordinary circumstances it just behaves as a complex variable, while in those cases where this would lead to infinities or where questions of convergence arise, one resorts to prescriptions analogous to the multiplication rules for the 'formal exponentials' $q^{H^i/4}$.

Besides the co-product, there is also a co-unit and an antipode for $U_q(\mathfrak{g})$, so that it is again a Hopf algebra. The co-unit is still defined as in equation (22.28), while the antipode is a q-deformation of (22.28), namely

$$s(H^i) = -H^i, \qquad s(E_\pm^i) = -q^{\pm 1/2} E_\pm^i \qquad (22.34)$$

and $s(e) = e$.

While the co-product (22.32) is obviously not co-commutative, it is still quasi-triangular. The universal R-matrix $\mathcal{R}$ which satisfies the relation (22.31) is again an infinite series in the basis elements of $U_q(\mathfrak{g}) \otimes U_q(\mathfrak{g})$; for $\mathfrak{g} = \mathfrak{sl}(2)$ it reads

$$\mathcal{R} = q^{H \otimes H/4} \sum_{n=0}^{\infty} f_n \left(q^{nH/4}(E_+)^n \right) \otimes \left(q^{-nH/4}(E_-)^n \right) \qquad (22.35)$$

with coefficients $f_n = q^{-n(n+1)/4}(q^{1/2} - q^{-1/2})^{2n} / \prod_{m=1}^{n}(q^{m/2} - q^{-m/2})$.

While at first sight the deformed algebra $U_q(\mathfrak{g})$ of a simple Lie algebra $\mathfrak{g}$ may look rather different from $U(\mathfrak{g})$, they are in fact closely related as long as the deformation parameter q is a formal variable or a complex number which is not a root of unity. In particular, the center of $U_q(\mathfrak{g})$ is spanned by appropriate q-deformations of the Casimir operators of $\mathfrak{g}$, and hence its elements are in one-to-one correspondence with the center of $U(\mathfrak{g})$. As a consequence, the representation theory of $U_q(\mathfrak{g})$ turns out to be isomorphic to that of $U(\mathfrak{g})$, and hence of $\mathfrak{g}$. For instance, the finite-dimensional modules of $U_q(\mathfrak{g})$ are unitary and fully reducible, and the irreducible finite-dimensional modules are parametrized by the dominant integral highest weights Λ of $\mathfrak{g}$. Also, the finite-dimensional modules can be decomposed into weight spaces precisely as for $\mathfrak{g}$-modules (compare equation (13.2)), with the same multiplicities as for the $\mathfrak{g}$-modules, and the tensor product multiplicities are identical as well. It is even possible to express the representation matrices for a highest weight representation of $U_q(\mathfrak{g})$ rather directly in a form close to those for the corresponding $\mathfrak{g}$-representation, namely by replacing some of the integers n that appear in the entries of the matrices for the Chevalley generators of $\mathfrak{g}$ in a suitable manner by so-called q-*integers* $[n]_q$. These are the complex numbers (respectively polynomials in the formal variable q) defined by

$$[n]_q := \frac{q^{n/2} - q^{-n/2}}{q^{1/2} - q^{-1/2}} = \sum_{m=1}^{n} q^{(n+1)/2 - m}. \qquad (22.36)$$

(The explicit formulæ, which are not particularly illuminating, can be found in [Curtright *et al.* 1991]. Also, when combined with the co-product (22.27) of $U(\mathfrak{g})$, these relations do *not* reproduce the co-product (22.32) of $U_q(\mathfrak{g})$, but rather another co-product that is related to (22.32) via a similarity transformation.)

Let us also note that when considering tensor products of $U_q(\mathfrak{g})$-representations, one can again apply the Brauer–Weyl duality between centralizer algebras (compare chapter 19). Instead of the discrete groups that were encountered for tensor products of $\mathfrak{g}$- (and hence $U(\mathfrak{g})$-) representations, one now gets certain q-deformations of these groups, or more precisely, of their group algebras. For example, for $\mathfrak{g} = \mathfrak{sl}(n)$ one now obtains, in place of the group algebras $\mathbb{C}S_\ell$ of the symmetric groups, the so-called Hecke algebras $H_\ell(q)$. The algebra $H_\ell(q)$ is generated by a unit element e and further invertible generators s_i, $i = 1, \ldots, \ell - 1$, modulo the relations

$$s_i s_i = (1 - q)\, s_i + q\, e\,, \qquad s_i s_{i+1} s_i = s_{i+1} s_i s_{i+1}\,,$$
$$s_i s_j = s_j s_i \quad \text{for } |i - j| \geq 2\,. \tag{22.37}$$

Thus $H_\ell(q)$ can in particular be regarded as a q-deformed Coxeter group.

If the deformation parameter q is an mth root of unity, i.e. $q^m = 1$, then the q-integer $[m]_q$ vanishes. As a consequence, entries of representation matrices which for generic q are finite can become zero or even infinite, and inner products of vectors in $U_q(\mathfrak{g})$-modules can degenerate to zero as well. (Also, the center of $U_q(\mathfrak{g})$ gets enlarged enormously, e.g. it contains certain powers of step operators.) This leads to drastic changes of the representation theory. One finds in particular that the only unitary finite-dimensional irreducible modules are those highest weight modules whose highest weight Λ is not only dominant integral, but in addition satisfies

$$0 \leq (\Lambda, \theta^\vee) \leq m - \mathfrak{g}^\vee\,. \tag{22.38}$$

Here θ is the highest root of $\mathfrak{g}$ and $\mathfrak{g}^\vee$ is the dual Coxeter number, and we assume that $m > \mathfrak{g}^\vee$. (Below we will in addition assume that m is the smallest power for which q^m is equal to 1. In this case q is called a *primitive mth root of unity*. For definiteness we take $q = \exp(2\pi i/m)$). Associated to the dominant integral weights which do not satisfy (22.38), there are non-unitary finite-dimensional modules which are reducible, but not fully reducible; they contain non-invariant submodules which for generic q would become irreducible. Furthermore, there are finite-dimensional 'cyclic representations' which do not possess any maximal weight vector, and which despite their finite-dimensionality are characterized by weights with continuous parameters.

Deformed enveloping algebras have been applied to the description of ro- **Information** tational and vibrational spectra in nuclei and diatomic molecules and of the shell structure of certain deformed nuclei; for a review, see [Bonatsos *et al.* 1995]. In these approaches one q-deforms an approximate symmetry in such a manner that the deformed structure provides a better approximation to the experimental data. The deformation parameter q is typically employed as a variable that can be tuned in such a way that the approximation of the data becomes optimal. An important tool is often a q-deformed harmonic oscillator algebra which provides a boson realization of $U_q(\mathfrak{sl}(2))$. Also, in this context one sometimes considers more general deformations which are no longer Hopf algebras so that the algebraic structure is much less rigid.

22.7 Group Hopf algebras

Besides the universal enveloping algebras, there are two other classes of Hopf algebras which already made their appearance in earlier chapters of this book. The first are the group algebras, and the second are algebras of functions on a group. Recall from chapter 19 that the group algebra $\mathbb{C}G$ of a finite group G is the finite-dimensional vector space $\mathbb{C}^{|G|}$, endowed with an associative multiplication that is obtained by extending the group product linearly (for a more precise description, see formula (19.15)). In a similar way one obtains a co-associative co-multiplication as well as a co-unit and an antipode on $\mathbb{C}G$, by setting

$$\Delta(e_\gamma) = e_\gamma \otimes e_\gamma, \qquad \varepsilon(e_\gamma) = 1 \tag{22.39}$$

and

$$\mathsf{s}(e_\gamma) = e_{\gamma^{-1}} \tag{22.40}$$

for all group elements $\gamma \in G$, and by extending these definitions by linearity and the required (anti-) homomorphism properties. In the relation (22.40) γ^{-1} is the ordinary group inverse of γ; the identity $(\gamma\gamma')^{-1} = \gamma'^{-1}\gamma^{-1}$ obeyed by inverses is the source of the *anti*-homomorphism property of the antipode (for the same reason the antipode of a general Hopf algebra is sometimes referred to as a 'generalized inverse', 'co-inverse' or 'quasi-inverse'). Just as for enveloping algebras, the co-product of a group Hopf algebra is co-commutative. Actually, almost any statement that applies to both universal enveloping algebras and group Hopf algebras is true for all other co-commutative Hopf algebras as well.

In principle these considerations can also be extended to (locally compact) infinite topological groups, and in particular to Lie groups, but the correct interpretation of the analogue of the space $\mathbb{C}^{|G|}$ is then rather intricate. Another possibility to study Lie groups G in this context is to interpret the group elements (in a neighborhood of the identity) as elements $\gamma = e^{\xi x}$ of some completion of the enveloping algebra $\mathsf{U}(\mathfrak{g})$, with $x \in \mathfrak{g}$. Applying the co-product (22.27) and the antipode (22.28) of $\mathsf{U}(\mathfrak{g})$ to such an element leads to

$$\Delta(e^{\xi x}) = e^{\xi x} \otimes e^{\xi x}, \qquad \mathsf{s}(e^{\xi x}) = e^{-\xi x}, \tag{22.41}$$

while the co-unit(22.28) yields $\varepsilon(e^{\xi x}) = 1$; thus they obey analogous relations as the group elements in the group algebra of a finite group. More generally, any element y of an arbitrary Hopf algebra $\mathfrak{A}$ which satisfies $\Delta(y) = y \otimes y$ and $y \circ \mathsf{s}(y) = e = \mathsf{s}(y) \circ y$ is called a *group-like element* of $\mathfrak{A}$. Group-like elements are in particular invertible, i.e. they are units of the Hopf algebra.

The group algebra $\mathbb{C}G$ of a discrete group can also be regarded as the zero-dimensional analogue of the algebra $L^2(G)$ of square integrable functions on a continuous group. The expression (21.54) of $L^2(G)$ as a sum of tensor products of irreducible modules continues to hold for discrete groups. In fact, the statement holds in the somewhat stronger version that as a G-module the algebra $\mathbb{C}G$ is isomorphic to the direct sum of all endomorphisms of all inequivalent irreducible G-modules: $\mathbb{C}G \cong \bigoplus_i \mathfrak{gl}(V_i)$. This shows in particular that these group algebras are semisimple algebras.

22.8 Functions on compact groups

Enveloping algebras and group algebras are co-commutative Hopf algebras. Another special, but important, type of Hopf algebras are those with a commutative multiplication. To obtain examples for such commutative Hopf algebras, we consider, for any compact topological group G, the vector space $\mathcal{F}G$ of continuous complex-valued functions on G. A natural product on $\mathcal{F}G$ is given by pointwise multiplication, i.e. the product $\varphi_1 \circ \varphi_2$ of two elements of $\mathcal{F}G$ is defined by $(\varphi_1 \circ \varphi_2)(\gamma) := \varphi_1(\gamma)\varphi_2(\gamma)$. Further, the function $\mathbf{1} \in \mathcal{F}G$ which maps any $\gamma \in G$ to the number 1 provides a unit element for this product. In a sense, this algebra structure merely reflects the structure of the 'target space' $\mathbb{C}$ of the maps $G \to \mathbb{C}$. Similarly, a co-algebra structure can be traced back to the structure of the other side of the maps, namely the group properties of G: By setting $\varepsilon(\varphi) := \varphi(e)$ and

$$(\Delta(\varphi))(\gamma \otimes \gamma') := \varphi(\gamma\gamma'), \qquad (\mathsf{s}(\varphi))(\gamma) := \varphi(\gamma^{-1}) \qquad (22.42)$$

for all $\gamma \in G$, one obtains a co-unit, co-product and antipode for $\mathcal{F}G$. Here e and γ^{-1} denote the unit and inverse in the group G, respectively. It is straightforward (see exercise 22.4) to verify that with these maps $\mathcal{F}G$ becomes a Hopf algebra. Also, by the definition of the product it is clearly commutative. Moreover, the co-product is co-commutative if and only if the group G is abelian.

In short, to any compact topological group one can associate a commutative Hopf algebra. It is an important result, known as the *Gelfand–Naimark theorem,* that this correspondence works also the other way round, i.e. every commutative Hopf algebra is isomorphic to the algebra $\mathcal{F}G$ of continuous functions on some compact topological group G. In fact:

> The structure theory of compact topological groups and of commutative Hopf algebras is completely equivalent.

In particular, one can describe the properties of a compact topological group entirely in terms of the associated Hopf algebra of continuous functions, without making explicit reference to the group as a topological space at all.

In view of the relations (22.22), (22.23) and (22.25) for the co-product, co-unit and antipode of a dual Hopf algebra it is apparent that the Hopf algebra dual to $\mathcal{F}G$ should be closely related to the group G, too. Indeed, if G is a connected and simply connected *Lie* group, then there is a (non-degenerate) duality between $\mathcal{F}G$ and the enveloping algebra $U(\mathfrak{g})$ of the Lie algebra $\mathfrak{g}$ of G, i.e. $\mathcal{F}G$ is isomorphic to a subspace of the dual space $U(\mathfrak{g})^*$. (The relevant subspace is contained in the so-called Hopf dual $U(\mathfrak{g})^\circ$ of $U(\mathfrak{g})$, which is defined as $U(\mathfrak{g})^\circ := \{\varphi \in U(\mathfrak{g})^* \mid \circ^*(\varphi) \in U(\mathfrak{g})^* \otimes U(\mathfrak{g})^*\}$; if $\mathfrak{g}$ is semi-simple, then the subspace is in fact equal to $U(\mathfrak{g})^\circ$. That one must take a subspace of $U(\mathfrak{g})^*$ is a consequence of the fact that for an infinite-dimensional Hopf algebra $\mathfrak{A}$ the dual of the co-product generically does not take values in $\mathfrak{A}^* \otimes \mathfrak{A}^*$, but in the larger space $(\mathfrak{A} \otimes \mathfrak{A})^*$.) This observation explains in particular to a large extent why for the purposes of this book we could focus our attention on algebras and did not have to study Lie *groups* in much detail.

When extending the functions in $\mathcal{F}G$ by linearity to the group algebra of a group G, i.e. setting $f(x) := \sum_{\gamma \in G} x_\gamma f(\gamma)$ for all finite sums $x = \sum_{\gamma \in G} x_\gamma e_\gamma$, one obtains a non-degenerate pairing between elements $f \in \mathcal{F}G$ and $x \in \mathbb{C}G$, and in fact one can show that $\mathcal{F}G$ and $\mathbb{C}G$ are dual as Hopf algebras. Furthermore, also the group Hopf algebras themselves can be considered as function algebras, namely by regarding the coefficients $x(\gamma) \equiv x_\gamma$ of $x \in \mathbb{C}G$ in the basis consisting of the group elements as functions on G. For discrete groups G the product in $\mathbb{C}G$ then reads

$$(x \diamond y)(\gamma') = \sum_{\gamma \in G} x(\gamma'\gamma^{-1}) y(\gamma). \tag{22.43}$$

For compact Lie groups, the summation gets replaced by integration with respect to the Haar measure (compare section 21.8), which yields the *convolution product*

$$(f_1 \bullet f_2)(\gamma') := \int_G d\mu_\gamma \, f_1(\gamma'\gamma^{-1}) \, f_2(\gamma) \tag{22.44}$$

of continuous functions on G. This continues to hold for groups which are non-compact, but still locally compact (in particular for finite-dimensional non-compact Lie groups), provided that one restricts to (a suitable completion of) the continuous functions with compact support.

22.9 Quantum groups

The deep insight expressed by the Gelfand–Naimark theorem is that topological groups can be described via commutative algebras. Turning things around, one may wonder what kind of 'geometrical' interpretation exists for non-commutative Hopf algebras, i.e. try to regard them as a kind of function spaces on suitable geometrical objects. In analogy with the commutative case, these objects are called *pseudogroups* or *quantum groups*. While it is not possible to describe them in a purely geometric language (e.g. interpret their elements as points), the algebraic properties of their 'function spaces' can be employed to analyze the analogues of various geometric and analytic concepts, such as vector bundles, differential calculus and integration, for these structures.

This is one of the basic ideas of what is nowadays known as non-commu- Information
tative geometry (for an introduction see e.g. [Coquereaux 1993]). The origin of this area of mathematics can be traced back to the foundations of
quantum mechanics: In classical mechanics, the observable quantities are
functions on phase space and hence form a commutative algebra. Upon
quantization, the observables become operators on a Hilbert space. So
they still form an algebra, but this is no longer commutative. A proper
treatment of these matters in the framework of relativistic field theory involves C^*-algebras (algebras with involution, i.e. 'hermitian conjugation',
and a compatible norm) and von Neumann algebras of bounded operators on Hilbert spaces, see e.g. chapter III.2 of [Haag 1992]. Accordingly
it should not come as a surprise that to a large extent non-commutative
geometry is formulated in the language of C^*-algebras.

In this context the analogue of a Lie group G with Lie algebra $\mathfrak{g}$ is the
quantum group or *deformed group* G_q that is obtained by identifying the
non-commutative Hopf algebra $\mathcal{F}G_q$ of 'functions on G_q' with the dual
$U_q(\mathfrak{g})^*$ of the deformed enveloping algebra $U_q(\mathfrak{g})$. The connections among
the various objects that are related to a Lie group G are summarized by
the following scheme:

group	functions	commutative Hopf algebra	dual space	co-commutative Hopf algebra
G	$\rightleftarrows$	$\mathcal{F}G \simeq U(\mathfrak{g})^*$	$\rightleftarrows$	$U(\mathfrak{g})$
				q-defor-mation $\downarrow\uparrow$ $q \to 1$
G_q	$\rightleftarrows$ "functions"	$\mathcal{F}G_q \simeq U_q(\mathfrak{g})^*$	$\rightleftarrows$ dual space	$U_q(\mathfrak{g})$
quantum group		non-commutative Hopf algebra		non-co-commuta-tive Hopf algebra

By abuse of terminology, the term 'quantum group' is not only used
for the 'geometrical' object G_q, but also for the algebra $U_q(\mathfrak{g})$, and more
generally for arbitrary quasi-triangular Hopf algebras. A related, and
more appropriate, term used to refer to $U_q(\mathfrak{g})$ is 'quantum algebra'.

22.10 Quasi-Hopf algebras

As remarked above, often the operation of forming tensor products of representations of an algebra $\mathfrak{A}$ is not commutative, but still has the equally
natural property of being commutative up to isomorphism. This led to
the notion of quasi-co-commutative (quasi-triangular) co-products. In the
same vein, tensor products of $\mathfrak{A}$-representations should in general not be
expected to be strictly associative, but (at best) only associative up to

isomorphism (compare e.g. equation (16.25)). Pulled back to the algebra $\mathfrak{A}$ itself, associativity of tensor products up to isomorphism corresponds to relaxing the co-associativity (22.14) of the co-product Δ to *quasi-co-asso-ciativity*, which means that the two sides of (22.14) are conjugate rather than identical, i.e. that there exists an invertible element $\mathcal{Q} \in \mathfrak{A} \otimes \mathfrak{A} \otimes \mathfrak{A}$

$$(\Delta \times id) \circ \Delta (x) = \mathcal{Q} \circ [(id \times \Delta) \circ \Delta (x)] \circ \mathcal{Q}^{-1} \tag{22.45}$$

for all $x \in \mathfrak{A}$. Provided that $\mathcal{Q}$ also satisfies certain compatibility require-ments, it is called a *co-associator* for $\mathfrak{A}$; the algebra $\mathfrak{A}$ is then called a *quasi-Hopf algebra*.

Among the latter compatibility conditions there is in particular the requirement that all ambiguities in multiple tensor products are already taken care of by putting brackets which indicate the order in which the multiplication is carried out. In particular, the two different possibilities – schematically displayed in figure (22.46) – of rearranging brackets in such a way that the four-fold tensor product $R_{U \otimes (V \otimes (W \otimes X))}$ gets transformed to $R_{((U \otimes V) \otimes W) \otimes X}$ should yield the same result.

$$\begin{array}{ccc}
& \boxed{U \otimes ((V \otimes W) \otimes X)} \longrightarrow \boxed{(U \otimes (V \otimes W)) \otimes X} \\
\nearrow & & \downarrow \\
\boxed{U \otimes (V \otimes (W \otimes X))} & & \\
\searrow & & \\
& \boxed{(U \otimes V) \otimes (W \otimes X)} \longrightarrow \boxed{((U \otimes V) \otimes W) \otimes X}
\end{array} \tag{22.46}$$

This property of tensor product representations translates to the relation

$$\begin{aligned}
(\mathcal{Q} \otimes e) &\circ [(id \times \Delta \times id)(\mathcal{Q})] \circ (e \otimes \mathcal{Q}) \\
&= [(\Delta \times id \times id)(\mathcal{Q})] \circ [(id \times id \times \Delta)(\mathcal{Q})]
\end{aligned} \tag{22.47}$$

for the co-associator (compare exercise 22.7); as the five factors in this formula correspond to the five edges in the pentagon-like diagram (22.46), this is called the *pentagon identity* for $\mathcal{Q}$. We have already encountered this relation in a different guise in section 16.7. There we were considering tensor products of representations of an ordinary enveloping algebra $\mathsf{U}(\mathfrak{g})$; the intertwiners F appearing in that context can be regarded as the repre-sentatives of $\mathcal{Q}$ in the relevant tensor product representations. Note that the co-product of $\mathsf{U}(\mathfrak{g})$ is co-associative, or in other words, $\mathsf{U}(\mathfrak{g})$ has trivial co-associator $\mathcal{Q} = e \otimes e \otimes e$. Nevertheless we had to deal with tensor prod-ucts which are associative only up to isomorphism, because we insisted on using a specific basis of the tensor product module $U \otimes V$ (namely the one corresponding to the decomposition of $U \otimes V$ into irreducible modules) which differs from the natural basis of the tensor product module, thereby providing an isomorphic, but not identical, realization of the tensor prod-uct representation. Indeed, it is possible to interpret the $6j$-symbols of $\mathfrak{g}$ as components of a non-trivial co-associator, such that when written out in components, the equation (22.47) reduces to the pentagon identity of section 16.7 for the $6j$-symbols (in particular for $\mathfrak{g} = \mathfrak{sl}(2)$ to the Biedenharn sum rule (16.38)).

Similarly, when the Hopf algebra is quasi-triangular, there is a compatibility condition involving both the co-associator $\mathcal{Q}$ and the universal R-matrix $\mathcal{R}$, which says that the two different rearrangements furnishing the transformation

$$
\begin{array}{ccccc}
\boxed{U \otimes (V \otimes W)} & \longrightarrow & \boxed{U \otimes (W \otimes V)} & \longrightarrow & \boxed{(U \otimes W) \otimes V} \\
\downarrow & & & & \downarrow \\
\boxed{(U \otimes V) \otimes W} & \longrightarrow & \boxed{W \otimes (U \otimes V)} & \longrightarrow & \boxed{(W \otimes U) \otimes V}
\end{array}
\tag{22.48}
$$

from $R_{U \otimes (V \otimes W)}$ to $R_{(W \otimes U) \otimes V}$ yield the same result, and analogously for the transformation from $R_{(U \otimes V) \otimes W}$ to $R_{V \otimes (W \otimes U)}$. These give rise to two (independent) *hexagon equations* for $\mathcal{Q}$ and $\mathcal{R}$, which in terms of suitable components of the intertwiners F and R again reproduce the hexagon equations that were already encountered in chapter 16 (i.e. for $\mathfrak{g} = \mathfrak{sl}(2)$ the Racah–Elliott relation (16.39)).

Let us also note that co-associativity together with the defining property (22.17) of the co-unit already imply the unitality $\Delta(e) = e \otimes e$ (22.10) for the co-product of a Hopf algebra. In contrast, for quasi-Hopf algebras the element $\mathcal{P} = \Delta(e)$ of $\mathfrak{A} \otimes \mathfrak{A}$ can be an arbitrary non-trivial projector, i.e. by (22.9) it is only required that $\mathcal{P}^2 = \mathcal{P}$. (Because of (22.17), $\Delta(e)$ cannot be zero in any bi-algebra.) In the description above, it was actually assumed that $\mathcal{P} = e \otimes e$. For $\mathcal{P} \neq e \otimes e$ the results continue to hold, but the universal R-matrix and co-associator $\mathcal{Q}$ can no longer required to be invertible; rather, the quantities $\mathcal{R}^{-1}$ and $\mathcal{Q}^{-1}$ must be interpreted as *quasi-inverses* of $\mathcal{R}$ and $\mathcal{Q}$, in the sense that $\mathcal{R}^{-1} \circ \mathcal{R} = \mathcal{P}$ and $\mathcal{R} \circ \mathcal{R}^{-1} = \pi_{12}(\mathcal{P})$, respectively $\mathcal{Q}^{-1} \circ \mathcal{Q} = (id \times \Delta)(\mathcal{P})$ and $\mathcal{Q} \circ \mathcal{Q}^{-1} = (\Delta \times id)(\mathcal{P})$.

22.11　Character rings

So far we have studied in this chapter structural features of universal enveloping algebras that are consequences of the tensor product of $\mathfrak{g}$-representations, as well as various generalizations thereof. We now slightly shift our focus and examine with which structure the set of irreducible representations, say of a simple Lie algebra, is endowed by the tensor product of representations.

To this end let us first have a look at finite abelian groups. The set $\widehat{G}$ of group characters ch of a finite abelian group G constitutes itself again an abelian group, with the product given by pointwise multiplication (see exercise 22.6); it has the same order (i.e. number of elements) $|G|$ as G. $\widehat{G}$ is called the *character group* of G. The characters, which are functions *ch*: G → ℂ, can be extended linearly to functions from the group algebra ℂG to ℂ. When doing so, Ĝ can be regarded as the set of algebra homomorphisms from ℂG to ℂ, and therefore the group algebra ℂĜ of the character group is nothing but the Hopf algebra dual to ℂG, and hence isomorphic to the function algebra *F*G. The isomorphism, i.e. the relation between characters *ch* ∈ ℂĜ and functions *f* ∈ *F*G, is just Fourier transformation, i.e. *f* ∈ *F*G is mapped to the element *f̃* ∈ ℂĜ, where

$$\tilde{f} = \sum_{ch\in\widehat{G}} \tilde{f}_{ch}\, ch \qquad \text{with} \qquad \tilde{f}_{ch} := \tfrac{1}{|G|}\sum_{\gamma\in G} ch(\gamma^{-1})\, f(\gamma)\,. \qquad (22.49)$$

In the special case of the cyclic group G = $\mathbb{Z}_N$ of order N, the group characters ch_i are given by

$$ch_j(m) = \exp\left(2\pi i\,\tfrac{im}{N}\right) \qquad (22.50)$$

for $j, m = 1, 2, \ldots, N$, so that (22.49) is just the Fourier decomposition of periodic functions in one variable.

For finite-dimensional compact Lie groups, an analogue of the Fourier decomposition (22.49) is provided by the formula (21.65). However, the characters of Lie group representations do not possess an inverse with respect to pointwise multiplication, and hence no longer form a group. Instead, according to the results of chapter 21 they span (modulo completion) the algebra of class functions on G, and in particular in the case of semisimple G one can consider the characters of finite-dimensional irreducible representations as a naturally distinguished basis of this space.

As far as the multiplication rules are concerned, we can work interchangeably with any of the following types of objects:

• the Lie group characters *ch*, i.e. functions G → ℂ;

• the Lie algebra characters $\mathcal{X}$, i.e. functions $\mathfrak{g}_\circ$ → ℂ;

• finite-dimensional representations of G respectively $\mathfrak{g}$, or more precisely, isomorphism classes of such representations (recall that isomorphic representations have the same character).

(This correspondence between characters and representations was e.g. the basis of the Racah–Speiser prescription for decomposing tensor products, compare section 15.7.) It is therefore natural to abstract from the particular meaning of the generators of the algebra as isomorphism classes of representations or functions on G respectively $\mathfrak{g}$ and consider the abstract algebra $\mathcal{A}(\mathfrak{g})$ which possesses the same multiplication rules. We will denote the generator of this abstract algebra that corresponds to an irreducible character $\mathcal{X}_\Lambda$ by φ_Λ. The character sum rule $\mathcal{X}_{\Lambda\times\Lambda'} = \mathcal{X}_\Lambda \mathcal{X}_{\Lambda'} = \sum_i \mathcal{L}_{\Lambda\Lambda'}{}^{\Lambda_i} \mathcal{X}_{\Lambda_i}$ (15.20) tells us that in the basis $\{\varphi_\Lambda\}$ of $\mathcal{A}(\mathfrak{g})$ the structure constants in this naturally distinguished basis of $\mathcal{A}(\mathfrak{g})$ are the tensor product multiplicities $\mathcal{L}_{\Lambda\Lambda'}{}^{\Lambda''}$ of $\mathfrak{g}$. Since these structure constants are all integers, it is often convenient to consider the ring spanned by the φ_Λ over the ring $\mathbb{Z}$ of integers instead of the algebra spanned over the field $\mathbb{C}$. (The fact that the structure constants are also non-negative cannot be directly implemented in the description of $\mathcal{A}(\mathfrak{g})$; note that on the set $\mathbb{Z}_{\geq 0}$ there is no inverse with respect to addition, so that $\mathbb{Z}_{\geq 0}$ is not a ring.) This ring, which for simplicity we will again denote by $\mathcal{A}(\mathfrak{g})$, is called the *character ring* or *representation ring* (sometimes also the Grothendieck ring) of the Lie algebra $\mathfrak{g}$, respectively the Lie group G.

The characteristic features of the ring $\mathcal{A}(\mathfrak{g})$ of a finite-dimensional semi-simple Lie algebra $\mathfrak{g}$ can be summarized as follows.

• $\mathcal{A}(\mathfrak{g})$ is a ring over $\mathbb{Z}$ with basis $\{\varphi_\Lambda\}$, where Λ runs over all dominant integral weights of $\mathfrak{g}$.

• The structure constants with respect to the distinguished basis $\{\varphi_\Lambda\}$ are *non-negative* integers.

• The product '$*$' of $\mathcal{A}(\mathfrak{g})$ is commutative and associative, and there is a unit $\varphi_\circ \equiv \varphi_{\Lambda=0}$ satisfying $\varphi_\circ * \varphi_\Lambda = \varphi_\Lambda$ for all Λ.

• There is a conjugation ' $+$', namely $(\varphi_\Lambda)^+ := \varphi_{\Lambda^+}$, where Λ^+ is the $\mathfrak{g}$-weight conjugate to Λ. This is an automorphism of order two, i.e. satisfies $(\varphi_\Lambda)^+ * (\varphi_{\Lambda'})^+ = (\varphi_\Lambda * \varphi_{\Lambda'})^+$ and $((\varphi_\Lambda)^+)^+ = \varphi_\Lambda$.

• The product $\varphi_{\Lambda^+} * \varphi_{\Lambda'}$ contains the unit, with multiplicity one, if and only if $\Lambda = \Lambda'$. In other words, the conjugation can be described as 'evaluation at the unit', $(\varphi_{\Lambda^+} * \varphi_{\Lambda'})|_{\varphi_\circ} = \delta_{\Lambda,\Lambda'}$.

In particular, conjugation preserves the natural basis.

Consider an infinite-dimensional ring which satisfies all these properties, and which in addition has a basis with the property that to any element of the basis one can associate a finite-dimensional vector space in such a way that the tensor product of these vector spaces reproduces the ring structure. To any such ring one can construct a topological group G (and, because of the Gelfand–Naimark theorem, also an associated Hopf algebra) such that the representation ring of G is isomorphic to the ring one started with. This is known as Tannaka–Krein reconstruction, or in the case of locally compact abelian groups as Pontryagin duality. The precise form of this duality is best formulated in the language of categories, compare e.g. section 5.1 of [Chari and Pressley 1994].

Information

To perform this construction of the group in practice, one must know the intertwiners between isomorphic vector spaces which arise in the decomposition of tensor products rather explicitly. In another version of the reconstruction theorem, due to Deligne, the requirement concerning the finite-dimensional vector spaces is replaced by the axiom that a certain number, the so-called rank of an object of the relevant category, is always integral.

In applications in quantum field theory (see e.g. chapters IV.2 and IV.4 of [Haag 1992]) this rank appears as the statistical dimension of a super-selection sector. In low-dimensional space-times the statistical dimensions are not necessarily integral (this leads to the concept of rational fusion rings, see the following section), so that the reconstruction theorem for *groups* no longer applies, and one must look for more general symmetry structures like e.g. quantum groups or quasi-Hopf algebras.

It can be shown that the representation rings of compact Lie groups are all finitely generated over $\mathbb{Z}$. We will make this somewhat more precise for some simple matrix Lie groups. In the case of SU(n), SO(n) with n odd, and of SP(n) the representation ring is isomorphic to the ring $\mathbb{Z}[x_1, x_2, \ldots, x_r]$ of polynomials in r = rank $\mathfrak{g}$ variables x_i with integral coefficients. In all these cases the isomorphism is given by mapping the fully antisymmetrized part of the ith tensor power of the defining representation $R_{\Lambda_{(1)}}$ to the variable x_i. (Note, however, that this isomorphism does not cover spinor representations of SO(n), and that SP(n) is not compact.) The representation ring of SO(n) with n even can be described as a module over a polynomial ring that is freely generated by two elements.

22.12 Rational fusion rings

Given a character ring $\mathcal{A}(\mathfrak{g})$, one may ask the question whether it can be consistently 'truncated' in the sense that there exist subrings which are closed under the product $*$ as well as under conjugation. Any such subring must contain the unit $\varphi_\circ$, and clearly the one-dimensional ring spanned by $\varphi_\circ$ provides such a truncation. But apart from this trivial example, there is only one further type of solution. To describe this, let us first recall that the tensor product of $\mathfrak{g}$-representations is graded by the conjugacy classes, compare the selection rule (15.17). The conjugacy classes are described by a discrete abelian group C. It follows that for any subgroup $C_\circ$ of C one obtains a subring of $\mathcal{A}(\mathfrak{g})$ by restricting the weights Λ to the conjugacy classes in $C_\circ$.

For instance, for $\mathfrak{g} = \mathfrak{sl}(2)$ one obtains this way the restriction to integer spin representations, for $\mathfrak{g} = \mathfrak{sl}(3)$ the restriction to 'zero triality', and for $\mathfrak{g} = \mathfrak{so}(n)$ the restriction to tensor representations. In terms of the Lie groups which have $\mathfrak{g}$ as their Lie algebra, the full character ring describes the representations of the simply connected covering group, while each truncated ring contains precisely the representations of a suitable non-simply connected subgroup of the covering group.

In particular, any non-trivial subring of $\mathcal{A}(\mathfrak{g})$ is still infinite-dimensional. However, there do exist finite-dimensional rings which share all the properties of $\mathcal{A}(\mathfrak{g})$ that we listed above, i.e. which are commutative associative rings over $\mathbb{Z}$ with a basis in which all structure constants are non-negative, which have a unit, and for which evaluation at the unit constitutes an automorphism of order two. Rather than delving into the structure theory of these finite-dimensional (also called *rational*) *fusion rings*, let us just present the two simplest examples. Both of them possess just one further generator ϕ besides $\varphi_\circ$, and the product of ϕ with itself reads

$$\phi * \phi = \varphi_\circ \qquad \text{respectively} \qquad \phi * \phi = \varphi_\circ + \phi. \qquad (22.51)$$

An easy consideration shows that despite all similarities the structure of such finite-dimensional fusion rings differs drastically from that of the rings $\mathcal{A}(\mathfrak{g})$. Recall that for the ring $\mathcal{A}(\mathfrak{g})$ the dimension sum rule $d_\Lambda d_{\Lambda'} = \sum_i \mathcal{L}_{\Lambda\Lambda'}^{\Lambda_i} d_{\Lambda_i}$ (15.21) holds. When imposing an analogous identity with positive numbers D on the fusion rings just mentioned, one observes the following. In the case of the first product of (22.51) one concludes that $D_\phi = 1$, which coincides with $D_{\varphi_\circ}$, while for the second product of (22.51) the result is even non-integral, namely $D_\phi = \frac{1}{2}(1 + \sqrt{5}) \simeq 1.618$.

In view of these strange properties of the putative 'dimension' D one might be tempted to conclude that such fusion rings are not of much interest. But this is not at all true, and indeed one can relate such rings in a sensible way to the character rings $\mathcal{A}(\mathfrak{g})$. Namely, one has to apply a generalized procedure of truncation in which not only the number of generators is reduced, but also some structure constants which involve the surviving generators are changed. This truncation can be performed in such a way that the resulting fusion ring is a homomorphic image of the character ring $\mathcal{A}(\mathfrak{g})$. In terms of the representation theory of the Lie algebra $\mathfrak{g}$ this manipulation does not possess any natural interpretation. However, there is an explanation in terms of the deformed enveloping algebras $\mathsf{U}_q(\mathfrak{g})$ with $q = \exp(2\pi i/m)$ a root of unity: The truncation can be performed in such a way that among all finite-dimensional representations precisely the unitary ones are left over, which then form a rational fusion ring $\mathcal{A}_q(\mathfrak{g})$. For example, the first of the rings described by (22.51) can be obtained as the ring $\mathcal{A}_q(\mathfrak{g})$ for $\mathfrak{g} = A_1$ and $m = 3$, while the second follows for $\mathfrak{g} = G_2$ and $\mathfrak{g} = F_4$ with $m = 5$ and $m = 10$, respectively. Moreover, there is a simple general formula for the numbers D as a product of q-numbers defined as in (22.36), namely

$$D(\varphi_\Lambda) = d_q(\varphi_\Lambda) := \prod_{\alpha > 0} \frac{[(\Lambda + \rho, \tilde{\alpha})]_q}{[(\rho, \tilde{\alpha})]_q}. \qquad (22.52)$$

Here the summation runs over all positive roots of $\mathfrak{g}$, ρ is the Weyl vector of $\mathfrak{g}$, and $\tilde{\alpha} = 2\alpha/(\theta, \theta)$ with θ the highest root. Thus (22.52) is just a q-deformed version of the Weyl formula (13.32) for the dimension of the $\mathfrak{g}$-module V_Λ; accordingly one refers to $d_q(\varphi_\Lambda)$ as the *q-dimension* or *quantum dimension* of the $\mathsf{U}_q(\mathfrak{g})$-module that corresponds to φ_Λ. A generator φ of unit quantum dimension, i.e. satisfying $D(\varphi) = 1$, is called a *simple current* of the fusion ring. (This terminology stems from conformal field theory, compare the remarks at the end of section 7.10).

In terms of the underlying (co-) algebra structure, the truncation of the representation theory just described translates into the procedure of forming the quotient of $U_q(\mathfrak{g})$ by a certain ideal $\mathcal{J}$. This is accompanied by a modification of the co-product (22.32) of $U_q(\mathfrak{g})$. The modified co-product $\tilde{\Delta}$ is related to the $U_q(\mathfrak{g})$-co-product Δ by $\tilde{\Delta}(x) = \tilde{\mathcal{P}} \circ \Delta(x)$, where $\tilde{\mathcal{P}} \equiv \tilde{\Delta}(e)$ is a non-trivial projector. Thus the quotient $\tilde{U}_q(\mathfrak{g}) = U_q(\mathfrak{g})/\mathcal{J}$ is no longer a Hopf algebra, but a quasi-Hopf algebra.

22.13 The extended Racah–Speiser formalism

The reader may already have recognized that the condition (22.38) for unitarity of $U_q(\mathfrak{g})$-modules is nothing but the condition (13.46) for the integrability of highest weight modules of the untwisted affinization $\mathfrak{g}^{(1)}$ of $\mathfrak{g}$ at level $k^\vee = m - g^\vee$. This suggests a close connection between the affine Lie algebra $\mathfrak{g}^{(1)}$ at level $k^\vee$ and the deformed enveloping algebra $U_q(\mathfrak{g})$, respectively its truncation $\tilde{U}_q(\mathfrak{g})$, with deformation parameter

$$q = \exp(2\pi i/(k^\vee + g^\vee)). \tag{22.53}$$

Indeed one can argue that the structure constants $\mathcal{N}_{\Lambda\Lambda'}^{\Lambda''}$ of the fusion ring of $\tilde{U}_q(\mathfrak{g})$ can be expressed through a formula that has a rather natural description in terms of the affine algebra $\mathfrak{g}^{(1)}$, namely as

$$\mathcal{N}_{\Lambda\Lambda'}^{\Lambda_i} = \sum_{\hat{w}\in\hat{W}} \mathrm{sign}(\hat{w})\, \mathrm{mult}_{\Lambda'}(\hat{w}(\Lambda_i + \rho) - \rho - \Lambda). \tag{22.54}$$

This result, known as the *Kac–Walton formula*, is an extension of the Racah–Speiser formula (15.23) for the tensor product multiplicities $\mathcal{L}_{\Lambda\Lambda'}^{\Lambda_i}$ of $\mathfrak{g}$. While in equation (15.23) the summation is over the Weyl group W of $\mathfrak{g}$, now one has to sum over the projection $\hat{W}$ of the Weyl group of $\mathfrak{g}^{(1)}$ to the weight space of $\mathfrak{g}$. The infinite group $\hat{W}$ is generated by the fundamental reflections $w_{(j)}$ of $\mathfrak{g}$ together with reflection about the hyperplane $(\Lambda, \theta^\vee) = k^\vee + 1 \equiv m - g^\vee + 1$ which implements the additional condition (22.38).

The formula (22.54) leads to a simple recipe for the calculation of the structure constants $\mathcal{N}_{\Lambda\Lambda'}^{\Lambda_i}$, namely an extension of the Racah–Speiser algorithm which consists of allowing also reflections with respect to the suitably shifted hyperplane that is perpendicular to the highest $\mathfrak{g}$-root θ. In short, this extended Racah–Speiser algorithm amounts to replacing the Weyl group of $\mathfrak{g}$ by the relevant affine Weyl group.

In the pictures (22.55) we apply this algorithm to the fusion product of two adjoint representations of $\mathfrak{g} = A_2$, at levels $k^\vee = 2$ and $k^\vee = 3$ of the affinization $A_2^{(1)}$. This amounts to supplementing the ordinary Racah–Speiser manipulations, which were already presented in the figure (15.33), by reflections about the line defined by $\Lambda^1 + \Lambda^2 = k^\vee + g^\vee \equiv k^\vee + 3$.

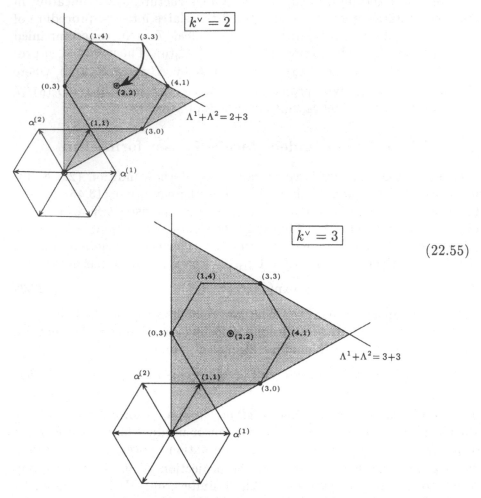

$$(22.55)$$

Note that the position of this hyperplane depends on the value of the level $k^\vee$; correspondingly, also the fusion rules depend on the level. From the figures (22.55) we read off that the fusion product reads $\varphi_{(1,1)} * \varphi_{(1,1)} = \varphi_{(0,0)} + \varphi_{(1,1)}$ at level 2, whereas $\varphi_{(1,1)} * \varphi_{(1,1)} = \varphi_{(0,0)} + 2\varphi_{(1,1)} + \varphi_{(3,0)} + 2\varphi_{(0,3)}$ at level 3. From the figures it is also evident that for higher level the fusion rules are equal to the Littlewood–Richardson coefficients. This is indeed a generic feature: For sufficiently high level the fusion rules become identical to the Littlewood–Richardson coefficients; however, the level for which this happens depends on the highest weights under consideration. The prescription for determining the structure constants of the fusion ring of $\tilde{U}_q(\mathfrak{g})$ from the tensor products of representations of the horizontal subalgebra $\mathfrak{g}$ of $\mathfrak{g}^{(1)}$ also implies that for any level $k^\vee$ there is a homomorphism from the character ring of $\mathfrak{g}$ to the fusion ring.

It must be emphasized, however, that the fusion ring of $\tilde{U}_q(\mathfrak{g})$ does *not* describe the tensor product of integrable $\mathfrak{g}^{(1)}$-representations. This follows e.g. from the observation that with respect to such tensor products the levels of the $\mathfrak{g}^{(1)}$-representations add up (additivity of quantum numbers), while all elements of the fusion ring of $\tilde{U}_q(\mathfrak{g})$ correspond to modules which all have one and the same level $k^\vee$.

Nevertheless the fusion ring $\mathcal{A}_q(\mathfrak{g})$ does arise in a situation characterized by the affine algebra $\mathfrak{g}^{(1)}$, namely in the two-dimensional conformal field theory based on $\mathfrak{g}^{(1)}$ which is known as a *WZW theory*. In this setting the fusion product '$*$' arises from an abstraction of the operator product algebra furnished by the quantum fields of this theory. Important information on the fusion ring is contained in a symmetric unitary matrix S which at the same time implements the modular transformation $\tau \mapsto -1/\tau$ (compare section 14.11) on the specialized characters $\tilde{\chi}_\Lambda(\tau)$ of integrable $\mathfrak{g}$-modules V_Λ and 'diagonalizes' the fusion product, which leads to the *Verlinde formula*

$$\mathcal{N}_{\Lambda\Lambda'}{}^{\Lambda''} = \sum_{\mu \text{ integrable}} S_{\mu,\Lambda} S_{\mu,\Lambda'} \overline{S}_{\mu,\Lambda''} / S_{\mu,k^\vee\Lambda_{(0)}} \tag{22.56}$$

for the structure constants of $\mathcal{A}_q(\mathfrak{g})$. Moreover, the quantum dimensions are given by $D_\Lambda = S_{\Lambda,k^\vee\Lambda_{(0)}} / S_{k^\vee\Lambda_{(0)},k^\vee\Lambda_{(0)}}$.

It is also instructive to compare (22.52) to the equation (18.33) for the 'asymptotic dimension' $\tilde{D}_\Lambda$ of an irreducible highest weight module with highest weight Λ (which in view of the Kac–Peterson formula (14.57) can be expressed in terms of the entries of the modular matrix S as $\tilde{D}_\Lambda = S_{\Lambda,k^\vee\Lambda_{(0)}}$). Namely, the quantum dimension is given by $d_q(\varphi_\Lambda) = D_\Lambda = \tilde{D}_\Lambda / \tilde{D}_{k^\vee\Lambda_{(0)}}$, with the deformation parameter q in (22.52) and the level $k^\vee$ in (18.33) related as in equation (22.53).

Fusion rings also arise in various other situations. They summarize the couplings between quantum fields in any arbitrary two-dimensional conformal field theory; they express the composition of superselection sectors in the C*-algebraic approach to relativistic quantum field theory; they describe the multiplication of (equivalence classes of) polynomials in certain quotients of polynomial rings; and they govern the statistical interactions of quasi-particles carrying both electric charge and magnetic flux in certain condensed matter systems. For an introduction to some of these issues, we refer to [Fuchs 1994] and to section 11.3 of [Chari and Pressley 1994].

Summary:

The question of whether for a given algebra $\mathfrak{A}$ it is possible to define the tensor product of representations leads to the notion of a co-product and a co-algebra. A bi-algebra is an algebra with a compatible co-algebra structure; a Hopf algebra is a bi-algebra with an antipode. Examples of Hopf algebras are (deformed) universal enveloping algebras, group Hopf algebras, the space of functions on a Lie group, and quantum groups.

The tensor product of representations also gives rise to the structure of fusion rings; for finite-dimensional simple Lie groups these are just the character rings. Finite-dimensional fusion rings are called rational. In the case of untwisted affine Lie algebras their structure constants can be computed with the extended Racah–Speiser algorithm.

Keywords:

Co-product, co-commutativity, co-associativity, co-unit, bi-algebra, Hopf algebra, antipode;
quasi-triangular Hopf algebra, universal R-matrix, deformed enveloping algebra, deformation parameter, Gelfand–Naimark theorem;
quantum group, quasi-co-associativity, quasi-Hopf algebra, pentagon and hexagon identities;
representation ring, character ring, (rational) fusion ring, quantum dimension, extended Racah–Speiser algorithm, Verlinde formula.

Exercises:

Use the unitality property (22.10) and the connecting axiom (22.7) to extend the definition (22.27) of the co-product to all of $U(\mathfrak{g})$. Show that this determines the co-product uniquely.

Exercise 22.1

Verify that the antipode of a Hopf algebra is an algebra anti-homomorphism.

Exercise 22.2

Show that the dual of a commutative Hopf algebra is co-commutative.

Exercise 22.3

For G a finite group, check that the algebra $\mathbb{C}G$ is a co-commutative Hopf algebra, and that $\mathcal{F}G$ is a commutative Hopf algebra.

Exercise 22.4

Write the co-product, unit, co-unit and antipode of the group algebra $\mathbb{C}G$ of a discrete group in a form that is appropriate to the interpretation of $\mathbb{C}G$ as a function algebra, analogously as is done for the product in the formula (22.43).

Exercise 22.5

Verify the group properties of the character group $\widehat{G}$ of a discrete abelian group G.

Exercise 22.6

To which arrows in the diagram (22.46) do the five individual factors in the pentagon identity (22.47) correspond?
Identify the maps which correspond to the six arrows in the diagram (22.48), and thereby write down the explicit form of the hexagon identity.
Compare the hexagon and pentagon identities with the formulæ (16.38) and (16.39) obtained in section 16.7 for the case of $\mathfrak{g} = \mathfrak{sl}(2)$.

Exercise 22.7

Epilogue

This book intends to provide some basic knowledge of Lie algebras and their representation theory, and to give the reader a flavor of the applications of these mathematical structures, in particular to the symmetries of physical systems. This final chapter is rather different in nature from all the others. Its aim is not to introduce any new material, but rather to present a list of various topics which are based on issues that were already treated in the book, but sometimes require also more advanced knowledge. To prepare the reader to address also such advanced questions, detailed suggestions for further reading are made.

1 Literature for further study

We have kept the treatment of symmetries and Lie algebras in this book rather self-contained. In fact, we are confident that to many readers the information supplied in our book will be sufficient for almost all practical needs. Nevertheless from time to time more detailed or advanced knowledge may be required. Even though there is a vast literature on the various aspects of Lie algebras, it can sometimes be difficult to find a useful reference. Therefore table XV presents for each chapter a few books which contain further information about the subjects that we addressed. As already stressed in the introduction, our choice of references neither gives credit to the original contributions, nor can it do justice to all of the relevant textbooks that are available. Rather, we present some sample literature which can serve as the source of further references. Most of this literature is, we hope, easily accessible, both bibliographically and intellectually.

There are also various aspects of Lie algebra theory which we touched only briefly or which we even could not treat at all. For some of these, we supply in the remaining sections a more detailed guide to the literature.

412

Table XV. *Literature*

chapter	suggested further reading
1	Arnold 1978, Barut–Rączka (chap. 13), Biedenharn–Louck 1981a, Marsden–Ratiu, Scheck, Tung (chap. 1,12,13)
2	Fulton–Harris (chap. 10,11), Kac–Raina (section 2.2)
3	Behrends, Fulton–Harris (chap. 15–18), Nachtmann (chap. 17), O'Raifeartaigh 1968
4	Humphreys 1972, Jacobson, Samelson
5	Barut–Rączka (chap. 8,10), Hamermesh (chap. 5)
6	Fulton–Harris (chap. 9,14), Helgason (chap. III), Moody–Pianzola, Samelson
7	Fulton–Harris (chap. 21), Helgason (chap. X), Kac (chap. 4)
8	Fulton–Harris (chap. 26), Gilmore, MacDonald 1978
9	Bourbaki 1989 (chap. III), Bröcker–tom Dieck, Gilmore, Marsden–Ratiu (chap. 9)
10	Bourbaki 1982 (vols. IV–VI), Fulton–Harris (app. D.4), Humphreys 1990, Moody–Pianzola (chap. 5)
11	Helgason (chap. X.4), Jacobson (chap. IX)
12	Fuchs 1992 (chap. 2), Kac, Kass *et al.*, Mickelsson, Pressley–Segal
13	Fuchs 1992 (chap. 1,2), Humphreys 1972, Kac, Kac–Raina
14	Dixmier, Jantzen, Kac, Moody–Pianzola (chap. 6), Weyl (chap. VII)
15	Hamermesh (chap. 5.6), Speiser
16	Biedenharn–Louck 1981b, Fano–Racah, Messiah (vol. 2, app. C)
17	Cvitanović, Hamermesh (chap. 10)
18	Dynkin, Gruber–Samuel
19	Barut–Rączka (§8.8, §10.2), Hamermesh (chap. 7), James–Kerber, MacDonald 1979, Weyl (chap. IV–VI)
20	Benn–Tucker, Naimark, Pressley–Segal (chap. 12), Tung (chap. 10); *for section 10*: Cornwell (Part D), Scheunert, Wess–Bagger
21	Bourbaki 1989 (§III.3.16), Halmos (chap. XI), Miller, Vilenkin–Klimyk
22	Chari–Pressley, Fuchs 1992 (chap. 4,5), Kassel, Lusztig

2 Computer packages

We have already stressed several times that many calculations involving Lie algebras and their representations which would be tedious to perform by hand can very efficiently be done on a computer. Often it is not difficult to implement the relevant formulæ into a computer program of ones own. But there are also several computer packages available which allow one to perform such calculations. As a general remark, we note that packages which provide their own environment, such as LiE or SCHUR, are typically faster than programs which are embedded in an environment (such as MAPLE or MATHEMATICA) with a much broader scope.

We mention a few packages that are known to us. For more extensive and up-dated information on these programs, we refer to the WWW page of this book at `http://norma.nikhef.nl/lie` (see also `http://www.cup.cam.ac.uk`). This page also contains updated information on how the programs are available.

• CHEVIE [Geck *et al.* 1996] is a computer algebra package for symbolic calculations with generic character tables of groups of Lie type, Weyl groups and Hecke algebras. It is based on the computer algebra systems GAP and MAPLE.

• ELIAS (Eindhoven Lie Algebra System) allows for analyzing the structure of finite-dimensional Lie algebras, e.g. the radical, Levi decomposition and derived series, as well as the centralizer and normalizer of any subalgebra. It will soon become available as a part of GAP.

• GAP (Groups, Algorithms, and Programming) [Schönert *et al.* 1994] is a system for computational discrete algebra, with particular emphasis on computational group theory. GAP contains e.g. routines for dealing with Weyl groups and Hecke algebras.

• LiE [van Leeuwen *et al.* 1992] is a software system for calculations in Lie algebras. It focuses on the representation theory of complex reductive Lie groups and algebras. LiE allows e.g. for the computation of the root system, the structure of the Weyl group and its action on weights, tensor product decompositions, branching rules, characters, centralizers of semisimple elements, and Kazhdan–Lusztig polynomials. There is also an unofficial variant called LIEGAP which is callable from GAP.

• LIE is also the name of a REDUCE package of functions for the classification of real finite-dimensional Lie algebras.

• In the Maple Share Library which is distributed with more recent releases of MAPLE, there are the packages COXETER and WEYL. The COXETER package allows to deal with root systems, reflection groups and characters, while the WEYL package contains procedures for manipulating weight vectors and computing weight multiplicities for irreducible representations of semisimple Lie algebras.

• The MATHEMATICA program LIE, available in the MathSource distribution, produces solutions of many linear and also non-linear (systems of) differential equations by making use of an underlying Lie algebra structure. As already mentioned in chapter 16, MATHEMATICA also contains the functions ClebschGordan, ThreeJSymbol and SixJSymbol which provide the Clebsch–Gordan coefficients, $3j$-symbols $6j$-symbols for $\mathfrak{sl}(2)$.

• SCHUR [Wybourne 1995] is an interactive program that can be used to calculate branching rules, Kronecker products, fusion rules of WZW theories, Casimir invariants, dimensions, Young diagrams and their hook lengths, etc. Its algorithms are based on the theory of partitions (which underlies the representation theory of the symmetric group), and in particular on Schur functions.

• SYMMETRICA is devoted to the representation theory and combinatorics of symmetric groups and related classes of groups. It allows to handle characters, matrix representations and various types of polynomials that are related to representations.

We also mention two more specific programs that are available from their authors.

• KAC [Schellekens 1995] is a C program designed to perform calculations with untwisted and twisted affine Lie algebras, including e.g. the Kac–Peterson formula for the modular matrix S and applications to coset conformal field theories.

• WEI [Fuchs 1991] is a FORTRAN 77 program that computes dimensions and weight systems of irreducible representations of finite-dimensional simple Lie algebras, tensor product decompositions of such representations and the corresponding fusion rules of WZW theories, and quantum dimensions.

3 Cohomology

Although in this book we have concentrated on the algebraic techniques that are relevant to symmetries, we still had to omit several important topics. One of these algebraic issues is Lie algebra cohomology. Detailed information on this subject can be found in chapter 6 of [de Azcárraga and Izquierdo 1995] and in chapter 21, §4 of [Barut and Rączka 1986].

To illustrate the basic idea of Lie algebra cohomology, let us sketch the so-called BRST approach (compare section 6.8 of [de Azcárraga and Izquierdo 1995]) for the case of a finite-dimensional Lie algebra $\mathfrak{g}$. We denote by $f^{ab}{}_c$ the structure constants with respect to a basis $\{T^a\}$ of $\mathfrak{g}$, and introduce so-called *anti-ghosts* β^a which transform in the adjoint representation of $\mathfrak{g}$ and so-called *ghosts* γ_a transforming in the co-adjoint representation. By definition, the ghosts and antighosts obey the anti-commutation relations

$$\{\gamma_a, \beta^b\} \equiv \gamma_a\beta^b + \beta^b\gamma_a = \delta_a{}^b, \qquad \{\gamma_a, \gamma_b\} = 0 = \{\beta^a, \beta^b\}.$$

Also, they span a (non-unitary) Fock space $\mathcal{F}$. For any $\mathfrak{g}$-module V the so-called BRST operator

$$Q := \sum_a \gamma_a T^a - \frac{1}{2} \sum_{a,b,c} f^{ab}{}_c \gamma_a\gamma_b\beta^c$$

acts on the tensor product space $V \otimes \mathcal{F}$.

The Jacobi identity of $\mathfrak{g}$ implies that Q is nilpotent, $Q^2 = 0$. As for any nilpotent operator one can consider the subspace of closed states in $V \otimes \mathcal{F}$, i.e. vectors $v \in V \otimes \mathcal{F}$ for which $Qv = 0$, modulo its subspace of exact states, i.e. vectors v' which are of the form $v' = Qw$ for some $w \in V \otimes \mathcal{F}$. This quotient vector space $H^*(\mathfrak{g}, V)$ is called the cohomology of $\mathfrak{g}$ with values in the module V.

One can check that when acting on $V \otimes \mathcal{F}$, the generators $\tilde{T}^a := T^a - f^{ab}{}_c \gamma_b\beta^c$ satisfy precisely the relations of $\mathfrak{g}$. Furthermore, one has $\tilde{T}^a = \{Q, \beta^a\}$, so that the action that is induced on $H^*(\mathfrak{g}, V)$ is trivial. Because of this property, cohomological methods have various applications in modern physics. Namely, whenever a symmetry of a physical system is in fact a gauge symmetry (cf. section 1.8), at some point in the analysis of the system one must 'divide out' the action of the symmetry. When the symmetry is realized in terms of a representation of a Lie algebra on a space V, this dividing out can be achieved precisely by the construction we sketched.

4 Associative algebras

After having studied this book the reader should also get easily acquainted
with algebraic structures other than Lie algebras. As an example we mention the theory of associative algebras, in particular the group algebras of
finite groups and a more detailed study of enveloping algebras. Besides
enveloping algebras of Lie algebras and the group algebras of finite groups
we have encountered in this book also Clifford algebras (compare chapter
20) as other examples of semisimple algebras.

   More information on semisimple algebras can be found in [Lang 1984].
For an introduction to finite groups and associative algebras, see [Curtis
and Reiner 1962], chapters 10 and 16 of [Kirillov 1975], chapters 2–5 of
[Ludwig and Falter 1988], and [Serre 1977]. The investigation of primitive
ideals of enveloping algebras leads to deep relations with geometry, e.g.
the structure of Schubert varieties, see [Borho 1986].

5 Lie groups and loop groups

We did not treat Lie *groups* in great detail. References for Lie groups
which are rather complementary to the present book are chapter III of
[Bourbaki 1989], [Gilmore 1974] and [Helgason 1978]. Loop groups and
their central extensions, which are infinite-dimensional Lie groups whose
Lie algebras are untwisted affine Lie algebras, are described in [Pressley
and Segal 1986].

   A Lie algebra theoretic treatment of a physical system is often particularly elegant when the system 'lives' on a group manifold. One example for
this phenomenon is provided by rigid bodies, compare exercise 9.9. Another class of examples is given by field theories for which the quantum
fields take values in a group manifold, like e.g. WZW theories (chapter
3 of [Fuchs 1992]), and more generally, nonlinear sigma models (compare
[Braaten *et al.* 1985]) on group manifolds.

6 Borel–Weil theory, coadjoint orbits

Lie algebra modules were in this book regarded as abstract vector spaces.
Sometimes it is however helpful to have also a geometrical realization of
these vector spaces at hand. Such a realization can be obtained by either
of two methods: Borel–Weil theory, or the method of coadjoint orbits. For
a review on Borel–Weil theory see [Bott 1988], and for analogous results
in the context of loop groups also chapters 2.8 and 11 of [Pressley and
Segal 1986]. Details about the method of coadjoint orbits can be found
in chapter 15 of [Kirillov 1975] and in section 2 of [Witten 1988].

   In Borel–Weil theory one considers the homogeneous space (compare section 11.10)
G/T, where T is a maximal torus of a finite-dimensional compact simple Lie group G.

This quotient is a smooth complex manifold; as a manifold it is isomorphic to $G_{\mathbb{C}}/B$, where $G_{\mathbb{C}}$ is the complexification of G and B a Borel subgroup, i.e. the exponentiation of a Borel subalgebra. Any $\mathfrak{g}$-weight λ gives rise to a complex line bundle over G/T. Namely, via exponentiation λ provides a map $e^{\lambda}\colon \mathfrak{g}_{\circ} \to S^1$, which after complexification can be extended to a map $e^{\lambda}\colon B \to \mathbb{C}^{\times}$. The total space of the line bundle is obtained from $G_{\mathbb{C}} \times \mathbb{C}$ by identifying $(\gamma\beta, \xi)$ and $(\gamma, e^{\lambda}(\beta)\xi)$ for all $\beta \in B$. The main result of Borel–Weil theory is that, if λ is a dominant integral weight, the space of holomorphic sections in this line bundle carries the irreducible representation with highest weight λ. The method of coadjoint orbits is based on the observation that any Lie group G has a natural coadjoint action on the dual space $\mathfrak{g}^*$ of its Lie algebra $\mathfrak{g}$. As it turns out, any orbit W of this action is endowed with a natural symplectic form Ω. One focuses on those orbits W over which one can construct a complex line bundle $\mathcal{L}_b$ which has the curvature form $i\Omega$. Using the technique of geometric quantization on W one attempts to relate an appropriate space of sections of $\mathcal{L}_b$ to an irreducible unitary representation of G. The coadjoint orbit approach is particularly useful for non-compact groups, where the Borel–Weil theory does not apply.

Related geometrical considerations have been used to apply tools from algebraic geometry to prove character formulæ for certain non-symmetrizable Kac–Moody algebras [Kumar 1987]. The coadjoint action of G on $\mathfrak{g}^*$ has also found applications in mechanics (see e.g. chapter 14 of [Marsden and Ratiu 1994]) and in optics (see chapter 19 of [Guillemin and Sternberg 1984]).

7 Fock spaces

Another tool for realizing $\mathfrak{g}$-modules explicitly are Fock spaces, which we already mentioned several times. They play a particularly important rôle in the canonical quantization of field theories. For detailed information we refer to [Ottesen 1995] and to section 2.2 of [Kac and Raina 1987].

Frequently the relevant Lie algebra is implemented on a Fock space in a non-unitary way, and to arrive at a unitary module of the algebra one has to identify a suitable quotient, similar as in the case of Verma modules. In the special case of modules of the Virasoro algebra with central charge $c <$ 1, this is known as the Coulomb gas approach or Feigin–Fuks construction, and in the case of affine Lie algebras with integral level as the Wakimoto construction. For some details, and applications in conformal field theory, see [Bouwknegt *et al.* 1993].

To any complex vector space V there are associated two types of Fock spaces, the symmetric or bosonic Fock space $\mathcal{F}_+$ and the antisymmetric or fermionic Fock space $\mathcal{F}_-$. These Fock spaces are constructed as the direct sums

$$\mathcal{F}_{\pm} := \bigoplus_{l \geq 0} V_{\pm}^{\otimes l},$$

where $V_{\pm}^{\otimes 0} \cong \mathbb{C}$, while $V_+^{\otimes l}$ and $V_-^{\otimes l}$ denote the totally symmetrized respectively antisymmetrized lth tensor power of V (compare section 19.1). As in section 2.5, on any Fock space one can define creation and annihilation operators, which generate the Fock space by acting on a vacuum vector (i.e. a non-zero vector in the subspace $V_{\pm}^{\otimes 0}$), and which obey canonical (anti-) commutation relations. Correspondingly the Fock space is the state space for a quantum field theory of free bosons and free fermions, respectively.

8 Catastrophes and A–D–E classifications

Another context in which, somewhat surprisingly, Lie algebraic structures occur is in the description of complex catastrophes. The symmetry
groups of such catastrophes are discrete groups generated by reflections.
In particular, the symmetries of the so-called simple singularities are the
Weyl groups of the simply laced simple Lie algebras. In particular, also
for these catastrophes there is an A–D–E classification. An introduction
to these matters can be found in [Arnold 1986] and, on a more technical
level, also in [Arnold *et al.* 1985].

A–D–E classifications also arise in a variety of other areas of mathematics and mathematical physics, such as the classification of the following
objects: Regular polyhedra in three-dimensional space [Slodowy 1983],
finite subgroups of SO(3), hypergeometric differential equations with finite monodromy group, singular fibers in elliptic pencils, caustics, wave
fronts; and finally, modular invariant partition functions of the affine Lie
algebra $A_1^{(1)}$ and of the $c = 1$ Virasoro algebra [Schellekens 1990]. These
A–D–E classifications typically emerge via a connection to additive functions (compare the end of section 7.9).

9 Conformal field theory and string theory

An area where the structures presented in this book are indispensable
tools is conformal field theory and its applications, both in string theory
and statistical mechanics. For an introduction to conformal field theory
based on affine Lie algebras, we refer to chapters 3 and 5 of [Fuchs 1992].
For applications in string theory, see chapters 2 and 4 of [Green *et al.*
1987] and chapter 11 of [Lüst and Theisen 1989].

The formalization of the notion of a chiral algebra in conformal field
theory gives rise to the mathematical structure of a vertex operator algebra, while the search for spectrum generating symmetries in string theory
leads to hyperbolic Kac–Moody algebras and Borcherds algebras; for further information, see e.g. [Frenkel *et al.* 1988], [Gebert 1993] and [Gebert
and Nicolai 1995]. More generally, from the space of physical states of
any vertex operator algebra one can recover the structure of a generalized
Kac–Moody algebra. Conversely, from any untwisted affine Lie algebra
one can construct a vertex operator algebra [Frenkel and Zhu 1992].

Among the various applications in this framework, let us mention that
the presence of null vectors in Verma modules can be used to derive differential equations for correlation functions, in particular the so-called
Knizhnik–Zamolodchikov equations; compare e.g. sections 3.4 and 3.5 of
[Fuchs 1992], [Feigin *et al.* 1995], and also (in particular for the relation
with quantum groups) section 16.2 of [Chari and Pressley 1994] and sec-

tion XIX.3 of [Kassel 1995]. Conformal field theories, in particular those related to affine Lie algebras, have connections to many areas of mathematics (partly already mentioned in section 14.11). An example is the denominator formula (14.45). Relations of that type, so-called Macdonald identities and Rogers–Ramanujan identities, play an important rôle in combinatorics and in the theory of modular functions, see e.g. chapters 12 and 13 of [Kac 1990] (original papers are e.g. [Lepowsky 1982], [Lepowsky and Milne 1978], [Lepowsky and Wilson 1984] and [Moody 1975]).

Also, various integrable field theories, such as Toda field theories, can be constructed from conformal field theories (via integrable perturbations, or via Hamiltonian reduction), and similarly, through the realization of affine Lie algebra representations in terms of free fermions one can construct so-called tau function solutions to integrable hierarchies. As literature we mention §14.11 of [Kac 1990] (some original references are: [Adler and van Moerbeke 1991], [Bergvelt and ten Kroode 1988], [Date *et al.* 1982], and [Olive and Turok 1983]).

10 *q*-anything

In section 22.6 we have encountered q-deformed enveloping algebras $U_q(\mathfrak{g})$ as examples of quasi-triangular Hopf algebras (quantum groups), and Hecke algebras as a q-deformation of group algebras of symmetric groups. Nowadays, q-deformations have been studied for many other algebraic structures (such as Lie superalgebras, affine Lie algebras, the Virasoro algebra, bosonic and fermionic oscillators, Poisson algebras, Lie bi-algebras, etc.) as well as for various geometric structures, and for special functions and the associated differential equations. q-analogues have even been proposed for Lagrangian field theories, in particular gauge theories. One might say that virtually anything in this world has been q-deformed (it should not come as a surprise that occasionally the resulting structures are of minor practical use). Just as for $U_q(\mathfrak{g})$, special phenomena typically arise when the deformation parameter q is a root of unity.

As general references we mention [Kulish 1992], [Fröhlich and Kerler 1993] and [Chari and Pressley 1994]. q-deformations of special functions are described in [Vilenkin and Klimyk 1991]. The universal R-matrix of a quasi-triangular Hopf algebra satisfies the Yang–Baxter equation, which plays a fundamental rôle in the theory of integrable systems [Baxter 1982]. Correspondingly there are deep links between quantum groups and integrable systems. These are described e.g. in [de Vega 1989] and [Majid 1990]. For references on q-gauge theories see e.g. [Sudbery 1996]. One potential application of q-deformations (compare e.g. [Fichtmüller *et al.* 1996]) is to provide a regularization for quantities which in a naive treatment would come out infinite, as happens for instance quite often in

Lagrangian quantum field theory.

With the help of q-deformations one can even address issues that are relevant for the 'classical' case $q = 1$ but which are difficult to study directly at $q = 1$. An example is provided by the so-called *crystal bases* for the finite-dimensional modules of $U_q(\mathfrak{g})$, which are distinguished by their simple behavior under forming tensor products. Surprisingly, these bases can also be given a well-defined meaning in the limit $q \to 0$. For more information, see [Kashiwara 1991] and chapter 14 of [Chari and Pressley 1994].

11 Quantum symmetry

In classical mechanics, finite symmetry transformations act on the configuration space or phase space of the system and therefore form a group (with respect to composition). In contrast, in *quantum* mechanics, where there is no longer a configuration space or phase space, more general symmetry structures than groups and the associated group algebras or Lie algebras can arise. We have already mentioned supersymmetries in section 20.9. Another type of generalized symmetries is encountered when one studies the superselection structure of quantum field theories; these symmetries are nowadays referred to as *quantum symmetries*. As references we mention the book [Haag 1992] as well as [Fredenhagen 1993] and [Mack and Schomerus 1993].

By the term statistics one summarizes the effects that arise from the exchange of any kind of objects. These effects strongly depend on the type of objects considered, and also on the dimension d of space-time. For $d \geq 3$ the statistical properties of elementary particles are governed by the representation theory of the infinite symmetric group S_∞, which leads in particular to the distinction between bosons and fermions. The associated quantum symmetries are compact groups G. In contrast, when d is smaller (or when one deals with quasiparticles rather than pointlike elementary particles), more general statistics are possible, such as anyons, and the rôle of S_∞ is taken over by the infinite *braid group* B_∞.

So far, however, it is not known what the analogue of the associated compact group G is, though several proposals have been made, such as certain quasi-triangular quasi-Hopf algebras (compare e.g. [Schomerus 1995] and [Fuchs *et al.* 1995]). Even though the precise structure of these symmetries is still under debate, what *is* generally accepted is that the representation theory of the quantum symmetry gives rise to the structure of fusion rings, and in particular, for finite number of superselection sectors, to rational fusion rings (compare section 22.12). The quantum dimensions that arise in this context play the rôle of the statistical dimensions of superselection sectors, which appear as amplification factors in various physical contexts.

Our list of more advanced topics is, of course, far from being exhaustive, and concerning our choice of references we refer once more to what have outlined in the introduction. Nonetheless, we hope that our remarks can serve as a stimulus for further study.

References

M. Adler and P. van Moerbeke: The Toda lattice, Dynkin diagrams, singularities and Abelian varieties. *Invent. math.* **103** (1991) 223

Y. Alhassid, F. Gürsey, and F. Iachello: Group theory approach to scattering. *Ann. Phys.* **148** (1983) 346

V.I. Arnold: *Mathematical Methods of Classical Mechanics* (Springer Verlag, New York 1978)

V.I. Arnold: *Catastrophe Theory* (Springer Verlag, Berlin 1986)

V.I. Arnold, S.M. Gusein-Zade, and A.N. Varchenko: *Singularities of Differentiable Maps* (Birkhäuser, Boston 1985)

N.M. Atakishiyev, W. Lassner, and K.B. Wolf: The relativistic coma aberration. II. Helmholtz wave optics. *J. Math. Phys.* **30** (1989) 2463

K. Bardakçi and M.B. Halpern: New dual quark models. *Phys. Rev.* **D 3** (1971) 2493

A.O. Barut and R. Rączka: *Theory of Group Representations and Applications*, second edition (World Scientific, Singapore 1986)

R.J. Baxter: *Exactly Solved Models in Statistical Mechanics* (Academic Press, New York 1982)

R.E. Behrends: Broken SU(3) as a particle symmetry. In: [Loebl 1968], p. 541

I.M. Benn and R.W. Tucker: *An Introduction to Spinors and Geometry with Applications in Physics* (Adam Hilger, Bristol 1987)

M.J. Bergvelt and A.P.E. ten Kroode: τ-functions and zero curvature equations of Toda-AKNS type. *J. Math. Phys.* **29** (1988) 1308

L.C. Biedenharn and J.D. Louck: *Angular Momentum in Quantum Physics – Theory and Application* (Addison–Wesley, Reading 1981a)

L.C. Biedenharn and J.D. Louck: *The Racah–Wigner Algebra in Quantum Theory* (Addison–Wesley, Reading 1981b)

D. Bonatsos, C. Daskaloyannis, P. Kolokotronis, and D. Lenis: Quantum algebraic symmetries in nuclei and molecules. In: *Collective Motion and Nuclear Dynamics*, A. Raduta and D. Bucurescu, eds. (World Scientific, Singapore 1995)

W. Borho: A survey on enveloping algebras of semisimple Lie algebras, I. *Canad. Math. Soc. Proc.* **5** (1986) 20

R. Bott: On induced representations. *Proc. Symp. Pure Math.* **48** (1988) 1

N. Bourbaki: *Groupes et Algèbres de Lie* (Masson, Paris 1982)

N. Bourbaki: *Lie groups and Lie algebras, Chapters 1–3*
(Springer Verlag, Berlin 1989)

P. Bouwknegt, J. McCarthy, and K. Pilch: Semi-infinite cohomology in conformal field theory and 2D gravity. *J. Geom. and Phys.* **11** (1993) 225

P. Bouwknegt and K. Schoutens: W symmetry in conformal field theory.
Phys. Rep. **223** (1993) 183

E. Braaten, T.L. Curtright, and C.K. Zachos: Torsion and geometrostasis in nonlinear sigma models. *Nucl. Phys.* B **260** (1985) 630

Th. Bröcker and T. tom Dieck: *Representations of Compact Lie Groups*
(Springer Verlag, New York 1985)

W. Buchmüller and W. Lerche: Geometry and anomaly structure of supersymmetric σ models. *Ann. Phys.* **175** (1987) 159

K.M. Case: Biquadratic spinor identities. *Phys. Rev.* **97** (1955) 810

V. Chari and A.N. Pressley: *A Guide to Quantum Groups*
(Cambridge University Press, Cambridge 1994)

J.-Q. Chen, P.-N. Wang, Z.-M. Lü, and X.-B. Wu: *Tables of the Clebsch–Gordan, Racah and Subduction Coefficients of SU(n) Groups*
(World Scientific, Singapore 1987)

S.-w. Chung, M. Fukuma, and A. Shapere: Structure of topological lattice field theories in three dimensions. *Int. J. Mod. Phys.* A **9** (1994) 1305

A.J. Coleman: Induced and subduced representations. In: [Loebl 1968], p. 57

R. Coquereaux: Non-commutative geometry: a physicist's brief survey.
J. Geom. and Phys. **11** (1993) 307

J.F. Cornwell: *Group Theory in Physics*, vol. III (Academic Press, London 1989)

L. Crane: Action of the loop group on the self dual Yang–Mills equation.
Commun. Math. Phys. **110** (1987) 391

M. Creutz: *Quarks, Gluons and Lattices* (Cambridge University Press, Cambridge 1983)

C.W. Curtis and I. Reiner: *Representation Theory of Finite Groups and Associative Algebras* (Wiley Interscience, New York 1962)

T. Curtright, G. Ghandour, and C. Zachos: Quantum algebra deforming maps, Clebsch–Gordan coefficients, coproducts, U and R matrices.
J. Math. Phys. **32** (1991) 676

P. Cvitanović: *Group Theory* (Nordita, Copenhagen 1984)

J. Daboul, P. Slodowy, and C. Daboul: The hydrogen algebra as centerless twisted Kac–Moody algebra. *Phys. Lett. B* **317** (1993) 321

E. Date, M. Jimbo, M. Kashiwara, and T. Miwa: Transformation groups for soliton equations VII – Euclidean Lie algebras and reduction of the KP hierarchy. *Publ. RIMS (Kyoto)* **18** (1982) 1077

J. de Azcárraga and J. Izquierdo: *Lie Groups, Lie Algebras, Cohomology and Some Applications in Physics* (Cambridge University Press, Cambridge 1995)

V.V. Deodhar, O. Gabber, and V.G. Kac: Structure of some categories of representations of infinite-dimensional Lie groups. *Adv. Math.* **45** (1982) 92

H.J. de Vega: Yang–Baxter algebras, integrable theories and quantum groups. *Int. J. Mod. Phys. A* **4** (1989) 2371

B. de Wit and H. Nicolai: On the relation between $d = 4$ and $d = 11$ supergravity. *Nucl. Phys. B* **243** (1984) 91

J. Dixmier: *Enveloping Algebras*
(American Mathematical Society, Providence 1996)

L. Dolan: Why Kac–Moody subalgebras are interesting in physics.
AMS Lectures in Appl. Math. **21** (1985) 307

E.B. Dynkin: Semi-simple subalgebras of semi-simple Lie algebras.
Amer. Math. Soc. Transl. **(2) 6** (1957) 111

F.J. Dyson: Missed opportunities. *Bull. Amer. Math. Soc.* **78** (1972) 635

J. Ellis, M.K. Gaillard, M. Günaydin, and B. Zumino: Supersymmetry and non-compact groups in supergravity. *Nucl. Phys. B* **224** (1983) 427

U. Fano and G. Racah: *Irreducible Tensorial Sets*
(Academic Press, New York 1959)

L. Fehér, L. O'Raifeartaigh, P. Ruelle, I. Tsutsui, and A. Wipf: On Hamiltonian reductions of the Wess–Zumino–Novikov–Witten theories.
Phys. Rep. **222** (1992) 1

B.L. Feigin, V.V. Schechtman, and A.N. Varchenko: On algebraic equations satisfied by hypergeometric correlators in WZW models. II.
Commun. Math. Phys. **170** (1995) 219

R.P. Feynman: *Statistical Mechanics. A Set of Lectures* (W.A. Benjamin, Reading 1972)

M. Fichtmüller, A. Lorek, and J. Wess: q-deformed phase space and its lattice structure. *Zeit. Physik C* **71** (1996) 533

K. Fredenhagen: Superselection sectors in low dimensional quantum field theory. *J. Geom. and Phys.* **11** (1993) 337

I.B. Frenkel, J. Lepowsky, and A. Meurman: *Vertex Operator Algebras and the Monster* (Academic Press, New York 1988)

424   *References*

I.B. Frenkel and Y. Zhu: Vertex operator algebras associated to representations of affine and Virasoro algebras. *Duke Math. J.* **66** (1992) 123

J. Fröhlich and T. Kerler: *Quantum Groups, Quantum Categories and Quantum Field Theory* [Lecture Notes in Mathematics 1542]
(Springer Verlag, Berlin 1993)

J. Fuchs: Computer program WEI (1991);
see the WWW page http://norma.nikhef.nl/lie

J. Fuchs: *Affine Lie Algebras and Quantum Groups*
(Cambridge University Press, Cambridge 1992)

J. Fuchs: Fusion rules in conformal field theory.
Fortschr. Phys. **42** (1994) 1

J. Fuchs, A.Ch. Ganchev, and P. Vecsernyés: Rational Hopf algebras: Polynomial equations, gauge fixing, and low-dimensional examples.
Int. J. Mod. Phys. A **10** (1995) 3431

W. Fulton and J. Harris: *Representation Theory, a First Course*
(Springer Verlag, New York 1991)

R.W. Gebert: Introduction to vertex algebras, Borcherds algebras and the monster Lie algebra. *Int. J. Mod. Phys.* A **8** (1993) 5441

R.W. Gebert and H. Nicolai: On E_{10} and the DDF construction. *Commun. Math. Phys.* **172** (1995) 571

M. Geck, G. Hiss, F. Lübeck, G. Malle, and G. Pfeiffer: CHEVIE – A system for computing and processing generic character tables.
Appl. Alg. in Engineering, Communication and Computing (1996)

R. Gilmore: *Lie Groups, Lie Algebras and Some of Their Applications*
(John Wiley, New York 1974)

P. Goddard, A. Kent, and D.I. Olive: Virasoro algebras and coset space models.
Phys. Lett. B **152** (1985) 88

P. Goddard and D.I. Olive: The magnetic charges of stable self-dual monopoles.
Nucl. Phys. B **191** (1981) 528

M.B. Green, J.H. Schwarz, and E. Witten: *Superstring Theory*, Vol. 1
(Cambridge University Press, Cambridge 1987)

B. Gruber and M.T. Samuel: Semisimple subalgebras of semisimple Lie algebras: The algebra A_5 (SU(6)) as a physically significant example.
In: [Loebl 1975], p. 95

V. Guillemin and S. Sternberg: *Symplectic Techniques in Physics*
(Cambridge University Press, Cambridge 1984)

R. Haag: *Local Quantum Physics* (Springer Verlag, Berlin 1992)

P.R. Halmos: *Measure Theory* (van Nostrand, Princeton 1950)

M.B. Halpern, E.B. Kiritsis, N.A. Obers, and K. Clubok: Irrational conformal field theory. *Phys. Rep.* **265** (1996) 1

M. Hamermesh: *Group Theory and its Application to Physical Problems* (Addison–Wesley, Reading 1962)

S.W. Hawking and G.F.R. Ellis: *The Large Scale Structure of Space-time* (Cambridge University Press, Cambridge 1980)

S. Helgason: *Differential Geometry, Lie Groups, and Symmetric Spaces* (Academic Press, New York 1978)

M. Henneaux and C. Teitelboim: *Quantization of Gauge Systems* (Princeton University Press, Princeton 1992)

J.E. Humphreys: *Introduction to Lie Algebras and Representation Theory* (Springer Verlag, Berlin 1972)

J.E. Humphreys: *Reflection Groups and Coxeter Groups* (Cambridge University Press, Cambridge 1990)

F. Iachello and A. Arima: *The Interacting Boson Model* (Cambridge University Press, Cambridge 1987)

K. Intriligator and N. Seiberg: Lectures on supersymmetric gauge theories and electric-magnetic duality.
In: Proceedings of the Cargèse Summer School on *Low Dimensional Applications of Quantum Field Theory* (Cargèse, France, July 1995); hep-th/9509066

C. Itzykson and J.-B. Zuber: *Quantum Field Theory* (McGraw-Hill, New York 1980)

N. Jacobson: *Lie Algebras* (Wiley Interscience, New York 1962)

G. James and A. Kerber: *The Representation Theory of the Symmetric Group* (Addison–Wesley, Reading 1981)

J.C. Jantzen: *Moduln mit einem höchsten Gewicht* [Lecture Notes in Mathematics 750] (Springer Verlag, Berlin 1979)

B.R. Judd: Group theory in atomic spectroscopy. In: [Loebl 1968], p. 469

V.G. Kac: *Infinite-dimensional Lie Algebras*, third edition (Cambridge University Press, Cambridge 1990)

V.G. Kac and A.K. Raina: *Highest Weight Representations of Infinite-dimensional Lie Algebras* (World Scientific, Singapore 1987)

V.G. Kac and M. Wakimoto: Modular and conformal invariance constraints in representation theory of affine algebras. *Adv. Math.* **70** (1988) 156

M. Kashiwara: On crystal bases of the Q-analogue of universal enveloping algebras. *Duke Math. J.* **63** (1991) 456

S. Kass, R.V. Moody, J. Patera, and R. Slansky: *Affine Lie Algebras, Weight Multiplicities, and Branching Rules* (University of California Press, Berkeley 1990)

C. Kassel: *Quantum Groups* (Springer Verlag, New York 1995)

T.W. Kephart and M.T. Vaughn: Tensor methods for the exceptional group E_6. *Ann. Phys.* **145** (1983) 162

R.C. King: Branching rules for classical Lie groups using tensor and spinor methods. *J. Phys. A: Math. Gen.* **8** (1975) 429

R.C. King and A.H.A. Al-Qubanchi: The Weyl groups and weight multiplicities of the exceptional groups. *J. Phys. A: Math. Gen.* **14** (1981) 51

A.A. Kirillov: *Elements of the Theory of Representations*
(Springer Verlag, Berlin 1975)

K. Koike and I. Terada: Young-diagrammatic methods for the representation theory of the classical groups of type B_n, C_n, D_n. *J. Algebra* **107** (1987) 466

P.P. Kulish (ed.): *Quantum Groups* [Lecture Notes in Mathematics 1510]
(Springer Verlag, Berlin 1992)

S. Kumar: A Demazure character formula in arbitrary Kac–Moody setting. *Invent. math.* **89** (1987) 395

S. Lang: *Algebra* (Addison–Wesley, Reading 1984)

J. Lepowsky: Affine Lie algebras and combinatorial identities.
Springer Lecture Notes in Mathematics **933** (1982) 130

J. Lepowsky and S. Milne: Lie algebraic approaches to classical partition identities. *Adv. Math.* **29** (1978) 15

J. Lepowsky and R. Wilson: The structure of standard modules, I: universal algebras and the Rogers–Ramanujan identities. *Invent. math.* **77** (1984) 199

E.M. Loebl (ed.): *Group Theory and its Applications* (Vol. I)
(Academic Press, New York 1968)

E.M. Loebl (ed.): *Group Theory and its Applications* (Vol. III)
(Academic Press, New York 1975)

W. Ludwig and C. Falter: *Symmetries in Physics* (Springer Verlag, Berlin 1988)

D. Lüst and S. Theisen: *Lectures on String Theory*
[Lecture Notes in Physics 346] (Springer Verlag, Berlin 1989)

G. Lusztig: *Introduction to Quantum Groups* (Birkhäuser, Boston 1993)

I.G. MacDonald: Algebraic structure of Lie groups.
In: *Representation Theory of Lie Groups*, M.F. Atiyah *et al.*, eds. (Cambridge University Press, Cambridge 1978), p. 91

I.G. MacDonald: *Symmetric Functions and Hall Polynomials*
(Oxford University Press, London 1979)

G. Mack and A. Salam: Finite-component field representations of the conformal group. *Ann. Phys.* **53** (1969) 174

G. Mack and V. Schomerus: A short introduction to quantum symmetry.
J. Geom. and Phys. **11** (1993) 361

S. Majid: Quasitriangular Hopf algebras and Yang–Baxter equations.
Int. J. Mod. Phys. A **5** (1990) 1

Yu.I. Manin: *Gauge Field Theory and Complex Geometry*
(Springer Verlag, Berlin 1988)

J.E. Marsden and T.S. Ratiu: *Introduction to Mechanics and Symmetry*
(Springer Verlag, New York 1994)

W.G. McKay and J. Patera: *Tables of Dimensions, Indices and Branching Rules
for Representations of Simple Lie Algebras* (Marcel Dekker, New York 1981)

A. Messiah: *Quantum Mechanics*, 11th printing
(North Holland Publ. Co., Amsterdam 1986)

J. Mickelsson: *Current Algebras and Groups* (Plenum, New York 1989)

M. Micu: Construction of invariants for simple Lie groups.
Nucl. Phys. **60** (1964) 353

W. Miller, Jr.: *Lie Theory and Special Functions*
(Academic Press, New York 1968)

R.A. Minard, G.E. Stedman, and A.G. McLellan: Reduction of angular momentum coupling trees and the polarization dependence of nonlinear optical
processes. *J. Chem. Phys.* **78** (1983) 5016

J.S. Mondragón and K.B. Wolf: *Lie Methods in Optics*
[Lecture Notes in Physics 250] (Springer Verlag, Berlin 1986)

R.V. Moody: Macdonald identities and Euclidean Lie algebras.
Proc. Amer. Math. Soc. **48** (1975) 43

R.V. Moody and A. Pianzola: *Lie Algebras With Triangular Decomposition*
(John Wiley, New York 1995)

G. Moore and N. Seiberg: Lectures on RCFT. In: *Physics, Geometry, and Topology*, H.C. Lee, ed. (Plenum, New York 1990), p. 263

O. Nachtmann: *Elementary Particle Physics: Concepts and Phenomena*
(Springer Verlag, Berlin 1990)

M.A. Naimark: *Linear Representations of the Lorentz Group*
(Pergamon Press, Oxford 1964)

M. Nakahara: *Geometry, Topology and Physics* (Adam Hilger, Bristol 1990)

B.G. Nickel: Evaluation of simple Feynman graphs. *J. Math. Phys.* **19** (1978)
542

H. Nicolai: Two-dimensional gravities and supergravities as integrable systems.
Springer Lecture Notes in Physics **396** (1991) 231

H. Nicolai: A hyperbolic Lie algebra from supergravity.
Phys. Lett. **B 276** (1992) 333

D.I. Olive and N. Turok: The symmetries of Dynkin diagrams and the reduction
of Toda field equations. *Nucl. Phys.* **B 215** (1983) 470

L. O'Raifeartaigh: Broken symmetry. In: [Loebl 1968], p. 469

L. O'Raifeartaigh: *Group Structure of Gauge Theories*
(Cambridge University Press, Cambridge 1986)

J.T. Ottesen: *Infinite-dimensional Groups and Algebras in Quantum Physics*
(Springer Verlag, Berlin 1995)

J.C. Parikh: *Group Symmetries in Nuclear Structure* (Plenum, New York 1978)

A.M. Perelomov and V.S. Popov: On eigenvalues of Casimir operators.
Sov. J. Nucl. Phys. **7** (1968) 290

K. Pilch and A.N. Schellekens: Formulae for the eigenvalues of the Laplacian on
tensor harmonics on symmetric coset spaces. *J. Math. Phys.* **25** (1984) 3455

A. Pressley and G.B. Segal: *Loop Groups* (Clarendon Press, Oxford 1986)

M. Reed and B. Simon: *Methods of Modern Mathematical Physics*
(Academic Press, New York 1972)

H. Samelson: *Notes on Lie Algebras*, second edition
(Springer Verlag, New York 1980)

F. Scheck: *Mechanics: From Newton's Laws to Deterministic Chaos*, second edition (Springer Verlag, Berlin 1994)

A.N. Schellekens: Conformal field theory for four-dimensional strings.
Nucl. Phys. B (Proc. Suppl.) **15** (1990) 3

A.N. Schellekens: Computer program KAC (1995);
see the WWW page http://norma.nikhef.nl/~t58/kac.html

A.N. Schellekens and S. Yankielowicz: Simple currents, modular invariants, and
fixed points. *Int. J. Mod. Phys. A* **5** (1990) 2903

M. Scheunert: *The Theory of Lie Superalgebras* [Lecture Notes in Mathematics
716] (Springer Verlag, Berlin 1979)

V. Schomerus: Construction of field algebras with quantum symmetry from local
observables. *Commun. Math. Phys.* **169** (1995) 193

M. Schönert *et al.*: *GAP – Groups, Algorithms, and Programming*
(RWTH University, Aachen 1994)

J.-P. Serre: *Linear Representations of Finite Groups*
(Springer Verlag, New York 1977)

P. Slodowy: Platonic solids, Kleinian singularities, and Lie groups.
Springer Lecture Notes in Mathematics **1008** (1983) 102

D.A. Smith: Chevalley basis for Lie modules.
Trans. Amer. Math. Soc. **115** (1965) 283

D. Speiser: Theory of compact Lie groups. In: Istanbul Summer School on *Group
Theoretical Concepts and Methods in Elementary Particle Physics*, F. Gürsey, ed.
(Gordon and Breach, New York 1964), p. 237

A. Sudbery: $SU_q(n)$ gauge theory. *Phys. Lett. B* **375** (1996) 75

K. Sundermeyer: *Constrained Dynamics, with Applications to Yang–Mills Theory, General Relativity, Classical Spin, Dual String Model* [Lecture Notes in
Mathematics 169] (Springer Verlag, Berlin 1982)

B.M. ter Haar Romeny, L.M.J. Florack, J.J. Koenderink, and M.A. Viergever:
Scale space: its natural operators and differential invariants.
Springer Lecture Notes in Computer Science **511** (1991) 239

M. Thoma and R.T. Sharp: Orbit-orbit branching rules for families of classical Lie algebra-subalgebra pairs. *J. Math. Phys.* **37** (1996) 4750

W.-K. Tung: *Group Theory in Physics* (World Scientific, Singapore 1985)

V.G. Turaev: *Quantum Invariants of Knots and 3-Manifolds* (de Gruyter, Berlin 1994)

C. Vafa and E. Witten: A strong coupling test of S-duality. *Nucl. Phys. B* **431** (1994) 3

M.A.A. van Leeuwen, A.M. Cohen, and B. Lisser: *LiE, A package for Lie Group Computations* (Computer Algebra Nederland, Amsterdam 1992)

N.Ja. Vilenkin and A.U. Klimyk: *Representation of Lie Groups and Special Functions* (Kluwer Academic Publishers, Dordrecht 1991)

S. Weinberg: *The Quantum Theory of Fields. Vol. 1: Foundations* (Cambridge University Press, Cambridge 1995)

J. Wess and J. Bagger: *Supersymmetry and Supergravity*, second edition (Princeton University Press, Princeton 1992)

H. Weyl: *The Classical Groups* (Princeton University Press, Princeton 1939)

E. Witten: Coadjoint orbits of the Virasoro group. *Commun. Math. Phys.* **114** (1988) 1

K.B. Wolf: Aberration classification and composition. In: *Quantum-like Models and Coherent Effects*, R. Fedele and P.K. Shukla, eds. (World Scientific, Singapore 1995), p. 420

B.G. Wybourne: Enumeration of group invariant quartic polynomials in Higgs scalar fields. *Australian J. Phys.* **33** (1980) 941

B.G. Wybourne: *Schur. An Interactive Programme for Calculating Properties of Lie groups and Symmetric Functions* (Mikołaja Kopernika University, Toruń 1995)

M. Yang and B.G. Wybourne: Extended Poincaré supersymmetry, rotation groups and branching rules. *J. Phys. A: Math. Gen.* **19** (1986) 2003

Index

Printed in the United States
By Bookmasters